Y0-BYK-868

METAL-LIGAND INTERACTIONS IN ORGANIC CHEMISTRY AND BIOCHEMISTRY

PART 2

THE JERUSALEM SYMPOSIA ON QUANTUM CHEMISTRY AND BIOCHEMISTRY

Published by the Israel Academy of Sciences and Humanities, distributed by Academic Press (N.Y.)

1st JERUSALEM SYMPOSIUM: *The Physicochemical Aspects of Carcinogenesis* (October 1968)
2nd JERUSALEM SYMPOSIUM: *Quantum Aspects of Heterocyclic Compounds in Chemistry and Biochemistry* (April 1969)
3rd JERUSALEM SYMPOSIUM: *Aromaticity, Pseudo-Aromaticity, Antiaromaticity* (April 1970)
4th JERUSALEM SYMPOSIUM: *The Purines: Theory and Experiment* (April 1971)
5th JERUSALEM SYMPOSIUM: *The Conformation of Biological Molecules and Polymers* (April 1972)

Published by the Israel Academy of Sciences and Humanities, distributed by D. Reidel Publishing Company (Dordrecht and Boston)

6th JERUSALEM SYMPOSIUM: *Chemical and Biochemical Reactivity* (April 1973)

Published and distributed by D. Reidel Publishing Company (Dordrecht and Boston)

7th JERUSALEM SYMPOSIUM: *Molecular and Quantum Pharmacology* (March/April 1974)
8th JERUSALEM SYMPOSIUM: *Environmental Effects on Molecular Structure and Properties* (April 1975)

VOLUME 9

METAL-LIGAND INTERACTIONS IN ORGANIC CHEMISTRY AND BIOCHEMISTRY

PART 2

PROCEEDINGS OF THE NINTH JERUSALEM SYMPOSIUM ON QUANTUM CHEMISTRY AND BIOCHEMISTRY HELD IN JERUSALEM, MARCH 29TH–APRIL 2ND, 1976

Edited by

BERNARD PULLMAN
Institut de Biologie Physico-Chimique
(Fondation Edmond de Rothschild) Paris, France

and

NATAN GOLDBLUM
The Hebrew University, Hadassah Medical School, Jerusalem, Israel

D. REIDEL PUBLISHING COMPANY
DORDRECHT-HOLLAND/BOSTON-U.S.A.

Library of Congress Cataloging in Publication Data

Jerusalem Symposium on Quantum Chemistry and Biochemistry, 9th, 1976.
Metal-ligand interactions in organic chemistry and biochemistry.

(The Jerusalem symposia on quantitative chemistry and biochemistry; v. 9)
Bibliography: p.
Includes index.
1. Ligands—Congresses. 2. Chemistry, Organic—Congresses.
3. Biological chemistry—Congresses. I. Pullman, Bernard, 1919–
II. Goldblum, Natan. III. Title. IV. Series.
QD474.J47 1976 547'.1'2242 76–44910

The set of two parts: ISBN 90–277–0771–5
Part 1: ISBN 90–277–0751–0
Part 2: ISBN 90–277–0754–5

Published by D. Reidel Publishing Company,
P.O. Box 17, Dordrecht, Holland

Sold and distributed in the U.S.A., Canada, and Mexico
by D. Reidel Publishing Company, Inc.
Lincoln Building, 160 Old Derby Street, Hingham,
Mass. 02043, U.S.A.

Printed in The Netherlands

TABLE OF CONTENTS

TABLE OF CONTENTS OF PART 2 VII

G. EISENMAN, J. SANDBLOM, and E. NEHER/Ionic Selectivity, Saturation, Binding, and Block in the Gramicidin A Channel: A Preliminary Report 1

A.P. THOMA and W. SIMON/Enantiomer-Selective Ion Carriers and Their Transport Properties in Bulk Membranes 37

L.K. STEINRAUF and M.N. SABESAN/Computer Simulation Studies on the Ion-Transporting Antibiotics 43

J. OTSUKA, Y. SENO, N. FUCHIKAMI, and O. MATSUOKA/Theoretical Calculation on Oxygen Binding to the Heme Iron in Myoglobin and Hemoglobin 59

A. DEDIEU, M.-M. ROHMER, and A. VEILLARD/Oxygen Binding to Iron Porphyrins. *Ab initio* Calculations 101

W.S. CAUGHEY, J.C. MAXWELL, J.M. THOMAS, D.H. O'KEEFFE, and W.J. WALLACE/Metal-Ligand Interactions in Porphyrins and Hemeproteins 131

M. CHEVION, W.E. BLUMBERG, and J. PEISACH/EPR Studies of Human Nitrosyl Hemoglobins and Their Relation to Molecular Function 153

R. BANERJEE, J.M. LHOSTE, and F. STETZKOWSKI/Carbon-13 NMR Studies of Human Hemoglobin and Some Other Hemoproteins 163

K.B. WARD and W.A. HENDRICKSON/Structures of the Non-heme Iron Proteins Myohemerythrin and Hemerythrin 173

D.H. HAYNES/Divalent Cation-Ligand Interactions of Phospholipid Membranes: Equilibria and Kinetics 189

T.O. HENDERSON/^{31}P Nuclear Magnetic Resonance Studies of Metal-Ligand Interactions in Human Blood Components 213

W.J. COOK and C.E. BUGG/Structures of Calcium-Carbohydrate Complexes 231

R.H. KRETSINGER/Why Does Calcium Play an Informational Role Unique in Biological Systems? 257

G. ZUNDEL and A. MURR/Influence of Transition Element Ions on the Hydrogen Bonds Formed by Their First Hydration Shell 265

M.A. WELLS and T.C. BRUICE/The Nucleophilicity of Metal Bound Hydroxide, Mechanisms of Displacement on Esters, and Transesterification Involving the Metal-Acyl Anhydride Bond 273

J. JAGUR-GRODZINSKI and Y. OZARI/Complexes of Cobalt with Polymeric Phosphoramides 285

J. KRESS, C. SOURISSEAU, and A. NOVAK/Infrared and Raman Studies of Some Organomagnesium and Allyl Compounds 299

B. BIRDSALL, J. FEENEY, and A.M. GUILIANI/Conformational Studies of Flexible Molecules Using Lanthanide Ions as NMR Probes 313

J. REUBEN/Metal Ions as Probes for NMR Studies of Structure and Dynamics in Systems of Biological Interest 325

G.H.F. DIERCKSEN and W.P. KRAEMER/SCF MO LCGO Studies on Electrolytic Dissociation: the System LiF.nH_2O (n=1,2) 331

B. ÅKERMARK, M. ÅLMEMARK, J.-E. BÄCKVALL, J. ALMLÖF, B. ROOS, and Å. STØGÅRD/*Ab initio* Calculations on Nickel-Ethylene Complexes 377

H. BASCH/Metal-Ligand Interactions and Photoelectron Spectroscopy 389

TABLE OF CONTENTS OF PART 1

PREFACE

A. PULLMAN/Search for General Features in Cation-Ligand Binding. *Ab initio*. SCF Studies of the Interaction of Alkali and Alkaline Earth Ions with Water

T.J. KISTENMACHER and L.G. MARZILLI/Chelate Metal Complexes of Purines, Pyrimidines and Their Nucleosides: Metal-Ligand and Ligand-Ligand Interactions

G.L. EICHHORN, J. RIFKIND, Y.A. SHIN, J. PITHA, J. BUTZOW, P. CLARK, and J. FROEHLICH/Recent Studies on the Effects of Divalent Metal Ions on the Structure and Function of Nucleic Acids

E. SLETTEN/Crystallographic Studies on Copper Complexes of Nucleic Acid Components

B. PULLMAN and A. PULLMAN/Quantum-Mechanical Studies on Cation Binding to Constituents of Nucleic Acids and Phospholipids

S. HANLON, B. WOLF, S. BERMAN, and A. CHAN/The Conformational Sensitivity of DNA to Ionic Interactions in Aqueous Solutions

G.V. FAZAKERLEY, G.E. JACKSON, J.C. RUSSELL, and M.A. WOLFE/ Interaction of Gadolinium (III), Manganese (II) and Copper (II) with Cyclic Nucleotides

W. LOHMANN, A. PLEYER, M. HILLERBRAND, V. PENKA, K.G. WAGNER, and H.A. ARFMANN/Influence of Heterocyclic Nitrogen Atoms on the Interaction between Purine Analogues and Metal Ions

B.M. RODE/The Investigation of Metal-Amide Interactions with the Aid of Quantum Chemical Model Calculations and Corresponding Model Experiments

C.N.R. RAO/Interaction of Alkali and Alkaline Earth Cations and Protons with Amides and Related Systems

D. BALASUBRAMANIAN and B.C. MISRA/Alkali Ion Binding to Polypeptides and Polyamides

M.H. BARON, H. JAESCHKE, R.M. MORAVIE, C. de LOZÉ, and J. CORSET/Infrared and Raman Studies of Interactions between Salts and Amides or Esters

B. SARKAR/Concept of Molecular Design in Relation to the Metal-Binding Sites of Proteins and Enzymes

L.H. JENSEN/Metal Complexes in Proteins

G. ROTILIO, L. MORPURGO, L. CALABRESE, A.F. AGRO, and B. MONDOVÌ/Metal-Ligand Interactions in Cu-Proteins

C.M. DOBSON and R.J.P. WILLIAMS/Nuclear Magnetic Resonance Studies of the Interaction of Lanthanide Cations with Lysozyme

M.M. WERBER, A. DANCHIN, Y. HOCHMAN, C. CARMELI, and A. LANIR/ Co(III)-ATP Complexes as Affinity Labeling Reagents of Myosin and Coupling Factor-1 ATPases

G.D. SMITH and W.L. DUAX/Crystallographic Studies of Valinomycin and A23187

M.R. TRUTER/Effects of Cations of Groups IA and IIA on Crown Ethers

R.M. IZATT, L.D. HANSEN, D.J. EATOUGH, J.S. BRADSHAW, and J.J. CHRISTENSEN/Cation Selectivities Shown by Cyclic Polyethers and Their THIA Derivatives

W. SAENGER, H. BRAND, F. VÖGTLE, and E. WEBER/ X-Ray Structure of a Synthetic, Non-Cyclic, Chiral Polyether Complex as Analog of Nigericin Antibiotics: Tetraethyleneglycol-bis (8-oxyquinoline)-ether·rubidium Iodide Complex

J.J. STEZOWSKI/Metal Ion - Ligand Interactions for the Tetracycline Derivative Antibiotics: a Structural Approach

IONIC SELECTIVITY, SATURATION, BINDING, AND BLOCK IN THE GRAMICIDIN A CHANNEL: A PRELIMINARY REPORT*

GEORGE EISENMAN, JOHN SANDBLOM and ERWIN NEHER
The Department of Physiology, UCLA Medical School, Los Angeles, California, U.S.A.; The Department of Medical Physics, University of Uppsala, Sweden; and the Max-Planck Institut für Biophysikalische Chemie, Göttingen, German Federal Republic

1. INTRODUCTION

An efficient channel should combine high transport rates with the ability to discriminate selectively between ion species (Bamberg, et al., 1976). For example, the Na^+ channel of nerve combines a high Na^+ to K^+ selectivity (20:1) with the ability to transport 10^8 Na^+ ions per sec (Hille, 1970). While the cation permeable channel produced by the neutral pentadecapeptide gramicidin A does have high transport rates (Hladky and Haydon, 1970; Hladky, 1973) and was initially reported by Mueller and Rudin (1967) to discriminate by a factor of 13 among the alkali cations, it has been concluded more recently to show little discrimination among monovalent cations, being not greatly different from the mobilities in free solution (Myers and Haydon, 1972; Hladky, 1973; and Bamberg et al., 1976); and it appeared that this, otherwise interesting, model for a biological channel might lack the important property of selectivity. However, Eisenman et al. (1974) found from membrane potential measurements that Tl^+ was 50 times more permeant than the like-sized K^+ ion, which demonstrated unambiguously that the gramicidin channel could discriminate selectively among monovalent cations. This observation was taken up by Neher (1975), who reported that although Tl^+ carried current through the gramicidin A channel better than Na^+ or K^+, it blocked Na^+ currents at unusually low concentrations. Andersen (1975) also confirmed the high Tl^+ selectivity and independently described blocking effects for K^+, Cs^+ and Tl^+. On examining this further, Eisenman et al. (1976a) discovered that the phenomenon of Tl^+ blocking of alkali cation currents was associated with a dependence of the permeability ratio on ion concentration, both phenomena being expected theoretically from the existence of a previously undescribed additional binding site for Tl^+ and K^+ in the channel. It should be noted that the high Tl^+ selectivity which must be a fundamental property of the amide carbonyl ligands of gramicidin is also characteristic of amide (and imide) carbonyl ligands for such typical cyclic octapeptides as cyclo $(Pro\text{-}Gly\text{-}Pro\text{-}d\text{-}Phenylala)_2$ (Eisenman et al., 1976), as well as

* Supported by: National Science Foundation Grant GB 30835, U.S. Public Health Services Grant NS 09931.

B. Pullman and N. Goldblum (eds.), Metal-Ligand Interactions in Organic Chemistry and Biochemistry, part 2, 1 - 36.

cyclo $(Pro\text{-}Gly)_4$ and cyclo $(Pro\text{-}d\text{-}Phenylala)_4$ (Eisenman, Kuv, Hint and Blout, unpublished observations).

Ionic selectivity, saturation, and block are also characteristic properties of the Na^+ channel of nerve (Hille, 1975b); and concentration dependence of the Na^+/K^+ permeability ratio has recently been described (Begenisich and Cahalan, 1975; Cahalan and Begenisich, 1976). binding of current-carrying monovalent cations has been implicated as the cause of the block by Hille, with Tl^+ being particularly strongly bound; and such binding has also been directly demonstrated by displacement of labelled tetrodotoxin in isolated preparations (Henderson et al., 1974; Reed and Raftery, 1976).

In view of the above phenomenological similarities between the biological Na^+ channels of unknown molecular structure and the artificial gramicidin A channel whose structure is reasonably well defined (Urry, 1971, 1972; Veatch and Blout, 1974a, b), an understanding of the origin of ion selectivity, saturation, binding, and block in the gramicidin A channel is clearly desirable. This paper, and the others of this series (Sandblom et al., 1976; Eisenman et al., 1976b; Neher et al., 1976), will show that the phenomena of blocking and of concentration dependent permeability ratios reflect two different aspects of the interactions incidental to the simultaneous occupancy by more than one cation in a four-site model for the gramicidin channel. The four-site simultaneously occupiable model we propose here extends previous treatments (Läuger, 1972; Hladky, 1972, 1973) to include the possibility of simultaneous occupancy by more than one cation and incorporates an additional binding site at the entrance to the channel so that there are two sites symmetrically located on each side. This extension causes certain parameters such as permeability ratios, which in the previous models were independent of ion concentrations, to become dependent upon these. Experimental measurements reported here on membrane potentials and single channel conductances in a variety of ionic solutions enable a preliminary assessment of the values of the key binding constants and rate constants of the model. The predictions of the model with these values for the constants will be shown here to account remarkably well for *all* of the as yet observed electrical behavior of the channel. This behavior includes such heretofore puzzling properties as permeability ratios higher then the conductance ratios and varying in opposite directions with increasing concentration (Myers and Haydon, 1972), the presence of a maximum instead of a simple saturation at high concentrations in the single channel conductance seen most strikingly with CsCl but also clearly present for KCl and NaCl (Hladky and Haydon, 1972, Figure 10), and the unusually low concentration at which Tl^+ was observed to inhibit Na^+ conductance (Neher, 1975).

2. GRAMICIDIN A CHANNELS EXHIBIT SELECTIVITY, BLOCK, SATURATION, BINDING AND CONCENTRATION DEPENDENT PERMEABILITY RATIOS WITH MONOVALENT CATIONS

The phenomena of selectivity, block, saturation, and binding in gramicidin A channels will be illustrated by the data of Table I and Figures 1 – 7; while the concentration dependence of the permeability ratios

TABLE I
Permeability ratios for gramicidin in glycerol mono-oleate-decane bilayers (after Eisenman et al., 1974, Table III)

	P_i/P_K
Li^+	0.083[a]
Na^+	0.33; 0.29[a]
K^+	1.0
Rb^+	1.03[a]
Cs^+	1.31[a]
Tl^+	50.0
NH_4^+	1.8[a]
H^+	16.0[a]
$CH_3NH_3^+$	0.24
$C_2H_5NH_3^+$	0.014
$(CH_3)_2NH_2^+$	0.0076
$(CH_3)_4N^+$	<0.00011[a]
$H_2NNH_3^+$	0.96
$HONH_3^+$	1.7
$Formamidinium^+$	1.55
$Guanidinium^+$	$\leq .027$ (pH 7.9)
$NH_2 \cdot guan^+$	0.02 (pH 7.2)
$OH \cdot guan^+$	≤ 0.24

Permeability ratios were measured in our usual way (Szabo et al., 1969) from membrane potential measurements in 2-cation mixtures of K^+ with 0.01 M solutions of the listed cations in the presence of a commercial mixture of gramicidin A (72% gramicidin A, 9% gramicidin B, 15% gramicidin C, according to Glickson et al., 1972) chosen to give a conductance of about $10^{-5}\ \Omega^{-1}\ cm^{-2}$ (usually about 10^{-9} M aqueous gramicidin)
a) Data of Myers and Haydon (1972, Table II) for bi-ionic potentials at 0.1 M.

will be demonstrated in Figures 8 and 9.

Table I summarizes the selectivity of gramicidin A channels, as manifested by permeability ratios from membrane potential measurements observed for macroscopically conducting membranes.

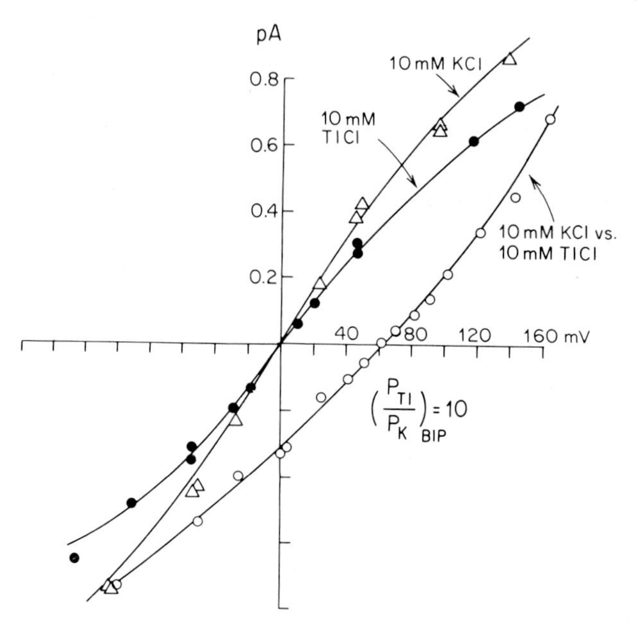

Fig. 1. Current-voltage relationship for single channels of purified valine gramicidin A in palmitoliden-lecithin/decane bilayers at 34.5 °C. (NB. These are the only data not at 22.5 °C in this paper). In this figure zero voltage has been assigned to that voltage which gives zero current after rupture of the membrane (the uncertainty in such an assignment is circumvented by performing the experiment in the manner of Figure 8). For the bionic interdiffusion case the current is carried by K^+ in the first quadrant and by Tl^+ in the third quadrant.

Note that substantial permeability selectivity exists for Tl^+, H^+, and certain polyatomic cations and even between such cations as Li^+ and Cs^+ (P_{Cs}/P_{Li} = 16). Figure 1 illustrates for a single channel the high Tl^+/K^+ permeability selectivity for a single channel despite the fact that the conductance selectivities (comparing pure Tl^+ with pure K^+ cases) are only slightly different. Figure 1 also illustrates the blocking effects seen in the conductance of K^+-Tl^+ mixtures. This will be seen by considering the figure in some detail. The filled circles and open triangles plot for single gramicidin A channels the observed current-voltage behavior when the membrane is interposed between symmetrical 10 mM solutions of TlCl and KCl, respectively; while the open circles plot the data for a membrane with 10 mM KCl on one side and 10 mM TlCl on the other. The permeability selectivity is manifested by the 60 mV membrane potential at zero-current in the bi-ionic case, corresponding to a bi-ionic permeability ratio $(P_{Tl}/P_K)_{BIP}$ of 10, calculated from the Goldman-Hodgkin-Katz equation at this temperature. Comparing the conductances in the pure Tl^+ and K^+ cases, it becomes apparent that the selectivity as seen in conductance ratio is different than that seen in the permeability ratio

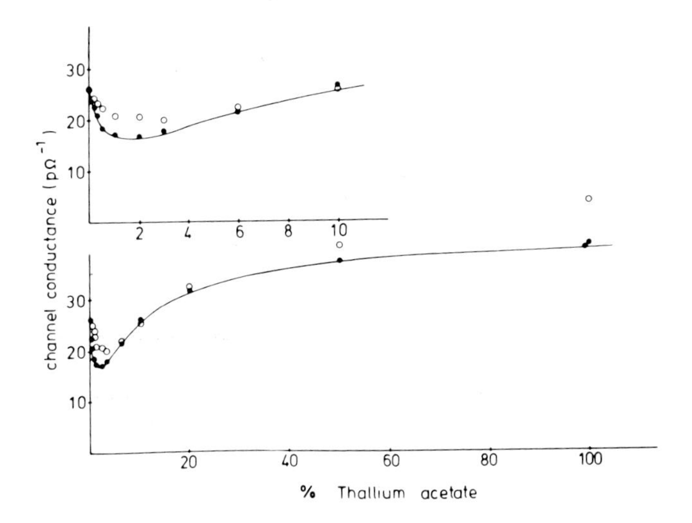

Fig. 2. Conductance of single gramicidin A channels vs. mole fraction of Tl^+ at 1M total concentration of TlAc plus NaAc symmetrically present on both sides of the membrane. Filled circles give the zero current conductance, open circles the conductance measured at ± 100 mV. The upper part expands the low Tl^+ concentration region around the conductance minimum. (After Neher, 1975, Figure 2). The temperature in this and all following figures 22.5 °C.

since the conductance selectivity, although small, is reversed, being slightly higher in the pure K^+ situation than in the pure Tl^+ case at all voltages.

The phenomenon of block is also illustrated in this figure, by noting that the slope of the current-voltage curve at zero current is lower in the bi-ionic situation then in either of the pure single-ion cases, indicating that the conductance is lower when the channel is interposed between Tl^+ and K^+ than when it is in the presence of either of these ions alone. This blocking effect is more clearly illustrated by measuring the conductance of a membrane interposed between identical mixtures of cations as illustrated in Figure 2 for Tl^+-Na^+ mixtures (Neher, 1975) by analogy to the experiment illustrated in Figure 3 first performed in glass membranes by Eisenman, Sandblom and Walker (1967). The experiments in Figures 2 and 3 are particularly simple in that the dependence of conductance on solution composition is clearly demonstrable without requiring any knowledge of the current-voltage characteristic of the system.

The phenomenon of saturation of single channel conductance with increasing electrolyte activity was first reported by Hladky and Haydon (1972), whose data are illustrated in Figure 4 for NaCl, KCl, and CsCl. This was interpreted by Hladky (1972) and Läuger (1973) as a natural consequence of assuming that the pore, for electrostatic reasons, was unlikely to be occupied by more then one cation at a time. Hladky (1972), however, clearly recognized that doubly occupied states could occur, the

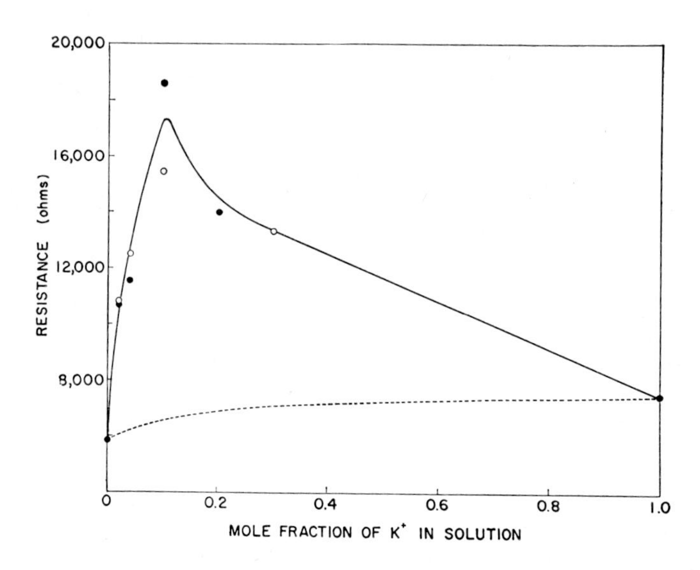

Fig. 3. Resistance of a 1 μ thick, completely hydrated K^+ selective glass membrane (interposed between identical solutions of varying Na^+-K^+ composition) as a function of the mole fraction of K^+ in solution. The dots and open circles refer to two series of measurements of the resistance, which was found to be time-independent and voltage-independent under these experimental conditions (After Eisenman et al., 1967, Figure 3).

pore being blocked in such states. Although the saturating aspects of the conductance were emphasized by these authors, the existence of a

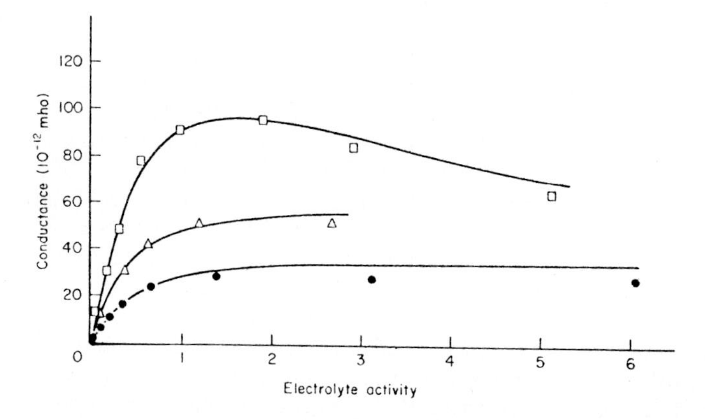

Fig. 4. Saturation of the single channel conductance of the gramicidin A channel as a function of electrolyte activity (M/L measured at 100 mV in NaCl (filled circles), KCl (open Triangle and CsCl (open boxes). (After Hladky, 1973, Figure 5).

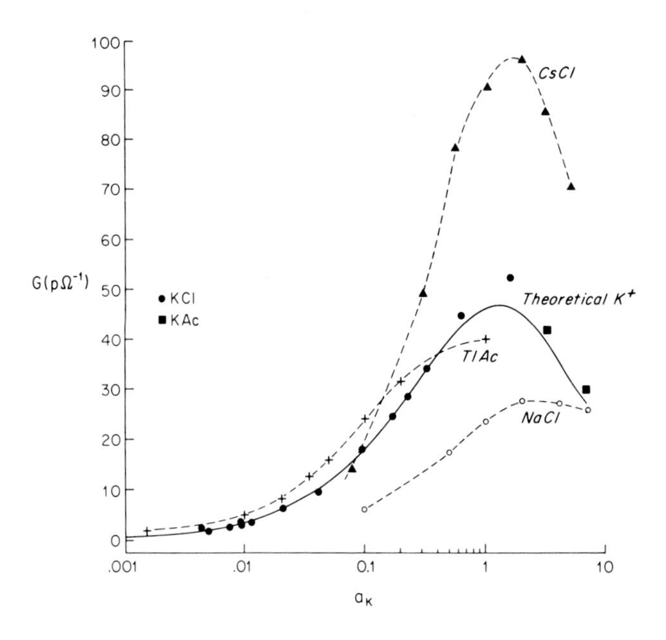

Fig. 5. Concentration dependence of single channel conductances of gramicidin A in GMO/hexadecane bilayers as a function of electrolyte activity. Our experimental data for TlCl, TlAc, KCl and Kac solutions (measured using applied voltages of ± 50 mV) are plotted by the indicated data points and their curves; and a theoretical curve for K^+ (calculated from Equation (19), using the parameters given in Table III and IV) is also plotted as the heavy curve, labelled theoretical K. This figure also present the data of Hladky and Haydon, 1972; and Hladky, 1973 for NaCl and CsCl. (Our data for KCl agree quite closely with theirs).

discrepancy from such simple behavior, most clearly seen in the decrease in CsCl conductance at very high concentrations, was recognized and explained as a consequence of double occupancy by Hladky. We have confirmed the existence of this phenomenon for K^+ solutions at very high concentrations, as illustrated by the KCl and KAc data points in Figure 5. In addition, as will be described in Section 4, we have been able to account for this as an expected property of our four-site model for the channel, which predicts the theoretically calculated curve for K^+ indicated by the heavy line in Figure 5.

Our membrane potential measurements also required the existence of an additional binding site at lower concentrations than had heretofore been studied. The existence of this site was verified (Eisenman et al., 1976a) and is demonstrated in Figures 6 and 7 by the two linear regions of different slopes for the single channel conductance-concentration dependence in Eadie-Hofstee type of Michaelis-Menten plots (cf. Zeffren and Hall, 1973, p. 67). Previous examinations by Hladky (1974) for Na^+ and by Neher (1975) for Tl^+ of such concentration-conductance dependences

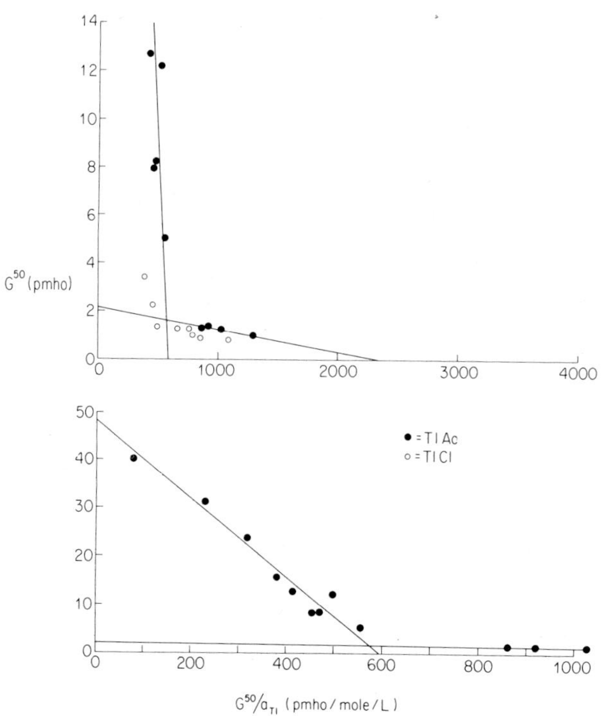

Fig. 6. Eadie-Hofstee type plot for conductance-concentration relationship for TlCl and TlAc for single channel conductances in GMO/hexadecane bilayers measured with ± 50 mV applied potentials. The ordinate is the observed conductance and the abscissa is the conductance divided by the activity. The lower portion of the figure presents an expanded view of the high concentration region. This is the region previously explored by Neher (1975). Our new finding is the additional linear region at very low concentrations seen best in the upper portion of the figure.

had demonstrated the linear region at high concentration, but had not been carried out at sufficiently low concentrations to detect the additional linear region illustrated to the right in Figures 6 and 7. The conductance-concentration dependence will be examined in more detail in Section 4.

The permeability ratio is expected to be independent of the ion concentration in a one-ion-saturable channel (Läuger, 1972; Hladky, 1973) and this appeared to be borne out by the data of Myers and Haydon for the group Ia cations except at high concentrations. In contrast we find for K^+-Tl^+ mixtures a strong concentration dependence of the permeability ratios becoming apparent at quite low concentrations of Tl^+. Typical observations are illustrated in Figures 8 and 9. Figure 8 illustrates how membrane potential measurements are made for a single channel and corre-

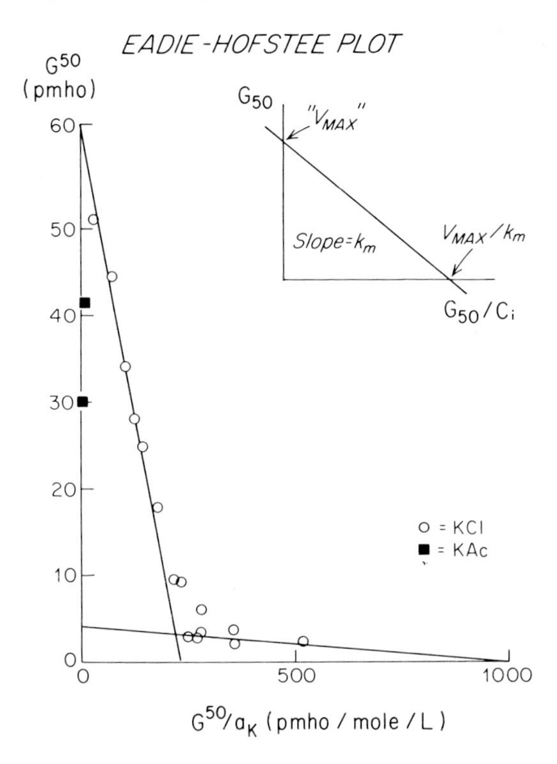

Fig. 7. Eadie-Hofstee type plot for KCl and KAc solutions. The single channel conductances in GMO/hexadecane bilayers were measured with ± 50 mV applied potentials. The insert presents schematically an Eadie-Hofstee plot with the terminology used for enzyme kinetics.

sponds to the left most data points of the lower portion of Figure 9, which has been carried out on grossly conducting membranes containing many channels. These two types of measurements should be, and are, identical to within experimental error (cf. the single channel data included in Figure 9); but the gross measurements can be made more easily and precisely.

The results are sufficiently complicated that it will be worthwhile examining Figure 9 in some detail. The upper portion of this figure illustrates the membrane potential observed for a membrane interposed between identical TlCl solutions when KNO_3 is varied on one side (see inset for experimental conditions); while the lower portion illustrates the behavior for symmetrical KCl solutions when Tl^+ is varied on one side. The points are measured experimentally, the curves are calculated according to the Goldman-Hodgkin-Katz equation (anion permeability assumed negligible) for the values of P_{Tl}/P_K indicated at the right.*

---

* These values are designated as apparent, $(P_{Tl}/P_K)_{App.}$ to distinguish them from (P_{Tl}/P_K) ratios calculated taking into account effects of ionic strength variation on activity coefficients.

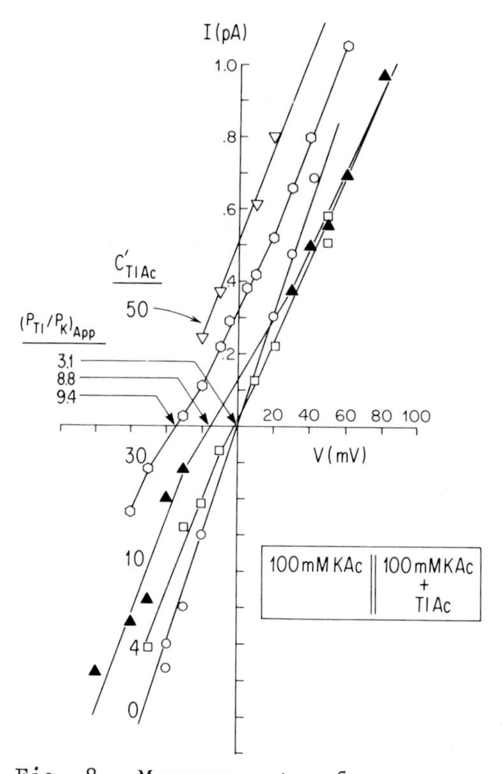

Fig. 8. Measurements of zero-current membrane potentials and calculations of permeability ratios for single channels of gramicidin A in GMO/hexadecane bilayers. The ordinate plots the current measured for the most frequently encountered channels at the applied voltages indicated along the abscissa. The solutions conditions are indicated in the insert. The numbers under the heading c'_{TlAc} indicate the TlAc concentration added unilaterally to the initially symmetrical 100 mM KAc solutions. All measurements were made on a single bilayer, and the changes in X intercepts with unilateral additions of Tl give directly the transmembrane potential for the single channel at zero current without any uncertainties as to junction potentials. The calculated values of apparent permeability ratio $(P_{Tl}/P_K)_{App}$ corresponding to the X intercepts are tabulated at the left. Also notice that a blocking effect of Tl^+ is apparent in the decrease in slope of (i.e. the conductance) at zero current on the addition of unilateral Tl^+. The values of the zero current intercepts have been plotted in the lower portion of Figure 9 for comparison with the many-channel data to be present there.

A strong dependence of P_{Tl}/P_K on bilateral Tl^+ concentration is seen in the upper part of the figure; for this ratio increases from 4 to 40 as the TlCl concentration is increased from 0.1 mM to 10 mM. Such a depen-

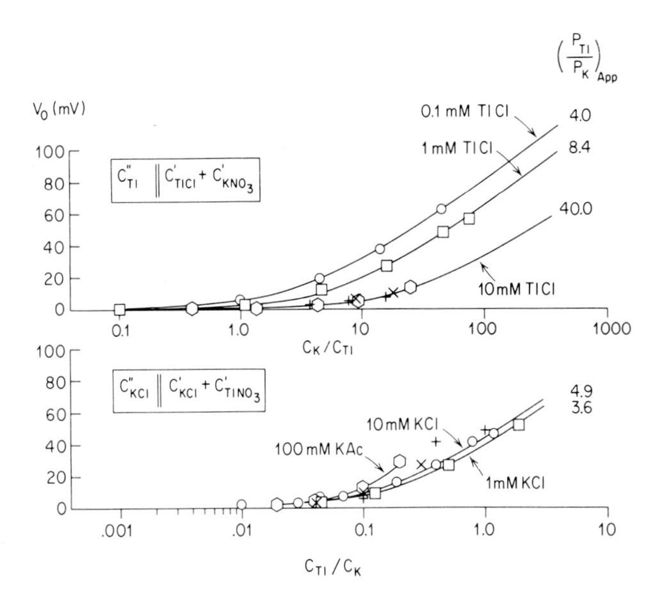

Fig. 9. Measurements of zero current membrane potentials and concentration dependence of permeability ratios in a many-channel (macroscopically conducting) situation. The observed membrane potential differences, plotted by open symbols, correspond to the membrane potentials measured in macroscopically conducting GMO/decane bilayers containing many gramicidin A channels simultaneously in the presence of approximately 10^{-9} M aqueous gramicidin A. For comparison with the many-channel data the x's and +'s correspond to measurements for single channels of valine gramicidin A in GMO/hexadecane bilayers (10 mM bilaterally symmetrical TlCl (× is KAc, + is KNO_3) in the upper portion of the figure) and 10 mM bilaterally symmetrical KCl (+) and 100 mM KAc (x) in the bottom portion of the figure.

dence is unprecedented in any prior studies on carriers (cf. Eisenman et al., 1973, p. 148) nor is it seen in nonactin controls carried out under the same ionic conditions.

In contrast, the lower portion of the figure demonstrates that there is much less change of P_{Tl}/P_{K^+} seen for different symmetrical K^+ levels for onesided variations of Tl^+, even though the one-sided Tl^+ concentration was varied over a range wider than encompassed in the two-sided variations of the upper portion of the figure. This is seen by the close fit of the data points to the theoretical Goldman-Hodgkin-Katz curves drawn for closely similar P_{Tl}/P_K values of 3.6 and 4.9 at 1 mM KCl and 10 mM KCl, respectively, and even the data obtained in 100 mM KAc (for solubility reasons) show only a small increase in P_{Tl}/P_K.

It is important to note that the close agreement between the Goldman-Hodgkin-Katz curves and the data points seen for all the chlorides verifies that in these experiments chloride permeability is negligible and also that the permeability ratios are independent of the transmembrane voltage.

2.1. *The Model*

All of the above electrical phenomena can be accounted for quantitatively by a theoretical analysis (Sandblom et al., 1976) of the 4-site model chematized in Figure 10. This model extends the one- or two- ion-satur-

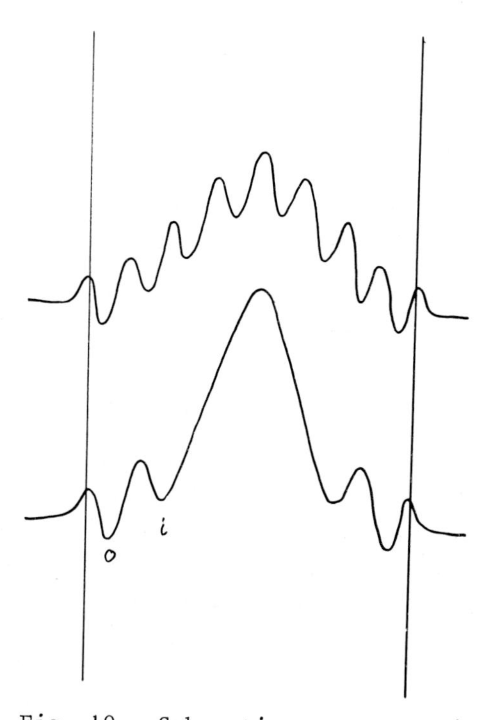

Fig. 10. Schematic representation of the gramicidin A channel as a sequence of potential energy minima (binding sites) separated by energy barriers. The upper portion represents a reasonably realistic schematic picture for a channel in which the binding sites are assumed to be chemically identical but to differ from each other energetically as a result of the addition of an electrostatic image force potential which is maximum at the center of the membrane. The lower portion of the figure further schematizes this situation as our four-site model by representing the possibilities for multiple occupancy of the channel by two binding sites on each side of the membrane which.are in equilibrium with the aqueous solution which they face (and with each other) while being separated by a central high (rate-limiting) barrier in the middle of the membrane.

able, two-compartment models of Läuger (1973) and Hladky (1974) by permitting multiple occupancy of the channel, with four binding sites explicitly represented in the lower portion of Figure 10 by the two binding sites, o and i, in each compartment on either side of a central high barrier. These sites are assumed each to be in equilibrium with the aqueous solution facing them (as well as with each other); and because of this, our model specifically excludes a single filing permeation mechanism. Indeed, the concepts of inner vs. outer do not necessarily refer to the distances from the membrane-solution interface, but could correspond, for example, to the radial distribution with the outer corresponding to an energy well for a binding site at the circumference of the pore, whereas the inner could correspond to the energy well at the center of the pore. Studies of the voltage-dependent properties of the system will be needed to choose between these different physical interpretations; and flux ratio-voltage studies will presumably be necessary to examine the consequences of single filing.

The upper portion of Figure 10 is presented to show more realistically th physical situations that our idealized model is intended to represent. It schematically represents the gramidicin A channel as a sequence of potential energy minima (binding sites) separated by energy barriers. The binding sites are assumed to be chemically identical, but their values are raised successively in moving toward the center of the membrane owing to the superposition of an electrostatic image force potential. The lower portion of the figure represents this schematically as two-binding sites, outer and inner.

In the following three sections, we will examine the behavior of the gramicidin A channel quantitatively proceeding from the most general properties to increasingly more specific ones. Thus, Section 3 characterizes permeability selectivity as seen in the membrane potential behavior, which depends only upon the heights of the peaks* in the energy diagram relative to the aqueous solution and the shifts of these peaks with various loadings. Membrane potential measurements alone neither need, nor yield, independent knowledge as to the depths of the wells (i.e. binding constants) or heights of barriers above wells (i.e. rate constants). Section 4 then deals with saturation and the binding properties for single salts, as seen in the single channel conductances. These phenomena are dependent upon the separate values of both of the wells and the heights of barriers above them, as well as how they shift with increasing loading for a single type of cation species. The detailed knowledge of the concentration dependence of conductance enables the levels of certain wells and their shifts with loading to be determined. Lastly, Section 5 examines the cross effects between two different ionic species, particularly as they are expected to be reflected in the single channel conductances in mixtures of cations. It is the shifts in energy profiles produced by these cross effects that lead to the concentration-dependent permeability ratios (through shifts in peaks) and to blocking

---

* The peak energy (relative to the aqueous solution) reflects the *product* of the binding constant and the rate constant, as is discussed by Hille with admirable clarity elsewhere (1975a).

(and enhancement) phenomena due to the separate shifts of wells and the heights of the barriers above them. Here, the data of Sections 3 and 4 are combined to predict the phenomena of blocking, and also of enhancement, and in this way demonstrate how blocking phenomena and concentration dependent permeability ratios represent separate aspects of the underlying ion interactions within the channel.

3. SELECTIVITY — AS SEEN IN THE PERMEABILITY RATIOS CHARACTERIZING THE MEMBRANE POTENTIAL IN THE PRESENCE OF MANY CHANNELS AS WELL AS FOR THE INDIVIDUAL CHANNEL

For the above described model our theoretical treatment (Sandblom et al. 1976) indicates that the membrane potential will be given generally by:

$$V_o = \frac{RT}{F} \ln \frac{P_x a'_x + P_y a'_y + P_A a''_A}{P_x a''_x + P_y a''_y + P_A a'_A} \tag{1}$$

In this equation a's are activities, the subscripts X and Y represent cations and A represents an anion, and the superscripts ' and " denote the solutions on the two sides of the mambrane. This equation is identical to the Goldman-Hodgkin-Katz equation and has been previously shown by Läuger (1973) and Hladky (1972, 1972) to be expected to hold under certain conditions for a one-ion-saturable, two lattice-site model for the gramicidin channel. In Läuger's (1973) treatment, which restricted to occupancy by no more than one ion at a time in the channel, the permeability ratio P_x/P_y is a constant, independent of ionic concentration; but Hladky (1972, p.103) explicitly recognized that the biionic potential could depend on the absolute concentration if a second ion entered the pore. In our treatment, which allows more than one cation at a time in the channel, this permeability ratio becomes a function of ion concentration (more precisely, activity) given by

$$\frac{P_x}{P_y} = \alpha \frac{(1 + \beta_x c'_x + \beta_y c'_y)(1 + \beta_x c''_x + \beta_y c''_y)}{(1 + \gamma_x c'_x + \gamma_y c'_y)(1 + \gamma_x c''_x + \gamma_y c''_y)} \tag{2}$$

which, for low ion concentrations reduces to Hladky's and Läuger's case for which (P_x/P_y) is given by the concentration independent limiting value, α. Equation (2) contains five independent constants (α, β_x, β_y, γ_x, γ_y), whose magnitudes can be measured experimentally from membrane potential measurements of the type presented in Figures 1, 8 and 9, from which the values given in Table II have been determined. The adequacy of this expression to describe the concentration dependence of P_{Tl}/P_K is illustrated in Figures 11 and 12 for a variety of experimental situations.

Figure 11 illustrates the satisfactory agreement between the theoretically expected (curves) and experimentally observed (points) values as a function of ionic activities calculated from the membrane potential

TABLE II
Experimentally measured values of the parameters determining the permeability ratio through Equation (2)

α	3.5
β_{Tl}	1.0 mM^{-1}
β_{K}	0.10 mM^{-1}
γ_{Tl}	0.5 mM^{-1}
γ_{K}	0.08 mM^{-1}

$x = Tl^{+}$, $Y = K^{+}$

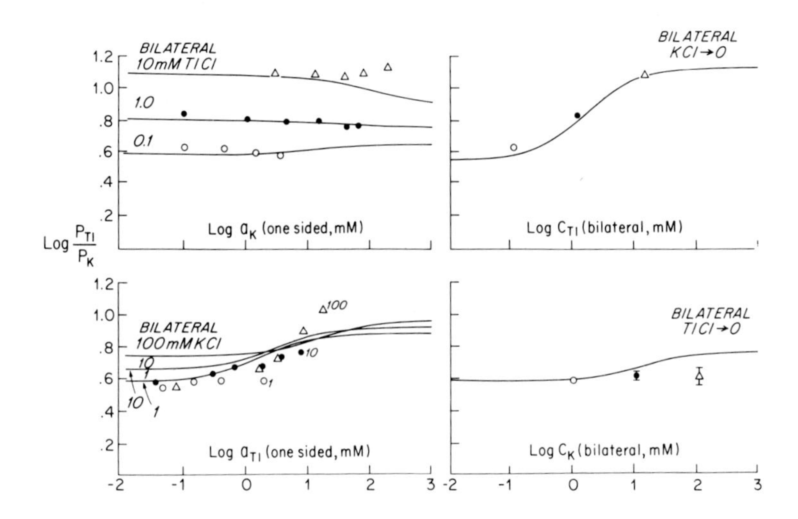

Fig. 11. Theoretically expected and experimentally observed dependence of the permeability ratio on ionic concentrations under a variety of solutions conditions. The points are permeability ratios computed directly from the experimental membrane potentials of Figure 9; while the curves are theoretically calculated according to Equation (2) using the values of the parameters given in Table II. In the right-hand figures the data points taken as the best experimental estimate for the indicated zero concentration limit are the average of the two lowest concentrations measured in the left-hand figures.

data of Figure 9 and Figure 12 illustrates this for the bi-ionic potential situation. In the experimental conditions of these figures anion permeability has been calculated to be negligible (Eisenman et al., 1976b), and the data have been corrected for the effect of variable ionic strength on the activity coefficients. The upper portion of Figure 11 presents the dependence of permeability ratios corresponding to the experiments of Figure 9 where Tl^{+} was varied bilaterally and K^{+} unilateral-

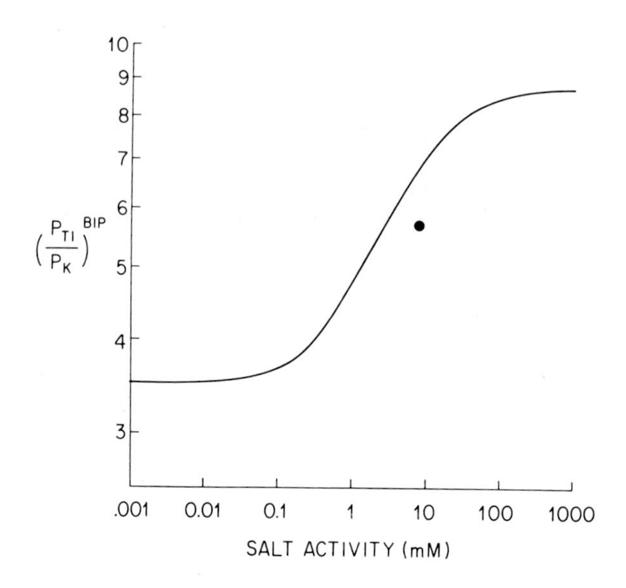

Fig. 12. Theoretically expected and experimentally observed permeability ratios for the bi-ionic situation, calculated according to Equation (6), with the parameters of Table II. The data point is the permeability ratio computed from the observed bi-ionic potentials at 10 mM concentration in many channel measurements of valine gramicidin A in GMO/decane bilayers.

ly; while the lower portion of Figure 11 gives the dependence of permeability ratios observed in the experiments of the lower portion of Figure 9 where K^+ was varied bilaterally and Tl^+ unilaterally. The theoretical curves have been drawn according to the parameters given in Table II ($X = Tl^+$, $Y = K^+$). Figure 12 presents the corresponding theoretical expectations for the bi-ionic situation, together with the data point observed experimentally at 10 mM in a many channel system. The agreement in Figures 11 and 12 between theory and experiment is satisfactory, particularly in view of the wide range of experimental conditions.*

Since Equation (2) seems to be verified, let us examine some of its properties and their implications.

In the limit of low salt concentration (c_x and $c_y \to 0$), Equation (2) reduces to

---

* Because we have yet to find an impermeant anion compatible with the solubility properties of Tl^+, our experiments have all been performed in the absence of any indifferent supporting electrolyte. Consequently, ionic strength effects produce changes in activity coefficients which are not always negligible; and the data presented in Figure 11 have been corrected for such effects as described elsewhere (Eisenman et al., 1976b).

$$\left(\frac{P_x}{P_y}\right)_{oo} = \left(\frac{P_x}{P_y}\right)_{c\to 0} = \alpha \tag{3}$$

where the subscripts o o signify that the outer binding sites are unoccupied in this limit.

In the case where the concentration of x becomes very large ($c_x \to \infty$) the permeability ratio becomes

$$\left(\frac{P_x}{P_y}\right)_{x_o x_o} = \left(\frac{P_x}{P_y}\right)_{c_x\to\infty} = \alpha\left(\frac{\beta_x}{\gamma_x}\right)^2 \tag{4}$$

where the subscript $x_o x_o$ indicates that in this limit both outer sites are loaded with species x. Here, besides α, the permeability ratio depends upon the ratio $(\beta_x/\gamma_x)^2$.

In the case where the concentration of y becomes very large ($c_y \to \infty$) and both outer sites are occupied by species y,

$$\left(\frac{P_x}{P_y}\right)_{y_o y_o} = \left(\frac{P_x}{P_y}\right)_{c_y\to\infty} = \alpha\left(\frac{\beta_y}{\gamma_y}\right)^2 \tag{5}$$

The corresponding equations for the bi-ionic case are, in general,

$$\left(\frac{P_x}{P_y}\right)^{BIP} = \alpha\,\frac{(1+\beta_x c)(1+\beta_y c)}{(1+\gamma_x c)(1+\gamma_y c)} \tag{6}$$

for low salt concentrations,

$$\left(\frac{P_x}{P_y}\right)^{BIP}_{o\ o} = \left(\frac{P_x}{P_y}\right)^{BIP}_{c\to 0} = \alpha \tag{7}$$

and for high salt concentrations,

$$\left(\frac{P_x}{P_y}\right)^{BIP}_{x_o y_o} = \left(\frac{P_x}{P_y}\right)^{BIP}_{c\to\infty} = \alpha\,\frac{\beta_x\beta_y}{\gamma_x\gamma_y} \tag{8}$$

where the subscripts in the latter equation indicate that at high concentrations in the bi-ionic situation one outer site is occupied by X and the other outer site by species Y.

The physical meaning of these formal limits of Equation (2) can now be scrutinized. The symbol β is used to indicate that the permeant species is X and the symbol γ is used to indicate that the permeant species is Y; whereas the subscripts indicate which species of ion is occupying the outer site. Thus β_x corresponds to the effect on P_x of loading a species X on the outer site on either side of the membrane, its magnitude giving the weighting factor for the effects of the concentration of X (notice that β_x appears only in the numerator and in front of the terms depending in the X concentrations, c'_x and c''_x). β_x appears symmetrically regardless of whether the concentration of X is varied in solution ' or ", as it should for a channel which is symmetrical with respect to both aqueous solutions.

β_y corresponds to the effect on P_x of loading a species Y on the outer site, and appears as the coefficient of those terms in the numerator which depend on the concentration of Y in the aqueous solutions. Similar considerations apply to the permeability of Y in the denominator where the symbol γ indicates that the permeant species is Y and the subscripts indicate the species of ion occupying the outer site. Thus, the coefficient γ_x gives the effect on P_y of loading a species X on the outer site and γ_y gives the effect on P_y of loading a species Y on the outer site.

Inspection of (2) or (6) shows that the formal requirement for a concentration dependence of the permeability ratio is that $\beta_x \neq \gamma_x$ and/or $\beta_y \neq \gamma_y$. Only if either, or both, of these conditions exist will the permeability ratio be a function of ionic concentration. Examining Table II and identifying X with Tl^+ and Y with K^+, we see that $\beta_{Tl} \neq \gamma_{Tl}$ whereas $\beta_K \simeq \gamma_K$ (our present data are not yet adequate to ascertain whether or not β_K is really different from γ_K). Thus the principle cause for the observed concentration dependence of permeability ratios is referrable to the difference in values of β_{Tl} and γ_{Tl}, which reflect the effects of loading Tl^+ on the outer sites. It seems physically reasonable that the terms reflecting the effects of loading K^+ on the external sites (β_K and γ_K) should be smaller and their possible differences perhaps less important.

The preceding discussion can be made more concrete, and the meaning of these parameters can be better grasped by presenting an energy profile diagram of the Eyring type (Eyring et al., 1949; Zwolinski et al., 1949 and Woodbury, 1971) so clearly discussed by Hille (1975a). Bezanilla and Armstrong (1972) and Hille (1975a) have indicated the particular significance of the peaks in an energy profile diagram in defining the permeability ratios, pointing out that the permeability for a given species depends only upon the energetic height of the internal peak, relative to the energy level of the aqueous solution. Hille has further noted that when the difference in peaks between two different species is constant, even if the individual peaks vary with an experimental variable, then the permeability ratios are expected to be constant, such a situation satisfying his constant offset energy peak condition.

An energy level diagram for the peak heights in the gramicidin A channel can be constructed from the present membrane potential data (using one additional parameter, measured in the next section for each ion, K^o_{Tl} and K^o_K, to assess their absolute, as well as relative, shifts with various loadings). Such a diagram is presented in Figure 13 where the peak values, and shifts in peaks, relative to the aqueous solution, are plotted for the present gramicidin A channel under the conditions of loading indicated along the abcissa. The ordinate plots the energy levels (in units of RT) calculated from the logarithms of the observed permeability ratio and its dependence upon concentration, as given by the parameters in Table II and the appropriate limits of Equation (2).

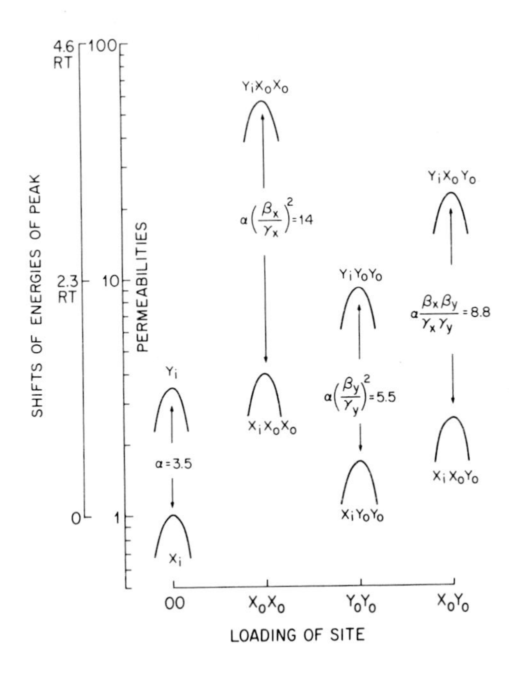

Fig. 13. Energy level diagram for the gramicidin A channel giving the values of the peaks (relative to aqueous solution) and shifts in peaks under the conditions of loading indicated along the abscissa based upon the parameters of Table II. The shifts of peaks relative to the unloaded state, and all referred to the aqueous reference state, are related to the α, β, and γ parameters governing permeabilities and the binding constants, K^o_x and K^o_y, for the external site as follows (recall Equation (13). The peaks for species X are shifted upwards by factors of 2 RT ln (K^o_x/β_x) for the $X_iX_oX_o$ state, by 2 RT ln (K^o/β_y) for the $X_iY_oY_o$ state, and by RT ln $(K^o_xK^o_y/\beta_x\beta_y)$ for the $X_iX_oY_o$ state. The peaks for Y are shifted upwards by 2 RT ln (K^o_x/γ_x) for the $Y_iX_oX_o$ state, by 2 RT ln (K^o_y/γ_y) for the $Y_iY_oY_o$ state, and by RT ln $(K^o_xK^o_y/\gamma_x\gamma_y)$ for the $Y_iX_oY_o$ state. Notice that the binding constants drop out of the differences between the peaks.

The abscissa lists four different states of loading of the outer sites, using the terminology discussed above, in which o o indicates no ions bound to the outer site, x_ox_o signifies both outer sites occupied by species X (Tl^+ in this case), etc.

The heavy arrows indicate the energetic distances between the peaks for species X and Y seen under the various loading conditions and are directly related to the corresponding permeability ratios. Thus, the shift in peak offset energies when both outer sites are loaded by Tl^+ (the shift in permeability ratio from 3.5 to 14) corresponds to the value $(\beta_X/\gamma_X)^2$ of Equation (4). Notice that the shift in peak offset energy in the bi-ionic situation is the arithmetic mean of the energies in the two symmetrically loaded situations (i.e. the geometric mean of the changes in permeability ratios).

4. SATURATION AND BINDING — AS SEEN IN THE INDIVIDUAL CHANNEL CONDUCTANCE IN SINGLE SALT SOLUTIONS

The dependence of the conductance of individual channels on the concentrations of TlCl and KCl (and KAc) in single salt solutions is illustrated in Figures 14 and 15. The points plot the experimentally observed data

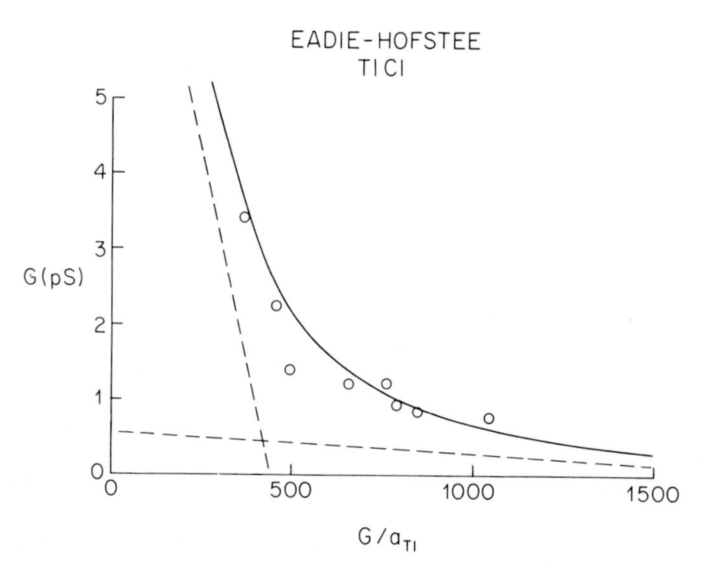

Fig. 14. Comparison of the experimentally observed (points) and theoretically calculated (curves) conductance for TlCl as a function of concentration according to Equation (10), (11), and (19) with the parameters given in Table III. The ordinate is the single channel conductance (measured using ± 50 mV applied voltage); the abscissa is conductance divided by Tl^+ activity.

(previously shown in Figures 6 and 7). For comparison the conductance theoretically expected from the four-site model is given by the continuous curves and straight line asymptotes drawn according to Equations (9 — 12).

The membrane conductance at zero current, G, in the presence of a single salt at the same concentrations on both sides of the membrane for

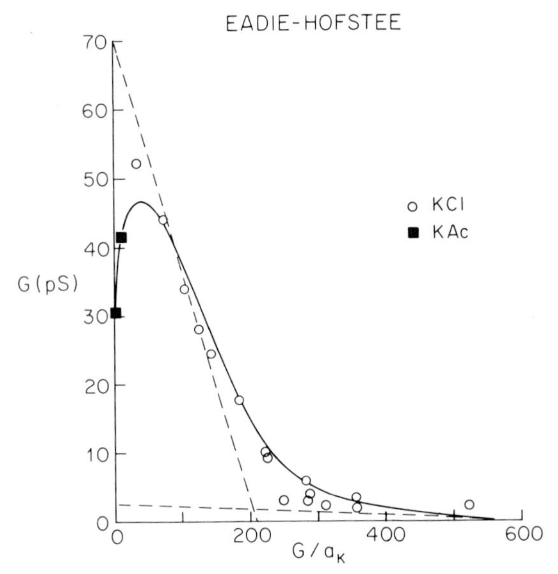

Fig. 15. Comparison of experimentally observed and theoretically expected conductance for KCl (circles) and KAc (boxes). The full curve is calculated according to Equation (19) and the asymptotes according to Equation (10) and (11) with the parameters given in Table III. The ordinate is the single channel conductance; the abscissa is conductance denoted by K^+ activity.

a four-site channel (Sandblom et al., 1976) is:

$$G = \frac{(p_1 c + 2p_2 c^2 + p_3 c^3)}{1 + K_1 c + K_2 c^2 + K_3 c^3 + K_4 c^4} \tag{9}$$

where c is the concentrations of species x (subscript omitted for simplicity), and the K's are combined binding constants (which are determined by the levels of the energy wells in Figure 10 and the shifts in these levels with concentration) and the p's are products of rate constants and binding constants (which are determined by the height of the central barrier and the shift in this height with concentration).* An alternative expression, giving the conductance in terms of permeabilities and experimentally measured binding constants, will be found in Eq. (19).

$$\begin{aligned} K_1 &= 2(k_{x_i:} + k_{x_o:}) \\ K_2 &= (2k_{x_o x_i:} + 2k_{x_o:x_i} + k_{x_i:x_i} + k_{x_o:x_o}) \\ K_3 &= 2(k_{x_o x_i:x_o} + k_{x_o x_i:x_i}) \\ K_4 &= (k_{x_o x_i:x_i x_o}) \end{aligned} \tag{9a}$$

---

* These constants are defined in terms of the more fundamental binding constants (K) and rate constants (v) characteristic of particular states of the system through eqs. (9a) and (9b) using terminology discussed in detail elsewhere (Sandblom et al., 1976).

$$P_1 = v_{x_i:} k_{x_i:}$$

$$P_2 = v_{x_i:}^{x_o:} k_{x_o x_i:} \tag{9b}$$

$$P_3 = v_{x_i:}^{x_o:x_o} k_{x_o x_i : x_o}$$

Briefly, the subscripts denote the states of occupancy of the channel, a colon representing the barrier. For example, $k_{x_i:}$ and $k_{x_o:}$, respectively denote the binding constants intrinsic to the inner and outer sites on one side of the membrane when no other sites are occupied. Similarly, $k_{x_o x_i:}$ is the binding constant when the outer and inner sites on the same side of the channel are both occupied by species x, $k_{x_o:x_i}$ is the binding constant when the outer site on one side and the inner site on the other side are occupied, etc. K_1 thus represent the binding constant to all singly occupiable states of the channel, K_2 represents the binding constant to all doubly occupiable states, K_3 to all triply, and K_4 to all quadruply occupiable states. The rate constants, v, are defined similarly. For example, $v_{x_i:}$ represents the rate constant for an ion x to jump from the inner site over the central barrier in an empty channel, $v_{x_i:}^{x_o:}$ is the rate constant when both the outer and the inner site on one side of the channel are occupied, et.

Equation (9) has three important limits. For low ion concentrations

$$G = G_x^o - \frac{G}{c} \frac{1}{K_x^o} \tag{10}$$

for high concentrations,

$$G = G_x^h - \frac{G}{c} \frac{1}{K_x^h} \tag{11}$$

and, for the very highest concentrations,

$$\log G = \log G_x^h - \log cK_x^\infty \tag{12}$$

The parameters in these equations (K_x^o, K_x^h, K_x^∞, G_x^o, G_x^h) are all experimentally measurable K_x^o and K_x^h are binding constants and G_x^o and G_x^h are maximum limiting conductances for the low (superscript o) and high (superscript h) concentration asymptotes on Eadie-Hofstee type plots such as in Figures 14 and 15 where the Y intercepts of the straight line asymptotes give the maximum limiting conductances and the slopes are inversely proportional to the binding constants. The binding constants K_x^∞ for the very highest concentration is defined through Equation (12) as the reciprocal of that value of the concentration at which the observed conductance equals G_x^h.

Figures 14 and 15 compare the experimantally observed conductances (data points) with the theoretically calculated continuous curves (for Equation (9) or (19)) and straight line asymptotes (from Equations (10) and (11)) according to the values of the parameters in Table III. Figure 14 shows clearly for TlCl, even though the data are restricted to relatively low concentrations because of the limited solubility of TlCl, the

TABLE III
Binding constants, maximum limiting conductances, and their products, as determined from the experimental data of Figures 14 and 15 for single salts of TlCl and KCl

Binding constants		Maximal limiting conductances (Y Intercepts)		Conductance·binding products (X Intercepts)	
K_{Tl}^o	3600 M^{-1}	G_{Tl}^o	0.53 $p\Omega^{-1}$	$G_{Tl}^o K_{Tl}^o$	1900 $p\Omega^{-1}M^{-1}$
K_K^o	260 M^{-1}	G_K^o	2.12 $p\Omega^{-1}$	$G_K^o K_K^o$	550 $p\Omega^{-1}M^{-1}$
K_{Tl}^h	45 M^{-1}	G_{Tl}^h	10.44 $p\Omega^{-1}$	$G_{Tl}^h K_{Tl}^h$	470 $p\Omega^{-1}M^{-1}$
K_K^h	2.9 M^{-1}	G_K^h	71.5 $p\Omega^{-1}$	$G_K^h K_K^h$	207 $p\Omega^{-1}M^{-1}$
K_{Tl}^∞	0 M^{-1}				
K_K^∞	0.22 M^{-1}				

two regions of low (o) and high (h) concentration dependences. (We do not analyze the data of Figure 6 for the more soluble TlAc here owing to anion effects seen with acetate which will be discussed elsewhere (Neher et al., 1976)).

Figure 15 is particularly interesting in that for K^+ it not only exhibits the two linear regions of concentration dependence corresponding to Equations (10) and (11) but also shows the decrease in conductance at the very highest concentrations expected from Equation (12). This is illustrated even more strikingly in Figure 16. As mentioned above in discussing Figure 5, a similar maximum conductance is clearly present in Hladky's and Haydon's conductance data for Cs^+, and is discernable also in their data for K^+ and Na^+ (recall Figure 6).

The experimentally measured constants G_x^o, G_x^h, K_x^o, K_x^h and K_x^∞ can be expressed in terms of the more fundamental rate constants and binding constants of the channel through the following relationships using a terminology which makes explicit the states of loading of the channels to which these constants refer. These relationships are (when the binding to the outer site is stronger than to the inner site, $k_{x_o} \gg k_{x_i}$):

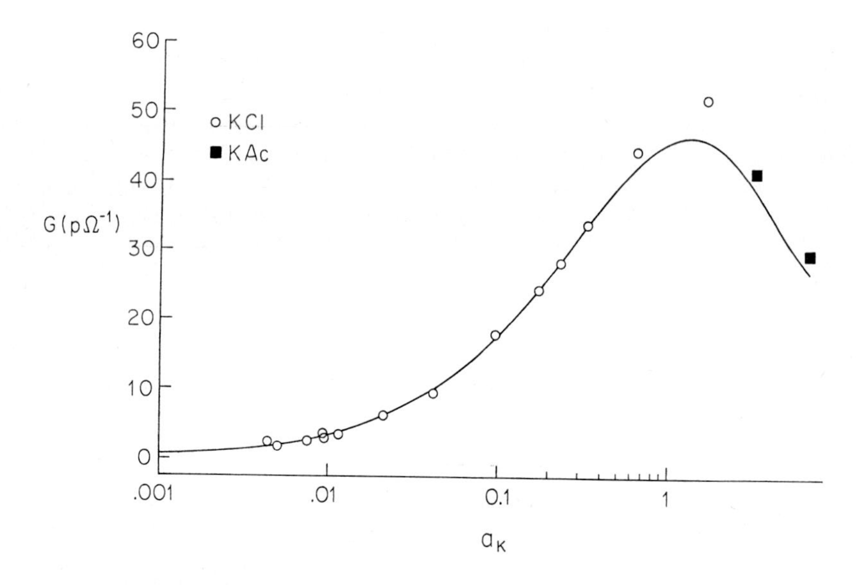

Fig. 16. Experimentally observed conductance for the individual gramicidin A channel as a function of K^+ activity compared with the theoretical expectations of the four-site model. The points are experimentally measured. The curve is drawn according to Equation (19) using the constants of Table II and III.

$$G_x^o = \frac{v_{x_i} k_{x_i}}{2k_{x_o}} \quad \text{and} \quad K_x^o = 2\, k_{x_o} \tag{13}$$

$$G_x^h = \frac{v^{x_o x_o}_{x_i}}{2} \quad \text{and} \quad K_x^h = 2k^{x_o x_o}_{x_i} \tag{14}$$

$$K_x^\infty = k^{x_o x_o x_i}_{x_i}/2 \tag{15}$$

The superscripts indicate the state of loading of the channel; while the subscripts indicate to which site the parameter refers. For example, $k^{x_o x_o}_{x_i}$ refers to the binding constant to the inner site (indicate by the subscript) when both outer sites (indicated by the superscripts) are occupied; while k_{x_i} and k_{x_o}, respectively, refer to the intrinsic binding constants to the inner site and the outer site in the absence of loading of any other site in the channel. Similarly, the rate constant $v^{x_o x_o}_{x_i}$ refers to the transition rate constant from the energy level of the inner site well when both outer sites are occupied across the height of the central peak when both outer sites are occupied; and v_{x_i} refers to the transition rate constant for an ion x to jump from the well depth of the inner site in an otherwise empty channel across the central barrier in an otherwise empty channel.

4.1. *Comparison With Potential Data*

Certain of the parameters assessed from conductance are, in terms of the theory, expected to be related to parameters measured in the previous section from membrane potentials. In particular, the parameter α should satisfy the condition

$$\alpha = \frac{G_x^o K_x^o}{G_y^o K_y^o} \tag{16}$$

and the values of β_{Tl} and γ_K should equal

$$\beta_{Tl} = \frac{K_T^o}{2}\sqrt{\frac{G_{Tl}^h K_{Tl}^h}{G_{Tl}^o K_{Tl}^o}} \tag{17}$$

and

$$\gamma_K = \frac{K_K^o}{2}\sqrt{\frac{G_K^h K_K^h}{G_K^o K_K^o}} \tag{18}$$

Table IV compares the values of those parameters, as measured by conductance, with those measured from membrane potentials; and it is clear that the agreement is satisfactory, which constitutes an important check on the adequacy of our model and the internal consistency of the data.

TABLE IV
Comparison of parameters measured from single channel conductance in single salts with those previously measured from membrane potentials in ionic mixtures at zero current

	From conductance	From potential
α	3.5	3.5
β_{Tl}	0.9 mM^{-1}	1.0 mM^{-1}
β_K	0.10 mM^{-1}	0.10 mM^{-1}
γ_{Tl}		0.5 mM^{-1}
γ_K	0.08 mM^{-1}	0.08 mM^{-1}

5. BLOCK — AS SEEN IN SINGLE CHANNEL CONDUCTANCE IN IONIC MIXTURES

The combination of parameters measured by membrane potential in Table II and single salt conductances in Table III are sufficient to predict the features to be expected for the conductance of a single channel exposed symmetrically to mixtures of K^+ and Tl^+; and Figure 17 presents the conductance expected theoretically (see Equation (19)), as a function of percent Tl^+ in Tl^+-K^+ mixtures at the indicated concentrations. The blocking implied by the strong minimum of conductance at 10 mM total

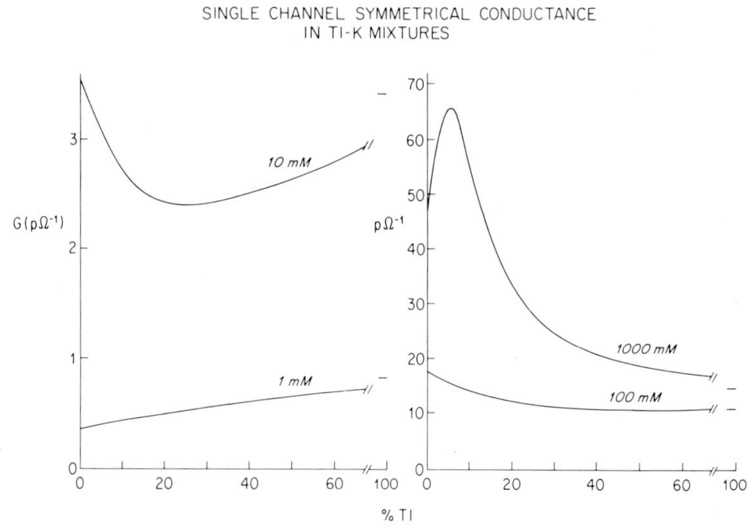

Fig. 17. Blocking and enhancement of conductance expected theoretically for the single channel in symmetrical Tl—K mixtures at the indicated total concentration. The abscissa gives the percent Tl^+ in the mixtures. Note the break in the abscissa at 70 %.

concentration is particularly noteworthy and is in accord with the conductance minimum in K^+-Tl^+ mixtures reported by Andersen (1975) and the minimum for Tl^+-Na^+ mixtures seen by Neher (1975) which can be seen in Figure 2. Indeed, the complex dependence on total concentration of the blocking and enhancement effects of Figure 17 doubtlessly underlies Andersen's (1975) report of the absence of a conductance minimum in Na^+-Tl^+ mixtures which is at variance with the data in Figure 2.

The theoretical expression which is the basis for combining the data of Sections 3 and 4 to yield the predictions of Figure 17 is:

$$G_o = \frac{P_x c_x + P_y c_y}{(1 + \frac{1}{2} K^o_x c_x + \frac{1}{2} K^o_y c_y)^2 + (\frac{1}{2} K^o_x c_x)^2 K^h_x c_x (1+K^{\infty}_x c_x) + (\frac{1}{2} K^o_y c_y)^2 K^h_y c_y (1+K^{\infty}_y c} \qquad (19)$$

The derivation of this expression is too lengthy to be given here; it will be found in Sandblom, Eisenman and Neher (1976).

6. DISCUSSION

6.1. *Anion Effects*

Although the gramicidin A channel is accepted as being negligibly permeant to anions (Myers and Haydon, 1972; Hladky, 1973; Bamberg et al., 1976) and anion permeabilities have indeed been found to negligible in the experiments described so far in this paper, we have detected a significant degree of anion permeation under experimental situations when both outer sites are loaded with cations, particularly Tl^+ (Eisenman et al., 1976a). This shows up as a difference in apparent permeability ratio depending on whether the measurement of membrane potential is made using a gradient of Tl^+ or a gradient of K^+. An example of this is illustrated in Table V and Figure 18. The experiment consists of measuring the apparent permeability ratio by making alternatively a small increment in K^+ or a small

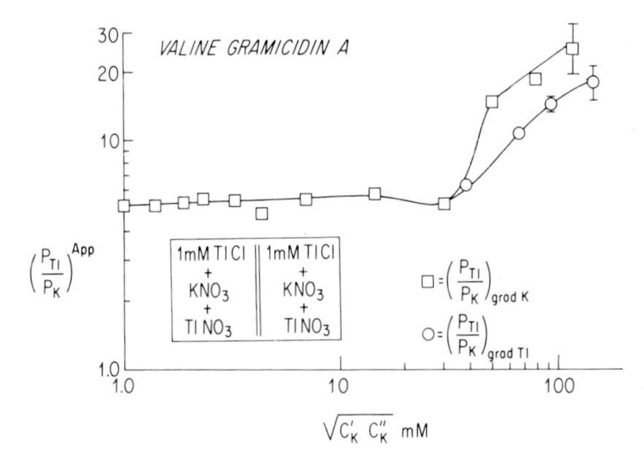

Fig. 18. The experiment of Table V demonstrating, in a many-channel system in GMO/decane bilayers, the difference in permeability ratios, seen at higher salt concentrations, when measured with a gradient of K^+ vs. with a gradient of Tl^+. Notice that the permeability ratios are essentially the same below 30 mM whether measured in a gradient of K or a gradient of Tl, but they then become different at higher concentrations. These findings can be accounted by an anion permeability becoming equal to K^+ permeability at higher concentrations, as discussed in the text.

increment in Tl^+ from a series of symmetrical starting mixtures of the two cations (cf. Table V). The data consist of the accurately measured small differences in membrane potential which result from appropriate increments of Tl^+ or K^+. Figure 18 presents the experimental results of Table V plotted as a function of $\sqrt{c'_K c''_K}$ (an equivalent plot can also be made as a function of $\sqrt{c'_{Tl} c''_{Tl}}$). The squares represent the apparent permeability ratio, $(P_{Tl}/P_K)^{App}$, measured in a gradient of K^+ and the circles represent $(P_{Tl}/P_K)^{App}$ measured in a gradient of Tl^+. The term 'apparent' is used here to indicate that these permeabil-

TABLE V

Solution concentrations: Side (')	Side ('')	V^{OBS} (mV)	$V-V^{Sym}_{Prec}$ (mV)	$V-V^{Sym}_{Fol}$ (mV)	ΔV_{corr} (mV)	$(P_{Tl}/P_K)^{App}$ grad K	grad Tl
1 mMTlCl	1 mMTlCl	0.15	symmetrical				
1 mMTlCl	1 mMTlCl + 1 mMKNO$_3$	4.05	3.9	3.6	4.33 – 4.63	5 - 5.4	
1 mMTlCl + 1 mMKNO$_3$	1 mMTlCl 1 mMKNO$_3$	0.55	symmetrical				
1 mMTlCl + 1 mMKNO$_3$	1 mMTlCl + 2 mMKNO$_3$	3.85	3.3		3.82	5.16	
1 mMTlCl + 1 mMKNO$_3$	1 mMTlCl + 3.5 mMKNO$_3$	7.75	7.2		8.43	5.35	
1 mMTlCl + 1 mMKNO$_3$	1 mMTlCl + 5.5 mMKNO$_3$	11.95	11.4		13.37	5.50	
1 mMTlCl + 2 mMKNO$_3$	1 mMTlCl + 5.5 mMKNO$_3$	8.75	8.2	7.7	9.67 - 10.17	5.11 - 5.56	
1 mMTlCl +3.5 mMKNO$_3$	1 mMTLCL + 5.5 mMKNO$_3$	5.15		4.1	5.55	4.70	
1 mMTlCl +5.5 mMKNO$_3$	1 mMTlCl + 5.5 mMKNO$_3$	1.05	symmetrical				
1 mMTlCl +5.5 mMKNO$_3$	1 mMTlCl + 9 mMKNO$_3$	7.25	6.2	5.9	6.81 - 7.11	5.18 - 5.72	
1 mMTlCl + 9 mMKNO$_3$	1 mMTlCl + 9 mMKNO$_3$	1.35	symmetrical				
1 mMTlCl + 9 mMKNO$_3$	1 mMTlCl + 24 mMKNO$_3$	16.95	15.6	15.2	17.51 - 17.91	5.51 - 5.97	
1 mMTlCl + 24 mMKNO$_3$	1 mMTlCl + 24 mMKNO$_3$	1.45	symmetrical				
1 mMTlCl + 24 mMKNO$_3$	1 mMTlCl + 39 mMKNO$_3$	10.05	8.6	8.9	10.27 - 10.57	4.68 - 5.71	
1 mMTlCl + 39 mMKNO$_3$	1 mMTlCl + 39 mMKNO$_3$	1.15	symmetrical				
1 mMTlCl + 39 mMKNO$_3$	3 mMTlNO$_3$ + 39 mMKNO$_3$	7.35	6.2	5.4	5.59 - 6.39		5.64-6.70
3 mMTlNO$_3$ + 39 mMKNO$_3$	3 mMTlNO$_3$ + 39 mMKNO$_3$	1.95	symmetrical				
3 mMTlNO$_3$ + 39 mMKNO$_3$	3 mMTlNO$_3$ + 69 mMKNO$_3$	7.45	5.5		7.69	14.38	
3 mMTlNO$_3$ + 69 mMKNO$_3$	3 mMTlNO$_3$ + 69 mMKNO$_3$	2.45	symmetrical				
3 mMTlNO$_3$ + 69 mMKNO$_3$	7 mMTlNO$_3$ + 69 mMKNO$_3$	10.95	8.5		8.69		10.69
7 mMTlNO$_3$ + 69 mMKNO$_3$	7 mMTlNO$_3$ + 69 mMKNO$_3$	2.35	symmetrical				
7 mMTlNO$_3$ + 69 mMKNO$_3$	7 mMTlNO$_3$ + 99 mMKNO$_3$	4.25	1.9	1.85	3.35 - 3.40	18 - 18.4	
7 mMTlNO$_3$ + 99 mMKNO$_3$	7 mMTlNO$_3$ + 99 mMKNO$_3$	2.6	symmetrical				
7 mMTlNO$_3$ + 99 mMKNO$_3$	11 mMTlNO$_3$ + 99 mMKNO$_3$	8.85	6.25	6.0	6.07 - 6.32		13.8 -15
7 mMTlNO$_3$ + 99 mMKNO$_3$	15 mMTlNO$_3$ + 99 mMKNO$_3$	13.75	11.15	10.9	11.12 - 11.37		14.71-15.3
11 mMTlNO$_3$ + 99 mMKNO$_3$	15 mMTlNO$_3$ + 99 mMKNO$_3$	7.55	4.95	4.7	4.7 - 4.95		12.9 -14.8
15 mMTlNO$_3$ + 99 mMKNO$_3$	15 mMTlNO$_3$ + 99 mMKNO$_3$	2.85	symmetrical				
15 mMTlNO$_3$ +159 mMKNO$_3$	15 mMTlNO$_3$ + 99 mMKNO$_3$	1.45	-1.4	-0.6	-2.58 to-3.38	20.5 - 30.8	

15 $mMTlNO_3$ + 159 $mMKNO_3$	15 $mMTlNO_3$ + 159 $mMKNO_3$	2.05	symmetrical			
21 $mMTlNO_3$ + 159 $mMKNO_3$	15 $mMTlNO_3$ + 159 $mMKNO_3$	-3.0	-5.05	-5.45	-5.3 av	17.4 av
21 $mMTlNO_3$ + 159 $mMKNO_3$	21 $mMTlNO_3$ + 159 $mMKNO_3$	2.45	symmetrical			

V^{OBS} is the observed potential between AgAgCl electrodes.

$V-V^{Sym}_{Prec}$ refers to the observed potential minus the potential observed in the symmetrical solution immediately preceding the measurements.

$V-V^{Sym}_{Fol}$ refers to the observed potential minus the potential observed in symmetrical solutions immediately following the measurement

ΔV_{corr} refers to the potential corrected for activity coefficient changes, due to ionic strength variations in the experiment.

The apparent permeability ratios were calculated according to the Goldman-Hodgkin-Katz equation and are presented in the appropriate column corresponding to whether they were measured in a gradient of potassium (grad K) or a gradient of thallium (grad Tl).

ity ratios have not been corrected for anion permeability effects. When such corrections are made, the data are well described by the present model (Eisenman et al., 1976b), but the corrections indicate that the nitrate and chloride anions must become nearly as permeant as K^+ at high bilateral salt concentrations.

A direct verification of this inferred induced anion permeability when both external sites are loaded with cations is given in Figure 19. Here, the acetate anion concentration is increased on both sides of a membrane, using tetramethylammonium as an cation, in order to see whether or not the membrane potential established by a constant gradient of Tl^+ (4.5 mM vs. 3.0 mM) is diminished on increasing the acetate concentration

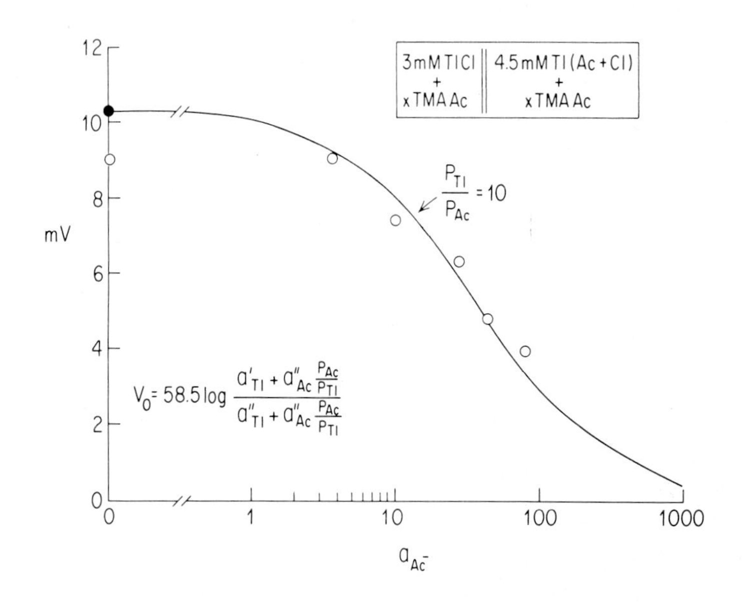

Fig. 19. Demonstration that acetate is one tenth as permeable as Tl^+ when both outer sites of the channel are loaded significantly with Tl^+.

in the manner expected if P_{Ac}/P_{Tl} were not negligible under these conditions. (Recall from the value of K^o_{Tl} = 3600 M^{-1} of Table III, that at greater than 3 mM Tl^+ concentration both outer sites are expected to be loaded nearly all of the time). The open circle on the ordinate, before adding any TMA Acetate, gives the dilution potential of 4.5 mM Tl^+ vs. 3.0 mM Tl^+ solutions. 9.4 mV is observed; whereas the theoretically expected value (if the membrane were impermeant to anions) is 10.3 mV (i.e. 58.5 Log_{10} (4.5/3), as indicated by the filled circle). This observation, by itself, suggests that the anions present initially (Ac^- plu Cl^-) are about 1/10 as permeant as Tl^+. The effect of adding tetramethylammonium acetate symmetrically to both sides of the membrane is plotted along the abscissa. If the acetate anion were negligibly permeable, the membrane potential would be expected to remain constant; but what is observed instead, as indicated by the data points, is a clear decrease of membrane potential with increasing acetate concentration, which is quite well-described by the curve drawn according to the Goldman-Hodgkin-Katz equation for a permeability ratio of thallium to acetate equal to 10. Since K^+ is only one tenth as permeable as Tl^+ under these conditions, this verifies that the acetate permeability is as large as potassium permeability at high Tl^+ loadings.

6.2. *Relation to Previous Studies on Gramicidin A*

A few comments on the relation of the present findings to previous results in the literature might not be inappropriate here. First, our data for both Tl^+ and K^+ indicate that the view advocated by both Hladky (1973) and Läuger (1973), that the gramicidin A channel can be occupied at most by one cation at a time needs to be modified. Our findings indicate that this channel can be occupied by several cations simultaneously, states of single occupancy, double occupancy, triple occupancy, and even quadruple occupancy being demanded by our data. The binding constants, previously inferred from the saturating conductance by Hladky (1973), by Läuger (1973) and by Neher (1975), are now seen to refer to the intermediate (triply occupied) state of loading of the channel. An additional binding site has been demonstrated here at very low salt concentrations both for Tl^+ and for K^+, with values for its binding constant of 3600 M^{-1} for Tl^+ and 260 M^{-1} for K^+. Moreover, at the very highest concentrations, a dramatic decrease in conductance has been observed for K^+ (recall Figure 16), which is completely accountable within the present theory as a consequence of full occupancy of all four binding sites, with subsequent loss of the conducting properties of the cahnnel which depends upon one internal site being availabel for occupancy. The binding constant for this latter case has been estimated to be 0.22 M^{-1} for K^+. A similar decrease of conductance with increasing salt concentration is also apparent in the literature data for Cs^+ and Na^+, as has been shown in Figure 5 and was indeed calculated by Hladky (1972, Appendix B, Equation (16)) to be a consequence of occypancy of both sites in his 2-site model.

The present four-site model, and its theory, also appears adequate to account for certain, heretofore unexplained, peculiaritues in permeability ratios and conductance ratios noted by Myers and Haydon (1972) who pointed out that the permeability ratios were always higher then conductance ratios and varied in the opposite direction with increasing concentration but did not provide an explanation of these differences other than to indicate that they were rather involved. That our four-site model can explain these observations is illustrated in Figure 20 for Tl^+ and K^+, which is the exact counterpart of Myers and Haydon's (1972) Fifure 3 for the K^+-Na^+ system. Figure 20 shows that the divergence between conductance ratios and permeability ratios for K^+ and Tl^+ and the detailed dependence of both of these quantities on salt concentration are completely accounted for in terms of the four-site model. It therefore seems likely that this is also the basis for the corresponding behavior seen for K^+ and Na^+ by Meyers and Haydon.

That the properties of the four-site model, demonstrated here for Tl^+ and K^+, will be applicable to the other monovalent cations seems

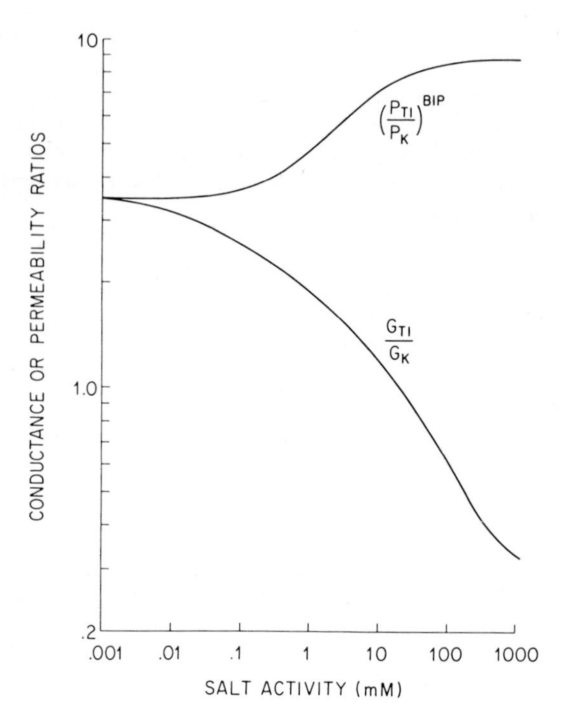

Fig. 20. Comparison of theoretically expected bi-ionic permeability ratios and conductance ratios for Tl^+ vs. K^+ as a function of salt concentration.

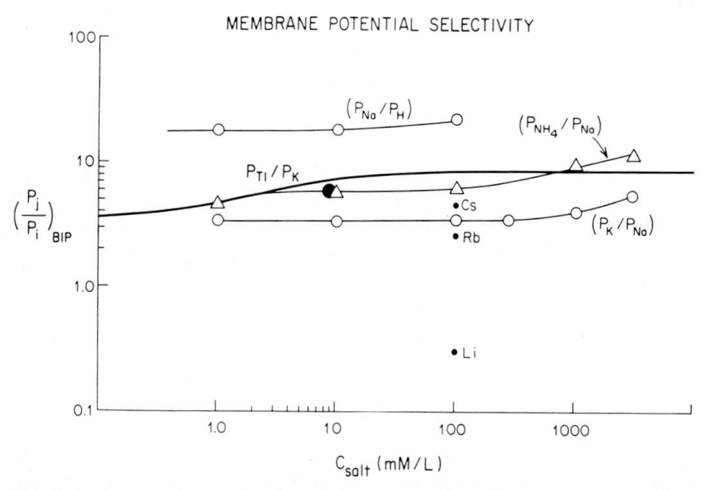

Fig. 21. Comparison of the concentration dependence of the theoretically expected bi-ionic permeability ratios for Tl^+ vs. K^+ in comparison with experimentally observed bi-ionic permeability ratios of Myers and Haydon (1972) for the indicated pair of cations (Cs, Rb, and Li are referred to Na).

apparent from the similarities, illustrated in Figure 21, between the concentration dependences of bi-ionic permeability ratios theoretically expected from the four-site model for thallium and potassium and those observed by Myers and Haydon (1972) for a variety of monovalent cations. It also seems likely that the four-site model can account for the blocking of monovalent conductances by divalent cations observed by Hladky (1972) and characterized by Bamberg and Läuger (unplublished results).

6.3. *Similarities to phenomena in biological channels*

Blocking in Na^+ channels and K^+ channels of nerve has been critically discussed by Hille (1975a) and by Armstrong (1975); and Hille (1975b) has shown how partial saturation for a one-ion-saturable channel gives a reasonable explanation for the simultaneous blocking and current carrying properties of cations, which is particularly noteworthy for the species Tl^+ and H^+. Hille (1975a, b) has also analyzed the interrelationships between selectivity, saturation and block within the conceptual framework of a one-ion-saturable model for the sodium channel.

A hint that the Na^+ channel might be occupiable by more then one cation at a time comes from a recent series of studies which have demonstrated a concentration dependence of P_K/P_{Na} for this channel (Begenisich and Cahalan, 1975; Cahalan and Begenisich, 1976). Such a concentration dependence of permeability ratios, by analogy to the situation in the gramicidin A channel, suggests that if the sodium channel were occupiable simultaneously by more then one cation the findings of both concentration dependent permeability ratios and blocking of conductance would be but two aspects of the same fundamental interactions between ions in a channel.

Similar phenomena, further suggesting interactions between ions in channels which can be occupied by more than one cation at a time, are beginning to appear in the biological literature. For example, blocking (and even enhancement) effects analogous to those of Figure 17 have been reported by Hagiwara and Takahashi (1974) for the membrane of the starfish egg; and Wiedner and Wright (1976) are finding concentration dependences of permeability ratios in the tight junctions of the gallbladder epithelium.

ACKNOWLEDGEMENTS

We thank Dr E. Gross of the National Institutes of Health for samples of purified valine gramicidin A made available directly to us, as well as via E. Grell and W.R. Veatch. We also acknowledge the generous support of the National Science Foundation (Grant GB 30835), and the U.S. Public Health Service (Grant NS 09931) which made this work possible.

REFERENCES

Andersen, O.S.: 1975, 'Ion-Specificity of Gramicidin Channels', Abstract, Internat. Biophys. Congress, Copenhagen, p. 112.

Bamberg, E., Kolb, H.A., and Läuger, P.: 1976, 'Ion Transport Through the Gramicidin A Channel', Reprinted from: The Structural Basis of Membrane Function (ed. by Y. Hatafo), Academic Press, New York, pp.

143-157.

Begenisich, T. and Cahalan, M.: 1975, 'Internal K^+ Alters Sodium Channel Selectivity', Abstract, Internat. Biophys. Congress, Copenhagen, p. 133.

Bezanilla, F. and C.M. Armstrong: 1972, 'Negative Conductance Caused by Entry of Sodium and Cesium Ions into the Potassium Channels of Squid Axons', J. Gen. Physiol. 60, 588-608.

Cahalan, M. and Begenisich, T.: 1976, 'Sodium Channel Selectivity: Dependence on Internal Permeant Ion Concentration', submitted to J.Gen. Physiol.

Eisenman, G., Krasne, S., and Ciana, S.: 1974, 'Further Studies on Ion Selectivity', Proceedings of International Workshop on Ion-Selective Electrodes and on Enzyme Electrodes in Biology and in Medicine, M. Kessler, L. Clark, D. Lübbers, I. Silver and W. Simon (Eds.), Urban and Schwarzenberg, Münich, Berlin, Vienna. In Press.

Eisenman, G., Deber, C., and Blout, E.R.: 1976 'Cyclic Octapeptides are Selective Carriers of Cations Across Lipid Bilayers', Biophys. Soc. Abstract, p. 81a.

Eisenman, G., Sandblo, J., and Neher, E.: 1976a, 'Evidence for Multiple Occupancy of Gramicidin A Channels by Ions', Biophys. Soc., Abstract, p. 81a.

Eisenman, G., Sandblom, J., Neher, E.: 1976b, 'Ionic Selectivity, Binding, and Block in the Gramicidin A Channel: II. Membrane Potentials in Ionic Mixtures and Concentration Dependent Permeability Ratios for K^+ and Tl^+', J. Memb. Biol., to be submitted.

Eisenman, G., Sandblom, J.P., and Walker, Jr., J.L.: 1967, Membrane Structure abd Ion Permeation 155, 965-974.

Eisenman, G., Szabo, G., Ciani, S., McLaughlin, S.G.A. and Krasne, S.: 1973, 'Ion Binding and Ion Transport Produced by Lipid Soluble Molecules', Prog. Surf. Meb. Sci. 6, 139-241.

Eyring, H., Lumry, R., and Woodbury, J.W.: 1949, 'Some Applications of Modern Rate Theory to Physiological Systems', Rec. Chem. Prog. 10, 100-114.

Glickson, J.D., Mayers, D.F., Settine, J.M., and Urry, D.W.: 1972, 'Spectroscopic Studies on the Conformation of Gramicidin A', Proton Magnetic Resonance Assignments, Coupling Constants, and H-D Exchange', Biochemistry 11, 477.

Hagiwara, S. and Takahashi, K.: 1974, 'The Anolomous Rectification and Cation Selectivity of the Membrane of a Starfish Egg Cell', J. Memb. Biol. 18, 61-80.

Henderson, R., Ritchie, J.M., and Strichartz, G.R.: 1974, 'Evidence That Tetrodotoxin and Saxitixin Act at a Metal Cation Binding Site in the Sodium Channels of Nerve Membrane', Proc. Natl. Acad. Sci. 71, 3936.

Hille, B.: 1970, 'Ionic Channels in Nerve Membranes', Prog. Biophys. Mol. Biol. 21, 3-32.

Hille, B.: 1975a, 'Ionic Selectivity of Na and K Channels of Nerve Membranes', in Membranes — A Series of Advances, voll. 3 'Dynamic Properties of Lipid Bilayers and Biological Membranes', G. Eisenman (Ed.), Marcel Dekker, New York, Ch. 4.

Hille, B.: 1975b, 'Ionic Selectivity, Saturation, and Block in Sodium Channels', J. Gen. Physiol. 66, 535-560.

Hladky, S.B.: 1972 'The Two-Site Lattice Made for the Pore. Appendix B', Ph. D. Dissertation, Cambridge, University, England.

Hladky, S.B.: 1974, 'Pore or Carrier? in B.A. Callingham (Ed.), Gramicidin A as a Simple Pore', Drugs and Transport Processes, University Park Press, pp. 193-210.

Hladky, S.B. and Haydon, D.A.: 1970, 'Discreteness of Conductance Change in Bimolecular Lipid Membranes in the Presence of Certain Antibiotics', Nature 225, 451-453.

Hladky, S.B. and Haydon, D.A.: 1972, 'Ion Transfer Across Lipid Membranes in the Presence of Gramicidin A. I. Studies of the Unit Conductance Channel', Biochim. Biophys. Acta 274, 294-312.

Läuger, P.: 1973, 'Ion Transport Through Pores: A Rate-Theory Analysis', Biochim. Biophys. Acta 311, 423-441.

Mueller, P. and Rudin, D.O.: 1967, 'Development of K^+-Na^+ Discrimination in Experimental Bimolecular Lipid Membranes by Macrocyclic Antibiotics', Biochem. Biophys. Res. Commun. 26, 398-404.

Myers, V.B. and Haydon, D.A.: 1972, 'Ion Transfer Across Lipid Membranes in the Presence of Gramicidin A. II. The Ion Selectivity', Biochim. Biophys. Acta 274, 313-322.

Neher, E.: 1975, 'Ionic Specificity of the Gramicidin Channel and the Thallous Ion', Biochem. Biophys. Acta 401, 540-544.

Neher, E., Sandblom, J., and Eisenman, G.: 1976, 'Ionic Selectivity, Binding and Block in Gramicidin A Channels: II. Saturating Behavior of Single Channel Conductances and Evidence for the Existance of Two Binding Sites for Monovalent Cations on Each Side of the Channel', J. Memb. Biol., to be submitted.

Reed, J.K. and Raftery, M.A.: 1976, 'Properties of the Tetrodotoxin Binding Component in Plasma Membranes Isolated from Electrophorus Electricus', Biochemistry, 15, 944-953.

Sandblom, J., Eisenman, G., and Neher, E.: 1976, Ionic Selectivity, Binding and Block in Gramicidin A Channels: I. Theory for the Electrical Properties of Ion Selective Channels Having Four Binding Sites and Multiple Conductance States', J. Memb. Biol., in press.

Szabo, G., Eisenman, G., and Ciani, S.: 1969, 'The Effects of the Macrotetralide Actin Antibiotics on the Electrical Properties of Phospholipid Bilayer Membranes', J. Memb. Biol. 1, 346.

Urry, D.W.: 1971, 'The Gramicidin A Transmembrane Channel: A Proposed π(L.D.) Helix', Proc. Natl. Acad. Sci. 68, 672-676.

Urry, D.W.: 1972, 'Protein Conformation in Biomembranes: Optical Rotation and Adsorption of Membrane Suspensions', Biochim. Biophys. Acta 265, 115-168.

Veatch, W.R., Fossel, E.T., and Blout, E.R.: 1974a, 'The Conformation of Gramicidin A', Biochemistry 13, 5249-5256.

Veatch, W.R. and Blout, E.R.: 1974b, 'The Aggregation of Gramicidin A in Solution', Biochemistry 13, 5257-5264.

Wiedner, G. and Wright, E.M.: 1976, 'Diffusion Potential Transients in the Rabbit Gallbladder', to be submitted to Pflug. Archiv.

Woodbury, J.W.: 1971, 'Eyring Rate Theory Model of the Current-Voltage Relationship of Ion Channels in Excitable Membranes', in J. Hirschfelder (ed.), Chemical Dynamics: Papers in Honor of Henry Eyring, John Wiley and Sons, Inc., New York.

Zeffren, E. and Hall, P.L.: 1973, The Study of Enzyme Mechanisms, John Wiley and Sons, New York.

Zwolinski, B.J., Eyring, H., and Reese, C.E.: 1949, 'Diffusion and Membrane Permeability', J. Physiol. Colloid Chem. 53, 1426-1453.

DISCUSSION

Rode: You suggest an energy profile for the polypeptide channel of the following shape (A)

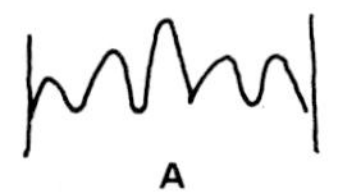

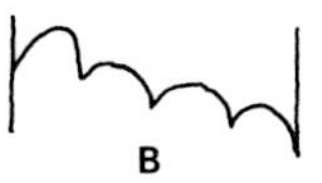

Do you think, that another shape (B), which would facilitate ion transpor through the channel, could also be possible? We have some evidence from calculations on energy surfaces for various ions in the field of various amides, that it should be possible to establish such an energy profile by suitable varying substitution in the peptide components.

Eisenman: By symmetry, the profile of your (B) type should be (C) because the gramicidin channel is a dimer.

C

Our data so far cannot exclude this possibility - although the interpretation of our parameters could be quite different in this case. Such a profile could put the important rate limiting steps at the interfaces rather than in the center. In principle the voltage dependence of the conductances should be able to decide between (A) and (C).

ENANTIOMER-SELECTIVE ION CARRIERS AND THEIR TRANSPORT PROPERTIES IN BULK MEMBRANES

A.P. THOMA and W. SIMON
Eidgenössische Technische Hochschule. Laboratorium für Organische Chemie. CH-8006 Zürich. Universitätstrasse 16, Switzerland

The overwhelming enantioner-selectivity of biochemical processes as well as the role of neuroregulators in neuronal functions stimulated efforts towards the design of chiral model systems capable to induce enantiomer-selective cation transport. Both electrically neutral [2] and charged ligands [4] may be used to achieve separation of enantiomers by the extraction from an aqeous to an organic phase. In view of an electrogenic ion transport electrically neutral ligands acting as ion carriers (ionophores) are of special interest. Such an ion transport by electrodialysis as well an enantiomer-selective electromotive behaviour of membranes has been described only recently [8].

Different chiral, neutral ligands which have been in succesful use in partition systems (4 in Figure 1) may fail for several reasons [6] if tested for chiral recognition in liquid membrane electrodes or electrodialytic racemate resolution. Thus we have been unable, up to now, to detect any enantiomer-selectivity attributable to the compounds 4 and valinomycin (Figure 1) in liquid membrane electrode measurements or of valinomycin in ion transport experiments. Using different model assumptions [6] the enantiomer-selective EMF-response of the cell

$$\text{Hg; } Hg_2Cl_2\text{, KCl(sat.)}/10^{-1}\text{ M } NH_4NO_3\text{/sample solution//liquid membrane// reference solution, AgCl; Ag} \quad (1)$$

may be described by

$$\Delta EMF = EMF_F - EMF_{\bar{F}} = \frac{RT}{F} \ln \left(\frac{K_{FL}}{K_{\bar{F}L}} \right) \quad (2)$$

with:

EMF_F, $EMF_{\bar{F}}$: potential differences measured over the cell (1) in equimolar solutions of the monovalent enantiomeric cations F and $\bar{F}$ respectively

$\frac{RT}{F}$: Nernst factor

K_{FL}, $K_{\bar{F}L}$: complex formation constants of the enantiomeric cations F and $\bar{F}$ with the chiral ligand L in the liquid membrane (1:1 stoichiometry)

B. Pullman and N. Goldblum (eds.), Metal-Ligand Interactions in Organic Chemistry and Biochemistry, part 2, 37-42.

and is correlated to the transport selectivity [5]

$$\frac{t_F}{t_{\bar{F}}} = \frac{K_{FL}}{K_{\bar{F}L}} \quad (3)$$

with t_F, $t_{\bar{F}}$: transport numbers of the enantiomeric cations F and $\bar{F}$ through the liquid membrane.

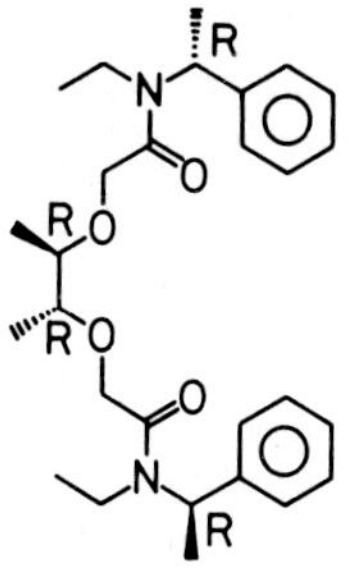

1

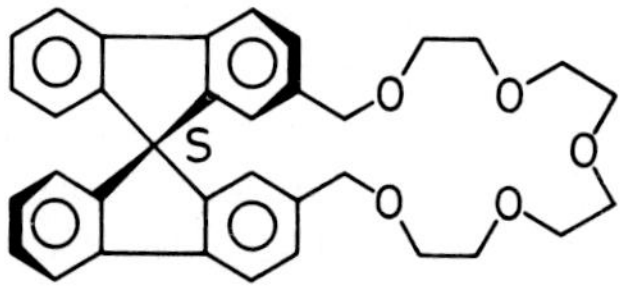

R S
2 3

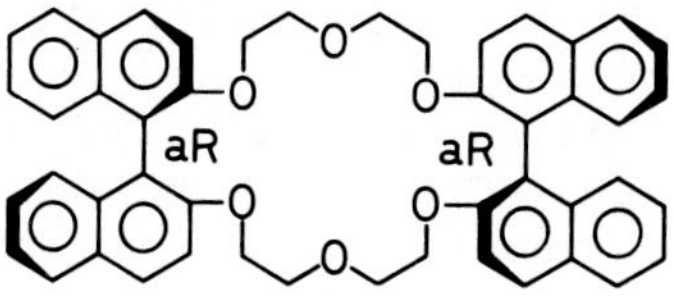

4

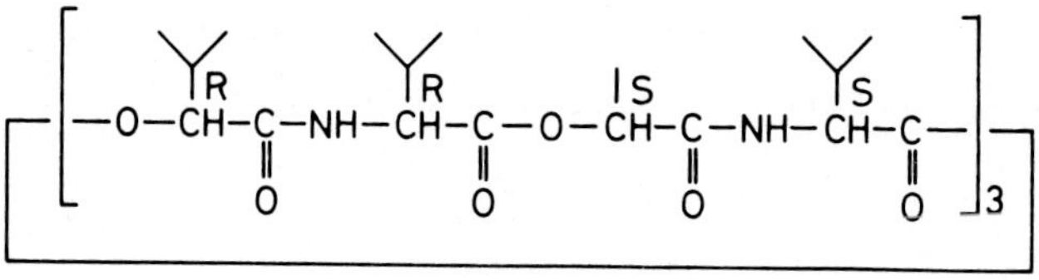

Fig. 1. Electrically neutral ligands studied. 1: see [8]; 2, 3: see [1] and [7]; 4: see [3]; bottom of the figure: Valinomycin.

For a more detailed discussion see [7]. The enantiomer-selective carrier-transport can be demonstrated conveniently by using a novel double-labelling technique (see Fig. 2) described earlier [8]. Using the micro-cell sh

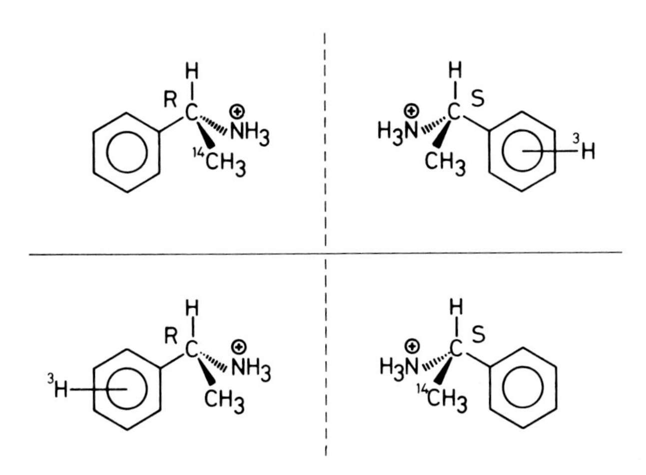

Fig. 2. Labelling of the α-phenyl-ethyl-ammonium ions used in the transport experiments [7].

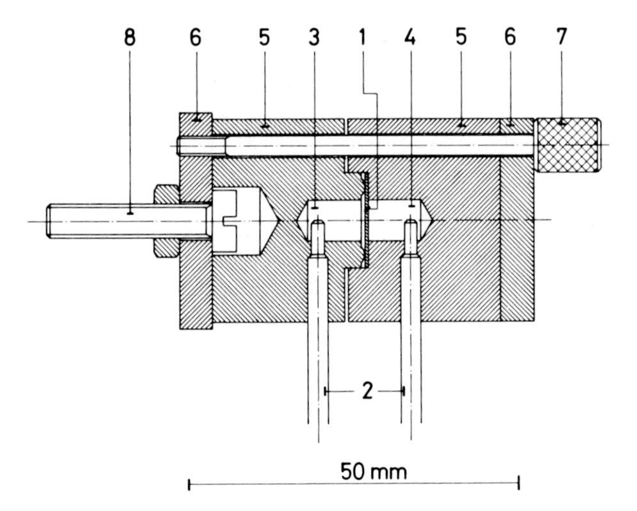

Fig. 3. Micro-cell used for electrodialysis,
1: Membrane — 2: Silver/silver chloride electrode: applied voltage: 1 to 10 V, average current: 1.5 μA, electrodialysis time: about 1 h — 3: Cathode compartment: half cell volue: 230 μl (200 μl 10^{-3} M racemic mixture, not labelled — 4: Anode compartment: half-cell volume: 230 μl (200 μl 10^{-3} M racemic mixture, radioactively labelled). Specific measured activities of α-phenyl-ethyl-ammonium chloride:

^{14}C-(R) : 9.4 × 10^{10} ipm mole^{-1} ^{3}H-(R) : 7.1 × 10^{10} ipm mole^{-1}
^{14}C-(S) : 9.4 × 10^{10} ipm mole^{-1} ^{3}H-(S) : 10.8 × 10^{10} ipm mole^{-1}

Scintillator: butyl-P B D (Ciba-Geigy AG, Basle). Liquid Scintillation Spectrometer: Mark I (Nuclear Chicago) — 5: Cell body (Teflon®) — 6: Cell support — 7: 4 tightening screws — 8: Mounting of the cell on wrist-shaker (continuous mixing of the halfcell contents achieved by shaking at a frequency of about 5 cps).

TABLE I
Enantiomer-selective transport af α-phenyl-ethyl-ammonium cations through liquid poly-(vinylchloride)-membranes using o-nitro-phenyl-octylether as membrane solvent; 95% confidence limits.

	t_R / t_S	
ligand	^{14}C-(R)-, ^{3}H(S)- racemic mixture	^{3}H-(R)-, ^{14}C-(S)- racemic mixture
1	1.067 ± 0.015	1.070 ± 0.026
2 (R)	1.124 ± 0.017	1.121 ± 0.012
3 (S)	$(1.093 \pm 0.027)^{-1}$	$(1.094 \pm 0.012)^{-1}$

in Figure 3 the results presented in Table I have been obtained. According to this table the ligands 1, 2 and 3 induce an enantiomer-selectivity of 7, 12 and 9% respectively. The results obtained with carrier 2 are in reasonable agreement with those for the enantiomer 3. EMF-measurements over cell (1) using α-phenyl-ethyl-ammonium chlorides as sample solutions led to the selectivities in the potentiometric chiral recognition given in Table II. The values obtained are in good agreement with the chiral recognition in the electrodialytic transport experiments (see Table I).

TABLE II
ΔEMF (mV) measured with enantiomer-selective liquid membrane electrodes using 0.1 M (R)- respectively (S)-α-phenyl-ethyl-ammonium chloride at 25 °C; o-nitro-phenyl-octylether was used as membrane solvent; 95% confidence limits.

ligand	$\Delta EMF = EMF_R - EMF_S$	calculated values of t_R/t_S (see Equation (2) and (3))
1	+ 2.5 ± 0.2[a]	1.102 ± 0.009
2 (R)	+ 2.9 ± 0.8[b]	1.118 ± 0.036
3 (S)	- 3.5 ± 0.7[b]	$(1.146 \pm 0.031)^{-1}$

[a] in potassium acetate/acetic acid-buffer (pH = 4.4)

[b] in KH_2PO_4 / K_2HPO_4-buffer (pH = 6.8)

ACKNOWLEDGEMENT

We thank Prof. Dr D.J. Cram for a generous supply of ligand 4. The present work was partly supported by the Schweizerischer Nationalfonds zur Förderung der wissenschaftlichen Forschung (project no. 2.1070.74).

REFERENCES

1. Bedeković, D. and Prelog, V.: 1976, in preparation.
2. Cram, D.J. and Cram, J.M.: 1974, Science, 803.
3. Helgeson, R.C., Timko, J.M., and Cram, D.J.: 1973, J. Amer. Chem. Soc. 95, 3023.
4. Lehn, J.M., Moradpour, A., and Behr, J.P.: 1975, J. Amer. Chem. Soc. 97, 2532.
5. Morf, W.E., Wuhrmann, P., and Simon, W.: 1976 Anal. Chem. 48, 1031.
6. Simon, W., Morf, W.E., and Meier, P.C.: 1973, Structure and Bonding, vol. 16, Springer Verlag, Berlin/Heidelberg/New York.
7. Thoma, A.P., Bedeković, D., Prelog, V., and Simon, W.: 1976, Helv. Chimica Acta, in press.
8. Thoma, A.P., Cimerman, Z., Fiedler, U., Bedeković, D., Güggi, M., Jordan, P., May, K., Pretsch, E., Prelog, V., and Simon, W.: 1975, Chimia 29, 344.

DISCUSSION

Saenger: (1) I would like to remark that separation of racemic mixtures was carried out about 20 years ago with cyclodextrins as host compounds. The separation was not as good as with your polyether molecules but it could be performed in aqeous solutions.

(2) How does the migration of the metal complex through the membrane occur? Does the complex diffuse as such or is the host molecule stationary in the membrane and the cation jumps from one host to the next?

Simon: (1) It was intended to achieve an electrogenic enantiomer-selective ion transport through membranes and to demonstrate an enantiomer-selective *electromotive* behaviour of such membranes and not only to separate racemic mixtures. The compounds you mention are not suitable for such membrane studies because of too small lipophilicity resp. too large water solubility.

(2) The host-guest complex is driven trough the bulk membrane by the electric potential gradient applied to the membrane. The host molecule is therefore not stationary (for similar bulk systems see Chimia (Switzerland) 27, 637 (1973)). Using valinomycin as carrier a contribution due to exchange of ligands during the transport process (carrier-relay mechanism) has been observed in somewhat similar bulk systems (Helv. Chim. Acta 53, 1605 (1970)).

Jagur-Grodzinski: Could you comment on the effect of counterions? There are no fixed ionic charges in your system and there will be a considerable tendency to drug a counterion into the membrane in order to preserve the macroscopic electroneutrality of the system.

Comment to Dr Saenger's question. In similar systems investigated by us, a neutral complex ($CO_2(NO)_2$ 2 DBCP) migrated through a membrane.

Simon: You are certainly right: In order to preserve the macroscopic electroneutrality in the bulk membranes, counterions have to enter the membrane phase during the carrier mediated cation transport. Using symmetrical cells, PVC membranes, and Cl^- as the counterions we have not been able so far to demonstrate an anion contribution to the transference number. The transference number for cations is however decreased well below 1.0 if lipophilic anions such as SCN^- are used. We anyway do not expect hydrophilic achiral anions, which are present in the membrane phase, to interfere with the enantiomer-selectivity based on the carrier complexation.

COMPUTER SIMULATION STUDIES ON THE ION-TRANSPORTING ANTIBIOTICS

L.K. STEINRAUF and M.N. SABESAN
Depts. of Biochemistry and Biophysics, Indiana University School of Medicine, Indianapolis, Indiana 46202, U.S.A.

1. INTRODUCTION

The ion-transporting antibiotics, of which about a dozen examples are known, were discovered more than twenty years ago and have had their physical, chemical and biological properties under intense investigation ever since. These compounds are characterized by the ability to bind a metal cation and render it soluble in non-polar solvents, and by this property to allow the bound cation to pass through lipid membranes in either biological or model systems. Some of the antibiotics display a high degree of specificity, being able to distinguish such similar cations as sodium and potassium. Binding of the cation is accomplished by the substitution of the hydration sphere of the ion by the oxygen atoms of the various alcohol, ether, carbonyl, or carboxylate groups of the antibiotic. The ion-transporting antibiotics have found extensive use, not as germ-killers, but as model systems for membrane transport and as modifiers of membrane regulated functions of cells and subcellular particles such as the mitochondrion.

The ion-transporting antibiotics can be classified according to several criteria. There are the carriers and the pore-formers; we shall consider here only the carriers. Among the carriers there are the C-types which are monocarboxylic acids that form a circle by head to tail hydrogen bonding to enclose the cation. No anion other than the carboxylate is necessary. The O-type carriers are neutral and cyclic, and an anion is necessary to preserve electroneutrality. O-type carriers may or may not use hydrogen bonding to maintain the geometry of the oxygen atoms binding the cation. The sphere of oxygen atoms surrounding the cation(s) may or may not include the anion(s).

One of the most rewarding methods of investigation of the ion-transporting antibiotics has been that of molecular structure determination by X-ray crystallography. However, even this method gives only one form of the probably many forms that the molecule can assume in solution or in membranes, albeit a form of low energy. It is possible by computer calculations to extend this single view of the molecule to other views such as those related by low energy rotations about single bonds. In this way the opening of the antibiotic to release the cation may be examined.

B. Pullman and N. Goldblum (eds.), Metal-Ligand Interactions in Organic Chemistry and Biochemistry, part 2 , 43-57.

Moreover, the cation may be changed and so the specificity to different cations may be examined, or other atoms may be added, deleted, or changed to determine their effects on ion-binding.

Our strategy has been to search for strain in the ion-binding form of the antibiotic as evidenced by abnormal bond distance, bond angles, or torsion angles, and then to examine bond rotations to try to find paths of cation release. We then require that any path should be enhanced by solvent interactions (principally hydrogen bond transfer to water molecules) that would be reasonably expected from the results of our calculated orientation of the antibiotic at a water-lipid interface.

2. METHODS

2.1. *Computer System*

The computer used for these studies was the PDP-15/30, running under the advanced Software System, which is file oriented and contains a very useful editor in addition to the assembler, the FORTRAN compiler, the loader, and other programs. The computer has 24 K of 18-bit 800 ns memory, paper tape reader and punch, four DecTape drives, a 1024 point resolution oscilloscope, a teletype, and an A/D converter. Our programs have been written in FORTRAN with assembly language subroutines.

2.2. *Computer Programming*

The BE (Bonding Energy) program was written by the authors to examine, manipulate, and display molecular structures. The maximum size of the molecule (about 100 atoms) is limited by the execution time of certain operations. The program usually starts with atomic coordinates from X-ray crystal structure determinations, but structures may be built atom-by-atom using the program. The program (Steinrauf, 1971) can produce bond distances, angles, and dihedral angles; it can insert, change, or delete atoms; it can produce the energy profile of rotation about a bond; it can display a ball-and-stick-model on the oscilloscope, slowly rotating to g[illegible] a 3-dimensional view of the current structure; it can refine the position [illegible] a single atom to an energy minimum with respect to other atoms; and it will find the orientation of minimum energy at a simulation of a lipid-water interface.

Energy for bond rotations was calculated by the sum of the potential energies between all of the atoms involved by the following formula:

$$E = \Sigma E_{ij} = \Sigma\left(\frac{A}{R_{ij}} + \frac{B}{R_{ij}^{3}} + \frac{C}{R_{ij}^{6}} + \frac{D}{R_{ij}^{12}}\right)$$

where i and j are two atoms of distance apart R_{ij}, and the coefficients A-D are the electrostatic contribution, the influence of the dielectric constant, the van der Waals attraction, and the core electron repulsion respectively. The coefficients A-D depend on the atom types; coefficient for the interactions between nineteen atom types, including the ions, are contained in the program. Some characteristics are given in Tables I and II.

TABLE I
Assumed distances of most favorable energy for cations to various types of oxygen atoms. The most favorable energy is assumed to be the same for each cation, except for sodium which is 80% of the others.

	Na^+	K^+	Rb^+	Ag^+	Tl^+
alcohol	2.58	2.64	2.75	2.45	2.65
ether	2.61	2.69	2.83	2.48	2.80
carbonyl	2.38	2.56	2.68	2.35	2.64
carboxylate	2.28	2.45	2.56	2.20	2.52
water	2.75	2.75	2.90	2.50	2.90

TABLE II
Fractions of an electron charge assumed to be on various atoms as would be seen by a univalent cation

Hydrogen of alcohol or carboxyclic acid	+ 0.2 e
Carbon single bonded to oxygen or nitrogen	+ 0.2 e
Carbon of carbonyl or carboxylic acid	+ 0.4 e
Nitrogen as in peptide	- 0.4 e
Oxygen of ester or carboxylic acid	- 0.2 e
Oxygen of ether or alcohol	- 0.4 e
Oxygen of any carbonyl	- 0.4 e
Oxygen of carboxylate (both)	- 0.7 e
Chloride ion	- 1.0 e
Monovalent cations	+ 1.0 e

These values have been extracted from the X-ray crystal structure determinations listed below. Although we have had considerable success in using this set of values, they should still be considered preliminary. In comparison to values given by others, as for example Pauling (1960), we are using longer bond distances for sodium ion and shorter for silver. Also we have found it necessary to reduce uniformly all bond energies involving sodium by 0.80 in order to preserve agreement with the known ion specificities of the antibiotics as given for example by Pressman (1968). Distances and energies for the 6–12 potentials were taken from Venkatachalam and Ramachandran (1967).

The refinement of the position of a single atom is usually confined to an ion with respect to its ligand atoms, and is accomplished by taking the gradient of the energy field at the point of the moving atom and then shifting the atom in the opposite direction of the gradient until a lower energy is encountered. Successive applications quickly find the local energy minimum.

The orientation of a molecule at a water-lipid interface is accomplished by stepping the molecule through the interface. The molecule is rotated about the two axes perpendicular to the interface to find the orientation of lowest energy for each step. The interface is assumed to be of a finite thickness specified by the user (usually 10 Å), and the energy of each atom in the interface is assumed to be a linear interpolation of the difference of energy of solution of the atom between pure water and pure lipid. These are given in Table III. The procedure allows each atom to be shielded from solvent by other atoms of the molecule.

TABLE III
Free energies of transfer used to calculate the preferred orientation of molecules at the water-membrane interface.

Group	$(\mu_W^o - \mu_M^o)$ in kcal/mole
methylene	0.88
methyl	2.10
alcohol	-1.34
carbonyl	-1.00
carboxylate	-2.00
peptide, nitrogen and hydrogen	-1.34

2.3. *X-Ray Crystal Structures*

Data for this study were taken from these crystal structure determinations: silver monensin (Pinkerton and Steinrauf, 1970), silver nigericin (Steinrauf et al., 1968), potassium nigericin (Geddes et al., 1975), sodium, potassium, and thallous dianemycin (Czerwinski and Steinrauf, 1971), valinomycin-potassium aurichloride (Pinkerton et al., 1969), valinomycin-rubidium aurichloride (Steinrauf, unpublished), uncomplexed beauvericin (Geddes, unpublished), and beauvericin barium picrate (Hamilton et al., 1975).

Use has also been made of the results of X-ray crystal structure determinations by others. We are grateful to Dr M.O. Chaney, Eli Lilly Research Laboratories, Indianapolis, Indiana for the atomic coordinates of the sodium complex of A-204A and to Dr J.F. Blout of Hoffman-La Roche Inc., Nutley, New Jersey for the coordinates of the silver complex of X-206. Other crystal structures include those of the silver and thallium complexes of grisorixin by Alleaume and Hickel (1970, 1972); the barium and silver complexes of X-537A by Johnson et al. (1970), and Maier and Paul (1971); nonactin potassium thiocyanate by Kilbourn et al. (1967); nonactin sodium thiocyanate by Dobler and Phizackerley (1974); and free monensin by Lutz et al. (1971). The crystal structure determination of silver nigericin has been repeated by Shiro and Koyama (1970) under the name of polyetherin A. It had been shown by Stempel et al. (1969) that nigericin, polyetherin A, and X-464 are in fact identical. In most of the above references a brief history of the antibiotics is included.

3. RESULTS

3.1. *Valinomycin – a perfect cage*

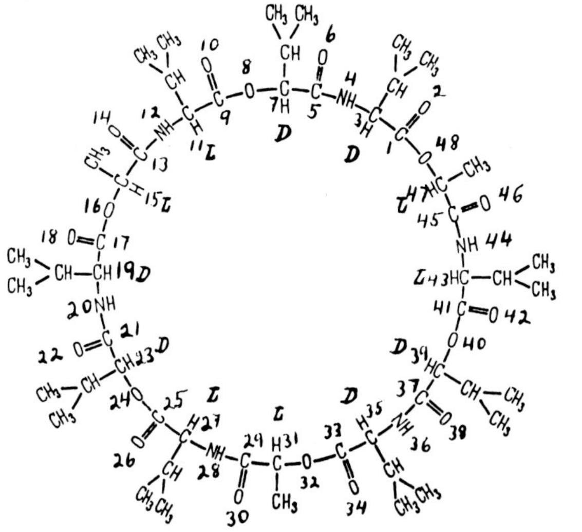

Fig. 1. The sequence of valinomycin.

The cyclic twelve residue polypeptide chain of valinomycin (Figure 1) is held in a rigid conformation by six hydrogen bonds. Since there is very little freedom in the structure, there is little ability to adopt to cations of different size. The change in the radius of the nearly octahedral cage of oxygen atoms surrounding the cation was observed to be only 0.037 Å from the potassium crystal structure to rubidium, and is accomplished entirely by rotations of the ester bonds. The energies and the positions of sodium, potassium, and rubidium ions were calculated inside the antibiotics as found in (1) valinomycin-potassium aurichloride and (2) in the isomorphous valinomycin-rubidium aurichloride. The resulting energies are given in Table IV and agree with the observed high specificity for potassium and rubidium and lower specificity for sodium. Both the observed and calculated positions of the cations are displaced toward the oxygens of the three L-valyl residues and away from the three-D-valyl residues. This displacement is greatest for sodium and least for rubidium; it is observed in the crystal structures to be 0.178 Å for potassium and 0.088 Å for rubidium.

There is little evidence for strain in this remarkably efficient structure. Each residue is in minimum energy conformation. The peptide groups average only 3° from planarity, the esters 8°. Only the six peptide hydrogen bonds have less than near-perfect geometry. It is therefore possible that the initial event in the opening of the structure involves the transfer of a peptide hydrogen to a water molecule at the membrane interface.

TABLE IV
Simulated binding energies of cations (kcal/mole) and distance of shift from the crystallographic position of the cation (Ångström units). Energy above, shift below

Crystal structure	Simulated cation				
	Na^+	K^+	Rb^+	Ag^+	Tl^+
K Valinomycin	-147 0.372	-169 0.332	-174 0.369		
Rb Valinomycin	-144 0.401	-166 0.224	-172 0.237		
Interior of Ba Beauvericin	-84	-76	-69		
Ag Monensin	-153 0.056	-146 0.057	-9 0.063	-167 0.056	
Ag Nigericin	-134 0.073	-145 0.990	-128 1.018	-151 0.036	
K Nigericin	-140 0.230	-151 0.148	-150 0.178	-139 0.324	
Na Dianemycin	-172 0.089	-161 0.081			-143 0.103
K Dianemycin	-171 0.104	-170 0.090			-143 0.120
Tl Dianemycin	-164 0.110	-163 0.086			-151 0.108

3.2. *Beauvericin – Inside or Outside?*

The only completed X-ray structure of beauvericin or enniatin (Figure 2) is that of the beauvericin-barium picrate dimer (Figure 3). Here three picrate ions form the bridging between two barium ions. The far side of each barium ion then bonds to the carbonyl oxygen of the three α-hydroxy isovaleryl residues. The complex ($Bv \cdot Ba \cdot Pic_3 \cdot Ba \cdot Bv$) has very close to 3-fold symmetry running through the two cations. Participation of anions in the complex is of considerable help in explaining the properties of beauvericin as has been discussed by Hamilton et al. (1975). Any attempt to give a complete picture requires at least one other complex of cation with beauvericin, probably (Bv·M·Bv), where M is either a mono or divale cation.

It is interesting to compare the structure of the barium beauverici complex with that of uncomplexed beauvericin which shows much less of 3- symmetry, and in the uncomplexed form the carbonyl oxygens of the hydrox residues are farther from each other by 0.56 Å and more irregularly spac On the side of the beauvericin distal to the barium the N-methyl groups and the carbonyl oxygens of the phenylalanyl residues form a six-membere circle with the N-methyls placed wider than the oxygens.

Computer calculations placing a rubidium ion at the center of the

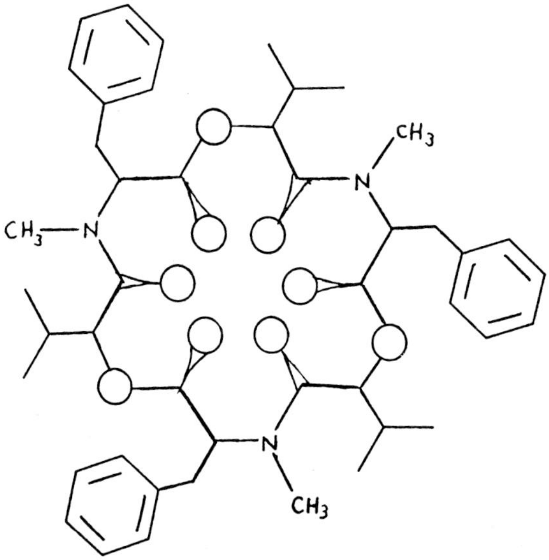

Fig. 2. The sequence of beauvericin.

phenylalanyl carbonyl oxygens produce cation to N-methyl distances of 3.95 Å or more, which are 0.4 Å greater than those found in other crystal structures from our laboratory. This indicated that cation binding from this direction is possible; however placing the cation at the center of all six carbonyl oxygens and refining gave reasonable bond distances and energy for sodium and too short distances and higher energy for potassium, while rubidium moved out to the same external position observed for barium. However, even for sodium ion the carbon-oxygen-cation bond angles would all be 92–100° as compared to 152–162° found for valinomycin. Also the carbonyl carbons would be 2.8–3.1 Å from the cation, while the shortest cation to ether carbon distances which we have observed are 3.3–3.4 Å in nigericin and dianemycin. Therefore, it seems unlikely that rubidium, potassium, or sodium could reside inside the oxygen cage of beauvericin, leaving only lithium or calcium as possibilities.

The residues of beauvericin are all in strained conformations. The major difference between the complexed and free forms is in the positions of the phenylalanyl residues. Obviously the benzene rings are contributing to the stability of these 'sandwiches'. We suggest that this stability can be disturbed at the water-membrane interface, particularly if the sandwiches contain anions which can be lost to the aqueous side of the interface.

The enniatins, a family of ion-transporting antibiotics similar to beauvericin (N-methyl-L-phenylalanyl is replaced by the corresponding valyl, leucyl, or isoleucyl residues) and having similar abilities for complexing with monovalent cations but relatively little affinities for divalent cations. We suggest that the enniatins, lacking the benzene rings, have little ability to incorporate anions in their complexes, and

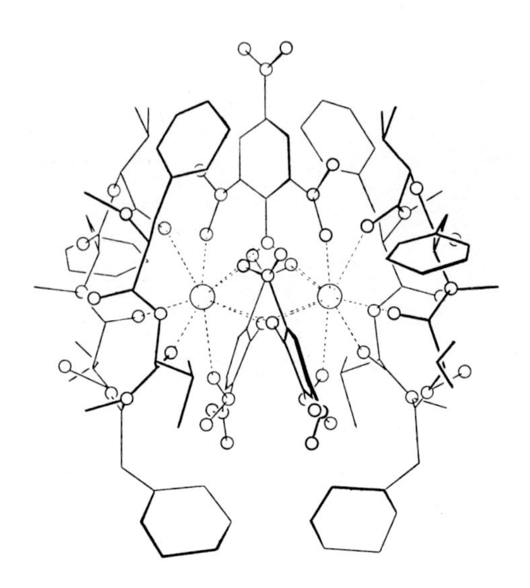

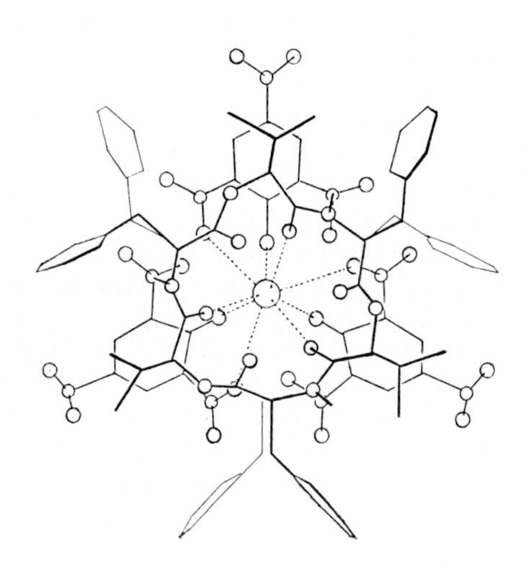

Fig. 3. Two views of the Bv·Ba·Pic_3·Ba·Bv complex from the crystal structure of beauvericin-barium picrate.

therefore cannot form complexes of low net charge as can beauvericin.

3.3. *Monensin: a Preference for Sodium*

The sequence of atoms of monensin but not the folding is given in Figure 4 along with the sequences of the other carboxylate antibiotics to be discussed. The carboxylate type of ion-transport antibiotic always contains one or more alcohol groups on the ring distal to the carboxylate which complete the enclosure around the bound cation by hydrogen bonding to the carboxylate. Flexibility of the molecule is restricted by the many 5- and 6-membered rings and also by the numerous methyl, methoxy, and other side groups. We have examined the steric factors involved in the rotations about single bonds of these molecules, and from our calculations we believe that each cation-binding form has each single bond in the lowest energy rotational conformation. The carbon-carbon single bonds between the rings specify the conformation of all of the important oxygen ligand atoms except the carboxylate. The methyl group on these rings considerably restricts rotation of the rings and the ethyl group of ring C is much more restrictive. These connecting single bonds are all very close to the staggered conformation. The average deviation from the perfect staggered dihedral angles is 5° and the greatest deviation is 12°. Of course the spiro connections absolutely prohibit rotation. The staggered conformation continues in the bonds of the carbon chain between the E ring and the carboxylate. There is always one exception here, and it is

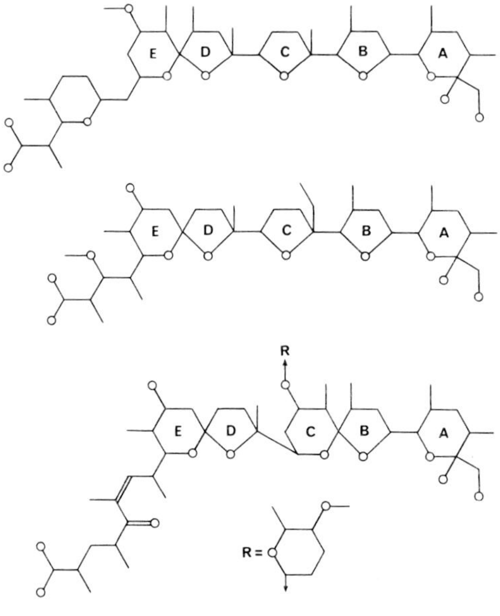

Fig. 4. The sequences of atoms in (top to bottom) nigericin, monensin, and dianemycin.

the same for monensin, nigericin and dianemycin, the second bond from the E ring. While this may be coincidence, it does suggest a mechanism for the unfolding of the antibiotic to allow passage of the cation. This will be demonstrated for monensin. When the second bond, which is 26° out of staggered conformation, is rotated by computer program back to the perfect staggered position, the carboxylate is moved away from the cation This is not unreasonable since there is no direct bonding between the carboxylate and the cation. One oxygen of the carboxylate is exposed to the solvent while the hydrogen bonds from the two alcohols of the A ring are transferred to the other carboxylate oxygen, but weakened in the process. Water molecules should now be able to hydrogen bond to the exposed carboxylate oxygen and compete for the alcohols allowing ring A to swing away.

When the cation inside monensin is changed by the computer, the larger cations press heavily on the primary alcohol of ring A. Rotation of the alcohol oxygen away from the cation relieves the force but weakens the hydrogen bond to the carboxylate and exposes the hydrogen atom to the solvent. There is therefore little adaptability toward cations of different size.

3.4. *Nigericin: a Taste for Potassium*

Since two crystal structures of nigericin have been done (silver and potassium), an experimental comparison of cation size is available. The sphere of coordinating oxygen atoms increases in size by 0.20 Å to accomm date the larger cation. This accommodation is by all of the bonds concerned to make the sphere larger. In order to maintain the hydrogen bonds to the carboxylate, the A ring, which does not bond to the cation, moves closer to the potassium. This flexibility is possible because nigericin has a carbon chain of 30 atoms while monensin has only 26. The relativel poor fit of our calculated position of the cation in the potassium nigericin structure is almost entirely due to the cation being pulled too nea the carboxylate. Either we have assigned an energy to the potassium-carboxylate bond that is too large, or the energy of this bond does not decrease with distance as rapidly as we have assumed.

3.5. *Dianemycin: Omnivorous*

Three crystal structures are available for dianemycin (sodium, potassium and thallium). Like nigericin, dianemycin has 30 carbon atoms in its chain; flexibility is however restricted by the second spiro system and by the double bonds between ring E and the carboxylate. Nevertheless, it might be expected that the sphere of oxygens would expand uniformly to accommodate larger cations. Such however is not the case; the accommodat is made by movement of the water molecule between the carboxylate and the alcohols of ring A. This movement of more that 0.3 Å is accomplished by some rearrangement of the hydrogen bonding for the thallium structure Dianemycin is known to complex with sodium, potassium, and rubidium ions almost equally well. Our potential energy parameters for rubidium and thallium are not significantly different, so it is interesting to see almost identical binding energies for each of the three ions in each of the three structures.

3.6. *The Membrane-Water Interface*

In water the ion-binding antibiotics have very weak binding capabilities and are very insoluble. On the other hand, the cation release cannot be expected to take place in any medium of low dielectric constant. Since as shown by Ashton and Steinrauf (1970), transport will take place in a system where the lipid bilayer is replaced by a centimeter thick layer of chloroform, the interface between water and membrane must be of importance in the release or uptake of cations. This situation is also amenable to computer simulation.

For each of the five antibiotics we have postulated a mechanism for release of the cation. For each antibiotic the mechanism depends on the antibiotic being properly oriented at the water-membrane interface. The final part of our simulation is to calculate the preferred orientation at the interface. The results of these calculations add strength to our proposed mechanisms by placing the carboxylates of monensin, nigericin, and dianemycin nearest the water phase. Valinomycin was found to have three nearly equal minima with the peptide hydrogen bonds pressed against the water phase. Such position would maximize the probability of hydrogen bond transfer to water. In Table V we have given the energy of the preferred orientation and the average energy of all orientations.

TABLE V
Free energy in kcal/mole for each antibiotic in the preferred orientation at the water-membrane interface and average free energy of all orientations

Antibiotic	Preferred orientation	Average of all orientations
Valinomycin K^+	-3.8	7.6
Uncomplexed Beauvericin	-0.2	9.9
Ag Monensin	-2.1	6.2
Ag Nigericin	1.1	7.9
K Dianemycin	-0.7	9.5

We have not attempted this simulation for the ($Bv\cdot Ba\cdot Pic_3\cdot Ba\cdot Bv$) complex because we are not satisfied with our potential energy parameters for the benzene and picrate rings. However, the simulation of uncomplexed beauvericin gives two nearly identical minima at 180° from each other, each having three carbonyl oxygens projecting into the water phase in an appropriate position to accept a cation.

Our simulation program on the PDP-15 will produce a stereo drawing of the oriented molecule on the oscilloscope. Photographs of these are given in Figure 5 to Figure 9.

We have plans to refine further the computer simulations. There is need to improve the potential energy parameters. In the simulation of K nigericin the cation is pulled too near the carboxylate, and in the simulation of K and Rb valinomycin the cation is not displaced enough toward the L-valyl residues. Our assignments of potential energy parameters

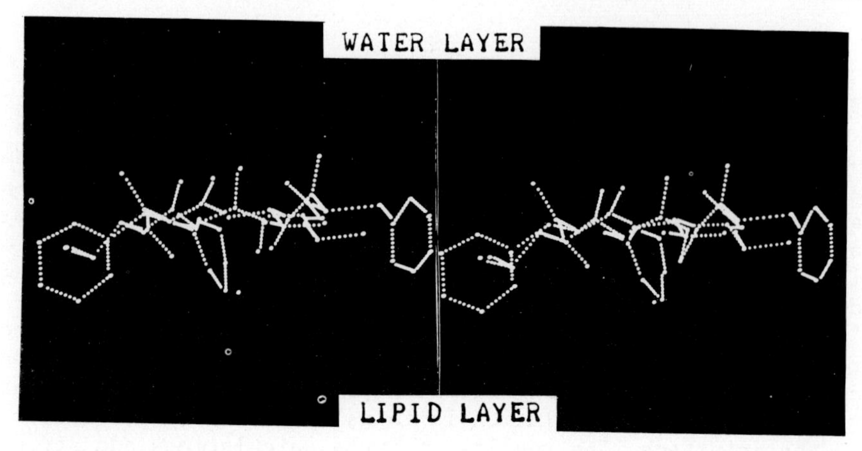

Fig. 5. Computer generated stereo drawing of free beauvericin as oriented at the simulated water-membrane interface. The interface is horizontal, the water direction is up and the lipid is down.

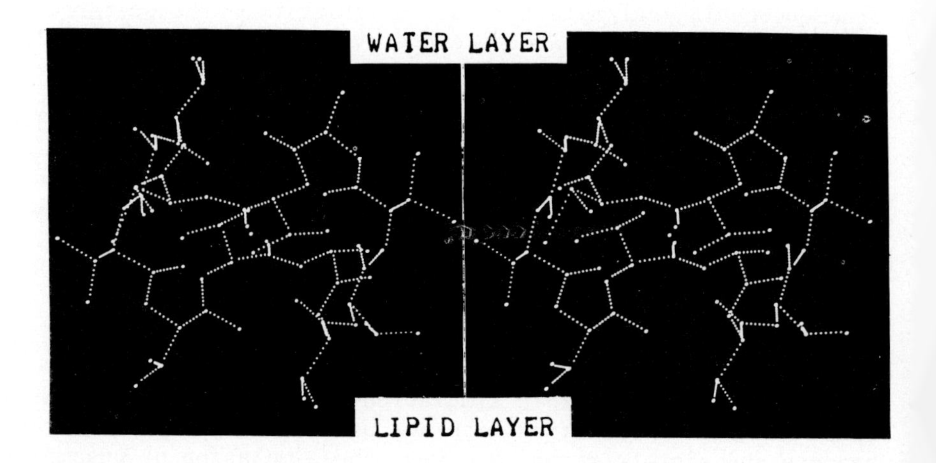

Fig. 6. Stereo drawing of K valinomycin.

for the cations in particular do not yet adequately describe the hydration of the ions in water, and we need to allow some bond rotations, particularly of side chains, during the orientation at the water-membrane interface.

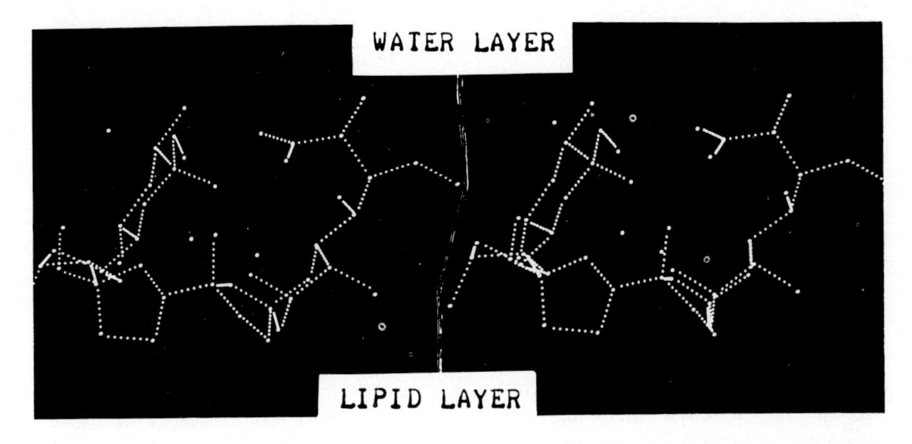

Fig. 7. Stereo of Ag monensin. Note the carboxylate oriented toward the water phase.

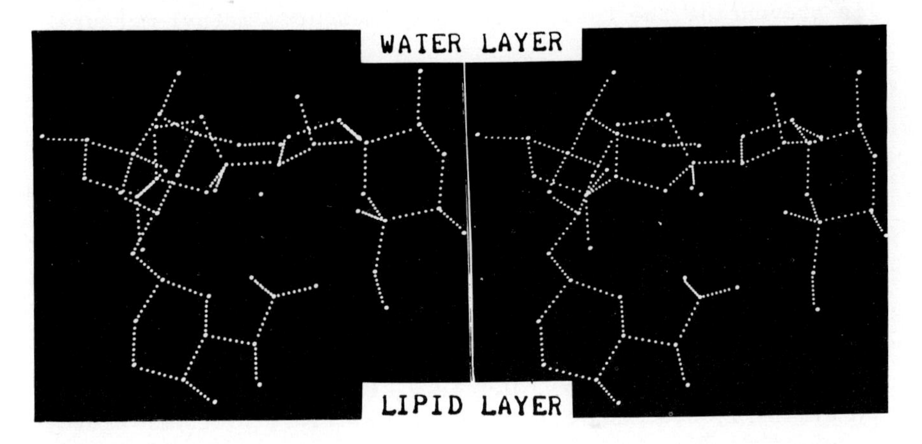

Fig. 8. Stereo of Ag nigericin.

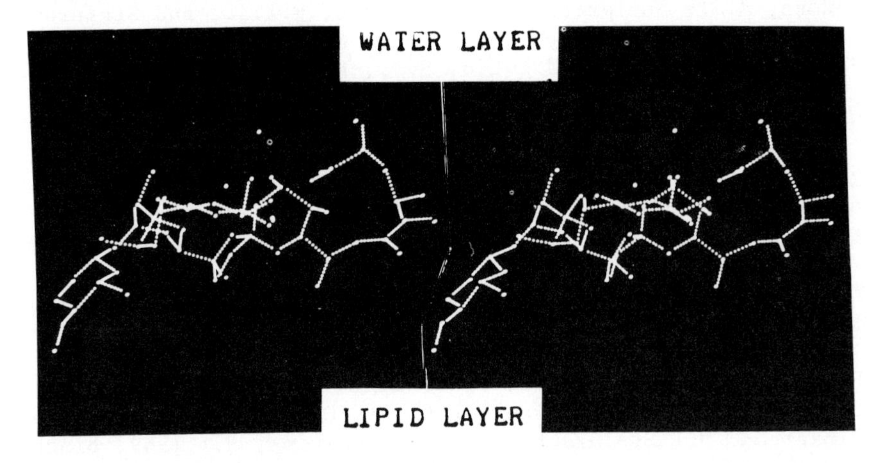

Fig. 9. Stereo of K dianemycin.

4. CONCLUSIONS

We have simulated the behaviour of some of the ion-transporting antibiotics toward the binding of cations, the rotation about single bonds, and the orientation at the water-membrane interface. From these results we have been able to postulate a mechanism of cation release for each antibiotic. These considerations have provided us with considerable insight into the function of various parts of the molecules. We have been able to postulate derivatives of the antibiotics which should have altered or enhanced properties, and several of these are being synthesized in the laboratory of Dr Roger Roeske of this department.

Such computer chemistry as we have described can provide a means of searching quickly for physical, chemical, or biological processes. As a further example this method has been used to conduct a survey for new chemotherapeutic agents for heavy-metal poisoning. From this survey has come a new class of compounds, the metal-binding steroids. Examples have now been synthesized and the compounds have been shown to be effective in preventing the toxic effects of mercury, methyl mercury, and lead in mice.

ACKNOWLEDGEMENTS

This work has been supported by NSF Grant GMS 74-22992 and PHS Grant R01ES942, and by computer time from Indiana University.

REFERENCES

Alleaume, H. and Hickel, D.: 1970, Chem. Comm., 1422-23.
Alleaume, H. and Hickel, D.: 1972, Chem. Comm., 175-6.
Czerwinski, E.W. and Steinrauf, L.K.: 1971, Biochem. Biophys. Res. Comm. 45, 1284-7.
Dobler, M. and Phizackerley, R.P.: 1974, Helv. Chim. Acta 57, 664-674.
Geddes, A.J., Sheldrick, B., Stevenson, W.T.J., and Steinrauf, L.K.: 1975, Biochem. Biophys. Res. Comm. 60, 1245-1251.
Johnson, S.M. Herrin, J., Liu, S.J., and Paul, I.C.: 1970, J. Amer. Chem Soc. 92, 4428-4435.
Hamilton, J.A., Steinrauf, L.K., and Braden, B.: 1975, Biochem. Biophys. Res. Comm., 64, 151-6.
Kilbourn, B.T., Dunitz, J.D., Pioda, L.A.R., and Simon, W.: 1967, J. Mol Biol. 30, 559-563.
Lutz, W.K., Winkler, F.K., and Dunitz, J.D.: 1971, Helv. Chim. Acta 54, 1103-8.
Maier, C.A. and Paul, I.C.: 1971, Chem. Comm., 181-2.
Pauling, L.: 1960, The Nature of the Chemical Bond, Cornell University Press, Ithaca, N.Y., pp. 505-562.
Pressman, B.C.: 1968, Fed. Proceedings 27, 1283-8.
Pinkerton, M. and Steinrauf, L.K.: 1970, J. Mol. Biol. 49, 533-546.
Pinkerton, M., Steinrauf, L.K., and Dawkins, P.: 1969, Biochem. Biophys. Res. Comm. 35, 512-8.
Shiro, M. and Koyama, H.: 1970, J. Chem. Soc., (B), 243-253.

Steinrauf, L.K.: 1971, DECUS Proceedings, Spring, pp. 49-51.
Steinrauf, L.K., Pinkerton, M., and Chamberlin, J.W.: 1968, Biochem. Biophys. Res. Comm. 33, 29-31.
Stempel, A., Westley, J.W., and Benz, W.: 1969, J. Antibiotics 22, 384-5.
Tanford, C.: 1973, The Hydrophobic Effect, John Wiley and Sons, New York.
Venkatachalam, C.M. and Ramanchandran, G.N.: 1967, in G.N. Ramanchandran (ed.), Conformation of Biopolymers, vol. I, Academic Press, New York and London, pp. 83-108.

DISCUSSION

A. Pullman: Did you also study with your simulation the capture of the cation?

Steinrauf: Not yet. We plan to follow the antibiotic after loss of the cation. This should give us some ideas about capture, and enable us to go to the uncomplexed form when available from X-ray crystallography. Three uncomplexed forms have been done, valinomycin, beauvericin, and monensin among our antibiotics.

THEORETICAL CALCULATION ON OXYGEN BINDING TO THE HEME IRON IN MYOGLOBIN AND HEMOGLOBIN

JINYA OTSUKA*
Dept. of Material Physics, Faculty of Engineering Science, Osaka University, Toyonaka, Osaka, Japan,

YASUNOBU SENO
Dept. of Physics, Faculty of Science, Nagoya University, Chikusa-ku, Nagoya, Japan,

NOBUKO FUCHIKAMI
Dept. of Physics, Faculty of Science, Tokyo Metropolitan University, Setagaya-ku, Tokyo, Japan,

and

OSAMU MATSUOKA
Laboratory of Applied Physics, The University of Electro-communication, Chofu-shi, Tokyo, Japan.

ABSTRACT. The binding of an oxygen molecule to the *heme iron* is studied by a theoretical calculation in the Heitler-London scheme with three ionic structures Fe^{2+}-O_2, Fe^{3+}-O_2^- and Fe^+-O_2^+. The electronic states of the iron atom and of the oxygen molecule are evaluated semi-empirically or empirically and only the interaction part between the iron atom and the oxygen molecule is calculated non-empirically. The calculations to determine the energy eigenvalues and eigenfunctions in the *heme*-O_2 system are carried out for both parallel and inclined arrangements of the O-O axis to the *heme* plane. In both arrangements, one of the singlet states decreases remarkably in its energy eigenvalue as the O_2 molecule approaches the *heme*. At the distance of 1.5 ~ 1.3 Å between the iron atom and one of the oxygen atoms, this singlet state is the lowest in energy if the values of state energies of the iron atom are chosen to be consistent with the other experimental results. In the parallel arrangement, the lowest state is composed mainly of $\psi(^3E; Fe^{2+})\phi(^3\Sigma_g^-; O_2)$ In the inclined arrangement, on the other hand, the electronic configuration of the lowest state is complicated, i.e. it is composed of $\psi(^3E; Fe^{2+})\phi(^3\Sigma_g^-; O_2)$, $\psi(^1A_1; Fe^{2+})\phi(^3\Sigma_g^-; O_2)$ and $\psi(^2E; Fe^{3+})\phi(^2\Pi_g; O_2^-)$. Although the present calculation does not make clear which of the arrangements is actually realized, the inclined arrangement seems to be preferable from the standpoint of the electronic configurations included in the lowest eigenstate.

---

* Present Address: Department of Applied Biological Science, Faculty of Science and Technology, Science University of Tokyo, Noda-shi, Chiba-ken, Japan.

B. Pullman and N. Goldblum (eds.), Metal-Ligand Interactions in Organic Chemistry and Biochemistry, part 2, 59-99.

1. INTRODUCTION

Myoglobin(Mb) and *hemoglobin(Hb)* belong to the so-called *hemoproteins*, which contain *hemes* as the prothetic groups, and their biological function is reversible oxygen binding. A *heme* is a metal complex compound, in which an iron atom is situated nearly at the center of the porphyrin ring. The 5th and 6th coordination positions of the *heme iron*, the direction of which is normal to the *heme* plane, are occupied by an intrinsic or extrinsic ligand, respectively. In *Mb* and *Hb*, the 5th coordination position is occupied by an intrinsic ligand identified as the nitrogen atom of histidine residue and the 6th coordination position is occupied by an extrinsic ligand such as oxygen molecule, O_2, and carbon monoxide, CO. A *Mb* molecule is a polypeptide chain with one *heme* and functions as storage of O_2 molecules especially in muscle cells. A *Hb* molecule is composed of four polypeptide chains, each of which contains one *heme*, and functions as carrier for the O_2 molecule from lung to various tissues. As is well known, O_2 molecules bind with *Hb* in a cooperative way, and recently the molecular mechanism of this cooperativity has been studied with growing interest, from the viewpoint of the conformational change of *Hb* molecule (Monod et al., 1965, Koshland et al., 1966; Perutz, 1970). The studies on the electronic structure of the *heme* and of the oxygenated *heme* may be important, because the change in the electronic state of the *heme* by oxygenation is considered as a trigger to cause the conformational change of the protein.

The purpose of the present work is a theoretical investigation into the nature of the binding of O_2 molecule to the *heme* by a quantum mechanical calculation. One theoretical calculation on the electronic structure of an oxygenated iron porphyrin complex has already been performed (Zerner et al., 1966). In this work, the molecular orbital energies of *oxyferroporphyrin monohydrate* are calculated by the extended Hückel method with self-consistent charge refinement. However, this treatment seems to be unsatisfactory for the problem considered here; exchange interaction cannot be taken into account explicitly in this method, and only one electronic configuration, in which both the iron atom and the O_2 molecule are singlet, is assumed as the binding state (singlet-singlet coupling). In the *heme*-O_2 system, other configurations, for example, triplet-triplet coupling, may also possibly produce the diamagnetism of this compound, since the ground state of a free O_2 molecule is triplet and the triplet state of the *heme* is also a low-lying state as will be mentioned later. In the present work, we intend to take carefully into account the various electronic states of the *heme* and of O_2 molecule.

As is well known, two calculation methods, the molecular orbital method and the Heitler-London method, are representative ones in the calculation of the electronic states of molecules. If a complete configuration interaction is considered, the two methods are equivalent to each other in their obtained results. Practically, however, a complete performance of the calculation of configuration interaction is usually very laborious. Therefore, it may be important to examine carefully which method is more suitable to the subject in question. Generally, the molecular orbital method is suitable to the system in which electrons move frequently from atom to atom. A good example of such a system is π-electrons in a conjugated double bonds system. On the other hand, in a hydrogen molecule the Heitler-London method is superior to the molecular orbital method as the first approach. In the bonding orbital of the molecular orbital method, both covalent state and ionic state are included with the same weight, but the result of more accurate calculation with configuration interaction shows that the contribution of the ionic state is much smaller in the ground state of a hydrogen molecule.

In the binding of the iron atom with O_2 molecule, many kinds of orbital couplings may possibly participate even if only the outer orbitals of the iron atom and of O_2 molecule are considered, since many multiplets arise from the iron atom in the *heme*, and low-lying excited states exist in O_2 molecule. Therefore, if we adopt the molecular orbital method, we must evaluate non-empirically many molecular orbital energies and then calculate the coulomb interaction between electrons, including the strong interaction within the iron atom and within O_2 molecule. It is not easy to obtain a satisfactory result along this line, since the interaction energy between the iron atom and O_2 molecule is small relative to the total energy of Fe—O_2 system, and very accurate evaluation of the strong interaction part is required. In many cases of π-electrons systems in hydrocarbons with conjugated double bonds, it may be allowed that the contribution of each intra atomic part of the constituent atom is regarded as the same and that the molecular orbital and their energy levels can be determined only by the symmetry of the system retaining some adjustable parameters, as long as much higher excitations such as $\sigma \to \pi$ transition are not discussed. And the method of evaluating these parameters and moreover the coulomb interaction between the π-electrons has been established (Pariser et al., 1953; Pople, 1953). This is not so in our *heme*-O_2 system. But, fortunately, information about the electronic structures of the iron atom in the heme has been accumulating as a result of many experimental measurements of magnetic susceptibility, EPR spectra, Mössbauer effect and optical spectrum (d→d transition), and by their theoretical analyses based on the ligand field theory. Thus, we adopt a Heitler-London scheme in which the electronic structures of the *heme* and of a free O_2 molecule are regarded as unperturbed parts and only the interaction part between the *heme* and the O_2 molecule is calculated non-empirically. Such a method of calculation, already proposed by Moffit, is called *method of atoms in molecule* (Moffit, 1951). In our case, we regard the *heme*-O_2 system as a *molecule*, and the *heme* and the O_2 molecule as *atoms* constituting the *molecule*. This treatment is based on the fact that the interaction energy (binding energy) is much smaller compared with the total energy of the *molecule* and the electronic structure of the *atoms* are fairly well conserved in the *molecule*. In this way, we intend to examine how the electronic structure changes as an O_2 molecule approaches the *heme*, and to investigate which of the unperturbed states contributes mainly to the binding of an O_2 molecule to the *heme*.

2. METHOD OF CALCULATION

2.1. *Treatment of the Electronic Structure of the Heme*

The study of the electronic structure of the *heme* itself continues to be a problem to be challenged at the present time. With respect to the interaction between the iron atom and the porphyrin ring, both π-π type and σ-σ type couplings are considered. Moreover, the d-electrons system is of open shell. Thus, many of the theoretical approaches hitherto carried out depend on their purposes whether they intend to explain the optical spectra originated from π-electrons in the porphyrin ring (Simpson 1949; Gouterman, 1959; Weiss et al., 1965), or whether they intend to explain the magnetic properties originated from the d-electrons in the iron atom (Griffith, 1957; Kotani, 1963; Otsuka, 1966, 1968). Although theoretical attempts to calculate the energy levels of *metalporphyrin complex* have been tried (Pullman et al., 1960; Ohno et al., 1963; Berthier et al., 1965; Zerner et al., 1966), their results do not attain to reproduce satisfactorily the energy levels, especially those to which d orbitels of the iron atom contribute mainly. Therefore, in the present work, the electronic structure of the heme is treated by the ligand field theory which attempts to explain the magnetic properties and the optical spectra of d → d transition in *metal-complex compounds*.

In the ligand field theory, d-electrons of the metal ion are assumed to be subject to the local electric field produced by the ligands which are neighbouring ions or atoms of the metal ion. However, in this theory, an explicit calculation of the effect of the local electric field is not carried out, but the feature of the splitting of the d-level is determined by the symmetry of the coordinations of the neighbouring ligands. Thus, the splitting energy values of the d-level are treated as adjustable parameters. Then, the coulomb interaction between the d electrons is calculated for a definite d^{ℓ} configuration. In a d^{ℓ} configuration, the coulomb interaction can be expressed with only three parameters, the so-called *Racah's* parameters A, B and C. *Racah's* parameter A appears only in diagonal elements with a common factor and the energy differences of eigenstates are independent of A value within a definite d^{ℓ} configuration. Usually, the values of B and C are taken to be slightly smaller than those of free metal ions (Tanabe and Sugano, 1954). In this way, the effect of covalency with the ligand orbitals is taken into consideration. The values of the splitting energy of d-level are determined by experimental data. Therefore, this method was originally to be valid only for the *metal-complex compounds* in which the covalency of the metal orbitals with the ligand orbitals is small. However, it has been discovered that this method is also useful to various kinds of *metal-complex compounds*, with its succesful explanations of their magnetic and optical properties if the values of B and C are chosen suitably. The treatment in the ligand field theory is discussed more fully by Griffith (1961).

Thus, in the present work, the eigenstates of the heme are constructed only from the d orbitals of the iron ions and their energy eigenvalues are evaluated reasonable by the values of the splitting parameters of the d-level through experimental data. This treatment may be justified by the following studies: (1) According to the result of the calculations on a *iron-porphyrin* system (Pullman et al., 1960; Ohno et al., 1963; Berthier et al., 1965; Zerner et al., 1966), mixing of the d orbitals in the iron atom into the *porphyrin* orbitals is not so remarkable. (2) An O_2 molecule interacts directly with the iron atom and not with the *porphyrin*, and higher excited states obtained by the transition such as

$d \rightarrow \pi$ $\pi \rightarrow d$ and $\pi \rightarrow \pi$ may not contribute much to the binding of O_2 molecule to the *heme*.

2.2. *Symmetry and Basic Functions*

Two simple models have been proposed to explain the nature of the chemical bond between the iron atom in the *heme* and O_2 molecule. One model was proposed by Pauling (1949), in which the O-O axis was at 60° to the *heme* plane and a σ bond was formed between the lone-pair orbital of the oxygen molecule and the empty octahedral hybrid orbital of the iron atom. Later, Griffith (1956) estimated the ionization potentials of the lone-pair orbital and of the π_g orbital in O_2 molecule, and comparing them, he proposed another model in which *heme*-O_2 was a π-complex with the O-O axis parallel to the *heme* plane. Recently, X-ray diffraction analysis on *oxy-myoglobin* (Watson and Nobbs, 1968) suggests that the O_2 molecule is bound to the iron atom in an orientation similar to that proposed by Pauling. With these situations in mind, we perform calculations for the two stereo-chemical arrangements; (1) parallel arrangement and (2) inclined arrangement.

The iron atom in the *heme* is situated nearly at the center of a *porphyrin ring*. Although non-planarity of the high spin *iron-porphyrin* has been pointed out by the recent X-ray analysis (Perutz, 1970), we assume an ideal geometry of the heme in which the iron atom lies in the plane and Fe—N bond length is 2.01 Å. The point group symmetry of the surroundings of the iron atom is almost C_{4v}. The chemical structure of *heme* is shown in Figure 1. In a field of the symmetry characterized by

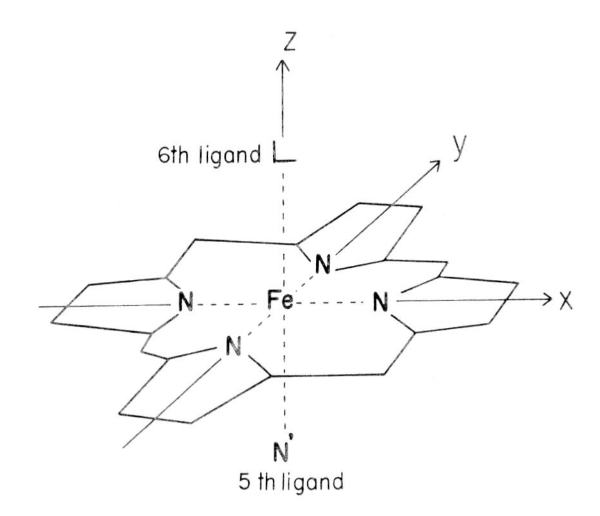

Fig. 1. The structure of the *heme*. In *myoglobin* and *hemoglobin*, the 5th coordination position is occupied by the nitrogen atom (N') of the histidine residue and the 6th coordination position is occupied by an extrinsic ligand (L) such as O_2 and CO. In the present paper, z axis is taken perpendicularly to the *heme* plane, and x and y axes are in the plane as shown in this figure.

the group C_{4v}, the d-level of the iron atom splits into four sub-levels differing in energy which are denoted by the symbols of the irreducible representations a_1, b_1, e and b_2. The explicit forms of the orbital functions for these levels are

$$a_1;\ u \propto 3z^2 - r^2$$
$$b_1;\ v \propto x^2 - y^2$$
$$e_x;\ \eta \propto xz$$
$$e_y;\ \xi \propto yz$$
$$b_2;\ \zeta \propto xy$$

where z axis is taken perpendicularly to the heme plane, and x and y axes are in the plane as shown in Figure 1. The energy differences between these sub-levels are expressed in terms of three parameters X, Y and Z as shown in Figure 2. When both Y and Z vanish, we have only two sub-levels, e_g and t_{2g}, separated by X, which corresponds to 10 Dq in the usua notation in cubic symmetry.

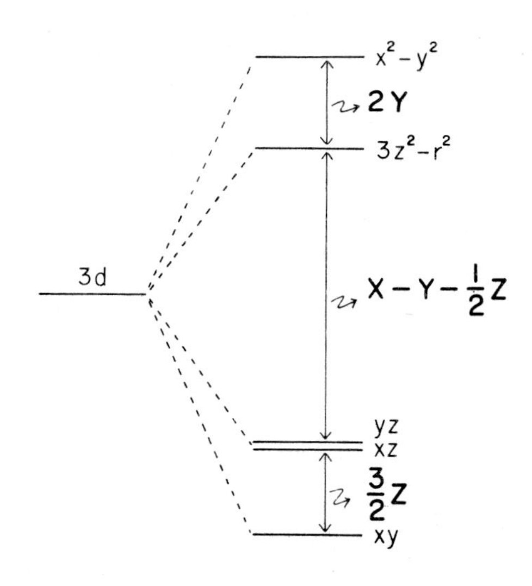

Fig. 2. The splitting of the d-level of the iron atom in the *heme*. The energy differences between the sub-levels are expressed in terms of three parameters X, Y and Z as shown in this figure.

With respect to O_2 molecule, the orbitals and the electronic states are labelled according to the axial symmetry of homopolar diatomic molecules ($D_{\infty h}$). In our treatment, the 1s orbital of each oxygen atom is shrunk to the nuclei and an O_2 molecule is treated as 12, 13, or 11 electrons system corresponding to ionic structure O_2, O_2^- or O_2^+, respectively. From the 2s and 2p atomic orbitals, are constructed eight molecular orbitals $2\sigma_g$, $2\sigma_u$, $3\sigma_g$, $\pi_u^\pm$, $\pi_g^\pm$ and $3\sigma_u$ in order of increasing orbital energy. In the *heme*-O_2 system considered here, however, it is convenient to use the following real forms of the doubly degenerate representations of $D_{\infty h}$ instead of $\Pi^+_{g,u}$, $\Pi^-_{g,u}$, $\Delta_g(2)$ and $\Delta_g(-2)$,

$$\Pi^{z'}_{g,u} = -\frac{1}{\sqrt{2}}(\Pi^+_{g,u} + \Pi^-_{g,u})$$

$$\Pi^{y'}_{g,u} = \frac{1}{\sqrt{2}}(-\Pi^+_{g,u} + \Pi^-_{g,u})$$

$$\Delta^a_g = \frac{i}{\sqrt{2}}(\Delta_g(2) + \Delta_g(-2))$$

$$\Delta^b_g = -\frac{i}{\sqrt{2}}(\Delta_g(2) - \Delta_g(-2))$$

where z' and y' show two directions perpendicular to O-O axis and making a right angle. In the parallel arrangement, in which the O-O axis is placed so as to be parallel to the heme plane, the point group symmetry of the total system *heme*-O_2 is C_{2v}. In this case, y' and z' directions can be taken to be consistent with y and z axes respectively, as shown in

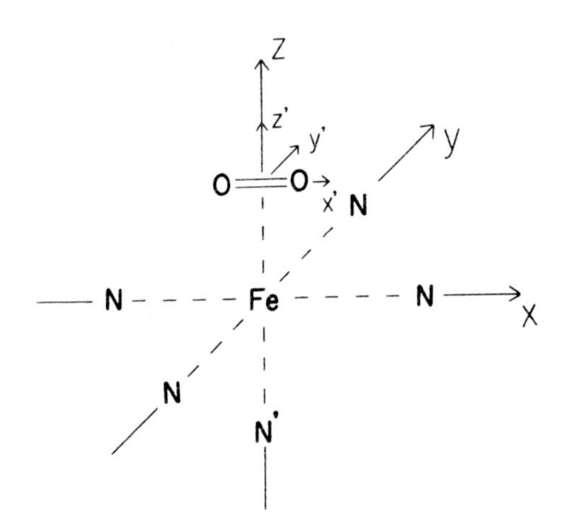

Fig. 3. The parallel arrangement of O_2 molecule to the heme plane. The O-O axis of the O_2 molecule is denoted by x', and y' and z' show two directions perpendicular to the O-O axis and making a right angle. In this arrangement, x', y' and z' directions can be taken to be consistent with x, y and z axes respectively as shown in this figure.

Figure 3. In this point group C_{2v}, there are four irreducible representations, A_1, A_2, B_1 and B_2. In this parallel arrangement, the bases of the irreducible representations of C_{4v} (*heme*) and of $D_{\infty h}$ (O_2 molecule) correspond to those of C_{2v} (*heme*-O_2) as follows,

C_{4v}	$D_{\infty h}$	C_{2v}
A_1, B_1	Σ_g^+, Π_u^z, Δ_g^a	A_1
A_2, B_2	Σ_u^-, Π_g^y	A_2
E_x	Σ_u^+, Π_g^z	B_1
E_y	Σ_g^-, Π_u^y, Δ_g^b	B_2

In the inclined arrangement, we take the stereochemical arrangement as shown in Figure 4. This arrangement is the same as the one proposed by Pauling (1949). In this case, point group symmetry of *heme*-O_2 system is C_s and there are two irreducible representations A' and A''. The bases which are symmetric with respect to xz plane belong to A' and those which are antisymmetric belong to A''. The bases of the irreducible representations of C_{2v}, whose symmetry is realized in the parallel arrangement, correspond to those of C_s as follows,

C_{2v}	C_s
A_1, B_1	A'
A_2, B_2	A''

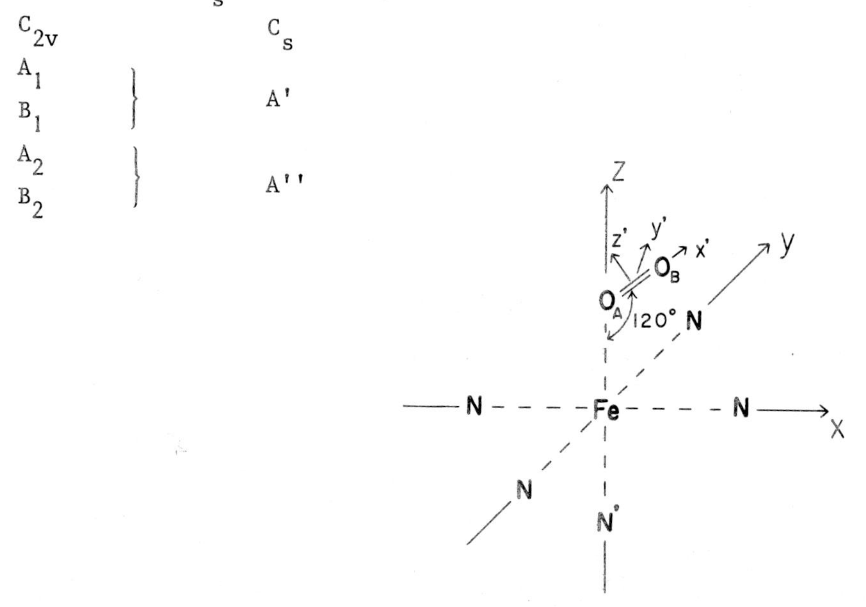

Fig. 4. The inclined arrangement of O_2 molecule to the heme plane. In this arrangement, the O-O axis (x' axis) makes a 60° angle with the z axis.

2.3. *Calculation Procedure*

First, we construct the basic functions of the *heme*-O_2 system which form the basis of the present calculation. A basic function $\Psi(^{2s+1}\Gamma)$ is assumed to be expressed as the anti-symmetrical product function of a wavefunction $\Psi(\alpha^{2s_1+1}\Gamma_1, \gamma_1)$ of the iron atom in the *heme* and of a wavefunction $\phi(^{2s_2+1}\Gamma_{2}, \gamma_2)$ of the O_2 molecule, i.e.,

$$\Psi(^{2s+1}\Gamma) = NA\{\psi(\alpha^{2s_1+1}\Gamma_1, \gamma_1)\cdot\phi(^{2s_2+1}\Gamma_2, \gamma_2)\}, \tag{1}$$

where N is the normalization constant, the symbol A means an antisymmetrizing operator for all electrons considered, $\gamma_i(i=1, 2)$ denotes the subspecies of the irreducible representation Γ_i. The spins S_1 and S_2 couple to the total spin S. In our treatment, $\psi(\alpha^{2s_1+1}\Gamma_1, \gamma_1)$ is chosen to be an electronic configuration of the iron atom of the strong field coupling scheme in the ligand field theory, i.e. in the form of $e^{\ell_1}b_2^{\ell_2}a_1^{\ell_3}b_1^{\ell-(\ell_1+\ell_2+\ell_3)}$ where $0 \leq \ell_1 \leq 4$, $0 \leq \ell_2 \leq 2$, $0 \leq \ell_3 \leq 2$ and ℓ is the total number of 3d electrons. The wavefunction of the O_2 molecule $\phi(^{2s_2+1}\Gamma_2,\gamma_2)$ is chosen to be the eigenfunction of a free O_2 molecule.

Next, we assume the eigenfunction Φ, which represents a stationary state of *heme*-O_2, can be expanded linearly in terms of the members of the basic functions Ψ_i,

$$\Phi = \sum_i C_i\Psi_i, \tag{2}$$

The finite expansion (2) is to represent a solution of the eigenvalue problem of the *heme*-O_2 system, namely,

$$H\Phi = W\Phi \tag{3}$$

where H is the Hamiltonian operator of the total system *heme*-O_2. Substituting (2) in (3), we find

$$\sum_i C_i H\Psi_i = W\sum_i C_i\Psi_i. \tag{4}$$

Multiplying both sides of this equation by Ψ_j^* from the left side and integrating over all space and spin coordinates, we then have

$$\sum_i H_{ji}C_i = W\sum_i S_{ji}C_i, \tag{5}$$

where we use the abbreviations

$$S_{ji} = \int\Psi_j^*\Psi_i d\tau = S_{ij}^* \tag{6}$$

$$H_{ji} = \int\Psi_j^* H\Psi_i d\tau = H_{ij}^* \tag{7}$$

The relations (6) and (7) define the matrix elements of energy matrix H and of overlap matrix S, respectively. Eliminating C_i from the system of Equation (5), we obtain the secular equation

$$\det(H - WS) = 0. \tag{8}$$

This equation gives the energy eigenvalues of *heme*-O_2 system. Next, let us consider the evaluation of a matrix element of total Hamiltonian H_{ji}. The explicit forms of Ψ_i and Ψ_j are;

$$\begin{aligned} \Psi_i &= NA\Sigma\psi_{iI}(q_1, q_2, \cdots\cdots q_\ell)\cdot\phi_{iII}(q_{\ell+1}, q_{\ell+2}, \cdots\cdots q_n)\}, \\ \Psi_j &= NA\{\psi_{jI}(q_1, q_2, \cdots\cdots q_{\ell'})\cdot\phi_{jII}(q_{\ell'+1}, q_{\ell'+2} \cdots\cdots q_n)\} \end{aligned} \tag{9}$$

$$(\text{generally } \ell \neq \ell')$$

where ψ shows the wavefunction of the iron atom, ϕ the wavefunction of the O_2 molecule, q_k the coordinate of the k-th electron including spin. The Hamiltonian of the total system may be resolved into the following according to the electronic configuration of $\psi_{iI}\cdot\phi_{iII}$.

$$H = H_I + H_{II} + V + U, \tag{10}$$

where $$H_I = \sum_{k=1}^{\ell} \{-\frac{1}{2}\Delta_k + v_1^c(k)\} + \sum_{k>h=1}^{\ell} \frac{1}{r_{hk}}, \tag{11}$$

$$H_{II} = \sum_{k=\ell+1}^{n} \{-\frac{1}{2}\Delta_k + v_2^c(k)\} + \sum_{k>h=\ell+1}^{n} \frac{1}{r_{hk}}, \tag{12}$$

$$V = \sum_{k=1}^{n} v_2(k) + \sum_{k=\ell+1}^{n} v_1(k) + \sum_{h=1}^{\ell} \sum_{h=\ell+1}^{n} \frac{1}{r_{hk}}, \tag{13}$$

and U shows the interaction between the core of the iron atom and that of the O_2 molecule. H_I and H_{II} are Hamiltonian of the iron atom in the he and that of the O_2 molecule, respectively, and V is the interaction part between the iron atom and the O_2 molecule. The matrix element of H, evaluated for the Hamiltonian resolved as in the Equation (10) will be denote by $\langle\Psi_j|H\Psi_i\rangle$. If ψ_{iI} and ϕ_{iII} are chosen to be eigenfunctions of H_I and H_{II}, respectively, these functions satisfy the following equations.

$$\begin{aligned} H_I\psi_{iI} &= W_{iI}\psi_{iI}, \\ H_{II}\phi_{iII} &= W_{iII}\phi_{iII}, \end{aligned} \tag{14}$$

where W_{iI} and W_{iII} are energy eigenvalues of unperturbed states of the *heme* and O_2 molecule, respectively. With the use of these relations, the matrix element $\langle\Psi_j|H\Psi_i\rangle$ is written as

$$\langle\Psi_j|H\Psi_i\rangle = (W_{iI} + W_{iII})\langle\Psi_j|\Psi_i\rangle + \langle\Psi_j|V\Psi_i\rangle + U\langle\Psi_j|\Psi_i\rangle. \tag{15}$$

On the other hand, $\langle\Psi_i|H\Psi_j\rangle$ is written as

$$\langle\Psi_i|H\Psi_j\rangle = (W_{jI} + W_{jII})\langle\Psi_j|\Psi_i\rangle + \langle\Psi_i|V\Psi_j\rangle + U\langle\Psi_i|\Psi_j\rangle. \quad (16)$$

Generally, $\langle\Psi_i|H\Psi_j\rangle$ is not necessarily equal to $\langle\Psi_i|H\Psi_j\rangle^*$. This indicates that the matrix of H constituted from the matrix elements evaluated in this way is not hermitian and the eigenvalues of this matrix are not necessarily real. In order to avoid this difficulty, we evaluate approximately the matrix element of H between j and i wavefunctions, H_{ji}, as follows.

$$H_{ji} = H^*_{ij} = \frac{1}{2}\{\langle\Psi_j|H\Psi_i\rangle + \langle\Psi_i|H\Psi_j\rangle\} \quad (17)$$

Finally, the following expression is obtained,

$$H_{ji} = \frac{1}{2}\{(W_{iI} + W_{iII}) + (W_{jI} + W_{jII})\}\langle\Psi_j|\Psi_i\rangle + \frac{1}{2}\{\langle\Psi_j|V\Psi_i\rangle + \langle\Psi_i|V\Psi_j\rangle\} + U\langle\Psi_j|\Psi_i\rangle. \quad (18)$$

To evaluate the elements of the overlap matrix and the interaction parts of the Hamiltonian, the following types of molecular integrals have to be computed,

zeroth order of overlap

- coulomb integral $(dd' \,;\, \chi\chi')$
- core integral $(\chi|v_1^c|\chi')$, $(d|v_2^c|d')$

first order of overlap

- hybrid integral $(dd' \,;\, d''\chi)$, $(\chi d' \,;\, \chi'\chi'')$
- resonance integral $(d|v_1^c|\chi)$, $(\chi|v_2^c|d)$
- overlap integral $(d|\chi)$

second order of overlap

- exchange integral $(\chi d \,;\, d'\chi')$

where d indicates a 3d atomic wavefunction of the iron atom and χ a molecular orbital of the O_2 molecule. With respect to the 3d orbital of iron atom, Clementi's Hartree-Fock atomic function (1965) of a free iron ion (Fe^{2+}, ^5D) is used. The radial part of this 3d function is approximately expanded in terms of eight Gaussian type orbitals,

$$R_{3d}(r) = \sum_{i=1}^{8} C_i N_i r^2 \exp(-\alpha_i r^2), \quad (19)$$

where

$$N_i = \frac{16}{\sqrt{15}}(2\pi)^{-1/4}\alpha_i^{7/4}. \quad (20)$$

For the molecular orbitals of O_2 molecule, we adopt the Schmidt orthogonalized orbitals constructed from the Gaussian type 2s and 2p functions of an oxygen molecule which are obtained by Huzinaga (1965), since the SCF molecular orbitels are almost equal to the Schmidt orthogonalized orbitals (1957).

All of the computations, the Gaussian expansion of 3d orbital functions, the evaluation of molecular integrals, the calculation of the matrix elements of the interaction part and the solution of the secular equations, are carried out by HITAC-5020E at the Computer Center, University of Tokyo and by FACOM 230-60 at the Data Processing Centre, Kyoto University.

3. UNPERTURBED STATES OF HEME AND OF OXYGEN MOLECULES

3.1. *Summary of the Electronic States of the Heme*

On the ferric forms of Mb and Hb, measurement of EPR spectra on their single crystal (Gibson et al., 1958) and temperature dependence of paramagnetic susceptibility at low temperatures (Tasaki et al., 1967) have revealed that the ground state of these compounds is sextet state $^6A_1$ (high spin state) and the spin degeneracy of this ground state is much removed, i.e., if the fine structure is expressed by a spin Hamiltonian DS_z^2, the value of D is 10 cm^{-1} in both Mb and Hb. Here, z axis is taken to be perpendicular to the *heme* plane and the anisotropy in the *heme* plane is hardly recognized. At room temperature, these hemoproteins are thermally in equilibrium between the compounds of high spin type and of low spin type (Beetlestone and George, 1961; George et al., 1964). As is clearly seen in the Tanabe Sugano diagram in cubic symmetry (Tanabe and Sugano, 1954) the energy of quartet state ($^4T_1$) is the lowest relative to the ground state where the high spin state is equal to the low spin state in energy. In tetragonal symmetry (C_{4v}), the quartet state is further split and the splitting may cause a large value of D. This is first pointed out by Kotani (1963). Otsuka (1968) has calculated the relation between the D value and the splitting parameters (Y and Z), assuming that the value of X is nearly equal to that which makes the energy of high spin state equal to that of low spin state. Then, the value of Y is estimated to be 4000 cm^{-1}, if the value of Z is estimated to be 1000 cm^{-1} from Griffith's analysis (1957) of EPR spectra of *hemoglobin azide* derivative. Along with these values, the value of X is chosen to be 28700 cm^{-1}.

On ferrous form of Mb and Hb, the temperature dependence of magnetic susceptibility has also been carried out and revealed the result that the fine structure of the ground quintet state is well expressed by a spin Hamiltonian DS_z^2 and the value of D is 5 cm^{-1} in both ferrous Mb and ferrous Hb (Nakano et al., 1971). The value of splitting parameter Z in ferrous form is estimated to be 300 cm^{-1}, if the coefficient of spin-orbit interaction is taken to be 200 cm^{-1}. The values of the other parameters X and Y are not determined directly in ferrous forms of Mb and Hb. Eaton and Charney (1969), however, have identified the d → d transition in ferrous cytochrome c (low spin compound) from absorption spectra and

CD spectra in the wave length region 5700 ~ 10 000 Å, and found that the value of X (in our notation) is between 17 000 and 19 000 cm^{-1}. This value of X gives almost equal energy to the singlet state ($^1A_1$) and the quintet state ($^5B_2$, or ^5E). On the basis of these facts, the value of X is assumed to be 17000 cm^{-1} in the present calculation. The parameter Y is treated as variable, being taken as 2000 cm^{-1}, 4000 cm^{-1} and so on.

Although the electrostatic interaction of d^{ℓ} configuration can be evaluated by the ligand field theory and electronic states can be determined, the result gives only the relative energy differences of various SΓ states within a definite d^{ℓ} configuration. In the present problem, the energy difference between $(3d)^5$ (Fe^{3+}) and $(3d)^6$ (Fe^{2+}) configurations and that between $(3d)^6$ and $(3d)^7$ (Fe^+) configurations in the *heme* are also needed. According to Moore's table (Moore, 1949), the ionization energy and the electron affinity of free Fe^{2+} ion are as follows.

$$E_0(3d^5,\ ^6S) - E_0(3d^6,\ ^5D) = 30.64 \text{ eV}. \quad (21\text{-}1)$$

$$E_0(3d^6,\ ^5D) - E_0(3d^7,\ ^4F) = 15.88 \text{ eV}. \quad (21\text{-}2)$$

Next, the charge distribution on the porphyrin ring should be taken into account. According to either the extended Hückel calculation on the iron *porphyrin complex* (Zerner et al., 1966) or the SCF-MO calculation on the π-electron system of the *porphyrin ring* (Weiss et al., 1965), the carbon atoms are almost electrically neutral, and negative charges are located mainly at the four nitrogen atoms. These negative charges decrease the ionization and the electron affinity of the iron atom. In the present calculation, we adopt an ionic model of the *heme* in which the negative charge -2 e is assumed to be distributed with the same weight on the four nitrogen atoms, i.e., one nitrogen atom is charged with -0.5 e. Thus, for the iron atom in the ferrous *heme*, the ionization energy (IP) and the electron affinity (EA) are evaluated approximately as the differences between the following energies,

$$E(3d^6,\ ^5D) = E_0(3d^6,\ ^5D) + \frac{2}{R_N}\times 6, \quad (22\text{-}1)$$

$$E(3d^5,\ ^6S) = E_0(3d^5,\ ^6S) + \frac{2}{R_N}\times 5, \quad (22\text{-}2)$$

$$E(3d^7,\ ^4F) = E_0(3d^7,\ ^4F) + \frac{2}{R_N}\times 7, \quad (22\text{-}3)$$

where R_N denotes the distance between the iron atom and the nitrogen atom of the *porphyrin*. If the distances R_N is taken to be 2.01 Å, then we have

$$IP = E(3d^5,\ ^6S) - E(3d^6,\ ^5D) = 16.31 \text{ eV}, \quad (23\text{-}1)$$

$$EA = E(3d^6,\ ^5D) - E(3d^7,\ ^4F) = 1.55 \text{ eV}. \quad (23\text{-}2)$$

If the energy $E(3d^6,\ ^5D)$ of the Equation (22-1) is taken as the origin of the energy levels of the *heme*, the matrix element of the intra-atomic

Hamiltonian of the *heme* $<3d^{\ell},\alpha\,^{2s+1}\Gamma|H_1|3d^{\ell},\beta\,^{2s+1}\Gamma>$ can be related by the electrostatic matrix element $(3d^{\ell},\alpha\,^{2s+1}\Gamma|H_1|3d^{\ell},\beta\,^{2s+1}\Gamma)$ in the ligand field theory in the following way.

$$<3d^{\ell},\alpha\,^{2s+1}\Gamma|H_1|3d^{\ell},\beta\,^{2s+1}\Gamma> = (3d^{\ell},\alpha\,^{2s+1}\Gamma|H_1|3d^{\ell},\beta\,^{2s+1}\Gamma) - \Delta(3d^{\ell})\delta_{\alpha\beta}, \quad (24)$$

$$\Delta(3d^6) = 15A - 35B + 7C, \quad (25\text{-}1)$$

$$\Delta(3d^5) = 10A - 35B - IP, \quad (25\text{-}2)$$

$$\Delta(3d^7) = 21A - 43B + 14C + EA, \quad (25\text{-}3)$$

where $\delta_{\alpha\beta} = 1$ for $\alpha = \beta$ and $\delta_{\alpha\beta} = 0$ for $\alpha \neq \beta$. The three terms (15A - 35B +7C), (10A - 35B) and (21A - 43B + 14C) are coulomb interaction energies among the 3d electrons in the ^5D state of Fe^{2+}, ^6S of Fe^{3+} and ^4F of Fe^+, respectively. Here, the orbital energy of d(xy) in the iron atom is assumed to be equal to that of d-orbital in a free iron atom, since d(xy) orbital stays away from the four nitrogen atoms of the porphyrin and from the nitrogen atom of the histidine residue (5th ligand). Thus, the orbital energy of d(xy) is taken as the origin, and the splitting parameters X, Y and Z are included in the matrix element $<3d^{\ell},\alpha\,^{2s+1}\Gamma|H_1|3d^{\ell},\alpha\,^{2s+1}\Gamma>$ or $(3d^{\ell},\alpha\,^{2s+1}\Gamma|H_1|3d^{\ell},\alpha\,^{2s+1}\Gamma)$.

The energy levels of low-lying terms in the *heme*, which are used in the present calculation, are shown in Figure 5. In the estimation of these energy values, the values of the parameters in the ligand field theory are taken as follows; X = 28700 cm^{-1}, Y = 4000 cm^{-1}, Z = 1000 cm^{-1}, B = 1015 cm^{-1} and C = 4800 cm^{-1} for the ferric form, and X = 17000 cm^{-1}, Y = 4000 cm^{-1}, Z = 300 cm^{-1}, B = 917 cm^{-1} and C = 4040 cm^{-1} for the ferrous form. Since the experimental data on Fe^+ in the *heme* is not available, the values of the parameters for Fe^+ are taken to be the same as for Fe^{2+} It should be noted that not only the low spin states (S = 0 in the ferrou form and S = 1/2 in the ferric form) but also the intermediate spin state (S = 1 in the ferrous form and S = 3/2 in the ferric form) are very low in energy. This is a characteristic feature of the electronic state of the iron atom in hemoproteins.

3.2. *Electronic State of O_2 molecule*

With respect to O_2 and O_2^+, the state energies are well established by numerous works on the absorption spectra, especially for O_2^+, of the Rydberg absorption series (Yoshino and Tanaka, 1968). The complete list of the references is given by Wallace (1962), and the tentative potential curve of O_2 molecule, together with nitrogen and nitric oxide, is propose by Gilmore (1965). On the other hand, in the quantitative estimation of the electron affinity of O_2, which corresponds to the energy of the groun state $^2\Pi_g$ of O_2^- relative to the ground state $^3\Sigma_g^-$ of O_2, there are differences. For example, Mulliken (1959) has suggested a value of 0.2 eV by his theoretical calculation, Curran (1961) has found a lower limit of 0.58 eV based on the adiabatic potential of O_2^- in electron-bombered O_3,

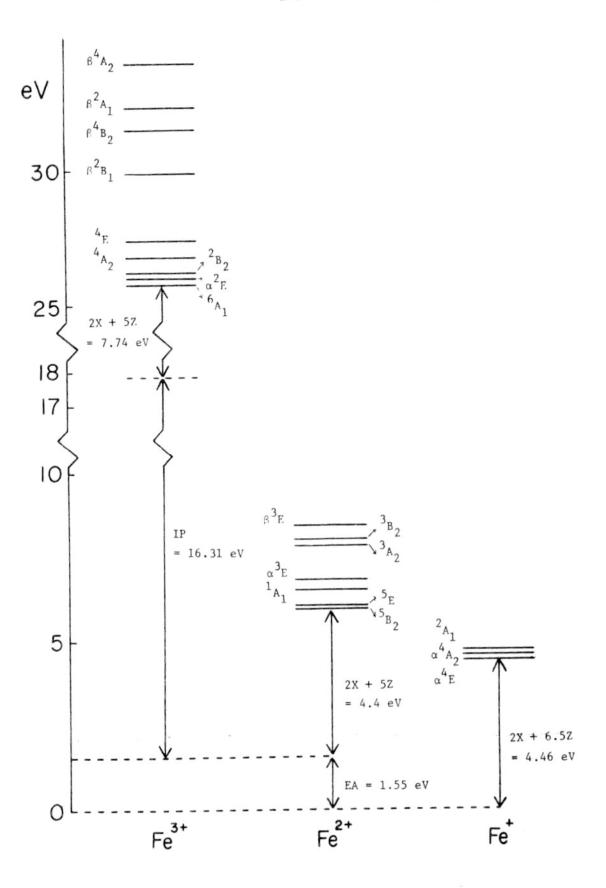

Fig. 5. The energy levels of low-lying terms of the iron atom in the *heme*. The notation of IP and EA_2 are the ionization energy and the electron affinity of the Fe^{2+} in the heme, respectively.

Phelps and Pack (1961, 1966) have obtained a value of 0.44 eV from a swarm experiment, and recently Stockdale et al. (1969) have determined the lower limit of the electron affinity to be 1.1 eV. However, the difference of ~0.5 eV may have no serious effect on the result of the present calculation, since the ionization energy of the ferrous iron atom in the *heme* is much larger than these values of the electron affinity of O_2. Therefore, we take the value of 0.44 eV for the electron affinity of O_2. The electronic states used in the present calculation are shown in Figure 6 with their state energies, where the state energy of $^3\Sigma_g^-(O_2)$ is taken to be zero.

3.3. *Basic Functions*

The basic function of the ionic structure Fe^{2+}-O_2, which we shall call normal configurations, are constructed from seven low-lying terms

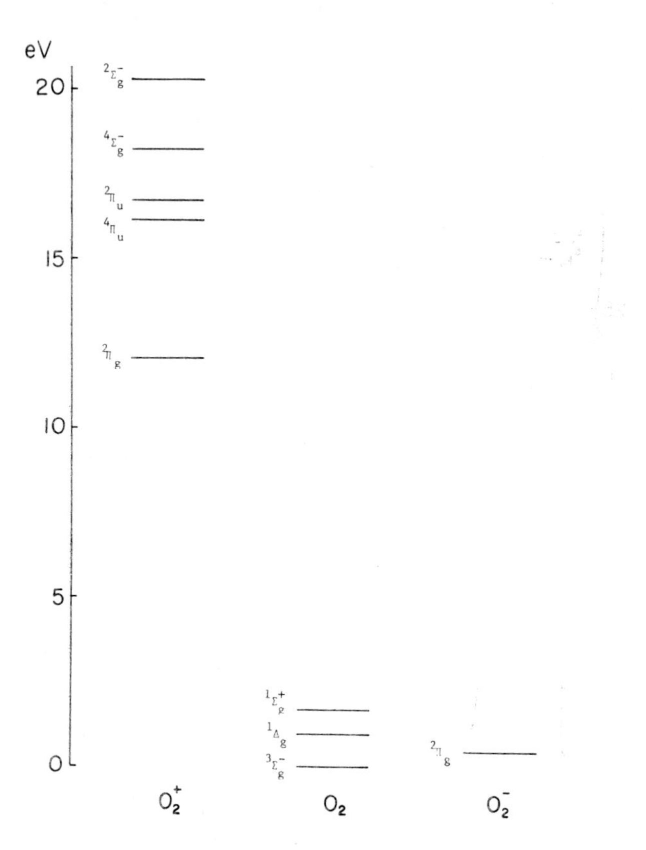

Fig. 6. The energy levels of low-lying terms of oxygen molecules. These values of the term energies are taken from experimental measurements (Wallace (1962), Phelps and Pack (1961 1962), Yoshino and Tanaka (1968)).

$^5B_2$, ^5E, $^1A_1$, $\alpha\,^3E$, $\beta\,^3E$, $^3A_2$ and $^3B_2$ of the *heme* and from three low-lying terms $^3\Sigma_g^-$, $^1\Delta_g$ and $^1\Sigma_g^+$ of O_2 molecule. Possible normal configurations which arise from the combinations of the above low-lying terms are listed in Table I. Another type of basic functions, which we shall call charge transfer configurations, are also taken into account in the present calculation. In principle, the charge transfer configurations considered here are those obtained from the normal configurations in Table I by one electron transfer from the iron atom to O_2 molecule or from O_2 molecule to the iron atom, with the limation that only the terms listed in Figure 5 and Figure 6 are used. However, with respect to the terms of Fe^{3+} and Fe^+, the following terms, which are excluded by the principle mentioned above but have the same coefficient of X in their expectation values as those in Figure 5, are also taken into consideration, since they are expected to interact considerably with those in Figure 5. Even with this limation, the number of the charge transfer configurations considered here becomes too many to be listed, and only possible electro

TABLE I
Normal configurations constructed from the low-lying terms of the *heme* (Fe^{2+}) and of O_2 molecule. The symmetries of these configurations in parallel and inclined arrangements are also shown.

	Parallel arrangement	Inclined arrangement
Singlet		
Ψ_1; $\psi(^1A_1)\phi(^1\Delta_g^a)$	$^1A_1$	$^1A'$
Ψ_2; $\psi(^1A_1)\phi(^1\Sigma_g^+)$		
Ψ_3; $\psi(\alpha\,^3E_y)\phi(^3\Sigma_g^-)$		
Ψ_4; $\psi(\beta\,^3E_y)\phi(^3\Sigma_g^-)$		
Ψ_5; $\psi(^3A_2)\phi(^3\Sigma_g^-)$	$^1B_1$	
Ψ_6; $\psi(^3B_2)\phi(^3\Sigma_g^-)$		
Ψ_7; $\psi(\alpha\,^3E_x)\phi(^3\Sigma_g^-)$	$^1A_2$	$^1A''$
Ψ_8; $\psi(\beta\,^3E_x)\phi(^3\Sigma_g^-)$		
Ψ_9; $\psi(^1A_1)\phi(^1\Delta_g^b)$	$^1B_2$	
Triplet		
Ψ_1; $\psi(^5E_y)\phi(^3\Sigma_g^-)$	$^3A_1$	$^3A'$
Ψ_2; $\psi(\alpha\,^3E_y)\phi(^3\Sigma_g^-)$		
Ψ_3; $\psi(\beta\,^3E_y)\phi(^3\Sigma_g)$		
Ψ_4; $\psi(\alpha\,^3E_y)\phi(^1\Delta_g^b)$		
Ψ_5; $\psi(\beta\,^3E_y)\phi(^1\Delta_g^b)$		
Ψ_6; $\psi(^5B_2)\phi(^3\Sigma_g^-)$	$^3B_1$	
Ψ_7; $\psi(^3A_2)\phi(^3\Sigma_g^-)$		
Ψ_8; $\psi(^3B_2)\phi(^3\Sigma_g^-)$		
Ψ_9; $\psi(^3A_2)\phi(^1\Delta_g^b)$		
Ψ_{10}; $\psi(^3B_2)\phi(^1\Delta_g^b)$		
Ψ_{11}; $\psi(\alpha\,^3E_x)\phi(^1\Delta_g^a)$		
Ψ_{12}; $\psi(\beta\,^3E_x)\phi(^1\Delta_g^a)$		
Ψ_{13}; $\psi(\alpha\,^3E_x)\phi(^1\Sigma_g^+)$		
Ψ_{14}; $\psi(\beta\,^3E_x)\phi(^1\Sigma_g^+)$		

TABLE I (Continued)

	Parallel arrangement	Inclined arrangement
Ψ_{15}; $\psi(^5E_x)\phi(^3\Sigma_g^-)$	$^3A_2$	$^3A''$
Ψ_{16}; $\psi(\alpha\,^3E_x)\phi(^3\Sigma_g^-)$		
Ψ_{17}; $\psi(\beta\,^3E_x)\phi(^3\Sigma_g^-)$		
Ψ_{18}; $\psi(\alpha\,^3E_x)\phi(^1\Delta_g^b)$		
Ψ_{19}; $\psi(\beta\,^3E_x)\phi(^1\Delta_g^b)$		
Ψ_{20}; $\psi(^3A_2)\phi(^1\Delta_g^a)$		
Ψ_{21}; $\psi(^3A_2)\phi(^1\Sigma_g^+)$		
Ψ_{22}; $\psi(^3B_2)\phi(^1\Delta_g^a)$		
Ψ_{23}; $\psi(^3B_2)\phi(^1\Sigma_g^+)$		
Ψ_{24}; $\psi(^1A_1)\phi(^3\Sigma_g^-)$	$^3B_2$	
Ψ_{25}; $\psi(\alpha\,^3E_y)\phi(^1\Delta_g^a)$		
Ψ_{26}; $\psi(\beta\,^3E_y)\phi(^1\Delta_g^a)$		
Ψ_{27}; $\psi(\alpha\,^3E_y)\phi(^1\Sigma_g^+)$		
Ψ_{28}; $\psi(\beta\,^3E_y)\phi(^1\Sigma_g^+)$		
Quintet		
Ψ_1; $\psi(^5E_y)\phi(^3\Sigma_g^-)$	$^5A_1$	$^5A'$
Ψ_2; $\psi(^5E_y)\phi(^1\Delta_g^b)$		
Ψ_3; $\psi(\alpha\,^3E_y)\phi(^3\Sigma_g^-)$		
Ψ_4; $\psi(\beta\,^3E_y)\phi(^3\Sigma_g^-)$		
Ψ_5; $\psi(^5B_2)\phi(^3\Sigma_g^-)$	$^5B_1$	
Ψ_6; $\psi(^5B_2)\phi(^1\Delta_g^b)$		
Ψ_7; $\psi(^5E_x)\phi(^1\Delta_g^a)$		
Ψ_8; $\psi(^5E_x)\phi(^1\Sigma_g^+)$		
Ψ_9; $\psi(^3A_2)\phi(^3\Sigma_g^-)$		
Ψ_{10}; $\psi(^3B_2)\phi(^3\Sigma_g^-)$		

TABLE I (Continued)

	Parallel arrangement	Inclined arrangement
Ψ_{11}; $\psi(^5E_x)\phi(^3\Sigma_g^-)$		
Ψ_{12}; $\psi(^5E_x)\phi(^1\Delta_g^b)$		
Ψ_{13}; $\psi(^5B_2)\phi(^1\Delta_g^a)$		
Ψ_{14}; $\psi(^5B_2)\phi(^1\Sigma_g^+)$	$^5A_2$	
Ψ_{15}; $\psi(\alpha^3E_x)\phi(^3\Sigma_g^-)$		
Ψ_{16}; $\psi(\beta^3E_x)\phi(^3\Sigma_g^-)$		$^5A''$
Ψ_{17}; $\psi(^5E_y)\phi(^1\Delta_g^a)$	$^5B_2$	
Ψ_{18}; $\psi(^5E_y)\phi(^1\Sigma_g^+)$		
septet		
Ψ_1; $\psi(^5E_y)\phi(^3\Sigma_g^-)$	$^7A_1$	$^7A'$
Ψ_2; $\psi(^5B_2)\phi(^3\Sigma_g^-)$	$^7B_1$	
Ψ_3; $\psi(^5E_x)\phi(^3\Sigma_g^-)$	$^7A_2$	$^7A''$

transfers are shown in this section. In the parallel, the following electron transfers are possible;

from the iron to oxygen

$\eta \rightarrow \pi_g^z$

$\zeta \rightarrow \pi_g^y$

from oxygen to the iron

$\pi_g^z \rightarrow \eta$

$\pi_g^y \rightarrow \zeta$

$3\sigma_g$, $\pi_u^z \rightarrow u, v$

$\pi_u^y \rightarrow \xi$

In the inclined arrangement, the following electron transfers are also possible in addition to those in the parallel arrangement;

from the iron to oxygen

$u, v \rightarrow \pi_g^{z'}$

$\xi \rightarrow \pi_g^{y'}$

from oxygen to the iron

$3\sigma_g$, $\pi_u^{z'} \rightarrow \eta$

$\pi_g^{z'} \rightarrow u, v$

$\pi_g^{y'} \rightarrow \xi$

$\pi_u^{y'} \rightarrow \zeta$

4. RESULT OF THE CALCULATION

4.1. *Parallel Arrangement*

The calculations in the parallel arrangement are carried out at three distances between the iron atom and the center of the O_2 molecule, 3.0 Å, 2.0 Å and 1.5 Å. The result of these calculations shows that the configuration interaction between the normal and the charge transfer configurations is small even at the short distance of 1.5 Å. That is, the obtained eigenstates are classified decidedly into two groups, one being composed mainly of the normal configurations and another mainly of the charge transfer configurations. The separation of the energy eigenvalues between these two groups is about 7~8 eV. The lower energy eigenvalues, whose eigenfunctions are composed mainly of normal configurations, are listed in Table II, where the sum of the unperturbed energy of the ground term $^5B_2$ of the *heme* and that of the ground term $^3\Sigma_g^-$ of O_2 is taken to be zero. The main feature of the result in this Table II are as follows.

(1) Among the singlet states, the energy eigenvalue of one of the $^1A_2$ states decreases remarkably in the region from 2.0 Å to 1.5 Å.

(2) At the distance of 1.5 Å, the energy eigenvalue of this $^1A_2$ state becomes lower than those of $^7B_1$, $^5B_1$ and $^3B_1$ states whose main components are constructed from the ground term of the *heme* ($^5B_2$) and from the ground term of O_2 ($^3\Sigma_g^-$).

(3) The energy eigenvalue of the lowest $^3A_1$ state is slightly smaller than that of the lowest eigenvalue of the $^1A_2$ state.

The eigenfunctions which correspond to the lowest eigenvalue in each $S\Gamma$ state at 1.5 Å in Table II are shown in Table III with the use of the value of the coefficient C_i defined in the Equation (1). In this table, the charge transfer configurations which contribute to the eigenfunction in Table II with coefficients of more than about 0.1 are listed.

In the lowest $^1A_2$ state, mixing of charge transfer configurations contributes considerably to decrease the energy eigenvalue of this state. These charge transfer configurations except $\psi(\alpha\,^4E_x)\phi(^4\Pi_u^y)$ are obtained by one electron transfer $\eta \rightarrow \pi_g^z$ or $\pi_g^z \rightarrow \eta$ from the normal configuration $\psi(\alpha\,^3E_x)\phi(^3\Sigma_g^-)$ or $\psi(\beta\,^3E_x)\phi(^3\Sigma_g^-)$. If the charge transfer configurations are not taken into account, the lowest eigenvalue of $^1A_2$ is calculated to be only -0.99 eV. The main configuration of this lowest $^1A_2$ state is $\psi(\alpha\,^3E_x)\phi(^3\Sigma_g^-)$, and the expectation value of this configuration is about 0.8 eV when the O_2 molecule is separated infinitely far from the *heme*. The decrease in the expectation value of this configuration is remarkable among all normal configurations considered in the present calculation. This is originated from negative exchange interaction. The term $\alpha\,^3E$ is an excited state of the *heme* in which two electrons with parallel spins are accomodated into η and u orbitals. The term $^3\Sigma_g^-$ is the ground state of a free O_2 molecule in which two electrons with parallel spins are accomodated into π_g^z and π_g^y orbitals. Since the overlap between the η orbital of the iron atom and the π_g^z orbital of the O_2 molecule is large in the present arrangement, the exchange interaction between the two electrons in η and π_g^z orbitals becomes a large negative value in the singlet configuration $\psi(\alpha\,^3E_x)\phi(^3\Sigma_g^-)$. Such a negative exchange interaction

TABLE II
The energy eigenvalues calculated in the parallel arrangement. The energy values are given in eV. The value of Y, one of the splitting parameters of d-level, is chosen to be 2000 cm^{-1}.

Singlet states

distance		1.5 Å	2.0 Å
$^1A_1$	ε_1	-2.048	0.597
	ε_2	-1.719	0.774
	ε_3	-0.842	1.683
	ε_4	-0.595	1.867
$^1B_1$	ε_1	-1.795	0.887
	ε_2	-1.335	1.417
$^1A_2$	ε_1	-3.460	0.302
	ε_2	-2.237	1.440
$^1B_2$	ε_1	-1.659	1.078

Triplet states

distance		1.5 Å	2.0 Å	distance		1.5 Å	2.0 Å
$^3A_1$	ε_1	-3.580	-0.343	$^3A_2$	ε_1	-3.300	-0.154
	ε_2	-2.168	0.562		ε_2	-2.417	0.380
	ε_3	-1.139	1.587		ε_3	-2.003	1.524
	ε_4	-1.050	1.666		ε_4	-0.634	1.588
	ε_5	0.459	2.794		ε_5	-0.320	1.740
$^3B_1$	ε_1	-3.290	-0.338		ε_6	0.483	2.226
	ε_2	-2.241	0.877		ε_7	0.631	2.716
	ε_3	-2.028	1.167		ε_8	0.868	2.814
	ε_4	-0.905	1.402		ε_9	1.959	3.151
	ε_5	-0.426	1.993	$^3B_2$	ε_1	-3.129	-0.037
	ε_6	-0.290	2.318		ε_2	-1.254	1.366
	ε_7	-0.013	2.404		ε_3	-0.213	2.339
	ε_8	0.793	2.484		ε_4	0.330	2.532
	ε_9	2.534	3.722		ε_5	1.015	3.504

Quintet states

distance		1.5 Å	2.0 Å	distance		1.5 Å	2.0 Å
$^5A_1$	ε_1	-3.163	-0.268	$^5A_2$	ε_1	-3.121	-0.193
	ε_2	-2.665	0.505		ε_2	-1.886	0.424
	ε_3	-1.390	0.836		ε_3	-1.503	0.830
	ε_4	-0.314	1.639		ε_4	-0.620	0.830

TABLE II (continued)

Quinted states (continued)

distance	1.5 Å	2.0 Å	distance	1.5 Å	2.0 Å
$^5B_1$ ε_1	-3.189	-0.275	ε_5	0.431	1.763
ε_2	-2.013	0.621	ε_6	0.538	1.782
ε_3	-1.777	0.840	$^5B_2$ ε_1	-2.124	0.447
ε_4	-1.693	1.362	ε_2	0.287	1.755
ε_5	-0.616	3.941			
ε_6	-0.030	3.941			

Septet states

distance	1.5 Å	2.0 Å
$^7A_1$ ε_1	-1.154	-0.118
$^7B_1$ ε_1	-3.260	-0.242
$^7A_2$ ε_1	-1.319	-0.099

TABLE III
The eigenfunctions for the lowest energy eigenvalue in each SΓ state. They are shown by the coefficients of the basic functions at the distance of 1.5 Å. In this table, tha basic functions with coefficients of smaller than about 0.1 are not listed, and the sum of the squares of these coefficients listed here is not necessarily to one

$^1A_1$ Ψ_i	Φ_1	$^1B_1$ Ψ_i	Φ_1
Fe^{2+}-O_2		Fe^{2+}-O_2	
$\psi(^1A_1)\phi(^1\Delta_g^a)$	0.6208	$\psi(^3A_2)\phi(^3\Sigma_g^-)$	0.8252
$\psi(^1A_1)\phi(^1\Sigma_g^+)$	-0.1745	$\psi(^3B_2)\phi(^3\Sigma_g)$	0.1583
$\psi(\alpha{}^3E_y)\phi(^3\Sigma_g^-)$	-0.4222		
$\psi(\beta{}^3E_y)\phi(^3\Sigma_g)$	-0.1321		
Fe^{3+}-O_2^-		Fe^{3+}-O_2^-	
$\psi(\alpha{}^2E_x)\phi(^2\Pi_g^z)$	0.3928	$\psi(\beta{}^2E_y)\phi(^2\Pi_g^y)$	0.0912
$\psi(\alpha{}^2B_2)\ (^2\Pi_g^y)$	-0.0876	$\psi(\gamma{}^2E_y)\phi(^2\Pi_g^y)$	-0.2740
$\psi(\beta{}^2B_2)\phi(^2\Pi_g^y)$	0.1445	Fe^+-O_2^+	
$\psi(\alpha{}^2A_2)\phi(^2\Pi_g^y)$	0.0321	$\psi(\alpha{}^4A_2)\phi(^4\Pi_u^y)$	-0.1379
		$\psi(\alpha{}^4A_2)\phi(^4\Sigma_g^-)$	0.1386

TABLE III (continued)

$^1A_1$		$^1B_1$	
Ψ_i	Φ_1	Ψ_i	Φ_1
$Fe^{+}-O_2^{+}$			
$\psi(^2A_1)\phi(^2\Pi_u^z)$	-0.1527		
$\psi(\alpha\,^4E_y)\phi(^4\Pi_u^y)$	0.0079		

$^1A_2$		$^1B_2$	
Ψ_i	Φ_1	Ψ_i	Φ_1
$Fe^{2+}-O_2$		$Fe^{2+}-O_2$	
$\psi(\alpha\,^3E_x)\phi(^3\Sigma_g^-)$	0.7577	$\psi(^1A_1)\phi(^1\Delta_g^b)$	0.7940
$\psi(\beta\,^3E_x)\phi(^3\Sigma_g^-)$	0.3049	$Fe^{3+}-O_2^-$	
$Fe^{3+}-O_2^-$		$\psi(\alpha\,^2E_x)\phi(^2\Pi_g^y)$	-0.4830
$\psi(\beta\,^2B_1)\phi(^2\Pi_g^y)$	0.1850	$\psi(\alpha\,^2B_2)\phi(^2\Pi_g^z)$	0.1085
$\psi(\beta\,^2A_1)\phi(^2\Pi_g^y)$	0.1526		
$\psi(\gamma\,^2A_1)\phi(^2\Pi_g^y)$	-0.0721		
$Fe^{+}-O_2^{+}$			
$\psi(^2A_1)\phi(^2\Pi_g^y)$	0.2431		
$\psi(^2B_1)\phi(^2\Pi_g^y)$	-0.0862		
$\psi(\alpha\,^4E_x)\phi(^4\Pi_u^y)$	-0.0847		

$^3A_2$		$^3B_2$	
Ψ_i	Φ_1	Ψ_i	Φ_1
$Fe^{2+}-O_2$		$Fe^{2+}-O_2$	
$\psi(\alpha\,^3E_x)\phi(^3\Sigma_g^-)$	0.6807	$\psi(^1A_1)\phi(^3\Sigma_g^-)$	0.7955
$\psi(\beta\,^3E_x)\phi(^3\Sigma g)$	0.2428		
$\psi(\alpha\,^3E_x)\phi(^1\Delta_g^b)$	-0.3243	$Fe^{3+}-O_2^-$	
		$\psi(\alpha\,^2E_x)\phi(^2\Pi_g^y)$	0.3926

TABLE III (continued)

$^3A_2$		$^3B_2$	
Ψ_i	Φ_1	Ψ_i	Φ_1
$Fe^{3+}-O_2^-$		$Fe^{+}-O_2^+$	
$\psi(\alpha^4E_y)\phi(^2\Pi_g^z)$	-0.1037	$\psi(^2A_1)\phi(^4\Pi_u^y)$	0.1365
$\psi(\beta^2A_1)\phi(^2\Pi_g^y)$	0.1528	$\psi(^2A_1)\phi(^4\Sigma_g^-)$	-0.1171
$\psi(\beta^2B_1)\phi(^2\Pi_g^y)$	0.1859		
$Fe^{+}-O_2^+$			
$\psi(^2A_1)\phi(^2\Pi_g^y)$	0.2521		

$^3A_1$		$^3B_1$	
Ψ_i	Φ_1	Ψ_i	Φ_1
$Fe^{2+}-O_2$		$Fe^{2+}-O_2$	
$\psi(^5E_y)\phi(^3\Sigma_g^-)$	0.8449	$\psi(^5B_2)\phi(^3\Sigma_g^-)$	0.8486
$Fe^{3+}-O_2^-$		$Fe^{3+}-O_2^-$	
$\psi(\beta^4A_2)\phi(^2\Pi_g^y)$	-0.1306	$\psi(\gamma^4E_y)\phi(^2\Pi_g^y)$	0.2173
$\psi(\beta^4B_2)\phi(^2\Pi_g^y)$	-0.1583		
$Fe^{+}-O_2^+$		$Fe^{+}-O_2^+$	
$\psi(\alpha^4A_2)\phi(^2\Pi_g^y)$	-0.2544	$\psi(\alpha^4E_y)\phi(^4\Pi_u^y)$	0.2419

$^5A_1$		$^5B_1$	
Ψ_i	Φ_1	Ψ_i	Φ_1
$Fe^{2+}-O_2$		$Fe^{2+}-O_2$	
$\psi(^5E_y)\phi(^3\Sigma_g^-)$	0.7736	$\psi(^5B_2)\phi(^3\Sigma_g^-)$	0.7228
$\psi(^5E_y)\phi(^1\Delta_g^b)$	-0.3590	$\psi(^5B_2)\phi(^1\Delta_g^b)$	-0.3538
		$\psi(^5E_x)\phi(^1\Delta_g^a)$	-0.1217
$Fe^{3+}-O_2^-$		$\psi(^5E_x)\phi(^1\Sigma_g^+)$	0.1123
$\psi(\beta^4A_2)\phi(^2\Pi_g^y)$	-0.1314		
$\psi(\beta^4B_2)\phi(^2\Pi_g^y)$	-0.1587	$Fe^{3+}-O_2^-$	

TABLE III (continued)

$^5A_1$	
Ψ_i	Φ_1
$Fe^+-O_2^+$	
$\psi(\alpha^4A_2)\phi(^2\Pi_g^y)$	-0.2546

$^5B_1$	
Ψ_i	Φ_1
$\psi(^6A_1)\phi(^2\Pi_g^z)$	-0.1721
$\psi(\gamma^4E_y)\phi(^2\Pi_g^y)$	0.2044
$Fe^+-O_2^+$	
$\psi(\alpha^4E_y)\phi(^2\Pi_g^y)$	0.2427

$^5A_2$	
Ψ_i	Φ_1
$Fe^{2+}-O_2$	
$\psi(^5E_x)\phi(^3\Sigma_g^-)$	0.6860
$\psi(^5E_x)\phi(^1\Delta_g^b)$	0.2656
$\psi(^5B_2)\phi(^1\Delta_g^a)$	-0.2230
$\psi(^5B_2)\phi(^1\Sigma_g^+)$	-0.2077
$Fe^{3+}-O_2^-$	
$\psi(^6A_1)\phi(^2\Pi_g^y)$	0.3874
$Fe^+-O_2^+$	
$\psi(\alpha^4E_y)\phi(^2\Pi_g^z)$	0.1638

$^5B_2$	
Ψ_i	Φ_1
$Fe^{2+}-O_2$	
$\psi(^5E_y)\phi(^1\Delta_g^a)$	0.6756
$\psi(^5E_y)\phi(^1\Sigma_g^+)$	0.5780
$Fe^+-O_2^+$	
$\psi(\alpha^4A_2)\phi(^2\Pi_g^z)$	0.3247

$^7A_1$	
Ψ_i	Φ_1
$Fe^{2+}-O_2$	
$\psi(^5E_y)\phi(^3\Sigma_g^-)$	0.9672
$Fe^+-O_2^+$	
$\psi(\beta^4E_y)\phi(^4\Pi_u^y)$	-0.1183
$\psi(\beta^4E_y)\phi(^4\Sigma_g^-)$	0.1072

$^7B_1$	
Ψ_i	Φ_1
$Fe^{2+}-O_2$	
$\psi(^5B_2)\phi(^3\Sigma_g^-)$	0.9316
$Fe^{3+}-O_2^-$	
$\psi(^6A_1)\phi(^2\Pi_g^z)$	-0.1337

TABLE III (continued)

	$^7B_1$	
	Ψ_i	Φ_1
	$Fe^+-O_2^+$	
	$\psi(^4B_2)\phi(^4\Pi_u^y)$	-0.1201
	$\psi(^4B_2)\phi(^4\Sigma_g^-)$	0.1044
	$\psi(\alpha^4E_y)\phi(^4\Pi_u^z)$	-0.1129

$^7A_2$	
Ψ_i	Φ_1
$Fe^{2+}-O_2$	
$\psi(^5E_x)\phi(^3\Sigma_g^-)$	0.8064
$Fe^{3+}-O_2^-$	
$\psi(^6A_1)\phi(^2\Pi_g^y)$	0.4171
$Fe^+-O_2^+$	
$\psi(\beta^4E_x)\phi(^4\Pi_u^y)$	0.1009

is characteristic only in the $^1A_2$ state and does not appear in any other singlet configuration. In Figure 7, these two orbitals, η and π_σ^z, are shown. This exchange effect in $^1A_2$ state contributes so much to the tota[l] energy that it cancels the excitation energy of the ^3E term in the free *heme*. Besides the negative exchange interaction, the diagonal elements of the Hamiltonian matrix of the normal configurations decrease as an O_2 molecule approaches the *heme*. This tendency is recognized in all norm[al] configurations. This is due to the charge distribution in our ionic mod-el. In our treatment, a positive charge (2 e) is located on the central iron atom and a negative charge (-2 e) is partitioned into four nitroge[n] atoms, i.e., -0.5 e is located on each nitrogen atom of the porphyrin ring. In an O_2 molecule, a negative charge appears at the midpoint of the two oxygen atoms owing to the large number of the electrons occupyi[ng] the bonding orbitals $2\sigma_g$, $3\sigma_g$ and $\pi_u^{y,z}$, and positive charges appear at the ends of the O_2 molecule. This charge distribution in the O_2 molecul[e] and that in the heme cause the large attractive coulomb interaction in the *heme*-O_2 system. This attractive coulomb interaction exceeds the ex-change repulsion even at 1.5 Å, and the energy values of diagonal eleme[nts]

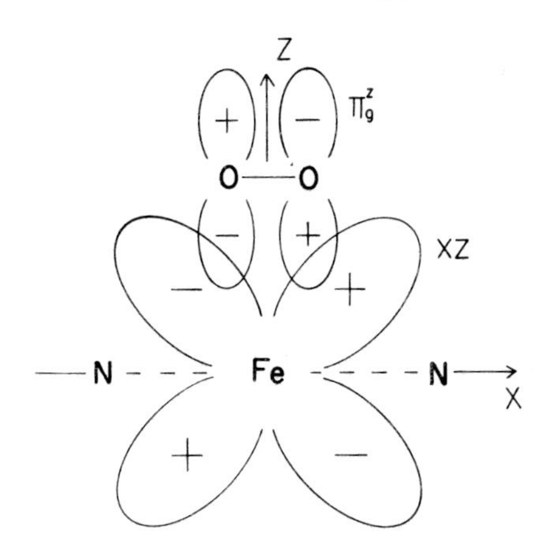

Fig. 7. The overlap between the xz orbital of the iron atom and the π_g^z orbital of the O_2 molecule. The value of this overlap is the largest in the parallel arrangement, and it causes the $^1A_2$ state to be the lowest in energy due to mixing of the charge transfer configurations between these orbitals and due to the attractive exchange interaction between the unpaired electron in the xz orbital and that in the π_g^z orbital.

decrease commonly by about 1 eV at this distance. When the diagonal elements are calculated on the smeared model, i.e., when negative charges on the *porphyrin* are gathered up to the central iron atom, then the diagonal elements of all normal configurations are of slightly positive values at 1.5 Å except for the normal configuration $\psi(\alpha\,^3E_x)\phi(^3\Sigma_g^-)$ of the $^1A_2$ state.

Next, our attention is given to $^1A_1$ state. This state is the lowest among the singlet states except for the $^1A_2$ state. At the distance of 2.0 Å, the main components of the lowest eigenstate of the $^1A_1$ state are two normal configurations $\psi(\alpha\,^3E_y)\phi(^3\Sigma_g^-)$ and $\psi(\beta\,^3E_y)\phi(^3\Sigma_g^-)$. At the distance of 1.5 Å, however, mixing of the other normal configurations $\psi(^1A_1)\phi(^1\Delta_g^a)$ and $\psi(^1A_1)\phi(^1\Sigma_g^+)$ increases remarkably. These two configurations can be rewritten by the two closed shell configurations, Ψ_z and Ψ_y, in the following way,

$$\psi(^1A_1)\phi(^1\Delta_g^a) = \frac{1}{\sqrt{2}}(\Psi_z - \Psi_y), \tag{26-1}$$

$$\psi(^1A_1)\phi(^1\Sigma_g^+) = \frac{1}{\sqrt{2}}(\Psi_z + \Psi_y), \tag{26-2}$$

where the explicit forms of Ψ_z and Ψ_y are given by

$$\Psi_z = A\{|\xi\bar{\xi}\eta\bar{\eta}\zeta\bar{\zeta}|\cdot|\pi_g^z\bar{\pi}_g^z\cdots\cdots|\}, \tag{27-1}$$

$$\Psi_y = A\{|\xi\bar{\xi}\eta\bar{\eta}\zeta\bar{\zeta}|\cdot|\pi_g^y\bar{\pi}_g^y\cdots\cdots|\}. \tag{27-2}$$

In the configuration Ψ_y, the charge transfer $\eta \to \pi_g^z$ and $\bar{\eta} \to \bar{\pi}_g^z$ are possible and the stabilization energy produced by these charge transfers is the largest among the four normal configurations of $^1A_1$ state. The result of the MO treatment by Zerner et al. (1966) shows that the lowest eigenstate is composed mainly of this configuration Ψ_y and of the charge transfer configurations arising from it, especially of $\psi(\alpha^2E_x)\phi(^2\Pi_g^z)$ of $Fe^{3+}-O_2^-$. In our ionic model, however, the coulomb interaction lowers the energy of the configuration Ψ_z more effectively, since the two electrons in π_g^z orbital are subjected to strong attractive field from the positive iron atom. Hence, it may be said that this mixing of Ψ_z suppresses the contribution of Ψ_y to the ground $^1A_1$ state. In practice, if we adopt the smeared *heme* model, the eigenfunction of the lowest $^1A_1$ state is calculated to be 0.58 $\{\psi(^1A_1)\phi(^1\Delta_g^a)\}$ - 0.41$\{\psi(^1A_1)\phi(^1\Sigma_g^+)\}$ - 0.08$\{\psi(\alpha^3E_y)\phi(^3\Sigma_g^-)\}$ - 0.03$\{\psi(\beta^3E_y)\phi(^3\Sigma_g^-)\}$ + 0.51$\{\psi(\alpha^2E_x)\phi(^2\Pi_g^z)\}$. This eigenfunction is nearly equal to that obtained by the MO treatment by Zerner et al. (1966). It is noteworthy, however, that the $^1A_2$ state is calculated to be lower than $^1A_1$ state even in this smeared *heme* model.

As seen Table II, the lowest state of $^3A_1$ is lower in energy than the lowest state of $^1A_2$ even at the distance of 1.5 Å. The main componen of the lowest $^3A_1$ state is a normal configuration $\psi(^5E_y)\phi(^3\Sigma_g^-)$, in which the negative exchange interaction between the electron in η orbital and that in π_g^z orbital appears in the same way as in the normal configuration $\psi(\alpha^3E_y)\phi(^3\Sigma^-)$ of the lowest $^1A_2$ state. The magnitude of this negative exchange interaction in $\psi(^5E_y)\phi(^3\Sigma_g^-)$ of $^3A_2$, however, is only one half o that in $\psi(\alpha^3E_y)\phi(^3\Sigma_g^-)$ of $^1A_2$. Moreover, mixing of the charge transfer configurations is more effective in the $^1A_2$ state than in the $^3A_1$ state, and the stabilization energy due to the charge transfers is larger in the $^1A_2$ state. Consequently, the $^1A_2$ state decreases in energy more remarkably than the $^3A_1$ state, as the O_2 molecule approaches the *heme*. As mentioned in section 3, we have no experimental information about the value of the parameter Y in ferrous Mb or Hb, although the value of this parameter is assumed tentatively to be 2000 cm^{-1} in the present calculation. For the larger value of Y, the term ^3E of the heme becomes lower in energy, while the term energy of ^5E remains almost the same, and the $^1A_2$ state is expected to be the lowest at the short distance of 1.5 Å. In fact, it is confirmed by additional calculations that $^1A_2$ state is lower in energy than $^3A_1$ state at 1.5 Å for Y > 4000 cm^{-1}.

Among the septet states, the $^7A_2$ state is especially low in energy. This is also due to the electron transfer from η to π_g^z orbital, that is, due to the mixing of $\psi(^6A_1)\phi(^2\Pi_g^y)$ of $Fe^{3+}-O_2^-$. If the value of X estimate in ferrous *heme* is used also in ferric *heme*, the energy difference betw the $^6A_1$ term of ferric *heme* and the ^5E term of ferrous *heme* is evaluate to be only about 16 eV and the $^7A_2$ state is lower in energy than the ^1A state. In the present calculation, the term energies of ferrous and ferric *hemes* are estimated with the different set of the splitting parameters determined experimentally for ferrous and ferric Mb (or Hb), respectively. In this case, the value of X in ferric *heme* is much larger than that in ferrous *heme* and the energy difference between the $^6A_1$ ter and the ^5E term is raised to about 20 eV. This difference of about 4 eV is critical to the production of the singlet ground state, $^1A_2$.

Other states are decidedly higher in energy and they are not discussed.

4.2. *Inclined Arrangement*

As seen in the parallel arrangement, the interaction between the O_2 molecule and the *heme* is too weak to obtain a reasonable result at the distance of 2.0 Å. In comparison with the parallel arrangement, the values of overlap integrals are more reduced in this inclined arrangement, since the distant oxygen atom O_B makes less of a contribution to them. With this fact in mind, we calculate the energy eigenvalues and eigenfunctions of the *heme*-O_2 system at three distances between the nearest oxygen atom O_A and the iron atom, 1.8 Å, 1.5 Å and 1.3 Å. The lower energy eigenvalues and the corresponding eigenfunctions thus calculated are listed in Table IV and Table V, respectively. In these eigenfunctions, normal con-

TABLE IV
The lower energy eigenvalues calculated in the inclined arrangement. The energy values are given in eV. The value of Y, one of the splitting parameters of d-level, is chosen to be 4000 cm^{-1}

Distance		1.3 Å	1.5 Å	1.8 Å
$^1A'$	ε_1	-8.673	-3.658	-0.409
	ε_2	-3.614	-1.141	0.343
	ε_3	-5.286	-1.368	0.817
	ε_4	-2.228	-0.347	0.953
	ε_5	-2.685	-0.136	1.530
	ε_6	2.106	2.280	2.196
$^1A''$	ε_1	-6.075	-2.153	-0.004
	ε_2	-2.048	-0.297	0.862
	ε_3	-1.594	0.359	1.676

figurations $\psi(^1A_1)\phi(^1\Delta_g^a)$ and $\psi(^1A_1)\phi(^1\Sigma_g^+)$ appear approximately in the forms of $\psi(^1A_1)\frac{1}{\sqrt{2}}\{\phi(^1\Delta_g^a) + \phi(^1\Sigma_g^+)\}$ and $\psi(^1A_1)\frac{1}{\sqrt{2}}\{\phi(^1\Sigma_g^+) - \phi(^1\Delta_g^a)\}$. Thus, they are denoted by Ψ_z', and Ψ_y', respectively, in the present section.

The lowest energy eigenvalue of the $^1A'$ state is outstandingly lower than that of the $^1A''$ state at each of the three distances. This leads us to the conviction that the binding state in the inclined arrangement is the $^1A'$ state. In the wave function of the lowest eigenstate $^1A'$, the main configurations $\psi(\alpha\,^3E_y)\phi(^3\Sigma_g^-)$ and Ψ_z', are included with almost the same weight. The former configuration represents the triplet-triplet binding which is a main component of the ground binding state $^1A_2$ in the parallel arrangement. The explicit form of the latter configuration is indicated by $A\{|\xi\bar{\xi}\eta\bar{\eta}\zeta\bar{\zeta}||\pi_g^{z}\bar{\pi}_g^{z}\cdots\cdots|\}$ and this configuration contributes greatly to the lowest $^1A_1$ state in the parallel arrangement. In the inclined arrangement, this configuration is coupled with the triplet-triplet configuration $\psi(\alpha\,^3E_y)\phi(^3\Sigma_g^-)$ due to the lower symmetry of the to-

TABLE V
The eigenfunctions for the energy eigenvalues in Table IV at 1.5 Å. They are shown by the coefficients of the basic functions. In this table, the basic functions with coefficients of smaller than about 0.1 are not listed, and the sum of the squares of the coefficients in each eigenstate is not necessarily equal to one

$^1A''$

Ψ_i	Φ_1	Φ_2	Φ_3
$Fe^{2+}-O_2$			
$\psi(\alpha\,^3E_x)\phi(^3\Sigma_g^-)$	0.8078	-0.1619	-0.1581
$\psi(\beta\,^3E_x)\phi(^3\Sigma_g)$	0.1527	-0.0976	0.8782
$\psi(^1A_1)\phi(^1\Delta_g^b)$	0.1477	0.8964	0.0741
$Fe^{3+}-O_2^-$			
$\psi(\alpha\,^2E_y)\phi(^2\Pi_g^{z'})$		0.1736	
$\psi(\alpha\,^2E_x)\phi(^2\Pi_g^{y'})$	0.3122		
$Fe^{+}-O_2^+$			
$\psi(\alpha\,^4E_x)\phi(^4\Pi_g^{y'})$			0.1732
$\psi(^2A_1)\phi(^2\Pi_g^{y'})$		0.2441	
$\psi(\beta\,^2E_x)\phi(^2\Pi_g^{y'})$	0.1938		
$\psi(\gamma\,^2E_x)\phi(^2\Pi_g^{y'})$			0.1573
$\psi(\delta\,^2E_x)\phi(^2\Pi_g^{y'})$			0.1007

$^1A'$

Ψ_i	Φ_1	Φ_2	Φ_3	Φ_4	Φ_5	Φ_6
$Fe^{2+}-O_2$						
$\psi(^1A_1)\phi(^1\Delta_g^a)$	0.4142	-0.4494			0.1329	0.6748
$\psi(^1A_1)\phi(^1\Sigma_g^+)$	0.4142	-0.4494			0.1329	-0.6748
$\psi(\alpha\,^3E_y)\phi(^3\Sigma_g^-)$	0.5328	0.6052	0.1579		-0.1902	
$\psi(\beta\,^3E_y)\phi(^3\Sigma_g)$		0.2503		-0.1400	0.8187	
$\psi(^3A_2)\phi(^3\Sigma_g^-)$			-0.0715	0.8964		
$\psi(^3B_2)\phi(^3\Sigma_g)$			0.8227	0.0737		

TABLE V (continued)

$^1A'$						
Ψ_i	ϕ_1	Φ_2	Φ_3	Φ_4	Φ_5	Φ_6
$Fe^{3+}-O_2^-$						
$\psi(\alpha\,^2E_x)\phi(^2\Pi_g^{z'})$						0.1250
$\psi(\gamma\,^2E_x)\phi(^2\Pi_g^{z'})$				0.0948		
$\psi(\alpha\,^2B_2)\phi(^2\Pi_g^{y'})$			0.3244			
$\psi(\alpha\,^2E_y)\phi(^2\Pi_g^{y'})$	0.2990					
$Fe^{+}-O_2^+$						
$\psi(\alpha\,^4E_y)\phi(^4\Pi_g^{y'})$					0.1639	
$\psi(^2A_1)\phi(^2\Pi_g^{z'})$		0.1117				0.1805
$\psi(\alpha\,^2A_2)\phi(^2\Pi_g^{y'})$				0.1542		
$\psi(\alpha\,^2B_2)\phi(^2\Pi_g^{y'})$			0.1888			
$\psi(\alpha\,^4A_2)\phi(^4\Pi_g^{y'})$				0.1679		
$\psi(^2A_1)\phi(^2\Pi_g^{z'})$	0.2770	-0.0904				
$\psi(^2B_1)\phi(^2\Pi_g^{z'})$					0.1744	
$\psi(\beta\,^2E_y)\phi(^2\Pi_g^{y'})$	0.1215	0.1607				
$\psi(\gamma\,^2E_y)\phi(^2\Pi_g^{y'})$					0.1486	

tal system *heme*-O_2 and both the two configurations contribute to the ground binding state. For the purpose of further investigation into the nature of the interaction between the heme and the O_2 molecule, the values of the diagonal elements of the interaction part of the Hamiltonian for the two configurations are listed in Table VI. In this table, v represent the total interaction energy, v_0 the zeroth order part of the interaction with respect to overlap, i.e., the coulomb energy, and v_2, which is equal to $v - v_0$, is regarded as approximately the second order of the interaction energy with respect to overlap. In the singlet-singlet configuration $\Psi_{z'}$, negative coulomb energy is predominant, because two electrons in $\pi_g^{z'}$ orbital feel strong attractive coulomb field from the core of the iron atom. Moreover, in this configuration, repulsive exchange energy v_2 takes a small value, since no electron occupies a u orbital which extends to the O_2 molecule and shares large overlapping with the closed shell orbitals of the O_2 molecule. In the triplet-triplet configuration $\psi(\alpha\,^3E_y)\phi(^3\Sigma_g^-)$, on the other hand, two types of attractive exchange interaction arise; one is between the two electrons occupying u and $\pi_g^{z'}$ orbitals and the other between the two electrons occupying ξ and

TABLE VI
The expectation values (eV) of the interaction part in total Hamiltonian in the configurations $\Psi_1(^1A')$ and $\Psi_{z'}(^1A')$, where
$\Psi_1(^1A') = \psi(\alpha\,^3E_y^-)\phi(^3\Sigma_g^-)$ and $\Psi_{z'} = \frac{1}{2}\{\psi(^1A_1)\phi(^1\Delta_g) + \psi(^1A_1)\phi(^1\Sigma_g^-)\}$

Distance	1.3 Å	1.5 Å	1.8 Å
$\langle\Psi_1(^1A')\|v\|\Psi_1(^1A')\rangle$	-2.674	-1.132	-0.250
$\langle\Psi_1(^1A')\|v_0\|\Psi_1(^1A')\rangle$	-4.799	-2.332	-0.612
$\langle\Psi_1(^1A')\|v_2\|\Psi_1(^1A')\rangle$	2.125	1.200	0.362
$\langle\Psi_{z'}(^1A')\|v\|\Psi_{z'}(^1A')\rangle$	-4.579	-2.748	-1.189
$\langle\Psi_{z'}(^1A')\|v_0\|\Psi_{z'}(^1A')\rangle$	-5.120	-2.759	-1.105
$\langle\Psi_{z'}(^1A')\|v_2\|\Psi_{z'}(^1A')\rangle$	0.541	0.011	-0.084

$\pi_g^{y'}$ orbitals. In this configuration, however, the electron in u orbital interacts strongly with the electrons in the closed shells of the O_2 molecule $2\sigma_g$, $2\sigma_u$, $3\sigma_g$ and $\pi_u^{z'}$. This interaction leads to considerable larger exchange repulsion and the resultant second order of exchange energy is positive in the configuration $\psi(\alpha\,^3E_y)\phi(^3\Sigma_g^-)$ as seen in Table VI. Thus, the interaction part of the expectation value of the configuration $\Psi_{z'}$ becomes lower than that of $\psi(\alpha\,^3E_y)\phi(^3\Sigma_g^-)$ as a whole. This energy difference of the interaction part between these two configuration is compensated by the difference of the intra-molecular part. The intra-molecular part is a sum of the corresponding term energies of the *heme* and of the O_2 molecule, i.e., $\varepsilon(\alpha\,^3E) + \varepsilon(^3\Sigma_g^-)$ for $\psi(\alpha\,^3E_y)\phi(^3\Sigma_g^-)$ and $\varepsilon(^1A_1) + \{\varepsilon(^1\Delta_g) + \varepsilon(^1\Delta_g^+)\}/2$ for $\Psi_{z'}$. For example, when the value of Y for Fe^{2+} is taken to be 4000 cm^{-1} and the distance R to be 1.5 Å, the difference of the expectation values of the total Hamiltonian between these two configuration is only slight, namely

$$\langle\psi(\alpha\,^3E_y)\phi(^3\Sigma_g^-)|H|\psi(\alpha\,^3E_y)\phi(^3\Sigma_g^-)\rangle - \langle\Psi_{z'}|H|\Psi_{z'}\rangle = -0.63 \text{ eV}. \quad (28$$

This is the reason why these two configurations have almost the same weight in the ground eigenstate $^1A'$.

With respect to the charge transfer configurations, three configurations $\psi(\alpha\,^2E_y;\ Fe^{3+})\phi(^2\Pi_g^{y'};\ O_2^-)$, $\psi(^2A_1;\ Fe^+)\phi(^2\Pi_g^{z'};\ O_2^+)$ and $\psi(\beta\,^2E_y;\ Fe^+)\phi(^2\Pi_g^{y'};\ O_2^+)$ contribute considerably to lower the energy of the ground state $^1A'$. The first configuration is obtainable from the normal configuration $\psi(\alpha\,^3E_y)\phi(^3\Sigma_g^-)$ by the electron transfer $u \to \pi_g^z$ and also from $\Psi_{z'}$, by the electron transfer $\xi \to \pi_g^{y'}$. The second configuration is obtainable from $\psi(\alpha\,^3E_y)\phi(^3\Sigma_g^-)$ by the electron transfer $\pi_g^y \to \xi$ and also from $\psi_{z'}$ by $\pi_g^{z'} \to u$. The last configuration is obtainable from

$\psi(\alpha^3E_y)\phi(^3\Sigma_g^-)$ by $\pi_g^{z'} \rightarrow u$. The considerable large contribution of these charge transfer configurations is to be expected from the large overlap integrals between the π_g type orbitals O_2 molecule and the 3d orbitals of the *heme* and also from the fact that the two terms of oxygen molecules included in these configurations are the lowest in energy in O_2^- and O_2^+ ionic states, respectively. The charge transfer configuration $\psi(\alpha^2E_y;$ $Fe^{3+})\phi(^2\Pi_g^{y'}; O_2^-)$, especially, is the lowest in energy among all charge transfer configurations and its expectation value decreases remarkably as the O_2 molecule approaches the *heme*. In order to evaluate the effect of non-orthogonality, we show the values of effective off-diagonal elements of the total Hamiltonian matrix between this charge transfer configuration $\psi(\alpha^2E_y; Fe^{3+})\phi(^2\Pi_g^{y'}; O_2^-)$ and the normal configuration $\psi(\alpha^3E_y)\phi(^3\Sigma_g^-)$ or $\Psi_{z'}$, in the following.

$$\langle\psi(\alpha^2E_y; Fe^{3+})\phi(^2\Pi_g^{y'}; O_2^-)|H|\psi(\alpha^3E_y)\phi(^3\Sigma_g^-)\rangle -$$
$$\langle\psi(\alpha^2E_y; Fe^{3+})\phi(^2\Pi_g^{y'}; O_2^-)|\psi(\alpha^3E_y)\phi(^3\Sigma_g^-)\rangle\langle\psi(\alpha^3E_y)\phi(^3\Sigma_g^-)|H|\psi(\alpha^3E_y)\phi(^3\Sigma_g^-)\rangle$$
$$= -\ 2.50\ \text{eV}. \qquad (29\text{-}1)$$

$$\langle\psi(\alpha^2E_y; Fe^{3+})\phi(^2\Pi_g^{y'}; O_2^-)|H|\Psi_{z'}\rangle -$$
$$\langle\psi(\alpha^2E_y; Fe^{3+})\phi(^2\Pi_g^{y'}; O_2^-)|\Psi_{z'}\rangle\langle\Psi_{z'}|H|\Psi_{z'}\rangle = -\ 1.90\ \text{eV}. \qquad (29\text{-}2)$$

where the distance R is taken to be 1.5 Å, This shows that the charge transfer configuration $\psi(\alpha^2E_y; Fe^{3+})\phi(^2\pi_g^{y'}; O_2^-)$ interacts more effectively with the normal configuration $\psi(\alpha^3E_y)\phi(^3\Sigma_g^-)$ than with $\Psi_{z'}$. At any rate, however, the lowest eigenfunction is composed mainly of the normal configurations even at the short distance of 1.3 Å. This tendency is recognized also in all the low-lying states listed in Table IV and Table V.

In the proposal of the inclined arrangement, Pauling (1949) assumed that the *heme* was of singlet state and that the valence state of the O_2 molecule was represented by sp^2 hybridized orbitals at each oxygen atom. Such a calculation that includes completely the valence state as proposed by Pauling becomes very laborious, since this valence state is composed of various states with much higher excitation energies in a free O_2 molecule. Therefore, we take into account only the low-lying states of the *heme* and of O_2 molecule also in the inclined arrangement. According to the present calculation, however, the diamagnetism of oxygenated *heme* can be explained even if such higher excited states of O_2 molecule are not considered. In this inclined arrangement, other states with higher spin multiplicities are decidedly higher in energy at 1.5 Å or 1.3 Å, as long as the value of Y is taken to be larger than 4000 cm^{-1}, and they are not discussed.

5. DISCUSSION AND CONCLUSION

In the present calculation for both parallel and inclined arrangements, the projection of the O-O axis to the *heme* plane is chosen so as to be parallel to the x axis which connects the two nitrogen atoms situated on opposite sides from each other in the porphyrin ring (Figure 3 and Figure 4). Since the nitrogen atoms in the porphyrin ring are negatively charged, it may be expected that the charge transfer configurations of the ionic structure Fe^{3+}-O_2^- are mixed more effectively into the ground binding state if the O-O axis is placed relative to the *heme* in another way, i.e., its projection to the *heme* plane makes an angle of 45° with the axis. Practically, however, energy eigenvalues and eigenfunctions depend only slightly on the orientation of the O-O axis in the *heme* plane (xy plane). This is because the distance between the two oxygen atoms of O_2 molecule is much shorter than that between the oxygen atom and the nitrogen atoms in the porphyrin ring and from the fact that the ionic structure Fe^{3+}-O_2^- is contained with small amount in the ground binding state. If the ideal C_{4v} symmetry is realized in the *heme* of Mb or Hb, the projection of the O-O axis to the *heme* plane may preferably make a 45° with the x axis. But, really, other effects such as the orientation of the *imidazole* plane of the proximal histidine (5th ligand) and of the distal histidine residue may affect the orientation of the O-O axis in the *heme* plane. Therefore, at the present time it may be impossible from the calculation to determine the orientation of the O-O axis in the *heme* plane. Thus, only the result of the calculation for the case when the projection of the O-O axis is chosen to be parallel to the x axis is given in the present paper. First of all, it is a matter in question whether an O_2 molecule binds with the *heme* in the parallel or inclined arrangement and which electronic state contributes mainly to the binding state.

Recently, experimental measurements have been carried out to inquire into the nature of the binding of an O_2 molecule to the *heme*, but their conclusions are not consistent. For example, from the Mössbauer spectroscopy on oxygenated Mb and oxygenated Hb, Lang et al. (1966) have indicated the possibility that the main ionic structure of this compound is Fe^{3+}-O_2^- in the parallel arrangement. Eicher and Trautwein (1969) have analyzed their own Mössbauer data on oxygenated Mb and concluded that the triplet-triplet coupling ($\psi(^3E)\phi(^3\Sigma_g^-)$ in our notation) of the ionic structure Fe^{2+}-O_2 is the most probable. Barlow et al. (1973) have inferred from the vibrational spectroscopy of the O_2 molecule bound to the *heme* that the O_2 molecule takes the ionic form of O_2^-. Within the framework of the present calculation, it is not easy to decide whether the O_2 molecule is bound with its O-O axis parallel or inclined to the *heme* plane. The result of the present calculation shows that the lowest energy eigenvalue $\varepsilon_1(^1A') = -3.66$ eV at the distance of 1.5 Å between O_A atom and the iron atom in the inclined arrangement is almost equal to the lowest eigenvalue $\varepsilon_1 = -3.64$ eV in the parallel arrangements at the distance of 1.5 Å between the center of the O_2 molecule and the iron atom. However, the following effects which are not considered in the present calculation seem to favour the inclined arrangement of the O_2 molecule. (i) In the present calculation, the used values of the splitting parameters of d-level are those determined experimentally on the unoxygenated form of Mb or Hb. But, recently, it has been suggested by X-ray diffraction analysis that the iron atom is out of the *heme* plane by about 0.7 Å in the unoxygenated form and it is changed so that it is almost in the *heme* plane by oxygenation (Perutz, 1970). Since the distance between the iron atom and the nitrogen atoms of the *porphyrin* ring is shortened by oxygenation, the energy level of $d(x^2-y^2)$ orbital may be higher in the oxygenated form than that used in the present calculation. This effect would lower the term energy of $^1A_1$ in the heme more than that used in the present calculation. Therefore, the ground state in the inclined arrangement to which the $^1A_1$ term contributes becomes lower in energy. (ii) Recently, it has also been suggested that the O_2 molecule bound to the *heme* is linked with the distal *histidine* through its protonated *imidazole* in native oxygenated form of Mb or Hb (Pauling, 1964; Yonetani et al., 1974). The result of the present calculation shows that the charge transfer configurations are mixed more effectively into the ground binding state in the inclined arrangement than in the parallel arrangement. As shown in §4, the charge transfer configuration $\psi(\alpha\,^2E_y;\ Fe^{3+})\phi(^2\pi_g^{y'};\ O_2^-)$ contributes particularly to the ground state in the inclined arrangement. Thus, if the protonated *imidazole* of the distal *histidine* is situated appropriately near to the bound O_2 molecule, a hydrogen bond may possible be formed between them. This is another factor which favours the inclined arrangement.

If the inclined arrangement is actually realized in the *heme*-O_2, the electronic structure of the ground binding state may be expected to be complicated. According to the result of the present calculation, the ground state is composed of the triplet-triplet configuration $\psi(^3E;\ Fe^{2+})\phi(^3\Sigma_g^-;\ O_2)$, the singlet-singlet configuration $\psi(^1A_1;\ Fe^{2+})\phi(^1\Delta_g;\ O_2)$ and charge transfer configurations such as $\psi(\alpha\,^2E;\ Fe^{3+})\phi(^2\Pi_g^{y'};\ O_2^-)$. Such a complicated electronic structure in the *heme*-O_2 system might reflect on the recent experimental results mentioned above.

It is ascertained theoretically by the present calculation that the high spin compound of Mb or Hb becomes diamagnetic by the binding with the paramagnetic molecule of O_2. In Mb and Hb, the splitting parameter X of d-level takes such a special value that the energy of low spin state $^1A_1$ is nearly equal to that of high spin state $^5B_2$ or ^5E. In this case, triplet states, ^3E and $^3A_2$, are also low in energy. The present calculation clarifies the meaning of this characteristic of the electroninc structure in the *heme*. Namely, these ^3E and $^1A_1$ contribute mainly to the binding of the *heme* with an O_2 molecule. This fact that the state contributing to the binding is not the ground state but low-lying excited states may be related closely with the conformational change of hemoglobin molecule induced by oxygenation. In Mb and Hb, the *porphyrin ring* is in van der Waals contacts with many side chains of the apo-protein part and the iron atom is also coordinated by the *imidazole* of the proximal *histidine* (5th ligand). When an O_2 molecule is bound to the iron atom, the electronic state of the *heme* changes from quintet state to triplet or singlet state. The electronic distribution of the triplet or singlet state spreads more to the direction of the *heme* plane than the quintet state. This may suggest the increase in the covalency of d(xz) and d(yz) orbitals of the iron atom with the π orbitals of the *porphyrin* ring. In fact, as mentioned already, it is observed by X-ray diffraction analysis (Perutz, 1970) that the iron atom is out of the *heme* plane with the direction of 5th ligand in the unoxygenated form and it is changed to be in the *heme* plane in the oxygenated form. Since the distance between the iron atom and the 5th ligand seems to be almost invariant by oxygenation (Perutz, 1970), this displacement of the iron atom may be a trigger to cause the conformational change of the protein. And this displacement originates from the change in the electronic state of the iron atom. If the singlet or triplet state is the ground state in the unoxygenated form of Hb, the change in the electronic distribution by oxygenation may not be so remarkable and the displacement of the iron atom may not be so large.

In the present calculation for either parallel or inclined arrangement, we cannot find any distance where energy eigenvalues become minimum. This may be partly because of our assumption that the core electron cloud of the iron atom is shrunk into a point charge at the iron nucleus but mainly from the wave functions used in the evaluation of the interaction part of the Hamiltonian. The wave functions used in the present calculation are those determined by the self-considtent Hartree-Fock method. Although the electron density near nucleus determined by this method is reliable, the marginal part of the wave function may not be reliable, since it does not severely influence the expectation value of energy. Thus, even if a plausible energy value of the system is obtained by a self-consistent procedure, it does not necessarily mean that more expanded wave functions are not needed. Therefore, it may be possibly considered that the wave functions used in the present calculation are not expanded enough to produce a correct value for the exchange repulsion between the iron atom and the O_2 molecule, and that the calculated energy value of the ground state in the present work is not a reliable one for the binding energy of an O_2 molecule to the *heme*.

REFERENCES

Barlow, C.H., Maxwell, J.C., Wallace, W.J., and Caughey, W.S.: 1973, Biochem. Biophys. Res. Comm. 55, 91.

Beetlestone, J. and George, P.: 1961, Biochemistry 3, 707.

Berthier, G., Millie, P., and Veiland, A.: 1965, J. Chim. Phys. 62, 8.

Clementi, E.: 1965, Table of Atomic Functions, A Supplement to IBM Journal of Research and Development 9, 2.

Eaton, W.A. and Charney, E.: 1964, J. Chem. Phys. 51, 4502.

Eicher, H. and Trautwein, A.: 1969, J. Chem. Phys. 50, 2540.

George, P., Beetlestone, J., and Griffith, J.S.: 1964, Rev. Mod. Phys. 36, 441.

Gibson, J.F., Ingram, D.J.E., and Schonland, D.: 1958, Discussions Faraday Soc. 26, 72.

Gilmore, F.R.: 1965, J. Quant. Spectrosc. Radiative Transfer 5, 369.

Gouterman, M.: 1959, J. Chem. Phys. 30, 1139.

Griffith, J.S. 1956, Proc. Roy. Soc. A235, 23.

Griffith, J.S.: 1957, Nature 180, 30.

Griffith, J.S.: 1961, The Theory of Transition-Metal Ions, Cambridge University Press, London.

Huzinaga, S.: 1965, J. Chem. Phys. 42, 1293.

Koshland, D.E., Nemethy, G., and Filmer, D.: 1966, Biochemistry 5, 365.
Kotani, M., Mizuno, Y., Kayama, K., and Ishhiguro, E.: 1957, J. Phys. Soc. Japan 12, 707.
Kotani, M.: 1963, Rev. Mod. Phys. 35, 717.
Lang. G. and Marshall, W.: 1966, Proc. Phys. Soc. 87, 3.
Moffit, M.: 1951, Proc. Roy. Soc. A210, 245.
Monod, J., Wyman, J., and Changeux, J.P.: 1965, J. Mol. Biol. 33, 283.
Moore, C.D.: 1958, Atomic Energy Levels: Circular of the National Bureau of Standards, U.S. Goverment Printing Office, Washington, vol. II, p. 49.
Mulliken, R.S.: 1959, Phys. Rev. 115, 1225.
Nakano, N., Otsuka, J., and Tasaki, A.: 1971, Biochim. Biophys. Acta 236, 222.
Ohno, K., Tanabe, Y., and Sasaki, F.: 1963, Theoret. Chim. Acta 1, 378.
Otsuka, J.: 1966, J. Phys. Soc. Japan 21, 596.
Otsuka, J.: 1968, J. Phys. Soc. Japan 24, 885.
Pack, J.L. and Phelps, A.V.: 1966, J. Chem. Phys. 44, 1840.
Pariser, R. and Parr, R.G.: 1953, J. Chem. Phys. 21, 466.
Pariser, R. and Parr, R.G.: 1953, J. Chem. Phys. 21, 767.
Pauling, L.: 1949, Haemoglobin: Sir Joseph Barcroft Memorial Symposium, London, Butterworth, p. 57.
Pauling, L.: 1964, Nature 203, 182.
Perutz, M.F.: 1970, Nature 228, 726.
Phelps, A.V. and Pack, J.L.: 1961, Phys. Rev. Letters 6, 111.
Pople, J.: 1953, Trans. Faraday, Soc. 49, 1375.
Pullman, B., Spanjaard, C., and Berthier, G.: 1960, Proc. Natl. Acad. Sci. U.S.A. 46, 1011.
Simpson, W.T.: 1949, J. Chem. Phys. 17, 1218.
Stockdale, J.A.D., Compton, R.N., Hurst, G.S., and Reinhardt, P.W.: 1969, J. Chem. Phys. 50, 2176.
Tanabe, Y. and Sugano, S.: 1954, J. Phys. Soc. Japan 9, 753.
Tanabe, Y. and Sugano, S.: 1954, J. Phys. Soc. Japan 9, 766.
Tasaki, A., Otsuka, J., and Kotani, M.: 1967, Biochim. Biophys. Acta 140, 284.
Wallace, L.: 1962, Astrophys. J. Suppl. 7, 165.
Watson, H.C. and Nobb, C.L.: 1968, 19 Colloquim der Gesellschaft für Biologische Chemie, Springer-Verlag, Berlin, New York, p.37.
Weiss, C., Kobayashi, H., and Gouterman, M.: 1965, J. Mol. Spectr. 16, 415.
Yonetani, T., Yamamoto, H., and Iizuka, T.: 1974, J. Biol. Chem. 249, 2168.
Yoshino, K. and Tanaka, Y.: 1968, J. Chem. Phys. 48, 4859.
Zerner, M., Gouterman, M., and Kobayashi, H.: 1966, Theoret. Chim. Acta 6, 363.

APPENDIX

The expression of the electronic configurations of the constituent wave functions, which appear in the low-lying eigenfunctions with coefficients of more than 0.1, and of low-lying terms shown in Figure 5 and Figure 6. In an O_2 molecule, the closed shell $(\pi_u^z)^2(\pi_u^y)^2(3\sigma_g)^2(2\sigma_u)^2(2\sigma_g)^2(1\sigma_u)^2(1\sigma_g)^2$ is not written down explicitly.

Fe^{2+}

$\psi(^1A_1) = |\xi\bar{\xi}\eta\bar{\eta}\zeta\bar{\zeta}|$ $\qquad \psi(\alpha\,^3E_x) = |u\xi\bar{\xi}\eta\zeta\bar{\zeta}|$

$\psi(\alpha\,^3E_y) = |u\xi\eta\bar{\eta}\zeta\bar{\zeta}|$ $\qquad \psi(\beta\,^3E_x) = |v\xi\bar{\xi}\eta\zeta\bar{\zeta}|$

$\psi(\beta\,^3E_y) = |v\xi\eta\bar{\eta}\zeta\bar{\zeta}|$ $\qquad \psi(^3B_2) = |u\xi\bar{\xi}\eta\bar{\eta}\zeta|$

$\psi(^3A_2) = |v\xi\bar{\xi}\eta\bar{\eta}\zeta|$ $\qquad \psi(^5B_2) = |uv\xi\eta\zeta\bar{\zeta}|$

$\psi(^5E_x) = |uv\xi\eta\bar{\eta}\zeta|$ $\qquad \psi(^5E_y) = |uv\xi\bar{\xi}\eta\zeta|$

Fe^{3+}

$\psi(\alpha\,^2E_x) = |\xi\bar{\xi}\eta\zeta\bar{\zeta}|$ $\qquad \psi(\alpha\,^2E_y) = |\xi\eta\bar{\eta}\zeta\bar{\zeta}|$

$\psi(\beta\,^2E_y) = \frac{1}{\sqrt{6}}\{2|u\xi\bar{\xi}\bar{\eta}\zeta| - |u\xi\bar{\xi}\eta\bar{\zeta}| - |\bar{u}\xi\bar{\xi}\eta\zeta|\}$

$\psi(\gamma\,^2E_x) = \frac{1}{\sqrt{6}}\{2|v\bar{\xi}\eta\bar{\eta}\zeta| - |v\xi\eta\bar{\eta}\zeta| - |\bar{v}\xi\eta\bar{\eta}\zeta|\}$

$\psi(\alpha\,^2B_2) = |\xi\bar{\xi}\eta\bar{\eta}\zeta|$ $\qquad \psi(\beta\,^2B_2) = \frac{1}{\sqrt{2}}\{|u\xi\bar{\eta}\zeta\bar{\zeta}| - |u\bar{\xi}\eta\zeta\bar{\zeta}|\}$

$\psi(\alpha\,^2A_2) = \frac{1}{\sqrt{2}}\{|v\xi\bar{\eta}\zeta\bar{\zeta}| - |v\bar{\xi}\eta\zeta\bar{\zeta}|\}$

$\psi(\alpha\,^2B_1) = \frac{1}{\sqrt{2}}\{|u\xi\bar{\xi}\zeta\bar{\zeta}| - |u\eta\bar{\eta}\zeta\bar{\zeta}|\}$

$\psi(\alpha\,^2A_1) = |u\xi\bar{\xi}\eta\bar{\eta}|$

$\psi(\beta\,^2A_1) = \frac{1}{\sqrt{2}}\{|u\xi\bar{\xi}\zeta\bar{\zeta}| + |u\eta\bar{\eta}\zeta\bar{\zeta}|\}$

$\psi(\gamma\,^2A_1) = \frac{1}{\sqrt{2}}\{|v\xi\bar{\xi}\zeta\bar{\zeta}| - |v\eta\bar{\eta}\zeta\bar{\zeta}|\}$

$\psi(\alpha\,^4E_x) = |u\xi\eta\bar{\eta}\zeta|$ $\qquad \psi(\alpha\,^4E_y) = |u\xi\bar{\xi}\eta\zeta|$

$\psi(\beta\,^4E_x) = -|v\xi\eta\bar{\eta}\zeta|$ $\qquad \psi(\beta\,^4E_y) = |v\xi\bar{\xi}\eta\zeta|$

$\psi(\gamma\,^4E_x) = |uv\eta\zeta\bar{\zeta}|$ $\qquad \psi(\gamma\,^4E_y) = |uv\xi\zeta\bar{\zeta}|$

$\psi(\alpha\,^4A_2) = |u\xi\eta\zeta\bar{\zeta}|$

$\psi(\beta\,^4A_2) = \frac{1}{\sqrt{2}}\{|uv\xi\bar{\xi}\zeta| + |uv\eta\bar{\eta}\zeta|\}$

$\psi(\alpha\,^4B_2) = |v\xi\eta\zeta\bar{\zeta}|$

$\psi(\beta\,^4B_2) = \frac{1}{\sqrt{2}}\{|uv\xi\bar{\xi}\zeta| - |uv\eta\bar{\eta}\zeta|\}$

$\psi(^6A_1) = |uv\xi\eta\zeta|$

Fe^{+}

$\psi(^2A_1) = |u\xi\bar{\xi}\eta\bar{\eta}\zeta\bar{\zeta}|$ $\qquad \psi(^2B_1) = |v\xi\bar{\xi}\eta\bar{\eta}\zeta\bar{\zeta}|$

$\psi(\alpha\,^2E_x) = |v\bar{v}\xi\bar{\xi}\eta\zeta\bar{\zeta}|$ $\qquad \psi(\alpha\,^2E_y) = |v\bar{v}\xi\eta\bar{\eta}\zeta\bar{\zeta}|$

$\psi(\beta\,^2E_x) = |u\bar{u}\xi\bar{\xi}\eta\zeta\bar{\zeta}|$ $\qquad \psi(\beta\,^2E_y) = |u\bar{u}\xi\eta\bar{\eta}\zeta\bar{\zeta}|$

$\psi(\gamma\,^2E_x) = \frac{1}{\sqrt{2}}\{|u\bar{v}\xi\bar{\xi}\eta\zeta\bar{\zeta}| - |\bar{u}v\xi\bar{\xi}\eta\zeta\bar{\zeta}|\}$

$\psi(\gamma\,^2E_y) = \frac{1}{\sqrt{2}}\{|u\bar{v}\xi\eta\bar{\eta}\zeta\bar{\zeta}| - |\bar{u}v\xi\eta\bar{\eta}\zeta\bar{\zeta}|\}$

$\psi(\delta\,^2E_x) = \frac{1}{\sqrt{6}}\{2|uv\xi\bar{\xi}\bar{\eta}\zeta\bar{\zeta}| - |u\bar{v}\xi\bar{\xi}\eta\zeta\bar{\zeta}| - |\bar{u}v\xi\bar{\xi}\eta\zeta\bar{\zeta}|\}$

APPENDIX (continued)

$\psi(\delta\,^2E_y) = \frac{1}{\sqrt{6}}\{2|uv\bar{\xi}\eta\bar{\eta}\zeta\bar{\zeta}| - |u\bar{v}\xi\eta\bar{\eta}\zeta\bar{\zeta}| - |\bar{u}v\xi\eta\bar{\eta}\zeta\bar{\zeta}|\}$

$\psi(\alpha\,^2A_2) = \frac{1}{\sqrt{2}}\{|u\bar{v}\xi\bar{\xi}\eta\bar{\eta}\zeta| - |\bar{u}v\xi\bar{\xi}\eta\bar{\eta}\zeta|\}$

$\psi(\alpha\,^2B_2) = |u\bar{u}\xi\bar{\xi}\eta\bar{\eta}\zeta|$

$\psi(\alpha\,^4E_x) = -|uv\xi\bar{\xi}\eta\zeta\bar{\zeta}|$

$\psi(\beta\,^4E_x) = -|u\bar{u}v\xi\eta\bar{\eta}\zeta|$

$\psi(\gamma\,^4E_x) = |uv\bar{v}\xi\eta\bar{\eta}\zeta|$

$\psi(\alpha\,^4A_2) = |uv\xi\bar{\xi}\eta\bar{\eta}\zeta|$

$\psi(\alpha\,^4E_y) = |uv\xi\eta\bar{\eta}\zeta\bar{\zeta}|$

$\psi(\beta\,^4E_y) = |u\bar{u}v\xi\bar{\xi}\eta\zeta|$

$\psi(\gamma\,^4E_y) = |uv\bar{v}\xi\bar{\xi}\eta\zeta|$

$\psi(^4B_2) = |u\bar{u}v\xi\eta\zeta\bar{\zeta}|$

O_2

$\phi(^3\Sigma_g^-) = |\pi_g^{z'}\pi_g^{y'}\cdots\cdots|$

$\phi(^1\Delta_g^a) = \frac{1}{\sqrt{2}}\{|\pi_g^{z'}\bar{\pi}_g^{z'}\cdots\cdots| - |\pi_g^{y'}\bar{\pi}_g^{y'}\cdots\cdots|\}$

$\phi(^1\Delta_g^b) = \frac{1}{\sqrt{2}}\{|\pi_g^{z'}\bar{\pi}_g^{y'}\cdots\cdots| - |\bar{\pi}_g^{z'}\pi_g^{y'}\cdots\cdots|\}$

$\phi(^1\Sigma_g^+) = \frac{1}{\sqrt{2}}\{|\pi_g^{z'}\bar{\pi}_g^{z'}\cdots\cdots| + |\pi_g^{y'}\bar{\pi}_g^{y'}\cdots\cdots|\}$

O_2^-

$\phi(^2\Pi_g^{z'}) = |\pi_g^{z'}\pi_g^{y'}\bar{\pi}_g^{y'}\cdots\cdots|$

$\phi(^2\Pi_g^{y'}) = |\pi_g^{z'}\bar{\pi}_g^{z'}\pi_g^{y'}\cdots\cdots|$

O_2^+

$\phi(^2\Pi_g^{z'}) = |\pi_g^{z'}\cdots\cdots|$

$\phi(^2\Pi_g^{y'}) = |\pi_g^{y'}\cdots\cdots|$

$$\phi(^2\pi_u^{z'}) = \frac{1}{2\sqrt{6}}\{2|\pi_g^{z'}\pi_g^{y'}\pi_u^{z'}\bar{\pi}_u^{z'}\bar{\pi}_u^{y'}\cdots\cdots| - |\bar{\pi}_g^{z'}\pi_g^{y'}\pi_u^{z'}\bar{\pi}_u^{z'}\pi_u^{y'}\cdots\cdots|$$

$$- |\pi_g^{z'}\bar{\pi}_g^{y'}\pi_u^{z'}\bar{\pi}_u^{z'}\pi_u^{y'}\cdots\cdots| + 3|\pi_g^{z'}\bar{\pi}_g^{z'}\pi_u^{z'}\pi_u^{y'}\bar{\pi}_u^{y'}\cdots\cdots|$$

$$+3|\pi_g^{y'}\bar{\pi}_g^{y'}\pi_u^{z'}\pi_u^{y'}\bar{\pi}_u^{y'}\cdots\cdots|\}$$

$$\phi(^2\Pi_u^{y'}) = \frac{1}{2\sqrt{6}}\{2|\pi_g^{z'}\pi_g^{y'}\bar{\pi}_u^{z'}\pi_u^{y'}\bar{\pi}_u^{y'}\cdots\cdots| - |\pi_g^{z'}\bar{\pi}_g^{y'}\pi_u^{z'}\pi_u^{y'}\bar{\pi}_u^{y'}\cdots\cdots|$$

$$- |\bar{\pi}_g^{z'}\pi_g^{y'}\pi_u^{z'}\pi_u^{y'}\bar{\pi}_u^{y'}\cdots\cdots| - 3|\pi_g^{z'}\bar{\pi}_g^{z'}\pi_u^{z'}\bar{\pi}_u^{z'}\pi_u^{y'}\cdots\cdots|$$

$$-3|\pi_g^{y'}\bar{\pi}_g^{y'}\pi_u^{z'}\bar{\pi}_u^{z'}\pi_u^{y'}\cdots\cdots|\}$$

$$\phi({}^2\Sigma_g^-) = \frac{1}{\sqrt{6}}\{2|\pi_g^{z'}\pi_g^{y'}\pi_u^{z'}\bar{\pi}_u^{z'}\pi_u^{y'}\bar{\pi}_u^{y'}3\bar{\sigma}_g 2\sigma_u 2\bar{\sigma}_u 2\sigma_g 2\bar{\sigma}_g \cdots|$$
$$-|\pi_g^{z'}\bar{\pi}_g^{y'}\pi_u^{z'}\bar{\pi}_u^{z'}\pi_u^{y'}\bar{\pi}_u^{y'}3\sigma_g 2\sigma_u 2\bar{\sigma}_u 2\sigma_g 2\bar{\sigma}_g \cdots|$$
$$-|\bar{\pi}_g^{z'}\pi_g^{y'}\pi_u^{z'}\bar{\pi}_u^{z'}\pi_u^{y'}\bar{\pi}_u^{y'}3\sigma_g 2\sigma_u 2\bar{\sigma}_u 2\sigma_g 2\bar{\sigma}_g \cdots|\}$$

$$\phi({}^4\Pi_u^{z'}) = |\pi_g^{z'}\pi_g^{y'}\pi_u^{z'}\bar{\pi}_u^{z'}\pi_u^{y'}\cdots\cdots|$$

$$\phi({}^4\Pi_u^{y'}) = |\pi_g^{z'}\pi_g^{y'}\pi_u^{z'}\pi_u^{y'}\bar{\pi}_u^{y'}\cdots\cdots|$$

$$\phi({}^4\Sigma_g^-) = |\pi_g^{z'}\pi_g^{y'}\pi_u^{z'}\bar{\pi}_u^{z'}\pi_u^{y'}\bar{\pi}_u^{y'}3\sigma_g 2\sigma_u 2\bar{\sigma}_u 2\sigma_g 2\bar{\sigma}_g \cdots|$$

DISCUSSION

A. Pullman: I am not quite clear as to the way in which you take into account the fact that the iron is bound to the heam group? Could you give some information on this point?

Otsuka: In our scheme, the eigenfunction of the heme-O_2 system is assumed to be expressed by a linear combination of the members of the basic functions, each of which is an anti-symmetrized production function of the eigenfunction of the unoxygenated heme and of the eigenfunction of the free O_2 molecule. We take into account the fact by determining the eigenvalues and eigenfunctions of the unoxygenated heme with the use of ligand field theory. Although the mixing of the iron orbitals and the porphyrin orbitals is not expressed formally, this may be allowed in the evaluation of the interaction between the iron atom in the heme and the O_2 molecule.

Veillard: You have reported total energies of -3.66 eV for the Pauling model and -3.64 eV for the Griffith model for a Fe—O distance of 1.5 Å (this value is certainly too small since the experimental Fe—O bond lenght in the dioxygen complex of the picket fence porphyrin is 1.75 Å). This corresponds to a difference in the relative stability of the two structures (end-on and perpendicular) of 0.5 kcal $mole^{-1}$ and one would expect an equilibrium between the two structures. This is at variance with (i) the experimental structure of the dioxygen complex of the picket-jence porphyrin (Collman, 1974) which shows that dioxygen is coordinated end-on; (ii) the results of our ab initio LCAO-MO-SCF calculations for the systems $FePO_2$ and $FePO_2NH_3$ (Fe=Fe(II), P = porphyrin) with the perpendicular structure found less stable than the end-on structure by 55 — 60 kcal $mole^{-1}$.

Otsuka: Strictly, our calculated energy eigenvalues are -3.658 eV for the Pauling model at a Fe—O distance of 1.5 Å and -3.460 eV (not -3.64 eV) for the Griffith model at a distance of 1.5 Å between the center of the oxygen molecule and the iron atom. Therefore, the difference is not so slight that one would expect an equilibrium between the two structures, although the result of our calculation has not been able to find a remarkable difference of the energy values between the two structures.

However, it should be also noted that the calculated energy eigenvalue in the present work is not a reliable one for the binding energy of an oxygen molecule to the heme, since exchange repulsions are not so effective even at the distance of 1.5 Å. This may be because of our assumption that the core electron cloud of the iron atom is shrunck into a point charge at the iron nucleus and also because of the wave functions used in the present calculation.

OXYGEN BINDING TO IRON PORPHYRINS. *AB INITIO* CALCULATIONS

A. DEDIEU, M.-M. ROHMER, and A. VEILLARD
Equipe de Recherche no. 139 du C.N.R.S., Université Louis Pasteur, B.P. 296/R8, 67000 Strasbourg, France

INTRODUCTION

Studies of synthetic oxygen carriers (Collman et al., 1975b; Basolo et al., 1975 and references therein) have significantly improved our understanding of the iron-dioxygen linkage in the oxygen carriers oxyhemoglobin and oxymyoglobin. However, a complete understanding of the coordinate link between iron and dioxygen (or carbon monoxide in the carbonyl complex) in the hemoproteins awaits further resolution of the following problems: (i) the alternative description in terms of a $Fe(II)-O_2$ or $Fe(III)-O_2^-$ configuration; (ii) the geometric structure of the iron-dioxygen unit; (iii) the stabilization of the coordinated dioxygen through hydrogen bonding to the imidazole of the distal histidine molecule in hemoglobin and myoglobin; (iv) the nature of the intramolecular motion (rotation about the iron-oxygen bond or inversion at the oxygen atom) interconverting the two types of coordinated dioxygen found in the picket-fence porphyrins (Collman et al., 1974); (v) the geometric structure of the iron-carbonyl unit in the carbonyl complexes, which could be either linear (as usually expected) or bent.

The original formulation by Pauling (1949) of the bonding of dioxygen to the heme in oxyhemoglobin assumed a σ donation from the dioxygen ligand to iron with a concomitant π back-donation as shown in *1*.

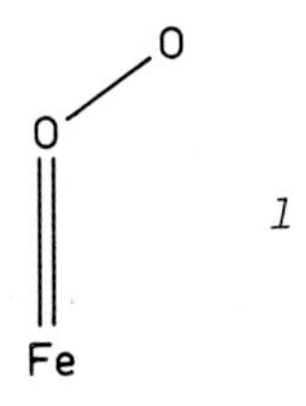

1

Since the metal-ligand transfers involve only a fraction of an electron, this is currently referred to-day as the $Fe(II)-O_2$ formulation. In this bent structure, the degeneracy of the π and $\bar{\pi}$ orbitals of the dioxygen ligand is removed and the two unpaired oxygen electrons are paired up

B. Pullman and N. Goldblum (eds.), Metal-Ligand Interactions in Organic Chemistry and Biochemistry, part. 2, 101-130.

into one of the antibonding π_g orbitals, thus accounting for the experimental diamagnetism of oxyhemoglobin (Pauling, 1936). To a first approximation, the low-spin Fe(II)ion has formally six electrons in the t_{2g} orbitals (assuming a local O_h symmetry) and the electronic structure of the iron-dioxygen unit in the Pauling formulation is $d^6\ \pi_g^2\ \pi_g^0$. A similar formulation Fe(II)-O_2 is implicit in the Griffith model with a perpendicular structure in place of a bent structure (Griffith, 1956; see also Klevan et al., 1973). The same formulation is found in the qualitative description of the bonding in oxyhemoglobin by Makinen et al. (1973). A different model has been proposed by Weiss (1964), with one electron transferred from iron to dioxygen and the binding described as a low-spin complex of iron (III) with a superoxide anion radical O_2^- (this is sometimes called the metsuperoxide model). This is a biradical model with two unpaired electrons, one localized on the iron and the other on dioxygen, the diamagnetism resulting from antiferromagnetic coupling of the unpaired spins as in *2*. The electronic structure may

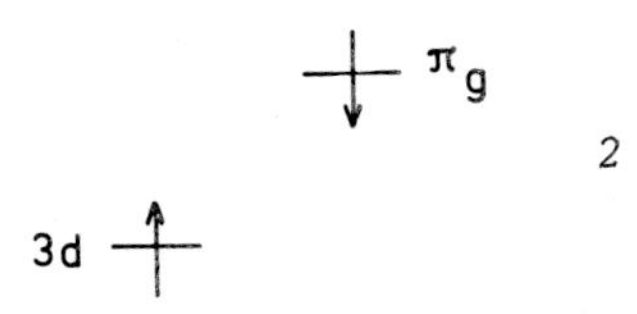

2

then be formally described as $d^5\ \pi_g^2\ \pi_g^1$. However one should realize that a transfer of electron density from iron to dioxygen, formulated as $Fe^{\delta+}-O_2^{\delta-}$, does not imply that the bonding should be described in terms of the Weiss model, but is compatible with the Pauling model with back-bonding donation exceeding the ligand to metal donation. The two-electron oxidative addition model of Gray (1971) *3* represents an extension of the

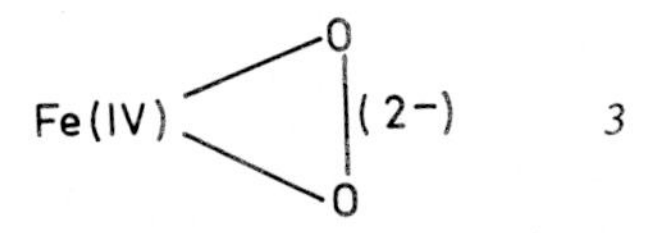

3

Weiss model, with two electrons transferred from iron to dioxygen resulting in a Fe(IV) complex with a peroxide ion O_2^{2-}. Support for the Weiss formulation Fe(III)-O_2^- comes from (i) the analogy with the well-documented Co(III)-O_2^- formulation for synthetic oxygen carriers of cobal (Basolo, 1975) (however the metal-dioxygen unit in the Co(II) complexes is not isoelectronic with the iron (II) dioxygen moiety, having one electron more); (ii) from the stabilization of the dioxygen complex of the tetraphenylporphyriniron(II) Fe(TPP)(B)$_2$ in a polar solvent (Basolo, 1975); (iii) from the infrared spectrum of oxyhemoglobin (Barlow et al., 1973; however see also Collman et al., 1975b); (iv) from the large quadrupole splittings observed in the Mössbauer resonance spectra of oxyhemoglobin (Lang et al., 1966); (v) from spectral evidence in support of the presence of Fe(III) in oxyhemoglobin (Peisach et al., 1968). However Gray (1971) has criticized the Weiss model on the basis that, given the proximity of the superoxide ligand O_2^- to the ligand NCS^- in the spectrochemical series, the low-spin character of oxyhemoglobin, which implies in the Weiss model a low-spin Fe(III) bound to an O_2^- ligand, is unexpected on the basis of the high-spin (S=5/2) character of met thiocyanate hemoglobin. Recent work on the dioxygen complex of a picket fence porphyrin, representing a synthetic model of the oxygen binding hemoproteins, favours the Pauling formulation as Fe(II)-O_2 (Collman et al., 1975b). Extended Hückel calculations for a model oxyhemoglobin compound account for the experimental quadrupole splittings of the Mössbauer resonance spectra with a ground state Fe(II)-O_2 configuration (Loew et al., 1975). However the possibility of a spin equilibrium between a singlet ground state corresponding to a dioxygen ferrous complex and a triplet state of slightly higher energy corresponding to a superoxide ferric complex (similar to the Weiss model but without compensation of spin) has been postulated in order to account for the paramagnetism of another dioxygen complex of the picket fence porphyrin (Collman et al., 1975b) anf for the Fe Kβ fluorescence emission spectrum of oxymyoglobin (Koster, 1975).

The geometric structure of the metal-dioxygen unit in the oxygen carriers has been a long pending problem. Different structural models have been successively proposed for dioxygen binding, including a linear M–O–O unit (Pauling et al., 1936), an end-on angular bond *1* (Pauling 1949, 1964) and a sideways perpendicular structure as in *3*, (Griffith, 1956). The dioxygen complex of the picket fence porphyrin has been recently characterized as a system with a Fe–O–O bent bond, the corresponding angle being 136° (Collman et al., 1974). However a perpendicular structure has been postulated for a dioxygen complex of a manganese(II) porphyrin (Weschler et al., 1975). It has been emphasized that the sideways, perpendicular structure corresponds to a formal coordination number of seven for the metal, a sterically unfavourable situation (Collman, 1975b) (among other things this would result from the fact that the Griffith model produces an unfavourably short N–O distance of less than 2.60 Å in oxyhemoglobin (Hoard, 1968)).

Pauling (1964) suggested originally that a stabilization of the dioxygen ligand in oxyhemoglobin might result from hydrogen bonding between the terminal oxygen atom O_β (bearing a formal negative charge) and the pyrrolic nitrogen atom of the imidazole group from the distal histidine as in *4*. However the formal charge separation postulated by Pauling in *4* appears rather dubious (calculations for the dioxygen complex of Co(acacen) indicate that the largest fraction of the formal negative charge associated with the dioxygen ligand is found on the O_α atom bound to cobalt (Dedieu et al., 1976)). On the other hand,

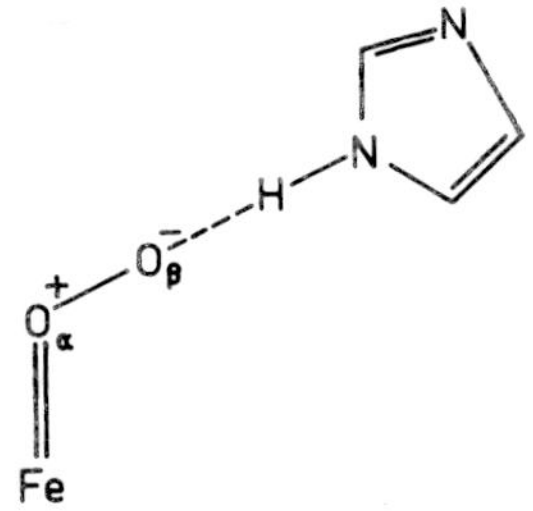

4 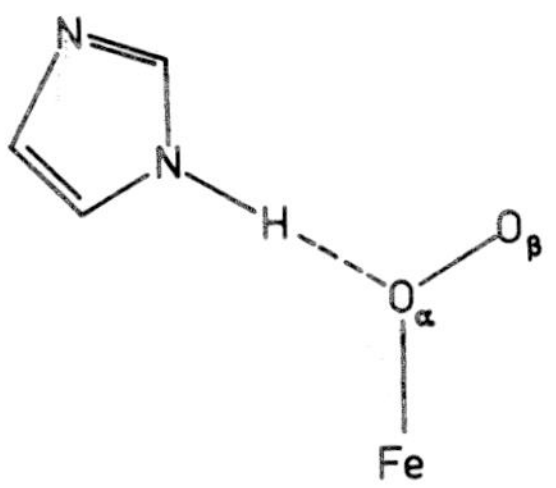 *5*

this O_α atom is at about 2.5 Å from the $N_{\varepsilon 2}$ atom of distal histidine, a situation which might lead to the hydrogen bonding represented in *5* (Antonini et al., 1971).

There are not many quantum mechanical calculations of oxygen binding to iron porphyrins. Zerner et al. (1966) carried out extended Hückel calculations for the oxygen complex of ferroporphyrin in the linear and perpendicular structures (with a water molecule as the sixth ligand). They concluded that the linear geometry should be unstable since the $1\pi_g$ orbitals of dioxygen were found below the normally occupied $e_g(d_\pi)$ and $b_{2g}(d_{xy})$ orbitals of iron (this would result in a formal structure $Fe(IV)-O_2^{2-}$ corresponding to the oxidative addition model of Gray(1971)). The system was found diamagnetic in the perpendicular structure, the degeneracy of the $1\pi_g$ orbital of dioxygen being lifted with the $1\pi_g(xy)$ orbital parallel to the porphyrin plane doubly occupied (the $1\pi_g(z)$ orbital perpendicular to the porphyrin plane mixes heavily with the metal $3d_{xz}$ orbital, an indication of $d\pi$-$p\pi^*$ backbonding). The same method has been applied by Loew et al. (1975) to the bent structure of oxyferroporphyrin (with an imidazole molecule as the sixth ligand), with a ground-state $Fe(II)-O_2$ configuration, in order to explain the large quadrupole splittings observed in the Mössbauer resonance spectra. Although $Fe(III)-O_2^-$ configurations were also considered, they were not necessary to accoun for both the magnitude and sign of the field gradient. A common feature of both extended Hückel calculations is the substantial negative charge (about 0.5 e) found on the dioxygen ligand. A number of calculations are restricted to the central part of the complex. In the extended Hückel calculations by Halton (1972, 1974), only the valence orbitals of the iron, oxygen and nitrogen atoms (of the pyrrole groups) were considered and an arbitrary addition of two electrons to the system was claimed necessary, in order to prevent the filling of the oxygen π orbitals at the expense of the iron 3d orbitals. Heitler-London calculations have been reported for the $Fe-O_2$ unit with the perpendicular structure (Seno et al., 1972). Goddard et al. (1975) carried out ab initio GVB-CI calculations for the $Fe-O_2$ unit with the bent and perpendicular structures, based on a biradical model where the configurations of the iron atom $(t_{2g})^5(e_g)^1$ and of dioxygen are triplet states, with a σ bond between iron and oxygen and a four-electron three-center π bond leading to a singlet ground-state. Although the idea of a biradical system is common to the models of Weiss and Goddard, the latter one corresponds to a $Fe(II)-O_2$ formulation on the basis of the reported population analysis.

We report here LCAO-MO-SCF ab initio calculations for a number of oxyferroporphyrins $FePO_2L$ (P=porphyrin, L=none, NH_3, imidazole) in order to investigate (i) the nature of the binding between iron and dioxygen; (ii) the possibility of hydrogen bonding between the dioxygen ligand and the distal imidazole; (iii) the barrier to rotation about the Fe-O bond; (iv) the role of the proximal imidazole ligand in oxyhemoglobin. Most of the calculations were carried out with an ammonia molecule as the fifth ligand in place of the proximal imidazole found in the natural oxygen carriers. However we have also carried one calculation with an imidazole ligand.

2. THE CALCULATIONS

Ab initio LCAO-MO-SCF calculations including all the electrons have been carried out for the systems of Table I with a gaussian basis set (10,6,4/7,3/3) contracted to [4,3,2/2,1/1] (minimal set except for the 3d functions which are split). The (10,6,4) basis set for the Fe atom is built from a (9,5,3) basis optimized for Fe^{2+} (Roos et al., 1971) incremented with one *s* function of exponent 0.20, one *p* function of exponent 0.25 and one *d* function of exponent 0.20 (these exponents are chosen so as to give a maximum of radial density about at midlength of the iron-ligand bonds (Veillard et al., 1976)). The (7,3) basis set for first-row atoms is the one of Roos et al, (1970) and the (3) basis set for hydrogen atoms is taken from Huzinaga (1965). The calculations were carried out with the Asterix system of programs (Benard et al., 1975a; Benard, 1976, see also Veillard, 1975). The open-shell treatment is based on the restricted Hartree-Fock formalism (Roothaan, 1960). All one- and two-electron integrals were computed with single-word accuracy on the Univac 1110 (word of 36 bits), with the few last SCF iterations carried out with double-word accuracy. A representative timing for the calculation on the iron(II)porphyrin FeP is 2 h 45 m of CPU time for the integral section on a Univac 1110 monoprocessor and 5 m of CPU time for each SCF iteration (in single precision).

The geometry of the iron-porphyrin core, with the iron atom in the porphin plane, was taken as the experimental geometry of the α,β,γ,δ-tetraphenyl porphinatoiron(II) (Collman et al., 1975a) (somewhat idealized) and is shown in Figure 1. For the Fe—O_2 moiety we have used the Fe—O bond length of 1.75 Å, the O—O bond length of 1.24 Å and the FeOO

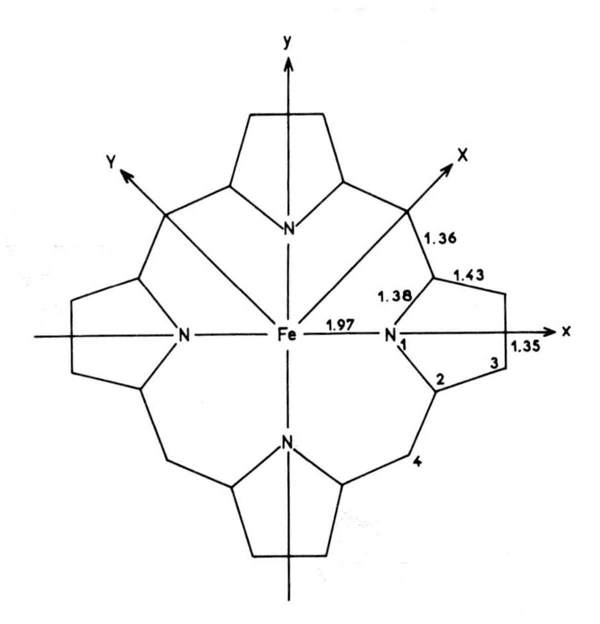

Fig. 1. Geometry of the iron-porphyrin core and the choice of axis.

TABLE I
Basis sets and SCF energies (in AU)

	Gaussian basis set	Contracted basis set	Number of electrons	Geometric structure	Molecular point group	Electronic configuration[a]	Total energy
Porphin dianion	420	132	162		D_{4h}		- 978.521
Fe(II)porphyrin(FeP)	472	157	186		D_{4h}		-2236.101
$FePO_2$	504	167	202	Linear	C_{4v}	$d^6(\pi_g^a)^2$	-2385.186
				Bent	C_s	$d^6(\pi_g^a)^2$	-2385.208
						$d^6(\pi_g^b)^2$	-2385.178
				Bent[c]	C_s	$d^6(\pi_g^a)^2$	-2385.199
				Perpen-dicular	C_{2v}	$d^6(\pi_g^a)^2$	-2385.022
						$d^6(\pi_g^b)^2$	-2385.108
$FePO_2(NH_3)$	529	175	212	Bent	C_s	$d^6(\pi_g^a)^2$ $a\,^1A'$	-2441.208
						$d^6(\pi_g^b)^2$ $a\,^1A'$	-2441.179
						$d_{xy}^1 d_{xz}^2 d_{yz}^2 (\pi_g^a)^2 (\pi_g^b)^1$ $^3A''$	-2441.217[b]
						ib. $^1A''$	-2441.217[d]
						$d_{xy}^2 d_{xz}^2 d_{yz}^1 (\pi_g^a)^2 (\pi_g^b)^1$ $^3A'$	-2441.232[b]
						ib. $b\,^1A'$	-2441.060[d]

[*TABLE I (Continued)*]

	Gaussian basis set	Contracted basis set	Number of electrons	Geometric structure	Molecular point group	Electronic configuration[a]	Total energy
				Perpen-dicular	C_s	$d^6(\pi_g^a)^2$	-2441.020
						$d^6(\pi_g^b)^2$	-2441.120
$FePO_2(NH_3)_2$	554	183	222	Bent	C_s	$d^6(\pi_g^a)^2$	-2497.117
$FePO_2Im$	596	196	238	Bent	C_1	$d^6(\pi_g^a)^2$	-2609.194
FePCO	504	167	200	Linear(α=180°)	C_{4v}	$d^6(\pi_g)^0$	-2348.4711
				Bent(α=172.5°)	C_s		-2348.4702
				Bent(α=165°)	C_s		-2348.4670

[a] Singlet state except otherwise stated, d^6 stands for $d_{xy}^2 d_{xz}^2 d_{yz}^2$.
[b] Triplet state.
[c] With the dioxygen ligand projecting along a Fe—N axis.
[d] Singlet state, computed from the SCF orbitals corresponding to the triplet state.

TABLE II
Coordinates (in Å) of the atoms for the distal NH_3 and the $Fe—O_2$ unit in the $FePO_2(NH_3)_2$ molecule.

	x	y	z
Fe	0	0	0
O_1	0	0	+1.75
O_2	+0.609	+0.609	+2.642
N_{distal}	-1.600	-0.500	+3.600
H_1	-0.952	-0.297	+2.850
H_2	-1.919	-1.450	+3.460
H_3	-2.403	+0.099	+3.460

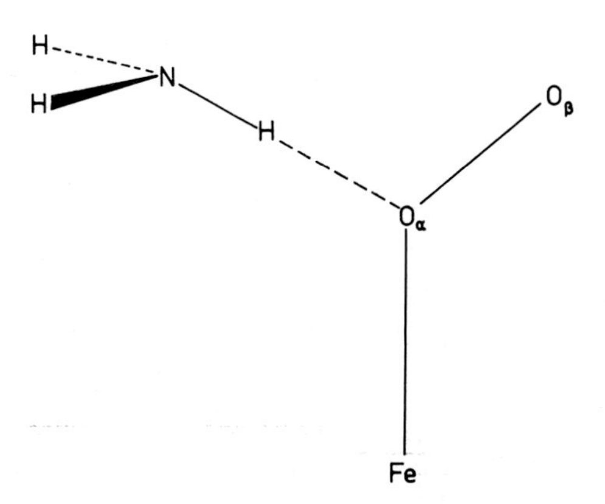

Fig. 2. The relative orientation of the oxygen and distal ammonia molecules in $FePO_2(NH_3)_2$.

angle of 136° corresponding to the experimental values for the dioxygen complex of the picket-fence porphyrin (Collman et al., 1975b). For the perpendicular structure of the $Fe—O_2$ unit we have kept the same Fe-O and O—O bond lengths used for the bent structure with the two Fe—O bond lengths equal. We have used the experimental geometries of the ammonia (Benedict et al., 1957) and imidazole (Martinez-Carrera, 1966) ligands with a Fe—N bond length of 2.12 Å for the ammonia ligand and 2.07 Å for the imidazole ligand (this is the experimental Fe(II)-N(imidazole) distance in the oxygen complex of the picket fence porphyrin (Collman, 1975 the bond length increment of 0.05 Å for the $Fe—N(NH_3)$ bond corresponds to an estimated $N(sp^3)-N(sp^2)$ difference by Little et al. (1974)). For the carbonyl complex, the C—O bond length of 1.16 Å and the Fe—C bond length of 1.77 Å are based on the values in Ru(TPP)(CO)(EtOH)(Bonnett et

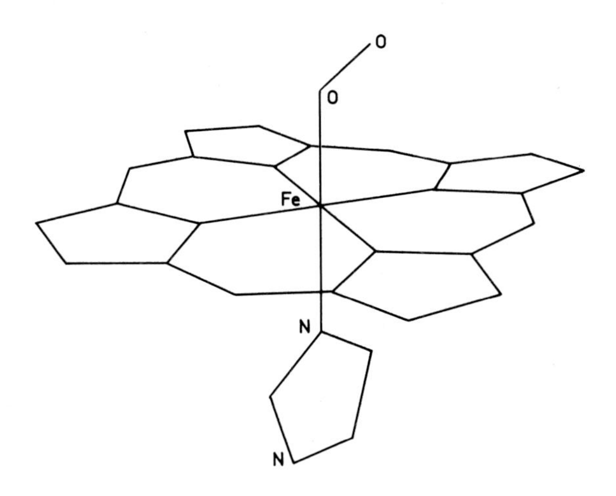

Fig. 3. The relative orientation of the imidazole and dioxygen ligands in $FePO_2Im$.

al., 1973). For the system $FePO_2(NH_3)_2$ with one additional NH_3 ligand to represent the distal imidazole of hemoglobin the nitrogen atom of the distal NH_3 ligand was positioned according to the X-ray data of Watson (1969) reported in Antonini et al. (1971) and the corresponding coordinates are given in Table II (the corresponding $N_{distal}-O_\alpha$ distance is equal to 2.50 Å, with the relative orientation of the oxygen and distal ammonia molecules shown in Figure 2). Our choice of axis is shown in Figure 1 with the porphyrin plane being the xOy plane, the four nitrogen atoms of the pyrrole rings being along the axis Ox and Oy. The Fe—O axis is the z axis and the O—O axis projects along the bissector OX of xOy (in what follows we sometimes refer for convenience to the axis OX and OY and to the linear combinations $3d_{Xz} = (1/\sqrt{2})(3d_{xz}+3d_{yz})$ and $3d_{Yz} = (1/\sqrt{2})(3d_{yz}-3d_{xz})$, since XOz will be the plane of symmetry for most of the systems studied) (for $FePO_2$ we have also carried one additional calculation with the dioxygen ligand projecting along Ox). For the $FePO_2Im$ molecule, the imidazole ligand is in the plane YOz (the relative orientation of the imidazole and dioxygen ligands is shown in Figure 3 and corresponds to one of the two situations found experimentally for the dioxygen complex of the picket fence porphyrin (Collman et al., 1975b)). The systems studied are assumed to be low-spin complexes (S=O) of iron(II) unless otherwise stated.

3. DISCUSSION

3.1. *The porphin dianion and the low-spin Fe(II) porphyrin*

We discuss briefly our results for the porphin dianion and the low-spin (S=O) planar iron(II) porphyrin (the ground state of the iron(II) porphyrin in the absence of axial ligands corresponds probably to an

TABLE III
The orbital energies (in AU) of the porphin dianion and of the low-spin (S=O) iron(II) porphyrin.

Porphin dianion		Fe(II)porphyrin	
$3a_{2u}(\pi)$ [a]	+0.014	$5a_{2u}(\pi)$	-0.316
$1a_{1u}(\pi)$	-0.045	$1a_{1u}(\pi)$	-0.317
$9b_{1g}(n)$	-0.106	$4e_g\ (\pi)$	-0.413
$3e_g\ (\pi)$	-0.115	$2b_{2u}(\pi)$	-0.424
$2b_{2u}(\pi)$	-0.121	$4a_{2u}(\pi)$	-0.442
$17e_u(n)$	-0.143	$3e_g\ (\pi)$	-0.469
$2a_{2u}(\pi)$	-0.167	$9b_{2g}(d_{xy})$	-0.518
$11a_{1g}(n)$	-0.187	$19e_u\ (n)$	-0.525
$2e_g\ (\pi)$	-0.201	$1b_{1u}(\pi)$	-0.536
$1b_{1u}(\pi)$	-0.258	$9b_{1g}(n)$	-0.537
$8b_{2g}(\sigma)$	-0.288	$2e_g\ (d_{xz},d_{yz})$	-0.554
$16e_u(\sigma)$	-0.310	$14a_{1g}(n)$	-0.588
$6a_{2g}(\sigma)$	-0.322	$8b_{2g}(\sigma)$	-0.593
$1b_{2u}(\pi)$	-0.326	$6a_{2g}(\sigma)$	-0.594
$15e_u(\sigma)$	-0.340	$18e_u\ (\sigma)$	-0.594
$8b_{1g}(\sigma)$	-0.348	$1b_{1u}(\pi)$	-0.613
$1e_g\ (\pi)$	-0.356	$17e_u\ (\sigma)$	-0.614
$1a_{2u}(\pi)$	-0.374	$8b_{1g}(\sigma)$	-0.620
$7b_{2g}(\sigma)$	-0.376	$7b_{2g}(\sigma)$	-0.637
$5a_{2g}(\sigma)$	-0.377	$13a_{1g}(\sigma)$	-0.650
		$1e_g\ (\pi)$	-0.653

[a] The notations π, n and σ denote respectively the porphin π MO's, the σ MO's which are made mostly of the nitrogen lone pairs and the other σ MO's.

TABLE IV
Gross orbital and atomic populations for the porphin dianion and the low-spin iron(II) porphyrin (the number in parenthesis refer to the porphin dianion, the numbering of the atoms is shown in Figure 1)

	Fe	N_1	C_2	C_3	C_4
s	6.17	3.48 (3.57)	2.96 (2.96)	3.06 (3.05)	3.03 (3.03)
p_x	4.13	1.49 (1.56)	0.97 (0.94)	1.08 (1.07)	1.09 (1.07)
p_y	4.13	1.06 (0.97)	0.98 (0.99)	1.06 (1.05)	1.09 (1.07)
p_z	4.04	1.37 (1.12)	1.02 (1.09)	1.03 (1.08)	1.02 (1.05)
$d_{x^2-y^2}$	0.35				
d_{xy}	1.91				
d_{xz}, d_{yz}	1.99				
d_{Xz}	1.99				
d_{Yz}	1.99				
d_{z^2}	0.05				
total	24.75	7.41 (7.23)	5.93 (5.97)	6.22 (6.25)	6.23 (6.23)

intermediate-spin S=1 or to a high-spin S=2 state (Collman et al., 1975a and references therein, Zerner et al., 1966)). We have reported the orbital energies for the high lying MO's in Table III and the results of the population analysis in Table IV. A pictorial representation of the high lying MO's of the iron(II) porphyrin is given in Figure 4.

The highest occupied orbital for the porphin dianion is the π a_{2u} orbital followed closely by the π a_{1u} orbital. The sequence of high-lying MO's is very similar to the one reported by Maggiora (1973) on the basis of a CNDO calculation. In the iron(II) porphyrin the π a_{2u} and a_{1u} orbitals become nearly degenerate and are separated from the next occupied orbital by a gap of about 0.1 AU (about 3 eV) (a similar gap has been emphasized by Maggiora (1973) for the magnesium porphines). No crossing of the $a_{2u}(\pi)$ and $a_{1u}(\pi)$ levels occurs upon going from the porphin dianion to the iron(II) porphyrin whereas such a crossing occured for the magnesium porphin as a consequence of a preferential stabilization of the $a_{2u}(\pi)$ orbital through its interaction with the 3p orbital of magnesium (Maggiora, 1973). The sequence of the high lying MO's are rather similar for the porphin dianion and the iron(II) porphyrin, the main differences being:

— a net stabilization of the orbitals of the iron(II) porphyrin due to the charge cancellation;

— a slight preferential stabilization of the $a_{2u}(\pi)$ orbital in the iron(II) porphyrin as a consequence of mixing with the iron $4p_z$ orbital;

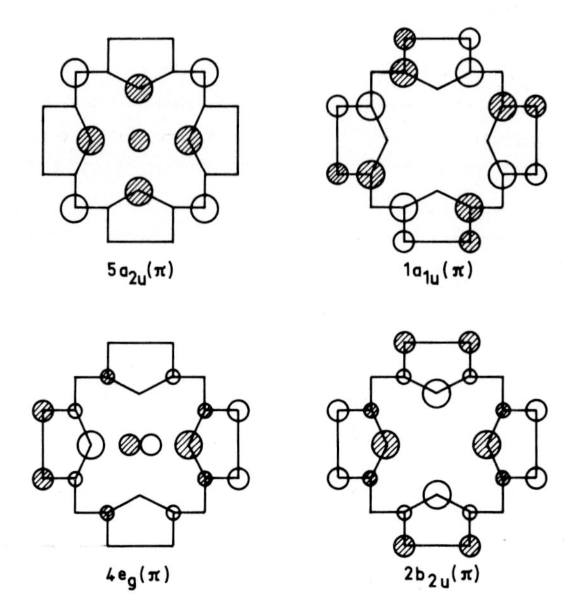

Fig. 4a.

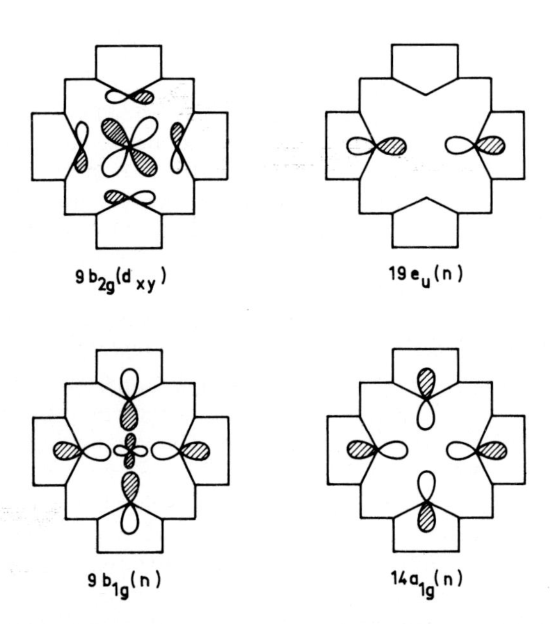

Fig. 4b. The high-lying MO's of iron(II) porphyrin. The size of the circles is approximately proportional to the basis functions coefficients. The view is looking down the positive z axis at the upper lobe of the p_π and d_π basis functions for the π MO's. The white circles indicate a positive value (for the upper lobe in the case of π functions) while the dark circles indicate a negative value.

— a preferential stabilization of the MO's b_{1g}, e_u and a_{1g} which are essentially built from the lone pairs of the pyrrole nitrogen, this stabilization being stronger for the b_{1g}(n) MO as a consequence of the bonding interaction with the iron $3d_{x^2-y^2}$ orbital (Figure 4).

The results of Table III have some obvious relationship to the analysis of the electronic spectra and photoelectron spectra of porphins and metalloporphyrins. The basic features of the visible and ultraviolet absorption spectrum of the closed-shell metal porphyrins have been rationalized with a six-orbital model by Gouterman and coworkers (Gouterman, 1959, 1961; Weiss, 1965; Caughey, 1965), one key point in this analysis being the near-degeneracy between the pairs (a_{1u}, a_{2u}) and (a_{2u}, b_{2u}) of the π MO's (however we postpone a detailed discussion of the electronic and photoelectron spectra of the metalloporphyrins since relaxation effects are expected to be relatively important for some of the excited or ionized states (Veillard et al., 1976)). The near-degeneracy of the highest filled orbitals a_{1u}(π) and a_{2u}(π) accounts for the striking differences observed in the EPR spectrum of porphyrin π cation radicals, with the possibility of either a $^2A_{2u}$ or a $^2A_{1u}$ ground state (Dolphin, 1974). The similarity between the sequences of MO's for the porphin dianion and the Fe(II) porphyrin accounts for the experimental observation that free base porphine and all the metallic derivatives have surprisingly similar photoelectron spectra (Khandelwal et al., 1975). A study of the redox potentials of metalloporphyrins led to the conclusion that, upon the introduction of a transition metal into a porphyrin, only the σ shell of the porphyrin (namely the nitrogen lone pairs) is appreciably shifted in energy (Fuhrhop, 1973), in essential agreement with the results of Table III. This is tantamount to the conclusion of Gouterman (1959) that the interaction between the metal d(π) and the porphyrin ring π orbitals is minimal. Since the d(π) orbitals of iron are fully populated in the low-spin iron(II) porphyrin, the charge transfer between the metal and the porphyrin π orbitals can only be either a dπ-pπ* back donation from the metal to the porphyrin ring or a donation from the pπ orbitals of the porphyrin ring to the $4p_z$ orbital of metal. The populations of the metal $3d_{xz}$ ($3d_{yz}$) and $4p_z$ orbitals are respectively 1.991 and 0.041 (Table IV), showing no appreciable π charge transfer between the metal and the porphyrin ligand. On the basis of the extended Hückel calculations of Zerner et al. (1966), Spiro et al. (1974) consider the low-spin iron (II) as a net π donor by about 0.1 electron. However we find that any comparison between our results and those of Zerner et al. is hampered by the fact that the only low-spin iron(II) systems considered by these authors are five- or six-coordinate with one or two molecules of water as additional ligands. Our conclusion that no appreciable π charge transfer occurs between the metal and the porphyrin ligand is similar to the one which we have reached previously for the synthetic oxygen carrier Co(acacen) (Dedieu et al., 1976). In both cases the metal atom has a formal oxydation number of two, a situation which is certainly not favourable for a back-bonding interaction dπ(M) → π*(ligand).

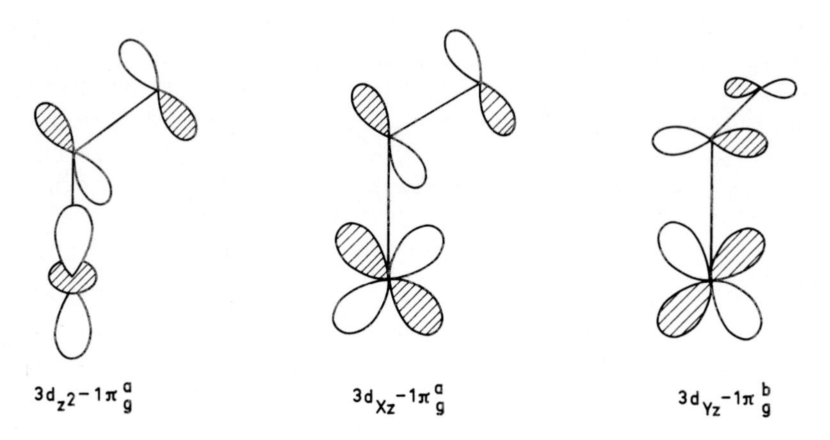

Fig. 5. The metal-dioxygen interactions in the bent structure.

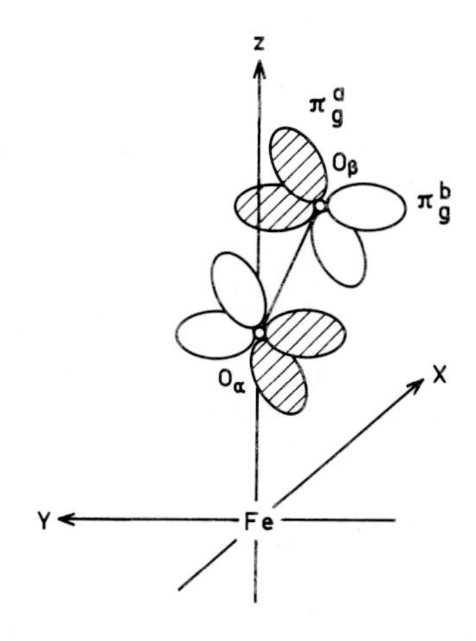

Fig. 6. The orbitals π_g^a and π_g^b of the dioxygen ligand in the dioxygen complex.

3.2. *The dioxygen complex*

We have previously discussed (Dedieu et al., 1976) the bonding in a metal-dioxygen complex in terms of the interactions between the orbitals of the metal and the orbitals of the dioxygen ligand. The bonding for the bent structure may be understood on the basis of the interactions of Figure 5, namely the ones between the $3d_{z^2}$, $3d_{xz}$ and $3d_{yz}$ (or $3d_{xz}$ and $3d_{yz}$) orbitals of iron and the $1\pi_g$ and $1\pi_g$ degenerate antibonding orbitals of dioxygen (we refer the reader to the work of Hoffman et al. (1974) on the coordination of the nitrosyl ligand for a detailed discussion of the interactions between a metal atom and a diatomic ligand in a linear or bent structure). These π_g and $\bar{\pi}_g$ orbitals which are degenerate for the molecule O_2 are no longer equivalent for the complex with a bent structure. We distinguish them through the use of the labels π_g^a and π_g^b (π_g^a is made of $2p_X = (1/\sqrt{2})(2p_x + 2p_y)$ and $2p_z$ orbitals of the oxygen atoms and is symmetrical with respect to the XOz plane which is a plane of symmetry for most of the systems of Table I, π_g^b is made of $2p_Y = (1/\sqrt{2})(2p_y - 2p_x)$ orbitals and is antisymmetrical with respect to the XOz plane, Figure 6). This results in a number of possible electronic configurations for the low-spin (S=O) ground state of the dioxygen complex, either of the closed-shell type as $d^6(\pi_g^a)^2(\pi_g^b)^0$ or $d^6(\pi_g^b)^2(\pi_g^a)^0$ or the open-shell type as $d^5(\pi_g^a)^2(\pi_g^b)^1$ or $d^5(\pi_g^b)^2(\pi_g^a)^1$ (d^6 stands for $d_{xy}^2\ d_{xz}^2\ d_{yz}^2$ and d^5 for one of the three possible configurations with one electron less) (we assume a low-spin ground-state configuration on the basis of the experimental diamagnetism of hemoglobin (Pauling et al., 1936) and of the dioxygen complex of the picket fence porphyrin (Collman et al., 1975b)). The first two configurations correspond to the Pauling model with a formal oxidation number of two for iron, hence to a Fe(II)-O_2 formal configuration while the latest correspond to the biradical model of Weiss with a Fe(III)-O_2^- formulation if one assumes antiparallel spins for the unpaired electrons.

According to the results of Table I, the lowest energy for $FePO_2NH_3$ corresponds to a triplet $^3A'$ with the *formal* configuration

$d_{xy}^2 d_{Xz}^2 d_{Yz}^1 (\pi_g^a)^2 (\pi_g^b)^1$. This configuration corresponds formally to the model of Weiss with the four-electron three-center π bond emphasized by Goddard (the choice between the Fe(II)-O_2 or the Fe(III)-O_2^- formulation is probably a semantic problem, since the formal charge of iron as given by the population analysis is rather close to one). Three other open-shell states are found at slightly higher energies, namely the singlet $b\,^1A'$ corresponding to the triplet $^3A'$ and another pair of triplet $^3A''$ and singlet $^1A''$ states corrsponding to the formal configuration $d_{xy}^1 d_{Xz}^2 d_{Yz}^2$ $(\pi_g^a)^2(\pi_g^b)^1$ (another pair of singlet and triplet states corresponding to the configuration $d_{xy}^2 d_{Xz}^1 d_{Yz}^2 (\pi_g^a)^2 (\pi_g^b)^1$ should have comparable energies). Then, at o.o24 AU (o.65 eV) above the lowest triplet, one finds the singlet closed-shell state $a\,^1A'$ corresponding to the model of Pauling with the configuration $d_{xy}^2 d_{Xz}^2 d_{Yz}^2 (\pi_g^a)^2$. However the correlation error bias these SCF calculations in favour of the open-shell configurations which have one electron pair less than the closed-shell configuration (roughly speaking, the difference in correlation energy between the closed-shell and open-shell configurations would be of the order of a pair correlation energy). Thus it is expected that the destabilization of the closed-shell configuration relatively to the open-shell configurations due to the correlation error would be of the order of 0.02-0.03 AU. This would be of the order of magnitude of the computed separation of 0.024 AU between the lowest triplet $^3A'$ and the singlet $a\,^1A'$ corre-

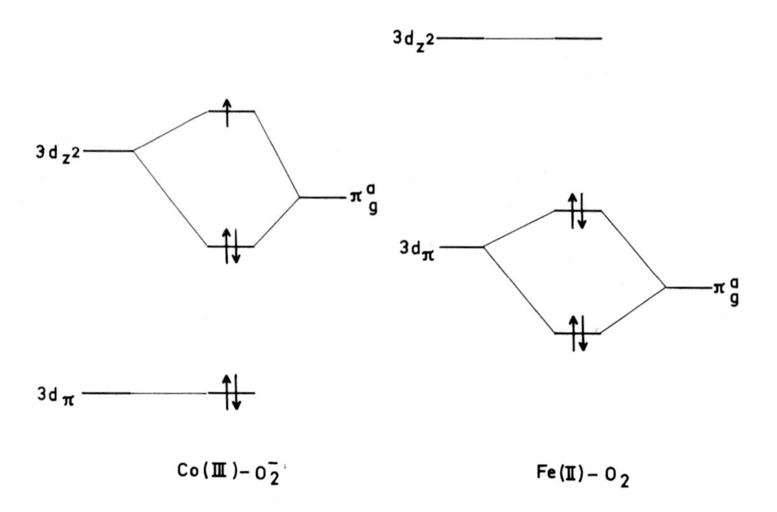

Fig. 7. The relative stabilities of the π_g^a, $3d_z2$ and $3d_\pi$ orbitals in the dioxygen complexes of Co(II)acacen and Fe(II) porphyrin and the corresponding interaction diagrams for the bent structures ($3d_\pi$ in the Fe-O_2 case stands for $3d_{xz}$).

sponding to the Pauling model with the closed-shell configuration. Thus our results are not at variance with the experimental diamagnetism of oxyhemoglobin and of the dioxygen complex of the picket fence porphyrin. In fact we may predict that the triplet state $^3A'$ should be very close in energy to the ground-state singlet, a prediction which seems corroborated by the existence of a paramagnetic dioxygen complex of the picket fence porphyrin (Collman et al., 1975b) and by the analysis of the Fe Kβ fluorescence emission spectrum of oxymyoglobin (Koster, 1975).

If we assume a low-spin ground state configuration, then the corresponding ground state should be the state $a\,^1A'$ with a closed-shell configuration $d^6(\pi_g^a)^2$ (the open-shell singlet states $b\,^1A'$ and $^1A''$ should be at higher energies than the closed-shell singlet $a\,^1A'$, since they are expected to have higher energies than the corresponding triplet states). The greater stability of the configuration $d^6(\pi_g^a)^2$ compared to the other closed-shell configuration $d^6(\pi_g^b)^2$ follows from the fact that the π_g^a orbital of dioxygen is stabilized relatively to the π_g^b orbital by the bonding interaction with the $3d_{xz}$ orbital of iron shown in Figure 5. With that respect, the situation is rather different from the one found in the dioxygen complex of Co(acacen) (Dedieu et al., 1976) where the stabilization of the π_g^a orbital was a consequence of a strong interaction with the metal $3d_{z^2}$ orbital, whereas no appreciable interaction was found between the $3d_\pi$ and π_g^a orbitals. This difference in bonding is a consequence of the essential difference in the ground-state configuration, which was of the type Co(III)-O_2^- for the dioxygen complex of Co(acacen) but which is of the type Fe(II)-O_2 for the dioxygen complex

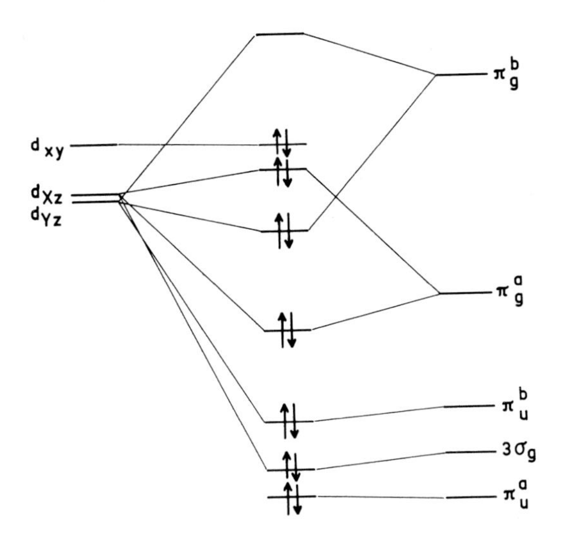

Fig. 8. A simplified interaction diagram between the metal 3d orbitals and the dioxygen π orbitals in the bent structure (the porphin π and σ levels are not represented).
Note. This interaction diagram is based on the sequence of energy levels of Table V and on the corresponding bonding and antibonding character of the orbitals. It turns out that a requisite to the construction of this interaction diagram is the assumption that the metal 3d orbitals on the left hand side and the pairs of dioxygen orbitals (π_g^a, π_g^b) and (π_u^a, π_u^b) on the right hand side are not degenerate (although they are degenerate for the free atom or the free ligand). This is merely a consequence from the fact that the SCF orbital energies of Table V do account for the bonding and antibonding interactions (as in simple Hückel theory) but also for the two-electron interactions. Since an interaction diagram represents just a convenient way of analyzing a wavefunction, we feel that this requirement is an acceptable one.

of iron(II) porphyrin. For a Co(III)-O_2 *formal* configuration, the π orbitals of dioxygen are relatively destabilized by the formal negative charge whereas the 3d orbitals of cobalt are relatively stabilized by the increased formal positive charge. Since the interaction between two orbitals depends among other things on the corresponding energy gap denominator (namely the magnitude of the interaction is inversely proportional to the difference of the orbital energies) (Salem, 1968), one may see from Figure 7 that the π_g^a-$3d_{z^2}$ interaction will be favoured in the Co(III)-O_2^- configuration whereas the π_g^a - $3d_{xz}$ will be favoured for the Fe(II)-O_2 configuration. A qualitative molecular orbital diagram has been proposed previously for the bent complex of dioxygen with cobalt(II) and iron(II) porphyrins (Wayland, 1974), and it is said that the bent diamagnetic complex contains essentially singlet O_2 binding

TABLE V
The energy levels for the high-lying molecular orbitals of $FePO_2NH_3$ in the bent structure

Nature and symmetry property[a]		Orbital energy (in AU)	Bonding (B) or antibonding (A) combination		Population analysis Iron	Dioxygen ligand
$\pi(a_{2u})$	S	-0.300				
$\pi(a_{1u})$	A	-0.309				
$\pi(e_g)$	A	-0.398				
$\pi(e_g)$	S	-0.400				
$\pi(b_{2u})$	A	-0.415				
$\pi(a_{2u})$	S	-0.430				
$\pi(e_g)$	S	-0.462				
$\pi(e_g)$	A	-0.462				
d_{xy}	S	-0.485				
d_{Xz}	S	-0.490	$\pi_g^a(O_2)$	A	1.06	0.55
d_{Yz}	A	-0.494	$\pi_u^b(O_2)$	A	1.48	0.14
$n(e_u)$	S	-0.508				
$n(e_u)$	A	-0.510				
$n(b_{1g})$	A	-0.520				
$\pi(b_{1u})$	S	-0.528				
$n(NH_3)$	S	-0.540				
$1\pi_g^a(O_2)$	S	-0.547	d_{Xz}	B	0.46	1.15

$1\pi_u^b(O_2)$	A	-0.711	d_{Yz}	B	0.12	1.12

$3\sigma_g(O_2)$		-0.752	d_{Xz}	B	0.05	1.45

$1\pi_u^a(O_2)$	S	-0.816				

[a] The irreducible representation between parenthesis refers to the parent MO of the iron(II) porphyrin of Table III. S and A notations stand for symmetric and antisymmetric with respect to the XOz plane. The dotted lines correspond to a number of σ MO's.

TABLE VI
Gross orbital and atomic populations for the iron and oxygen atoms of $FePO_2(NH_3)$ (upper line for the bent and perpendicular structures, the numbers in parenthesis refer to the perpendicular structure) and $FePO_2Im$ (lower line)

	Fe	O_α	O_β
s	6.16 (6.17)	3.80 (3.86)	3.86 (3.86)
	6.15	3.80	3.87
p_x, p_y	4.11 (4.13)	1.36 (1.57)	1.27 (1.57)
	4.11	1.36	1.27
p_z	4.07 (4.09)	1.53 (1.11)	1.52 (1.11)
	4.07	1.53	1.53
$d_{x^2-y^2}$	0.33 (o.38)		
	0.33		
d_{xy}	1.91 (1.91)		
	1.91		
d_{xz}, d_{yz}	1.93 (1.80)		
	1.93		
d_{Xz}	1.95 (1.62)		
	1.96		
d_{Yz}	1.91 (1.98)		
	1.90		
d_{z^2}	0.23 (0.31)		
	0.22		
Total	24.78 (24.73)	8.05 (8.11)	7.93 (8.11)
	24.77	8.04	7.93

Fe(II) by the empty d_{z^2}. Although this description would probably hold for the complex of dioxygen with a cobalt(II) porphyrin (where the bonding should be rather similar to the one in the complex of Co(acacen) with a Co(III)-O_2^- formal ground state (Hoffmann, 1970)), it does not correspond to the results of our calculation for the complex of dioxygen with iron(II) porphyrin. The energy levels corresponding to the high-lying MO's of $FePO_2NH_3$ for the bent structure are reported in Table V and a simplified interaction diagram between the metal 3d orbitals and the dioxygen π orbitals is shown in Figure 8.

We have reported in Table VI the gross orbital and atomic populations (iron and oxygen atoms only) for the bent structure of the system $FePO_2(NH_3)$. The electron distribution at the iron atom departs slightly from the symmetric low-spin ferrous configuration $(d_{xy})^2(d_{xz})^2(d_{yz})^2$. The population 0.23 of the $3d_{z^2}$ orbital is rather low (much lower than the corresponding population of 0.5 — 0.7 found for the dioxygen complex of Co(acacen) with various fifth ligands, Dedieu et al., 1976) and this reflects the absence of any appreciable $3d_{z^2}$-$1\pi_g^a$ interaction. Some ap-

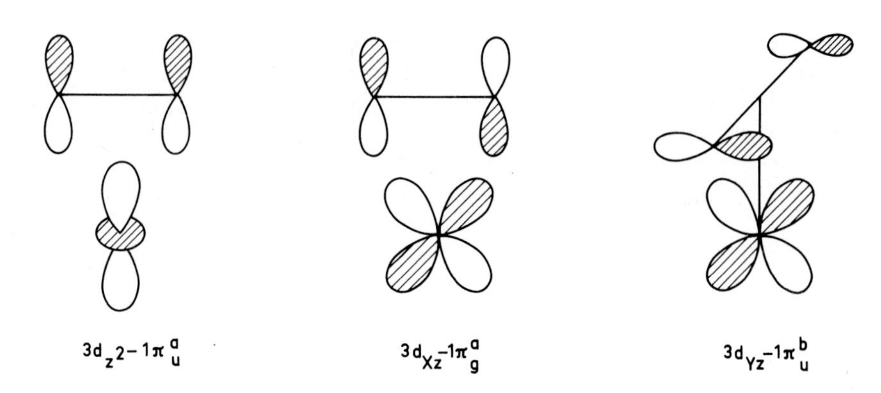

Fig. 9. The metal-dioxygen interactions in the perpendicular structure.

preciable $d\pi$-$p\pi$ backbonding between the $3d_{yz}$ orbital and the π_g^b orbital as in Figure 5 is also evident from the population of 1.91 for the $3d_{yz}$ orbital (again this contrasts with the situation in the dioxygen complex of Co(acacen) with a population close to 2.00 for the d_π orbitals, a consequence of the Co(III)-O_2^- configuration which is rather unfavourable for backbonding). A salient feature of Table VI is the quasi-neutrality of the dioxygen ligand (contrary to the conclusion of an extended Hückel calculation where the dioxygen ligand carries a formal negative charge of -0.52, Loew et al, 1975), thus supporting the description of the iron-dioxygen unit as Fe(II)-O_2. These results are in agreement with two experimental observations: (i) the observed ν_{O_2} at 1385 cm^{-1} in the dioxygen complex of the picket-fence porphyrins is only 100 cm^{-1} below that of singlet oxygen, supporting a description of the nature of dioxygen as coordinated singlet oxygen (Collman et al., 1975); (ii) the Fe—O bond length of 1.75 Å in the same dioxygen complex is appreciabl shorter than the Co—O bond length of 1.86 Å in the dioxygen complexes of Co(II), a difference which has been considered as a possible indication of backbonding in the Fe(II) complexes (Collman et al., 1975).

The ground state configuration for the perpendicular structure is the configuration $d^6 (\pi_g^b)^2 (\pi_g^a)^0$. The same change in the ground-state configuration from the bent to the perpendicular structure has been found for the dioxygen complex of Co(acacen) (Dedieu et al., 1976). Bonding in the perpendicular structure may be described in terms of the interactions $3d_{z^2}-1\pi_u^a$, $3d_{xz}-1\pi_g^a$ and $3d_{yz}-1\pi_u^b$ of Figure 9. Examination of the wavefunction (Table VII and Figure 10) indicates that the largest metal-dioxygen interaction for the perpendicular structure is the $3d_{xz}$ - $1\pi_g^a$ interaction (which is the most favourable one in terms of overlap). This results in a strong stabilization of the $3d_{xz}$ orbital and a strong destabilization of the $1\pi_g^a$ orbital with respect to the $1\pi_g^b$ orbital as shown in Figure 10, thus leading to a ground-state configuration $(\pi_g^b)^2 (\pi_g^a)^0$. One marked difference in the electronic structure of the

TABLE VII
The energy levels for the high-lying molecular orbitals of $FePO_2NH_3$ in the perpendicular structure

Nature and symmetry property[a]		Orbital energy (in AU)	Bonding (B) or antibonding (A) combination	Population analysis Iron	Dioxygen ligand
$\pi(a_{2u})$	S	-0.301			
$\pi(a_{1u})$	A	-0.312			
$\pi(e_g)$	A	-0.398			
$\pi(e_g)$	S	-0.408			
$\pi(b_{2u})$	A	-0.411			
$\pi(a_{2u})$	S	-0.429			
$\pi_g^b(O_2)$	A	-0.459			
$\pi(e_g)$	S	-0.464			
$\pi(e_g)$	A	-0.464			
n	A	-0.512			
n	S	-0.518			
$\pi(b_{1u})$	S	-0.530			
d_{Xz}	S	-0.542	$\pi_g^a(O_2)B$	1.24	0.42
d_{xy}	S	-0.542			
$n(b_{1g})$	A	-0.549			
n	S	-0.555			
d_{Yz}	A	-0.573	$\pi_u^b(O_2)A$	1.57	0.20

$\pi_u^a(O_2)$	S	-0.683			

$\pi_u^b(O_2)$	A	-0.792	d_{Yz} B	0.19	1.12

[a] The irreducible representation between parenthesis refers to the parent MO of the iron(II) porphyrin of Table III. S and A notations stand for symmetric or antisymmetric with respect to the *XOz* plane. The dotted lines correspond to a number of σ MO's of the porphyrin core.

bent and perpendicular geometries, as judged from Tables V and VII and from Figures 8 and 10, corresponds to the relative location of the metal 3d and dioxygen π_g^a orbitals. In the bent structure, the π_g^a orbital is lower than the $3d_{Xz}$ orbital and appears *stabilized* through its interaction

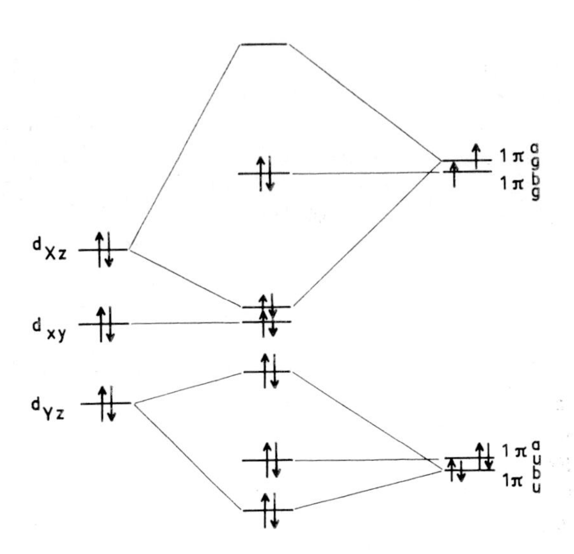

Fig. 10. A simplified interaction diagram between the metal 3d orbitals and the dioxygen π orbitals in the perpendicular structure (the porphin π and σ levels are not represented)

with $3d_{xz}$. In the perpendicular structure, the π_g^a orbital is higher than the $3d_{xz}$ orbital and is destabilized through its interaction with $3d_{xz}$ This difference reflects a situation rather comparable to the one of Figure 7, with the dioxygen ligand more negative in the perpendicular structure (see Table VI) as a consequence of an increased backbonding and the π orbitals of dioxygen destabilized relatively to the 3d orbital Another interaction, although somewhat smaller, is the four-electron destabilizing interaction $3d_{yz}$. $-1\pi_u^b$ (cf. below). The importance of the $3d_{xz}-1\pi_g^a$ interaction has already been emphasized in the extended Hückel calculation of Zerner et al. (1966), where the metal $3d_{xz}$ orbital has so mixed with the O_2 $1\pi_g$ MO directed into the porphin plane that no MO of the complex can be accurately described as a $3d_{xz}$ metal orbital. As a consequence this calculation yielded the same ground state configuration $(\pi_g^b)^2(\pi_g^a)^0$.

From the results of Table I, the perpendicular structure is found less stable than the bent one by 55 kcal mole^{-1} for $FePO_2NH_3$ and 63 kcal mole^{-1} for $FePO_2$. We consider these values as indicative since they woul change slightly with further theoretical refinements at the SCF level such as the basis set extension and the geometry optimization. However this energy difference between the bent and perpendicular structures is large enough for the perpendicular structure to be ruled out for the dioxygen complexes of iron(II) porphyrins. We have previously reached a similar conclusion for the perpendicular structure of the dioxygen complex of Co(acacen) and we have ascribed the destabilization of the perpendicular structure to the four-electron destabilizing interaction

(Salem, 1961) $3d_{yz}-1\pi_u^b$. There are two reports in the literature of a perpendicular structure for a metalloporphyrin. The ESR spectrum of single crystal oxycobaltmyoglobin has been interpreted on the basis of a sideways perpendicular structure (Chien et al., 1972), however this conclusion has been criticized (Brown et al., 1975). A perpendicular structure has been tentatively assigned to a dioxygen complex of a manganese(II) porphyrin (Weschler et al., 1975). However this complex would be either a high-spin (S = 5/2) or an intermediate spin (S = 3/2) system and the occupations of the 3d orbitals of the metal should be rather different from the corresponding ones for the low-spin iron(II) porphyrin, hence some significant differences may be expected in the nature and the degree of the ligand-metal interactions (for instance the four-electron destabilizing interaction $3d_{yz}-1\pi_u^b$ may become a three-electron interaction). For this reason any conclusion drawn for a Mn(II) dioxygen complex from our calculation for the Fe(II) dioxygen complex would appear as premature.

3.3. *The interaction between the dioxygen ligand and the distal imidazole molecule in the natural oxygen carriers*

The system $FePO_2(NH_3)_2$ represents a model system to investigate the occurence of an hydrogen bond between the dioxygen ligand and the distal imidazole molecule in the molecules of oxymyoglobin and oxyhemoglobin. Since a calculation with two imidazole molecules (one proximal and one distal) would have slightly exceeded our present computational possibilities, we resort to a model system where the imidazole groups are represented by two ammonia molecules. With the coordinates given in Table II, the nitrogen-oxygen distance of 2.50 Å fulfills probably the geometric requirements for an hydrogen bond. The computed energy of -2497.117 AU for this system should be compared to the sum -2497.156 AU of the energies of $FePO_2(NH_3)_{prox.}$ (namely -2441.208 AU, Table I) and of NH_3 calculated with the same geometry and the same basis set (-55.948 AU from an independant calculation). Thus the interaction between the $FePO_2NH_3$ complex and the distal ammonia molecule is found *repulsive* and the corresponding destabilization energy amounts to 24 kcal mole^{-1}. Our model calculation relies essentially on two approximations (we do not consider the fact of using a fixed geometry as an approximation, since we want to investigate the possibility of hydrogen bonding in a given system, namely oxymyoglobin, with the corresponding experimental geometry). The first approximation is the replacement of the distal imidazole molecule by an ammonia molecule with probably a different hydrogen bonding ability. We do not know how the result of the calculation is affected by this approximation, however, it is dubious that the error could be larger than 24 kcal mole^{-1}, in other words that it might change the interaction from repulsive to attractive. The second approximation is the use of a minimal basis set at the SCF level, which is known to overestimate the energy of an hydrogen bond (hence the attractive character) because of the basis set superposition effect (namely when the constituent units are brought together, basis functions on the one can make up for the deficiencies in those on the other, thus favouring the hydrogen bonded system over the separated constituents) (Joesten et al., 1974). Thus it seems highly dubious that further theoretical refinements in this calculation could change the interaction from repulsive (by 24 kcal mole^{-1}) to attractive. Then one may probably conclude rather safely that this calculation does not support the often postulated stabilization of the coordinated dioxygen through hydrogen bonding to the distal imidazole. This conclusion is certainly a reasonable one since the dioxygen ligand in the complex is found nearly neutral and the hypothesis of an hydrogen bond was implicitly based on the idea that the dioxygen ligand assumes either a superoxide O_2^- or a charge separated $O^+ - O^-$ configuration. Collman (1975c) concluded that it is not necessary to invoke hydrogen bonding from the distal imidazole to dioxygen in the biological systems, since the picket fence porphyrin is a good model for myoglobin (in the sense that it reproduces the thermodynamic constants of the biological systems) but lacks that feature. Hydrogen bonding between the dioxygen ligand and the distal histidyl residue has been postulated in oxycobaltmyoglobin and oxycobalthemoglobin to interpret the sharpened EPR adsorption upon measurements in D_2O (Yonetani et al., 1974), however the situation in the dioxygen complexes of cobalt(II) porphyrins might be a completely different one since their Co—O_2 unit has been described as Co(III)-O_2^- (Hoffmann et al., 1970).

3.4. *The rotation about the Fe—O_2 bond*

We have reported in Table I the computed energy for the two different conformations *6* and *7* of $FePO_2$ with respect to rotation about the Fe—O bond (the first one may be labelled as staggered and the other as eclipsed). The staggered one which we have considered so far corresponds to the energy minimum, with the O—O bond projecting along the bissector of the angle NFeN. The eclipsed one corresponds to an energy maximum

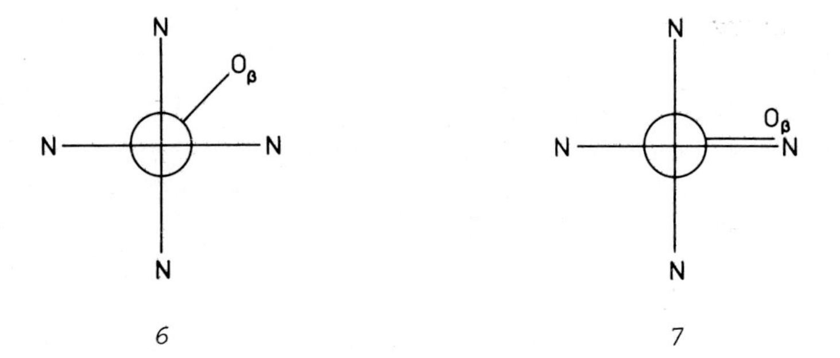

with the O—O bond projecting along a Fe-N axis. A value of 5.6 kcal mole^{-1} is computed for the barrier to the rotation about the Fe—O axis, of fourfold symmetry. A rotation barrier of less than 5 kcal mole^{-1} has been reported from an extended Hückel calculation (Loew et al., 1975). This relatively low barrier to rotation is in good agreement with the fourway statistical disorder in the structure of the dioxygen complex of the picket fence porphyrin, with the terminal oxygen atom occupying any of the four possible positions such as the one in *6* (Collman et al., 1974). However the same result might be achieved also by inversion at the bound oxygen O_1 rather than by rotation about the Fe-O bond, with a linear Fe—O—O unit as the transition state. We find (Table I) a value of 13.8 kcal mole^{-1} for the corresponding barrier, the transition state being represented by the singlet $^1A_1$ of electronic configuration $(\pi_g^a)^2$ (the ground state for the linear structure should be a triplet $^3A_2$ with the configuration $(\pi_g^a)^1(\pi_g^b)^1$ since π_g^a and π_g^b are degenerate).

3.5. *The imidazole ligand*

We have carried out only one calculation with an imidazole molecule as the axial ligand, with a bent structure, and we have assumed the same ground state $d^6(\pi_g^a)^2(\pi_g^b)^0$ which we have found when the axial ligand is an ammonia molecule. The purpose of this calculation was twofold: (i) to ascertain the effect of having replaced the imidazole ligand of the biological oxygen carriers by an ammonia molecule which lacks the π electron system of imidazole; (ii) to ascertain the role of the imidazole ligand in both the natural and synthetic oxygen carriers, either as a π electron donor or as a π electron acceptor (cf. below). We have reported in Table VI the gross orbital and atomic populations for the iron and oxygen atoms of $FePO_2Im$, they may be compared directly with the corresponding results for $FePO_2(NH_3)$. The populations of Table VI for the orbitals of the iron and oxygen atoms turn to be nearly identical (within the reported accuracy) for the ammonia and imidazole ligands, with the following implications (we do not report the populations on the atoms of the porphyrin ring, which are also identical to within two figures). First, our approximation of dealing with systems where the ligand imidazole has been replaced with an ammonia molecule is probably reasonable in the light of the nearly identical populations obtained with these two differents ligands. Next, this implies that the electronic behaviour of the imidazole ligand should not be very different from the one of the ammonia ligand, which is limited to its σ-donor ability. The population of the π-electron system of imidazole in $FePO_2Im$ is found equal to 6.007 showing no significant donor or acceptor properties at the π level. We have reached previously the same conclusion for the dioxygen adduct of Co(acacen)Im, with a population of 6.004 for the π-electron system of imidazole (Dedieu et al., 1976). On these basis, the imidazole ligand seems to behave as a pure σ-donor. π-donor ability has been postulated previously for the imidazole ligand. According to Stynes et al. (1975) the π donor properties of the imidazole molecule as axial ligand play a dominant role in oxygen binding to cobalt: imidazole as a *good π donor* strengthens the $Co-O_2$ bond by *increasing* the electron density available on the cobalt atom for backbonding to dioxygen. Chang et al. (1973) suggested that, in the iron porphyrins, the large π basicity of imidazole reinforces the iron to dioxygen backbonding. However Walker (1973) emphasizes the π *acceptor* character by considering that the aromatic amines such as pyridine and imidazole stabilize the cobalt-dioxygen bond through π backbonding from cobalt to the amine, thus *decreasing* the electron density on the cobalt atom (this should strengthen the Co(III)-O_2^- bond). We believe that the π-electron donor or acceptor ability of the imidazole ligand has probably been overemphasized (a dioxygen complex of the protoheme IX exists with an axial ligand the tert-butylamine molecule, of σ-donor character only (Wagner et al., 1974)).

3.6. *The structure of the Fe—CO unit in the iron(II) carbonyl-porphyrin complex*

We have investigated the structure of the Fe—CO unit in the carbonyl complex of iron(II) porphyrin by varying the FeCO angle. The results of Table I indicate that the FeCO unit should be linear (a parabolic fit of the three energy values of Table I gave a minimum for an angle of 179.2°). The RuCO bond in Ru(CO)(TPP)(EtOH) is essentially linear with a bond angle of 175.8° (Bonnett et al., 1973, these authors have questioned the results of Cullen et al., 1972, who found a Ru—C—O bond angle of 153°). Collman et al. (1975) mention a linear FeCO unit for the carbonyl complex of the picket fence porphyrin. The closely related complex Fe($C_{22}H_{22}N_4$)(NH_2NH_2)(CO) has an essentially linear structure with a FeCO angle of 178° (Goedken et al., 1973). Thus both the theoretical and experimental results lead to the conclusion that the MCO unit should be linear in the carbonyl complexes of iron(II) or ruthenium(II) porphyrins. Wayland et al. (1974) arrived to a similar conclusion on the basis of a qualitative molecular orbital diagram. An X-ray structure of the carbonyl complex of the hemoglobin erythrocruorin showed a carbon monoxide molecule inclined to the heme plane, the angle in the FeCO unit being 145° ± 15°, however this bending has been ascribed to the protein interactions (Huber et al., 1970). Trautwein et al. (1974) estimated a value of 135° for the Fe—C—O angle in the carbonyl complex of myoglobin on the basis of the agreement between the electric field gradients found experimentally from Mossbauer spectroscopy and calculated from extended Hückel calculations, however their corresponding total energy has a minimum for a linear geometry. Finally, on the basis of the isotopic shifts found in the infrared study of the carbonyl complex of hemoglobin, Alben et al. (1968) favoured a linear unit with the oxygen atom of the carbonyl coordinated to the iron atom. We believe that this structure should be energetically unfavourable since SCF calculations for the two species NiCO and NiOC indicate that coordination at the carbon atom is favoured by more than 10 kcal $mole^{-1}$ (Veillard et al., 1976). The σ and π populations of the carbonyl ligand in the complex amount respectively to 9.838 (indicative of σ-donation from the carbony and 4.146 (indicative of π back-donation to the carbonyl), the net balance being a slight donation of 0.016 e from the carbonyl to the iron(II) porphyrin.

4. CONCLUSION

Through *ab initio* SCF calculations with a minimal basis set, we find that the singlet ground state of the dioxygen complex of iron(II) porphyrin corresponds probably to the Pauling model with a formal configuration $d^6(\pi_g^a)^2(\pi_g^b)^0$ of the Fe(II)-O_2 type. However a triplet state with a formal configuration $d_{xy}^2 d_{xz}^2 d_{yz}^1 (\pi_g^a)^2(\pi_g^b)^1$ should be rather close in energy to the singlet ground state. The bent structure proposed by Pauling for the Fe—O_2 unit is found more stable than the perpendicular structure of Griffith by about 55 kcal mole^{-1}. Thus, on energy grounds, the perpendicular structure appears as extremely unfavourable. Stabilization of the coordinated dioxygen through hydrogen bonding to the distal imidazole in the hemoproteins is not supported by a model calculation where the proximal and distal imidazole molecules of the natural oxygen carriers are represented by two ammonia molecules. Rotation about the Fe—O axis, with a computed barrier of 5.6 kcal mole^{-1}, is an easier process than inversion at the bound oxygen atom, with a linear Fe—O—O unit as the transition state and a computed barrier of 13.8 kcal mole^{-1} The imidazole ligand of the natural oxygen carriers seems to function as a pure σ-donor and we believe that its π-electron donor or acceptor ability has been overemphasized. Finally the FeCO unit in the carbonyl complex of iron(II) porphyrin is found linear. These are the first *ab initio* calculations reported for this type of molecules. Their limitations are obvious: use of a minimal basis set (except for the d functions of the metal) at the SCF level together with fixed geometries (either experimental or assumed). However, besides showing the technical feasibility of *ab initio* calculations for systems with 200 — 300 electrons, they are probably somewhat more accurate than the previous semiempirical studies.

ACKNOWLEDGEMENTS

Calculations have been carried out at the Centre de Calcul du C.N.R.S. (Strasbourg-Cronenbourg) and we are grateful to the staff of the Centre for their assistance and cooperation. This work has been supported through the A.T.P. N° 2240 of the C.N.R.S.

REFERENCES

Alben, J.O. and Caughey, W.S.: 1968, Biochem. 7, 175.

Antonini, E. and Brunori, M.: 1971, Hemoglobin and myoglobin in their reactions with ligands, North-Holland Publ., Amsterdam, p. 85 and following.

Barlow, C.H., Maxwell, J.C., Wallace, W.J., and Caughey, W.S.: 1973, Biochem. Biophys. Res. Commun. 55, 91.

Basolo, F., Hoffman, B.M., and Ibers, J.A.: 1975, Acc. Chem. Res. 8, 384.

Benard, M., Dedieu, A., Demuynck, J., Rohmer, M.M., Strich, A., and Veillard, A.: Asterix: a System of Programs for the Univac 1110, unpublished work.

Benard, M.: 1976, J. Chim. Phys., 413.

Benedict, W.S. and Plyler, E.K.: 1957, Can. J. Phys. 35, 1235.

Bonnett, J.J., Eaton, S.S., Eaton, G.R., Holm, R.H., and Ibers, J.A.: 1973, J. Amer. Chem. Soc. 95, 2141.

Brown, L.D. and Raymond, K.N.: 1975, Inorg. Chem. 14, 2595.
Caughey, W.S., Deal, R.M., Weiss, C., and Gouterman, M.: 1965, J. Mol. Spectrosc. 16, 451.
Chang, C.K. and Traylor, T.G.: 1973, J. Amer. Chem. Soc. 95, 8477.
Chien, J.C.W. and Dickinson, L.C.: 1972, Proc. Natl. Acad. Sci. U.S.A. 69, 2783.
Collman, J.P., Gagne, R.R., Reed, C.A., Robinson, W.T., and Rodley, G.A.: 1974, Proc. Natl. Acad. Sci. U.S.A. 71, 1326.
Collman, J.P., Hoard, J.L., Kim, N., Lang, G., and Reed, C.A.: 1975a, J. Amer. Chem. Soc. 97, 2676.
Collman, J.P., Gagne, R.R., Reed, C.A., Halbert, T.R., Lang, G., and Robinson, W.T.: 1975b, J. Amer. Chem. Soc. 97, 1427.
Collman, J.P., Brauman, J.I., and Suslick, K.S.: 1975c, J. Amer. Chem. Soc. 97, 7185.
Cullen, D., Meyer, E., Srivastara, T.S., and Tsutsui, M.: 1972, Chem. Comm., 584.
Dedieu, A., Rohmer, M.M., and Veillard, A.: 1976, J. Amer. Chem. Soc. 98, 3717.
Dolphin, D. and Felton, R.H.: 1974, Acc. Chem. Res. 7, 26.
Fuhrhop, J.-H., Kadish, K.M., and Davis, D.G.: 1973, J. Amer. Chem. Soc. 95, 5140.
Goddard, W.A. and Olafson, B.D.: 1975, Proc. Natl. Acad. Sci. U.S.A. 72, 2335.
Goedken, V.L., Molin-Case, J., and Whang, Y.: 1973, J.C.S. Chem. Comm., 337.
Gouterman, M.: 1959, J. Chem. Phys. 30, 1139.
Gouterman, M.: 1961, J. Mol. Spectrosc. 6, 138.
Gray, H.B.: 1971, Structural Models for Iron and Copper Proteins Based on Spectroscopic and Magnetic Properties in R.F. Gould (ed.), Bioinorganic Chemistry, Adv. Chem. Ser., 100, 365.
Griffith, J.S.: 1956, Proc. Roy. Soc. Ser. A, 235, 23.
Halton, M.P.: 1972, Theoret. Chim. Acta 24, 89.
Halton, M.P.: 1974, Inorg. Chim. Acta 8, 131.
Hoard, J.L.: 1968, in A. Rich et al. (eds), Structural Chemistry and Molecular Biology, W.H. Freeman, San Francisco, P. 573.
Hoffman, B.M. and Petering, D.H.: 1970, Proc. Natl. Acad. Sci. U.S.A. 67, 637.
Hoffmann, R., Chen, M.M.L., Elian, M., Rossi, A.R., and Mingos, D.M.P.: 1974, Inorg. Chem. 13, 2666.
Huber, R., Epp, O. and Formanek, H.: 1970, J. Mol. Biol. 52, 349.
Huzinaga, S.: 1965, J. Chem. Phys. 42, 1293.
Joesten, M.D. and Schaad, L.J.: 1974, Hydrogen bonding, M. Dekker, New York, p. 93.
Khandelwal, S.C. and Roebber, J.L.: 1975, Chem. Phys. Let. 34, 355.
Klevan, L., Peone, J., and Madan, S.K.: 1973, J. Chem. Educ. 50, 670.
Koster, A.S.: 1975, J. Chem. Phys. 63, 3284.
Lang, G. and Marshall, W.: 1966, Proc. Phys. Soc. 87, 3.
Little, R.G. and Ibers, J.A.: 1974, J. Amer. Chem. Soc. 96, 4440 and 44
Loew, G.H. and Kirchner, R.F.: 1975, J. Amer. Chem. Soc. 97, 7388.
Maggiora, G.M.: 1973, J. Amer. Chem. Soc. 95, 6555.
Makinen, M.W. and Eaton, W.A.: 1973, Polarized Single Crystal Absorptio

Spectra of Carboxy- and Oxyhemoglobin in A.D. Adler (ed.), The Chemical and Physical behaviour of Porphyrin Compounds and Related Structures, Ann. N.Y. Acad. Sci. 206, 210.
Martinez-Carrera, S.: 1966, Acta Cryst. 20, 783.
Pauling, L. and C.D. Coryell: 1936, Proc. Natl. Acad. Sci. U.S.A. 22, 210.
Pauling, L.: 1949, Hemoglobin, Butterworth, London, p. 57.
Pauling, L.: 1964, Nature, 203, 182.
Peisach, J., Blumberg, W.E., Wittenberg, B.A., and Wittenberg, J.B.: 1968, J. Biolog. Chem. 243, 1871.
Roos, B. and Siegbahn, P.: 1970, Theoret. Chim. Acta 17, 209.
Roos, B., Vinot, G., and Veillard, A.: 1971, Theoret. Chim. Acta 20, 1.
Roothaan, C.C.J.: 1960, Rev. Mod. Phys. 32, 179.
Salem, L.: 1961, Proc. Roy. Soc. Ser. A. 264, 379.
Salem, L.: 1968, J. Amer. Chem. Soc. 90, 543.
Seno, Y., Otsuka, J., Matsuoka, O., and Fuchikami, N.: 1972, J. Phys. Soc. Japan 33, 1645.
Spiro, T.G. and Strekas, T.C.: 1974, J. Amer. Chem. Soc. 96, 338.
Stynes, D.V., Stynes, H.C., James, B.R., and Ibers, J.A.: 1973, J. Amer. Chem. Soc. 95, 1796.
Trautwein, A., Maeda, Y., Harris, F.E., and Formanek, H.: 1974, Theoret. Chim. Acta 36, 67.
Veillard, A.: 1975, The Logic of SCF Procedures, in G.H.F. Diercksen et al. (eds.), Computational Techniques in Quantum Chemistry and Molecular Physics, D. Reidel, Dordrecht, p. 201.
Veillard, A. and Demuynck, J.: 1976, Transition Metal Compounds in H.F. Schaeffer (ed.), Modern Theoretical Chemistry II Electronic Structure: Ab Initio Methods, Plenum Publishing Corporation, in the press.
Veillard, H. and Veillard, A.: 1976, to be published.
Wagner, G.C. and Kassner, K.J.: 1974, J. Amer. Chem. Soc. 96, 5593.
Walker, F.A.: 1973, J. Amer. Chem. Soc. 95, 1154.
Watson, H.C.: 1969, The Stereochemistry of the Protein Myoglobin in B.J. Aylett et al. (eds.), Progress in Stereochemistry, Butterworths, London, 1969, p. 299.
Wayland, B.B., Minkiewicz, J.V., and Abd-Elmageed, M.E.: 1974, J. Amer. Chem. Soc. 96, 2795.
Weiss, C., Kobayashi, H., and Gouterman, M.: 1965, J. Mol. Spectrosc. 16, 415.
Weiss, J.J.: 1964, Nature 202, 83.
Weiss, J.J.: 1964, Nature 203, 183.
Weschler, C.J., Hoffmann, B.M., and Basolo, F.: 1975, J. Amer. Chem. Soc. 97, 5278.
Yonetani, T., Yamamoto, H., and Iizuka, T.: 1974, J. Biol. Chem. 249, 2168.
Zerner, M., Gouterman, M., and Kobayashi, H.: 1966, Theoret. Chim. Acta 6, 363.

DISCUSSION

CAUGHEY: This is very interesting work. For the experimentalist interested in using such theoretical results for the purpose of predicting the influence of cis effects (i.e., porphyrin structure and porphyrin ring-protein interactions), trans effects (i.e., basicity and steric accessibility of trans ligand), and the environment (protein and/or ligand) on O_2 binding to heme proteins can you provide information on the degree of uncertainty in these results. Specifically, how precisely can you evaluate the change in electronegativity of the iron (e.g. as felt by the porphyrin) on going from CO to NO to O_2 as ligands? And, how reliable is the finding that the bound O_2 is neutral?

VEILLARD: A comparison of the charge distribution with the two ligands CO and O_2 will be presented subsequently, together with a tentative analysis of the trans effect (including for instance a comparison of the charge distribution in $FePO_2$ and $FePO_2NH_3$). Regarding the question of how reliable is the conclusion that the bound O_2 is quasi-neutral, minimal basis set calculations usually tend to overemphasize the polarity of the charge distribution, whereas extended basis set calculations tend to level off the charge distribution. Thus it is expected that the near-neutrality of the bound O_2 will remain in more accurate calculations.

A. Pullman: Concerning correlation effects, it is very unlikely that it is relevant in any of the areas where Dr Veillard drew conclusions. He was extremely careful to avoid drawing conclusions in the domains where his theoretical approach would not be applicable.

As to the possible inversion of the orbitals by going to a split basis set, I do not think that even this would affect the general conclusion which in fact was essentially based on the values of the *total* energy and he used the orbital diagram essentially to explain the results in familiar terms. In fact even if the orbitals were inversed, the total energies might very well be in the same order and would simply show that the orbital model is too crude to be reliable.

METAL-LIGAND INTERACTIONS IN PORPHYRINS AND HEMEPROTEINS

WINSLOW S. CAUGHEY, JOHN C. MAXWELL, JOHNNY M. THOMAS,
DAVID H. O'KEEFFE*, and WILLIAM J. WALLACE
Dept. of Biochemistry, Colorado State University, Fort Collins, Colo. 80523, U.S.A.

1. INTRODUCTION

In hemoglobin and in myoglobin, heme B (Figure 1) is surrounded by protein (globin) with a passageway from one side of the iron to the outer surface of the protein through which potential ligands in the external

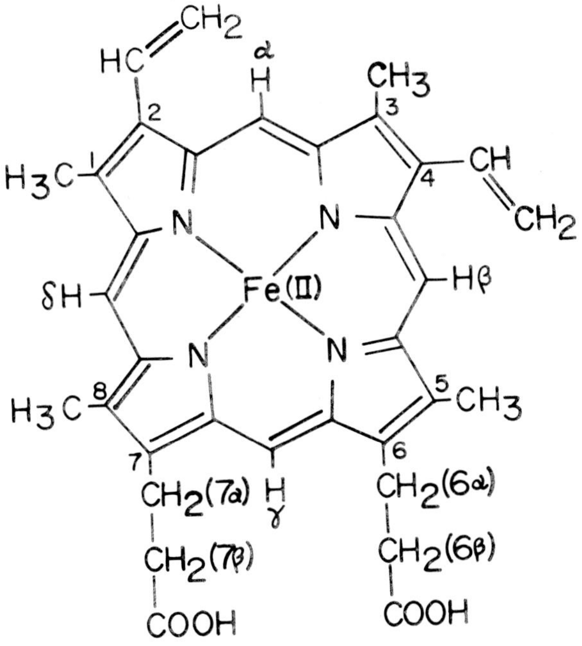

Fig. 1. Heme B (Protoheme).

* Present address: Department of Biochemistry, The University of Texas Health Science Center at Dallas, Dallas, Texas 75235.

B. Pullman and N. Goldblum (eds.), Metal-Ligand Interactions in Organic Chemistry and Biochemistry, part 2, 131–152.

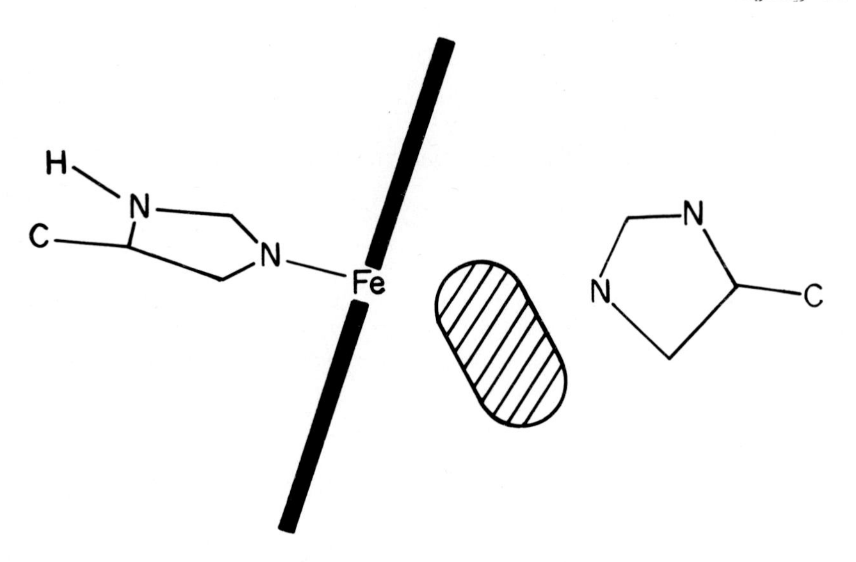

Fig. 2. Schematic representation of the oxygen binding site in hemoglobins and myoglobins. The hatched area shows where the O_2 (or another ligand such as CO, NO, or N_3) must be.

medium can pass to bind at iron. X-ray crystallographic and other studies revealed that several hemoglobins and myoglobins possess a ligand binding site as shown in Figure 2 (Caughey et al., 1975a) in which the external ligand binds between iron and a distal histidine; a proximal histidine can bind to iron on the other side of the porphyrin plane. Structural and mechanistic aspects of binding by such ligands as O_2, CO, NO, and R-NC to iron(II) and OH^-, HO_2^-, CN^-, N_3^-, and F^- to iron(III) have been of great interest recently. Of special concern for the better understanding of the physiological functions of these proteins are the effects of protein structure upon ligand binding and the effects of ligand binding on the properties of the proteins.

Among hemeproteins, in general, the porphyrin structure, the amino acid sequences, and the conformations of protein can all vary. As the protein environment about the heme moiety changes from protein to protein the bonding (e.g., covalent, van der Waals, electrostatic) between heme and protein, and the extent to which the external medium (solvent) approaches the heme may also vary. Studies with several types of protein-free metal porphyrins and with a few proteins have shown the interdependence of porphyrin, metal, axial ligand and medium (protein and/or solvent) in the determination of properties and in the control of reactions (Caughey et al., 1973; 1975b). Recent findings in several such studies utilizing infrared, electronic, and magnetic resonance spectroscopic probes are discussed here.

2. NICKEL(II) PORPHYRINS

The interrelationships between porphyrin, metal, and axial ligands (pri-

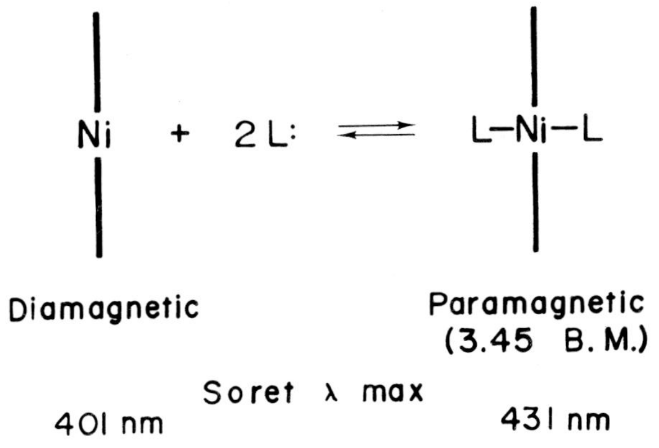

Ni + 2L: ⇌ L–Ni–L

Diamagnetic — Paramagnetic (3.45 B.M.)

Soret λ max

401 nm — 431 nm

Fig. 3. Equilibrium between planar and tetragonal nickel(II) porphyrins. Nitrogen bases (amines, pyridine, etc.) can serve as axial ligands (L:). The λ_{max} values are for protoporphyrin complexes.

marily of the sigma-donor type) were perhaps first clearly demonstrated with nickel(II) porphyrins (Caughey et al., 1962; 1966; McLees and Caughey, 1968). Nitrogen bases (e.g., pyridine, piperidine, alkylamines, etc.) bind weakly to the diamagnetic planar complex to form a paramagnetic (3.45 B.M.) tetragonal complex of nickel(II) in an equilibrium represented in Figure 3. The equilibrium constant for this reaction is very dependent upon the structures of both the porphyrin and the ligand. Thus, with the common ligand piperidine, the less basic the porphyrin nitrogens (Table I), the greater will be the equilibrium constant (e.g., at 25 $^{\circ}$C in chloroform solutions: diethyl 9×10^{-3}, dihydrogen 2×10^{-2}, divinyl 4×10^{-2} and diacetyl 0.6; McLees and Caughey, 1968). Whereas with a common porphyrin the more basic is L:, the more the six-coordinate species will be favored compared with the four-coordinate. These are basically electronic effects and hold only in the absence of steric effects which can be largely avoided in these systems by suitable choices of porphyrin and ligand. The binding of two nitrogen bases as axial ligands renders the nickel ion less effective as an electron acceptor from the porphyrin nitrogens and is accompanied by a large shift in electronic spectrum (e.g., from 400 to 430 nm in Soret band maximum) (Figure 3). A red shift in Soret band is what we have come to expect when the apparent electronegativity of the central metal ion is reduced and the magnitude of the shift (30 nm) is consistent with a marked reduction in electronegativity (Caughey, 1973). As the electron acceptance by the central metal ion increases, there also results a reduction in ring current field effects in PMR spectra and in diamagnetic susceptibilities (Caughey and Koski, 1962; Caughey et al., 1967; 1973), and electron delocalization onto substituents at the periphery of the porphyrin ring is also reduced (Fuchsman et al., 1975). These and other data clearly show marked effects of porphyrin, metal, and axial ligand on each other.

TABLE I
Influence of Porphyrin Structure on ν_{CO} in Hemes, Hemoglobins, and Myoglobins

2,4-Substituent on the deuteroheme	pK_3 [a]	pyridine[b] Fe^{II} CO	ν_{CO}, cm^{-1} hemoglobin A[c] CO	myoglobin[c,d] CO
acetyl	3.3	1984	1957	—
vinyl	4.8	1977	1951	1944
hydrogen	5.5	1975	1949	1941
ethyl	5.8	1973	1947	1939

[a] An expression of the basicity of the metal-free porphyrin-related to the equilibrium between neutral porphyrin (PH_2) and monoprotonated species (PH_3^+).
[b] Alben and Caughey (1968). Recorded in bromoform with .12 M pyridine.
[c] In the case of the acetyl, hydrogen, and ethyl 2,4-substituents, the proteins are reconstituted from the globin and the appropriate heme. Sperm whale myoglobin was used.
[d] McCoy and Caughey (1971).

3. CARBONYLS

The heme carbonyls, low-spin iron(II) compounds, are well studied systems in which porphyrin structure, the ligand trans to CO, and the medium all affect the iron-CO bond. Changes in porphyrin structure which make the nitrogens of the metal-free porphyrin more basic increase the strength of the bonding between iron and CO and the increased bond strength is reflected indirectly in lower ν_{CO} values (Table I) and in increasing affinities for CO (Alben and Caughey, 1968; Caughey, 1973). These effects are constistent with strong π-donation to CO from iron (Caughey et al., 1973). Similar effects of porphyrin structure on ν_{CO} are observed in reconstituted proteins (hemoglobins or myoglobins) so it is clear that interactions between the protein and the heme do not materially affect the ability of substituents at the periphery of the porphyrin ring to influence CO bonding. This conclusion contrasts with suggestions that steric effects dominate differences in ligand affinities among the reconstituted proteins (Yonetani et al., 1974). When the basicity of the ligand trans to CO is altered in simple heme-CO complexes, the ν_{CO} responds in a way that is entirely consistent with the effects produced by changes in the 2,4-substituents on the porphyrin (Caughey et al., 1973) in that increasing the electron density at iron allows the π transfer to CO to increase and ν_{CO} to decrease. Changes in ligands in the low-spin iron(II) hemes also affect ring current fields in PMR spectra and the electronic spectra (with the visible region more sensitive than the Soret region). As seen in Table II, with diamagnetic iron(II) porphyrins the Soret band is rela-

TABLE II
Soret Band Maxima for Heme B with Different Axial Ligands

Axial Ligands	λ_{max}, nm
dipyridine	418
dihydrazine	417
pyridine carbonyl	419
hydrazine carbonyl	417
1-methylimidazole nitrosyl	418
bis-(methylisonitrile)	433
μ-hydrazido-bis-carbonyl	408

TABLE III
Soret Band Maxima for Native and Reconstituted Hemoglobins A with CO, NO, or O_2 Bound

2,4-Substituent on the deuteroheme	λ_{max}, nm			
	deoxyHb	HbCO	HbNO	HbO_2
ethyl	421[a]	410	409	404
hydrogen	420[a]	409	407	402
vinyl	430[a]	420	418	416
acetyl	442	433	430	425

[a] Sugita and Yoneyama (1971).

tively insensitive to exchange among sigma-donor and π-acceptor type ligands compared to the shifts between diamagnetic and paramagnetic nickel. Exceptions are the bis-isonitrile and the dimeric carbonyl with a bridging hydrazine group.

4. NITROSYLS

The bonding of NO to hemes and hemeproteins is qualitatively similar to the bonding of CO to the same hemes and hemeproteins. However, Soret band maxima for the NO complexes are slightly blue shifted compared with the corresponding CO complexes in accord with a small increase in electronegativity of iron when NO replaces CO (Table III). Furthermore the trend in the positions of the band maxima through a series of porphyrins is the same for the NO complexes as for the CO complexes. Just as the ν_{CO} around 1950 cm^{-1} for CO complexes indicates π transfer from metal to ligand, the ν_{NO} value near 1600 cm^{-1} for iron(II)-NO complexes indicates appreciable transfer of electron density through the π system from iron into the N—O bond (Table IV). As free gases, $^{14}N^{16}O$ and $^{15}N^{16}O$ exhibit ν_{NO} values of 1876 and 1843 cm^{-1}, respectively. Cis and trans effects that arise from the substituents on the porphyrin

TABLE IV
Influence of Porphyrin Structure and Trans Ligand on N—O Stretching Frequencies[a]

	$\nu_{^{14}N^{16}O}$	$\nu_{^{15}N^{16}O}$
1-Methylimidazole heme B nitrosyl (in 1-methylimidazole)	1618 cm^{-1}	—
Pyridine heme B nitrosyl (in pyridine)	1633	—
Heme B nitrosyl (five-coordinate) (in $ClCH_2$-CH_2Cl)	1669	—
Hemoglobin A nitrosyl (stripped)	1615	1587
Hemoglobin A nitrosyl (with inositol hexaphosphate)	1615, 1668	1587, 1635
Hemoglobin A nitrosyl (reconstituted with 2,4-diacetyldeuteroheme)		1591

[a] Maxwell and Caughey (1976) with the exception of the reconstituted hemoglobin ν_{NO}.

and in the ligand bound to iron trans to the NO, respectively, follow the pattern established for CO complexes. Thus, increased electron density on iron promotes π transfer to NO with a corresponding lowering of ν_{NO} (Figure 4). Particularly interesting in this respect is the large frequency shift that occurs when five-coordinate (FeNO) is transformed into six-coordinate (LFeNO). This frequency shift provides a convenient means by which the five- and six-coordinate complexes can be readily distinguished. The binding of inositol hexaphosphate (IHP) between portions of polypeptide chain from each β subunit at the center of the hemoglobin A tetramer (Arnone and Perutz, 1974) causes a shift in infrared frequency from 1615 cm^{-1} to 1668 cm^{-1} for two of the four bound NO ligands that is indicative of cleavage of the proximal histidine to iron(II) bond in these hemes (Maxwell and Caughey, 1976). The two IHP-altered hemes also gave infrared evidence for an altered interaction between distal histidine and bound NO. ESR and ultraviolet-visible spectra also support two of the nitrosyl hemes becoming five-coordinate while the other two remain unchanged as a result of IHP binding. Infrared spectra reveal quite a different picture for NO bonding to iron(III) species. Thus, in iron(III) peroxidase, $\nu_{^{15}N^{16}O}$ is found at 1865 cm^{-1} (22 cm^{-1} higher frequency than is found in the free gas) which is consistent with net donation of electron density from NO to iron(III). Thus, a change in oxidation state from iron(II) to iron(III) is responsible for a marked change in iron-NO bonding.

That reactions and properties of groups at the periphery of the porphyrin ring can be significantly influenced by the oxidation state

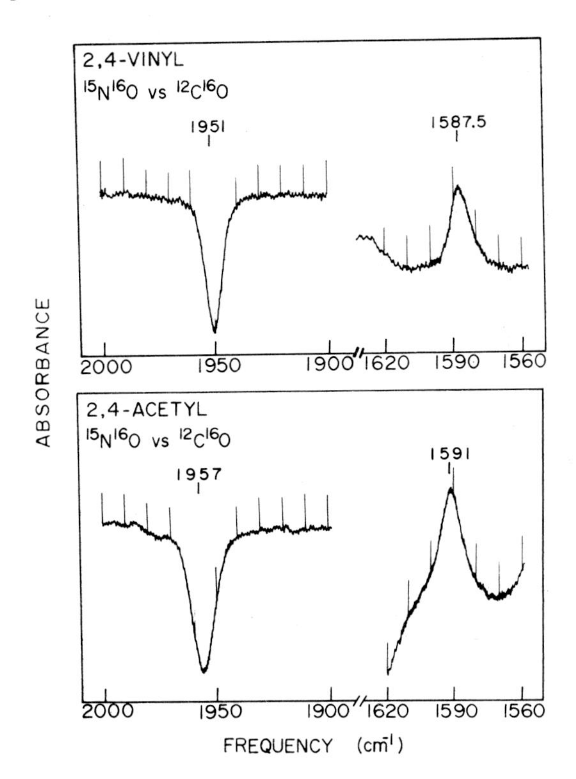

Fig. 4. Hemoglobin difference infrared spectra for a nitrosyl in the sample beam and a carbonyl in the reference beam. Upper trace: native hemoglobin A, ν_{CO} = 1951 cm^{-1}, ν_{NO} = 1587.5 cm^{-1}. Lower trace: hemoglobin A reconstituted with 2,4-diacetyl-deuteroheme, ν_{CO} = 1957 cm^{-1}, ν_{NO} = 1591 cm^{-1}.

and the axial ligands is interestingly demonstrated by the data of Table V. In hemoglobin A reconstituted 2,4-diacetyldeuterohemin, the ν_{CO} for the acetyl groups are fairly reliably identified as being at 1637 cm^{-1} in the carbonyl but at 1649 cm^{-1} in the nitrosyl species. This shift in ν_{CO} is reasonably explained as due to easier electron transfer from porphyrin onto the acetyl carbonyl group in HbCO than in the HbNO. This is further evidence for a less electronegative iron in HbCO than in HbNo. Other iron and nickel complexes of Table V provide further support for such effects (Fuchsman et al., 1975).

5. OXYGENYLS

Dioxygen binding is also similar to that for CO and NO but greater blue-shifts in Soret band maxima are seen with dioxygen (Table III) suggesting greater π transfer through iron onto the bound dioxygen. The order of blue shifts in Soret band frequency from CO to NO to O_2 species indicate

TABLE V
Ligand Binding and Oxidation State Efffects Upon the C—O Stretch of Acetyl Groups in 2,4-Diacetyldeuteroporphyrins[a]

Species	ν_{CO}, cm^{-1}
Metal-free porphyrin	1661
Iron(III) chloride	1665
Iron(III) bromide	1664.5
μ-oxo-bis-hemin	1665.5
Iron(II) dipyridine	1647
Nickel(II) (four coordinate)	1659
Nickel(II) dipyridine	1647
Reconstituted hemoglobin A carbonyl	1637
Reconstituted hemoglobin A nitrosyl	1649

As the electronegativity of iron decreases, electron delocalization from porphyrin onto the acetyl carbonyl becomes easier with an accompanying decrease in C—O stretch frequency.

[a] The hemoglobins were reconstituted with the diacid species, the remaining compounds were dimethyl esters. These esters have been discussed by Fuchsman et al. (1975); both $CHCl_3$ and pyridine were used as solvents but all values above were corrected for solvent effects to be comparable to $CHCl_3$ solutions.

TABLE VI
Infrared Data for Bound O_2 in Hemoglobins and Myoglobins

Oxygenated Species	ν_{O_2}, cm^{-1}	
	$^{16}O_2$	$^{18}O_2$
Human Red Blood Cell[a]	1107	1065
Hemoglobin A (HbA)	1106	1065
HbA reconstituted with:		
Deuteroheme	1106	1065
2,4-diethyldeuteroheme	1106	1065
2,4-diacetyldeuteroheme	1106	1065
Cobalt(II) deuteroporphyrin IX[b]	1105	1065
Myoglobin (Mb), bovine heart[c]	1103	1065

[a] Barlow et al. (1973).

[b] Maxwell and Caughey (1974).

[c] Maxwell et al. (1974).

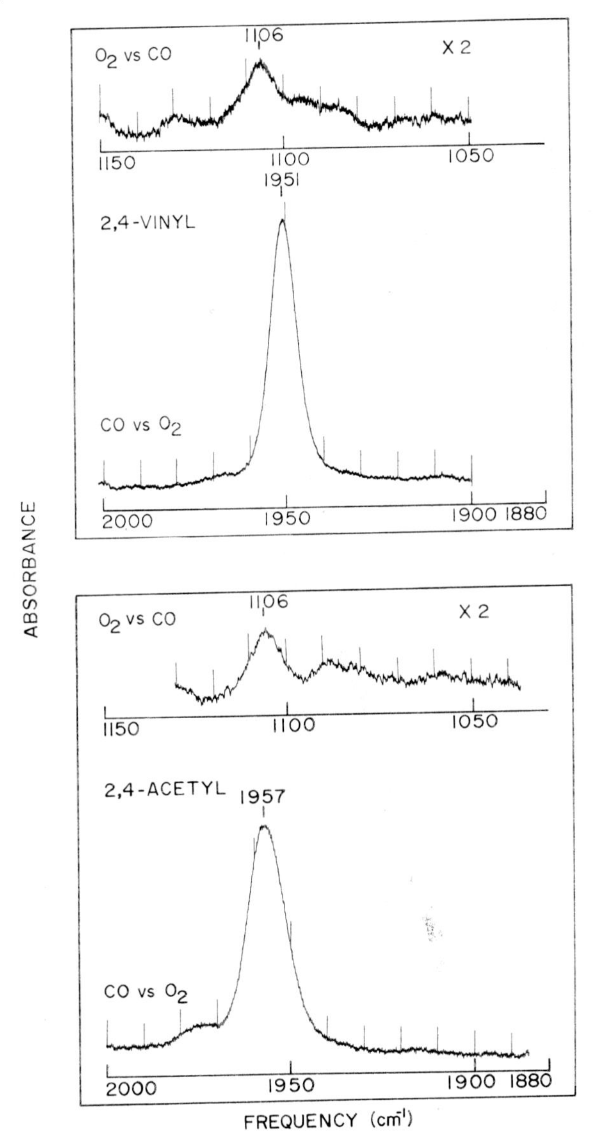

Fig. 5. Hemoglobin difference infrared spectra for oxygenyl and carbonyl complexes. Upper traces: native hemoglobin A, ν_{O_2} = 1106 cm^{-1}, ν_{CO} = 1951 cm^{-1}. Lower traces: hemoglobin A reconstituted with 2,4-diacetyldeuteroheme, ν_{O_2} = 1106 cm^{-1}, ν_{CO} = 1957 cm^{-1}.

this is also the order of increasing donation from iron(II) onto the axial ligand. The ν_{O_2} found near 1105 cm^{-1} for oxyhemoglobins and oxymyoglobins (Table VI; Figure 5) is indicative of a bent-end-on con-

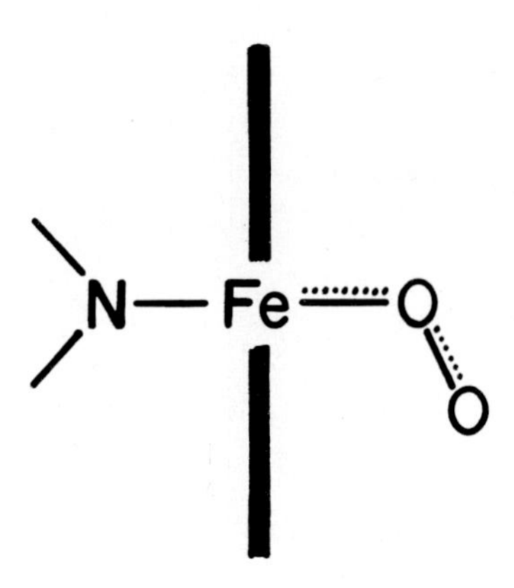

Fig. 6. Representation of oxygenyl bonding between Fe^{II} and O_2 as in oxyhemoglobin or oxymyoglobin.

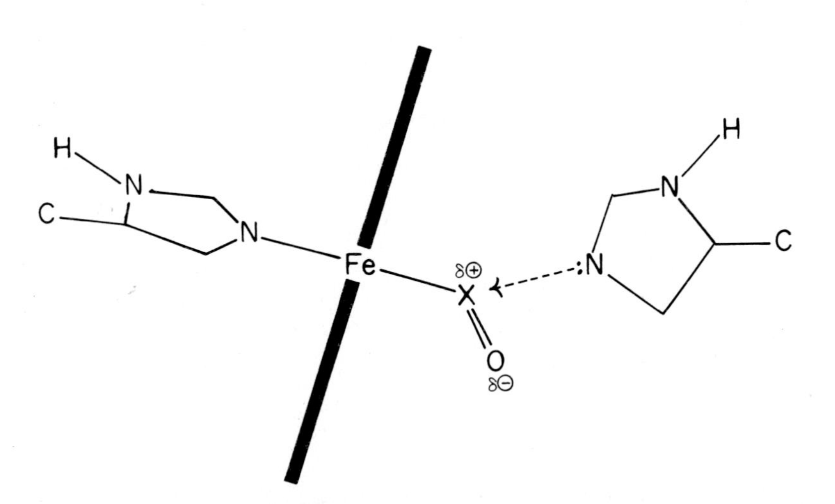

Fig. 7. A schematic representation of the nature of ligand binding to hemoglobins or myoglobins with both proximal and distal histidines present. X may be carbon, nitrogen, or oxygen. The partial positive charge on X may be stablized by interaction with the distal histidine wherein a proton-free imidazole nitrogen acts as electron donor and X as acceptor.

figuration (i.e., Fe—O$_{\diagdown O}$). The shift in frequency from the free gas value (1555 cm^{-1}, Raman) shows that electron-density has been added to the O—O bond as a consequence of binding at iron(II). The ν_{O_2} is slightly sensitive to protein structure, but is not affected nearly as much by a change in porphyrin structure or by a change in metal as is noted with carbonyls and nitrosyls. ν_{O_2} is typically rather insensitive to changes in other ligands (Vaska, 1976). The ν_{O_2} frequency of 1105 cm^{-1} is also not inconsistent with bound dioxygen being O_2^-, superoxide ion. However, for the reason that superoxide is displaced much less readily by other ligands than are anions such as F^-, N_3^-, CN^-, and OH^-, as well as for thermodynamic and other reasons, it must be concluded that appreciable covalent (multiple) bond character exists between iron(II) and O_2 (Caughey et al., 1975; Wallace et al., 1974). Oxygenated hemeproteins of the hemoglobin type may thus be well designated as *oxygenyl* species in analogy to carbonyls and nitrosyls and, the bonding in oxygenyl hemoglobin may be represented as in Figure 6. The polarity which will develop on the bound O_2 can influence interactions with the distal histidine; the general case of O_2, NO, and CO bonding in hemoglobins and myoglobins with both proximal and distal histidines present may be represented as in Figure 7.

6. AZIDES

Azidohemins, in the solid and in solvents that do not provide a suitable sixth ligand (e.g., chloroform or benzene) exist as five-coordinate, high-spin species not unlike the chlorides or acetates in terms of many physical properties (Caughey, 1973). With a suitable ligand (e.g., pyridine or 1-methylimidazole) present, an equilibrium between five-coordinate (high-spin) and six-coordinate (low-spin) forms result (McCoy and Caughey, 1970) (Figure 8). Infrared bands provide a ready means for the identification of the two forms. These equilibria are highly sensitive to temperature, porphyrin structure, and sixth ligand. The six-coordinate species is favored with increasing basicity of the sixth ligand, with decreasing basicity of the porphyrin, and with decreasing temperature (Figure 8).

Similar equilibria have been observed in infrared spectra in methemoglobin and metmyoglobin azides where bands at 2045 cm^{-1} and 2025 cm^{-1} for high- and low-spin forms, respectively, are found (McCoy and Caughey, 1970). In the azido proteins, increasing the temperature also increases the proportion of high-spin form (Alben and Fajer, 1973). Recently, we noted that upon addition if inositol hexaphosphate to stripped azido methemoglobin A, the intensity of the 2045 cm^{-1} band increases compared with the 2025 cm^{-1} band. Typically, in stripped azido metHb A, the 2045 cm^{-1} band was about 8% the intensity of the 2025 cm^{-1} band; whereas with IHP present, it was 14%. This IHP effect is thus consistent with the presence of five-coordinate and six-coordinate species in equilibrium; the proximal histidine is better able to serve as ligand trans to azide with IHP absent — a finding not unlike that seen with the nitrosyls (Maxwell and Caughey, 1976). An obvious question here is how far away from the iron does the proximal histidine need to be for the infrared and magnetic data to indicate five- coordination? In other words, how weak does the proximal histidine to iron

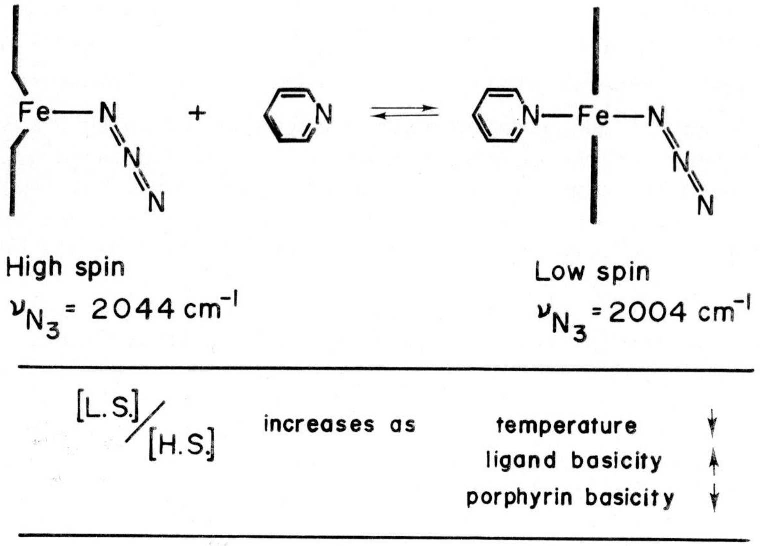

High spin
ν_{N_3} = 2044 cm^{-1}

Low spin
ν_{N_3} = 2004 cm^{-1}

[L.S.]/[H.S.] increases as temperature ↓, ligand basicity ↑, porphyrin basicity ↓

Fig. 8. Equilibrium between five- and six-coordinate species of azidohemin B in pyridine.

bond become in the high-spin species? These data certainly suggest (as did infrared and ESR data for nitrosyls) that this bond is completely broken when highlspin azide is observed. It was concluded earlier that bonding between the proximal histidine and iron in the metmyoglobin and methemoglobin fluorides must be either absent, or extremely weak (Caughey 1973), and that the iron would lie appreciably out of the porphyrin plane towards the fluoride as noted for hemin B chloride (Koenig, 1965).

The azide stretching frequency of the high-spin form of azidohemin B in pyridine compares favorably with the frequencies found in the protein suggesting the distal histidine in the protein serves the same role as does solvent pyridine. However, the frequency of the low-spin form of the hemin (ca. 2000 cm^{-1}) is much lower than that found in the low-spin form of the protein (ca. 2025 cm^{-1}) suggesting that distal histidine can no longer interact with bound azide in the same way as in the high-spin form or in pyridine solution. Thus, the two forms of HbN_3 and MbN_3 may differ not only in terms of the spin state of the iron, with the high-spin state five-coordinate and the low-spin state six-coordinate, but also in positioning the azide ligand such that one form (high-spin) can interact with distal histidine while the other form (low-spin) cannot.

7. CYANIDES

The cyanides of hemins and hemeproteins have long been known as conveniently prepared, entirely low-spin, iron(III) species. When hemin B chloride and potassium cyanide are dissolved together in aqueous

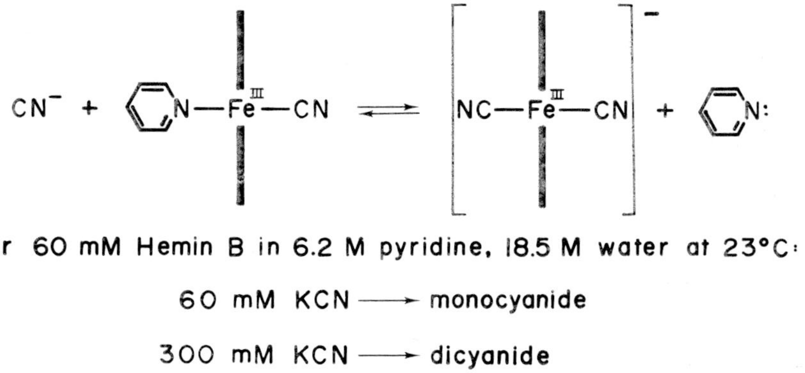

$$CN^- + \text{py}N\text{—}Fe^{III}\text{—}CN \rightleftharpoons [NC\text{—}Fe^{III}\text{—}CN]^- + \text{py}N:$$

For 60 mM Hemin B in 6.2 M pyridine, 18.5 M water at 23°C:

60 mM KCN ⟶ monocyanide

300 mM KCN ⟶ dicyanide

Fig. 9. Equilibrium between mono- and di-cyanides of hemin B in aqueous pyridine.

pyridine, the dipyridine, pyridine cyano, and dicyano hemin species are all present in the equilibrium mixture (Caughey et al., 1973). The relative proportions of each of the species is governed by the pyridine/cyanide concentration ratio. Hence, by manipulating this ratio, each of the species can, in turn, be made the dominant component in the mixture. Conditions for obtaining essentially pure mono- or di-cyano species are shown in Figure 9. The electronic spectrum is markedly influenced by the axial ligands (Table VII), and, following the reasoning that the frequency shifts reflect a change in effective electronegativity of iron(III), it is concluded that replacement of pyridine ligands successively by cyanide ligands makes the iron(III) increasingly less effective as an electron acceptor from the porphyrin as the formal charge on iron changes from positive to neutral to negative. It may be noted (Table VII) that the methemoglobin and metmyoglobin cyanides, wherein the proximal histidine is trans to cyanide, exhibit Soret and visible frequencies similar to those for the hemin B monocyanide, but unlike those for hemin B dicyanide. Also, the C—N stretching frequencies for the proteins are very close to ν_{CN} for the monocyanide (2125.5 cm^{-1}) but differ significantly from the dicyanide (2112.5 cm^{-1}). The basicity of the porphyrin has a small, but significant effect upon ν_{CN} in both mono- and di-cyanides (Table VII). The effect on ν_{CN} of changes in porphyrin structure and in trans ligand can be rationalized in terms of a simple model in which two forms, (a) $M\text{—}C{\equiv}N$ and (b) $\overset{+}{M} = C = \overset{-}{N}$, contribute to the structure in varying degrees. The more basic is the porphyrin, the more (b) is favored over (a). Similarly, replacement of neutral pyridine by the negatively charged cyanide ions as trans ligand will increase the contribution of (b) compared to that of (a). The effects of porphyrin structure upon azide frequencies are even less than with the cyanides (Table VIII); effects of comparable or lesser magnitude have been observed for reconstituted methemoglobin A azides (O'Toole, 1972). Whereas the presence or the absence of a trans ligand affects ν_{N_3} by

TABLE VII
Electronic and Infrated Spectral Data for Hemin B and Hemeprotein Cyanides

	λ_{max}, nm		ν_{CN}, cm^{-1}
Dicyanohemin B[a]	433	554	2112.5
Pyridine, cyanohemin B[a]	417.5	542	2125.5
Cyanomethemoglobin A	419	540	2125.0[b]
Cyanomethmyoglobin	422[c]	540[c]	2125.0[b]
Dipyridine hemin B[a]	410	530	—

[a] In aqueous pyridine.

[b] McCoy and Caughey (1970).

[c] Tamura et al. (1973).

TABLE VIII
Influence of Porphyrin and Trans Ligand Structures on Cyano and Azido Infrared Band Frequencies

	2,4-substituent on deuterohemin		
Axial ligands	H—	—CH=CH$_2$	CH$_3$—C(=O)—
pyridine-Fe^{3+}-cyanide	2124.5 cm^{-1}	2125.5 cm^{-1}	2127.5 cm^{-1}
cyanide-Fe^{3+}-cyanide	2112.0	2112.5	2114.0
(pyridine) Fe^{3+}-azide[a] (high spin)	2043.0	2042.5	2042.0
pyridine-Fe^{3+}-azide[b] (low spin)	2005.0	2004.5	2004.0
1-methylimidazole-Fe^{3+}-azide[b, c]	—	2000	—

[a] In pyridine solution, but presumably five-coordinate.

[b] Presumably six-coordinate. The frequency difference of 4.5 cm^{-1} between pyridine and 1-methylimidazole species of hemin B cannot be attributed solely to a trans effect. Rather, the difference could result largely from the change in solvent (i.e., pyridine vs. 1-methylimidazole); solvent effects of this magnitude have been observed in ν_{NO} and ν_{CO} (Maxwell and Caughey, 1976).

[c] In 1-methylimidazole solution.

ca. 40 cm^{-1} and solvent effects can be significant, ν_{N_3} appear only subtly influenced by porphyrin or trans ligand basicity in porphyrin or hemeprotein systems. Thus, with azides infrared spectroscopy is less useful as a probe of porphyrin and trans ligand structure than it is for spin-state and medium. With the cyanides, both visible-Soret and infrared spectra indicate that cyano pyridine hemin B in aqueous pyridine is a close model of the methemoglobin and metmyoglobin cyanides whereas the dicyanide complex of hemin B is a less satisfactory model.

TABLE IX
PMR Chemical Shifts for Hemin Cyanides[a]

Axial ligands	2,4-Substituent	Ring Substitutent Resonances					
		1,3,5,8-CH_3		α-CH_2	β-CH_2	α,β,γ,δ-H	
Pyridine, cyano	hydrogen	-20.53 -16.62	-17.66 -14.85	-8.66 -7.34	-0.70 -0.53	-4.84 -3.29	-3.75 -2.91
	vinyl	-19.95 -15.45	-19.43 -14.12	-8.00 -6.76	-0.65	-5.10 -2.98	-3.21 -2.56
	acetyl	-26.25 -12.33	-19.98 - 4.88	-9.09 -6.88	-1.31 -0.87	— -0.69	-1.76 -0.33
Dicyano	hydrogen	-18.47 -14.51	-15.97 -12.51	-7.89 -6.81	-0.76 -0.60	-3.57 -0.76	-1.28 -0.51
	vinyl	-17.63 -13.24	-17.13 -11.69	-7.41 -7.12	-0.66	-3.42 -9.12	-0.59 +0.70
	acetyl	-20.13 - 9.69	-19.68 - 3.88	-7.88 -6.05	-1.61 -1.28	— +2.87	+0.18 +6.46

[a] In pyridine d_5, D_2O (7:3 v/v) at 21 °C

Changes in axial ligand were shown several years ago to modify the paramagnetic shifts seen in PMR spectra for high-spin hemins (Caughey and Johnson, 1969) and for low-spin dipyridine hemins (Hill and Morallee, 1970, 1972; La Mar and Walker, 1973). With the cyanides, as can be seen in Table IX, the magnitudes of the chemical shifts are sensitive both to the number of cyanides (1 or 2) bound to same hemin and to the porphyrin structure. The diacetyl hemin is strikingly different from the dihydrogen and divinyl hemins, especially in the monocyanides. For the diacetyl compound, the methyl group resonances range over 21 ppm for the monocyanide (with one at -26) and range somewhat less, 16 ppm, for the dicyanide. With hemin B and deuterohemin the range of shifts is narrower. The chemical shifts for the ring methyls are of special interest in relation to PMR data for the hemeprotein cyanides. Values for the 1,3,5,8-methyls range, in methemoglobins from -22.6 upfield,

TABLE X
PMR Chemical Shifts for Porphyrin Methyl Groups in Hemeprotein, and Hemin Cyanides

	ppm to lower field			
Metmyoglobin[a]	-26.9	-18.2	-13.2	
Metdeuteromyoglobin[b]	-24.5	-18.4	-12.5	-11.6
Metmesomyoglobin[c]	-25.0	-22.0	-12.7	-11.3
Methemoglobin A[d]	-22.7α	-21.7β	-16.6α	-15.7β
Pyridine cyano deuterohemin (2,4-hydrogen)	-20.53	-17.66	-16.62	-14.85
Pyridine cyano hemin B (2,4-vinyl)	-19.95	-19.43	-15.45	-14.12
Pyridine cyano diacetyldeuterohemin	-26.25	-19.98	-12.33	- 4.88

[a] Mayer et al. (1974).

[b] Shulman et al. (1969).

[c] Shulman et al. (1971).

[d] Ogawa et al. (1972). The α and β designate from which chain of the tetramer the resonances are assigned. Thus, four resonances, two each from the α and β chains have not been identified.

TABLE XI
Paramagnetic Shifts for Hemin Monopyridine Monocyanides With Diamagnetic Dipyridine Hemes as Reference

2,4-Substituent on Deuterohemin	Shifts in ppm for Ring Methyl Groups on the Porphyrin			
Hydrogen	-17.02	-14.15	-13.11	-11.34
Vinyl	-16.40	-15.88	-11.90	-10.57
Acetyl	-22.65	-16.38	- 8.73	- 1.28

in metmyoglobin from -26.9 upfield, in hemin B monocyanide from -19.5 to -14.1, and in hemin B dicyanide from -17.6 to -11.7 (Table X). The monocyanide values appear near those for methemoglobin but differ from the metmyoglobin values, especially the resonance at -26.9.

The actual magnitudes of paramagnetic shifts, i.e., the difference in chemical shift between the iron(III) monocyanide and the diamagnetic iron(II) dipyridine for ring methyl proton resonances, are listed in Table XI. These extremely wide variations in paramagnetic shift among hemins of different structure suggest an explanation for the differences between the metmyoglobin and the methemoglobin cyanide PMR data. The metmyoglobin shifts must differ from those of methemoglobin solely due to differences in protein structure. The protein of the myoglobin interacts with its hemin B, presumably through donor-acceptor interactions, to cause one methyl (in one pyrrole moiety) to experience a spin density of the magnitude found in hemin B only when a vinyl was replaced by the much more strongly electron-withdrawing acetyl group. Asymmetric substitution of groups that strongly interact with the porphyrin ring (by resonance and inductive mechanisms) result in markedly different paramagnetic shifts at equivalent positions around the ring. It may be noted that the iron(III) cyanide of the parent compound porphin, that is devoid of either *beta* or *meso* substituents, experiences only one resonance for the *beta*-protons and one for the *meso*-protons (Wuthrich, 1970).

A first suggestion in explanation of the -26.9 ppm methyl resonance for metmyoglobin is that an acceptor on the protein affects the π-system of the pyrrole ring to which that methyl group is attached in a way not unlike the effect of the electron-withdrawal from the periphery by an acetyl group. Present theory (Schulman et al., 1971) does not appear to handle these problems well. However, we can suggest that the changes in the coupling between substituents on the porphyrin ring and the inner π-systems may be more variable, and thus a more important factor, than has been appreciated. These data also suggest new ways of viewing hemin cyanide paramagnetic shifts.

Since the metal eg orbitals mix primarily with two porphyrin π orbitals (the highest filled, 3_{eg}, or the lowest unoccupied, 4_{eg}), electron spin density may be delocalized via ligand-to-metal charge transfer ($L_{3_{eg}} \rightarrow M_{eg}$) or metal-to-ligand charge transfer ($M_{eg} \rightarrow L_{4_{eg}}$). Simple Hückel molecular orbital calculations (Lonquet-Higgins et al., 1955) indicate that ligand-to-metal charge transfer would result in large downfield contact shifts for ring methyl proton resonances and little or none for *meso*-proton resonances. In contrast metal-to-ligand charge transfer would result in smaller downfield contact shifts for the ring methyl resonances and large upfield contact shifts for the *meso* protons.

It has been suggested that for low-spin hemin complexes the ligand-to-metal charge transfer spin delocalization pathway dominates (Schulman et al., 1971; Hill and Morallee, 1972; LaMar and Walker, 1973). And the paramagnetic shifts observed for the *meso*-proton resonances were considered to be pseudo contact (through space) in nature as calculated from the aniosotropic g-tensor determined from low temperature EPR spectra. However, results from recent ^{13}C-NMR spectral studies of several hemin cyanides have indicated a much smaller pseudo contact contribution to the prarmagnetic shift of all ring substituent resonances (Wuthrich and Baumann, 1973a, b). Since both the ring methyls *and* the *meso* proton resonances experience large paramagnetic shifts in the mono- and di-cyanides investigated, both charge transfer spin delocalization pathways may be operative here.

As judged from averaged paramagnetic shifts, the $L_{3eg} \rightarrow M_{eg}$ spin delocalization pathway contributes less and the $M_{eg} \rightarrow L_{4eg}$ pathway more as the 2,4-substituent is changed from hydrogen to vinyl to acetyl (cis effect). A similar effect is observed upon axial ligand replacement, i.e. exchanging the trans pyridine for a second cyano also results in increased $M_{eg} \rightarrow L_{4eg}$ charge transfer. This increase in $M_{eg} \rightarrow L_{4eg}$ spin delocalization is caused by a change in the iron to porphyrin nitrogen bonding since as the electron-withdrawing ability of the 2,4-substituent increases, hydrogen < vinyl < acetyl, the basicity of the porphyrin decreases thereby weakening the iron-porphyrin nitrogen bonding and strengthening the iron-axial ligand bonding. An increase in the energy of the d_{xz} and d_{yz} metal e_g orbitals therefore occurs in this same order such that their mixing with the 4_{eg} porphyrin orbital becomes energetically more favorable than with the 3_{eg} orbital. A similar effect occurs as the basicity of the axial ligand(s) increases.

Hopefully, mention of these problems may entice theoreticians in attendance at this Symposium to the bait presented by these intriguing problems. It is becoming ever clearer that detailed insight into metal-ligand interactions lies at the heart of an understanding of hemeprotein function. Courage to take up the pursuit of such insight comes from the increasing awareness of the great promise from applications of the several independent spectroscopic and other physical probes. Such probes are particularly valuable when they can be refined on well-defined small molecule systems of relevance to hemeproteins and the results of this refinement applied meaningfully in chemical terms to an investigation of the more complex protein systems and even intact tissue. The studies briefly outlined here provide further evidence for the interdependence of protein, porphyrin, metal, and axial ligand and for the need for careful discrimination as to the relative importance of the several different factors that can control the properties and function of hemeproteins.

ACKNOWLEDGEMENT

This research was supported in part by a grant (HL-15980) from the United States Public Health Service.

REFERENCES

Alben, J.O. and Caughey, W.S,: 1968, An Infrared Study of Bound Carbon Monoxide in the Human Red Blood Cell, Isolated Hemoglobin, and Heme Carbonyls, Biochemistry 7, 175-183.

Alben, J.O. and Fager, L.Y.: 1972, Infrared Studies of Azide Bound to Myoglobin and Hemoglobin. Temperature Dependence of Ionicity, Biochemistry 11, 842-847.

Arnone, A. and Perutz, M.F.: 1974, Structure of Inositol Hexaphosphate - Human Deoxyhemoglobin Complex, Nature (London) 249 (5452), 34-36.

Barlow, C.H., Maxwell, J.C., Wallace, W.J., and Caughey, W.S.: 1973, Elucidation of the Mode of Binding of Oxygen to Iron in Oxyhemoglobin by Infrared Spectroscopy, Biochem. Biophys. Res. Commum. 55, 91-95.

Caughey, W.S.: 1973, Iron Porphyrins —Hemes and Hemins, in G.L. Eichorn (ed.), Inorganic Biochemistry, Elsevier, Amsterdam, 1973, pp. 797-831.

Caughey, W.S., Barlow, C.H., Maxwell, J.C. Volpe, J.A., and Wallace, W.J

1975a, Reactions of Oxygen with Hemoglobin, Cytochrome *c* Oxidase and Other Hemeproteins, Ann. N.Y. Acad. Sci. 244, 1-9.

Caughey, W.S., Barlow, C.H., O'Keeffe, D.H., and O'Toole, M.C.: 1973, Spectroscopic Studies of Cis and Trans Effects in Hemes and Hemins, Ann. N.Y. Acad. Sci., 206, 296-309.

Caughey, W.S., Deal, R.M., McLees, B.D., and Alben, J.O.: 1962, Species Equilibria in Nickel(II) Porphyrin Solutions: Effect of Porphyrin Structure, Solvent and Temperature, J. Amer Chem. Soc. 84 (9), 1735-1736.

Caughey, W.S., Fujimoto, W.Y., and Johnson, B.P.: 1966, Substituted Deuteroporphyrins II. Substituent Effects on Electronic Spectra, Nitrogen Basicities and Ligand Affinities, Biochemistry 5, 3830-3843.

Caughey, W.S., and Johnson, L.F.: 1969, Effects of Axial Ligand Upon Paramagnetic Shifts in Proton NMR Spectra of High-Spin Deuterohemins, Chem. Commum. 1362- 1363.

Caughey, W.S. and Koski, W.S.: 1962, Nuclear Magnetic Resonance Spectra of Porphyrins, Biochemistry 1, 923-931.

Caughey, W.S., Smythe, G.A., O'Keeffe, D.H., Maskasky, J.A., and Smith, M.L.: 1975b, Heme A of Cytochrome *c* Oxidase. Structure and Properties: Comparisons with Hemes B,C,S and Derivatives, J. Biol. Chem. 250, 7602-7622.

Caughey, W.S., York, J.L. and Iber, P.K.: 1967, Aspects of the Porphyrin Ring System, in A. Ehrenberg et al. (eds.), Magnetic Resonance in Biological Systems, Pergamon Press, Oxford, 1967, pp. 25-34.

Fuchsman, W.H., Weng, S.-H., and Caughey, W.S.: 1975, An Infrared Probe of Iron Porphyrin Oxidation State and Axial Ligation, Bioinorg. Chem. 4, 353-359.

Hill, H.A.O. and Morallee, K.G.: 1970, The Transmission of Electronic Effects in Paramagnetic Metal Complexes: the Proton Magnetic Resonance Spectra of Bis-pyridinoiron(III) Protoporphyrin IX Complexes, Chem. Commun., 266-267.

Hill, H.A.O. and Morallee, K.G.: 1972, Nuclear Magnetic Resonance Spectra of Bis(pyridinato) iron(III) — Protoporphyrin IX Complexes, J. Amer. Chem. Soc. 94, (3) 731-738.

Koenig, D.F.: 1965, The Structure of α-Chlorohemin, Acta Cryst. 18, 663-673.

LaMar, G.N., and Walker, F.A.: 1973, Proton Nuclear Magnetic Resonance and Electron Spin Resonance Investigation of the Electronic Structure and Magnetic Properties of Synthetic Low-Spin Ferric Porphyrins, J. Amer. Chem. Soc. 95, 1782-1796.

Lonquet-Higgins, H.C., Rector, C.W., and Platt, J.R.: 1955, Molecular Orbital Calculations on Porphine and Tetrahydroporphine, J. Chem. Phys. 18, 1174-1181.

Maxwell, J.C. and Caughey, W.S.: 1974, Infrared Evidence for Similar Metal - Dioxygen Bonding in Iron and Cobalt Oxyhemoglobins, Biochem. Biophys. Res. Commun. 60, 1309-1314.

Maxwell, J.C. and Caughey, W.S.: 1976, An Infrared Study of NO Bonding to Heme B and Hemoglobin A. Evidence for Inositol Hexaphosphate Induced Cleavage of Proximal Histidine to Iron Bonds, Biochemistry 15, 388-396.

Maxwell, J.C., Volpe, J.A., Barlow, C.H., and Caughey, W.S.: 1974,

Infrared Evidence for the Mode of Binding of Oxygen to Iron of Myoglobin from Heart Muscle, Biochem. Biophys. Res. Commun. 58, 166-171.

Mayer, A., Ogawa, S., Shulman, R.G., Yamane, T., Cavaleiro, J.A.S., Gonsalves, A.M.A.R., Kenner, G.W., and Smith, K.M.: 1974, Assignments of the Paramagnetically Shifted Heme Methyl Nuclear Magnetic Resonance Peaks of Cyanometmyoglobin by Selective Deuteration, J. Mol. Biol. 86, 749-756.

McCoy, S. and Caughey, W.S.: 1970, Infrared Studies of Azido, Cyano, and Other Derivatives of Metmyoglobin, Methemoglobin, and Hemins, Biochemistry 9, 2387-2393.

McCoy, S. and Caughey, W.S.: 1971, Infrared Evidence for Two Types of Bound CO in Carbonyl Myoglobins in Chance et al. (eds), Probes of Structure and Function of Macromolecules and Membranes vol. II: Probes of Enzymes and Hemoproteins, Academic Press, Inc., New York, 1971, pp. 289-293.

McLees, B.D. and Caughey, W.S.: 1968, Substituted Deuteroporphyrins V. Structures, Stabilities and Properties of Nickel(II) Complexes with Axial Ligands, Biochemistry 7, 642-652.

Ogawa, S., Shulman, R.G., and Yamane, T.: 1972, High Resolution Nuclear Magnetic Resonance Spectra of Hemoglobin. I. The Cyanide Complexes of α and β Chains, J. Mol. Biol. 70, 291-300.

O'Toole, M.C.: 1972, Infrared Study of Ligands Bound to Hemoglobin, Myoglobin, and Cytochrome *c* Oxidase, Ph. D. Dissertation, Arizona State University, Tempe, Az.

Shulman, R.G., Glarum, S.H., and Karplus, M.: 1971, Electronic Structure of Cyanide Complexes of Hemes and Hemeproteins, J. Mol. Biol. 57, 93-115.

Shulman, R.G., Wuthrich, K., Yamane, T., Antonini, E., and Brunori, M.: 1969, NMR of Reconstituted Myoglobins, Proc. Nat. Acad. Sci. U.S.A. 63 623-628.

Sugita, Y. and Yoneyama, Y.: 1971, Oxygen Equilibrium of Hemoglobins Containing Unnatural Hemes, J. Biol. Chem. 246, 389-394.

Tamura, M., Asakura, T., and Yonetani, T.: 1973, Heme Modification Studies of Myoglobin. I. Purification and Some Optical and EPR Characteristics of Synthesized Myoglobins Containing Unnatural Hemes, Biochim. Biophys. Acta 295, 467-479.

Vaska, L.: 1976, Dioxygen-Metal Complexes: Toward a Unified View, Acc. Chem. Res., in press.

Wallace, W.J., Maxwell, J.C., and Caughey, W.S.: 1974, The Mechanisms of Hemoglobin Autoxidation. Evidence for Proton-Assisted Nucleophilic Displacement of Superoxide by Anions, Biochem. Biophys. Res. Commun. 57, 1104-1110.

Wuthrich, K.: 1970, Structural Studies of Hemes and Hemoproteins by Nuclear Magnetic Resonance Spectroscopy, Structure and Bonding 8, 53-121.

Wuthrich, K. and Baumann, R.: 1973a, Hyperfine Shifts of the Carbon-13 NMR in the Low-Spin Iron(III)-Porphyrin Complexes, Helv. Chim. Acta 56, 585-96.

Wuthrich, K. and Baumann, R.: 1973b, Recent Developments in the Investigation of the Paramagnetic Centers in Low-Spin Ferric Hemoproteins.

Carbon-13 Hyperfine Shifts in Iron Porphyrin Complexes, Ann. N.Y. Acad. Sci. 222, 709-721.

Wuthrich, K., Shulman, R.G., Yamane, T., Wyluda, B.J., Hugli, T., and Gurd, F.R.N.: 1970, High Resolution Proton Magnetic Resonance Studies of Cyanoferrimyoglobins and Alkylated Derivatives from Different Species, J. Biol. Chem. 245, 1947-1953.

Yonetani, T., Yamamoto, H., and Woodrow, G.V., III: 1974, Studies on Cobalt Myoglobins and Hemoglobins. I. Preparation and Optical Properties of Myoglobins and Hemoglobins Containing Cobalt Proto-, Meso-, and Deuteroporphyrins and Thermodynamic Characterization of Their Reversible Oxygenation, J. Biol. Chem. 249, 682-690.

DISCUSSION

A. Lanir (Tel Aviv Univ.)*:* In the ^{13}C-NMR spectrum of ^{13}C—O bound to hemoglobin, two resonance peaks had been identified. I wonder if you can comment, whether the differences are due to changes in the bond between the iron and carbon-monoxide or to other subtler interactions between ligand and some groups around the heme pocket?

Can you say that the surroundings of the oxygen bound to hemoglobin are somewhat different in the α and β subunits?

Can you see such small differences by following IR spectrum of CO-hemoglobin?

Caughey: The two resonances peaks in the ^{13}C-NMR spectra of Hb ^{13}CO are reasonably assigned to α and β subunits. However the reasons for the chemical shift difference are by no means clear. There has not been sufficient data from simple heme carbonyls or from a variety of hemoprotein structures to permit an evaluation of the relative importance of trans ligands and environmental factors. Hopefully, with the availability of more data, the interpretation of ^{13}C-NMR data for bound ^{13}CO can be made more meaningful. Presently, interpretations of C—O strech frequencies are much more informative. With rabbit hemoglobins both IR and ^{13}C-NMR approaches have proven useful in the detection of differences in O_2 binding sites.

The infrared spectra for bound CO molecules provide a very sensitive probe for subtle differences in binding site. The C—O stretch is sensitive to Fe-CO bonding, to medium and to steric factors that may interfere with the otherwise preferred linearity of the Fe-C-O bonds. Hemoprotein carbonyls have ranged from about 1900 cm^{-1} to 1970 cm^{-1}. A range nearly as great has been seen among abnormal hemoglobins. However, in human hemoglobin A the ν_{CO} for the two subunits appear to differ by only about 2 cm^{-1} and is therefore difficult to resolve.

Ward: You mentioned that, unlike the case for CO and N_3^-, for oxy Hb the O—O stretching frequency is relatively insensitive to substitution at the 2 and 4 positions on the heme group. Have you an explanation for this effect?

Caughey: The stretching frequencies for CO and NO are indeed quite sensitive to both cis effects (due to changes in porphyrin) and trans effects (due to changes in trans ligand) whereas the O—O stretch is relatively much less sensitive to such effects. These differences are typically found in CO, NO, and O_2 coordination complexes and suggest that the O—O bond is less sensitive to cis and trans effects despite observations that on and off *rates* for O_2 binding may be quite sensitive to such effects.

Veillard: We have carried out a model calculation on the system $FePO_2(NH_3)_{proxim}(NH_3)_{distal}$ (P = porphyrin, $NH_{3\ proxim}$ and $NH_{3\ distal}$ represent the proximal and distal imidazole molecules of hemoglobin) in order to investigate the occurrence of a hydrogen bond between the dioxygen ligand and the distal imidazole molecule. The computed interaction is found repulsive (by 24 Kcal/mole), in agreement with your conclusion concerning the absence of hydrogen bonding. However the donor-acceptor interaction which you postulate between the $X^{\delta+} - O^{\delta-}$ ligand and the pyridine-type nitrogen to the distal imidazole seems hard to reconcile with (i) the X-ray crystallographic data of Watson which correspond approximately to a distance of 2.5 Å between the X_α atom of a diatomic ligand XO and the nearest nitrogen atom of the distal imidazole; (ii) the fact that, according to our calculation for the dioxygen complex of the Fe(II) porphyrin, the O_α atom appears slightly negative (computed charge 8.05).

Caughey: It is interesting that your calculations do not predict hydrogen bonding between distal histidine and bound dioxygen to be likely in accord with our earlier conclusions based largely on the nature of pH effects upon the infrared spectra of bound ligands. However I would add a note of caution to the use of NH_3 in place of histidine imidazoles since marked effects on bound ligand have been observed due to changes in either trans ligand or solvent. Furthermore, protein structure can significantly affect the accessibility of proximal histidine to iron or of the distal histidine to bound O_2.

The evidence *against* hydrogen bonding between distal histidine and O_2 does appear stronger than does the evidence *for* donor-acceptor interactions. However, I am reluctant to discard the possibility of bonding interactions of the donor-acceptor type between distal histidine and bound O_2 (at the αO atom) or another ligand. First, the distance (2.5 Å) you mention would not appear well established for the MbO_2 or HbO_2 cases. Second, development of a positive charge on the αO atom is not unreasonable particularly in view of the polarizing effect of an approac pyridine-like imidazole nitrogen. Third, teleological reasoning suggests the distal histidine, while not essential for reversible oxygen binding does have a role since it appears to be invariant among a wide variety of species (e.g., birds, mammals, fish, reptiles).

EPR STUDIES OF HUMAN NITROSYL HEMOGLOBINS AND THEIR RELATION TO MOLECULAR FUNCTION*

M. CHEVION**, W.E. BLUMBERG, and J. PEISACH
Depts. of Pharmacology and Molecular Biology, Albert Einstein College of Medicine of Yeshiva University, Bronx, N.Y. 10461, U.S.A. and Bell Laboratories, Murray Hill, N.J. 07974, U.S.A.

Nitric oxide binds to ferrous heme proteins forming nitrosyl heme compounds. Unlike oxy- and carbonmonoxy hemoglobin these compounds have an odd electron spin ($S = 1/2$) and their electronic structure can be investigated in detail using EPR [1] spectroscopy. Low temperature EPR spectra of nitrosyl derivatives of heme proteins are particularly interesting in that a superhyperfine structure (SHF) is often resolved at the part of the spectrum arising from g_z which is in the direction normal to the heme plane [1, 2, 3]. The SHF is sensitive to the interactions of the unpaired electron spin originating in the NO with nitrogen nuclei located along the heme normal. Such SHF patterns would consist of three, five or nine lines depending upon whether there is interaction of the unpaired electron with one, two equivalent or two inequivalent nitrogen nuclei, respectively.

It has been suggested using kinetic [4, 5], NMR [5] and CD [6] techniques that nitrosyl hemoglobin can undergo a change in quaternary structure which correlates with the lowering of affinity for NO upon the

* The portion of this investigation carried out at the Albert Einstein College of Medicine was supported in part by U.S. Public Health Service Grant HL-13399 from the Heart and Lung Institute and by National Cancer Institute Contract NO1-CP-55606 to J. Peisach. This is Communication No. 353 from the Joan and Lester Avnet Institute of Molecular Biology.

** Permanent Address: Department of Cellular Biochemistry, The Hebrew University and Hadassah, Medical School, Jerusalem, Israel.

[1] Abbreviations used are:

CD - Circular dichroism
DSS - 2,2-dimethyl-2-silapentane-5-sulfonate
EPR - Electron paramagnetic resonance
IHP - Inositol hexaphosphate
NMR - Nuclear magnetic resonance
SHF - Superhyperfine

B. Pullman and N. Goldblum (eds.), Metal-Ligand Interactions in Organic Chemistry and Biochemistry, part 2, 153-162.

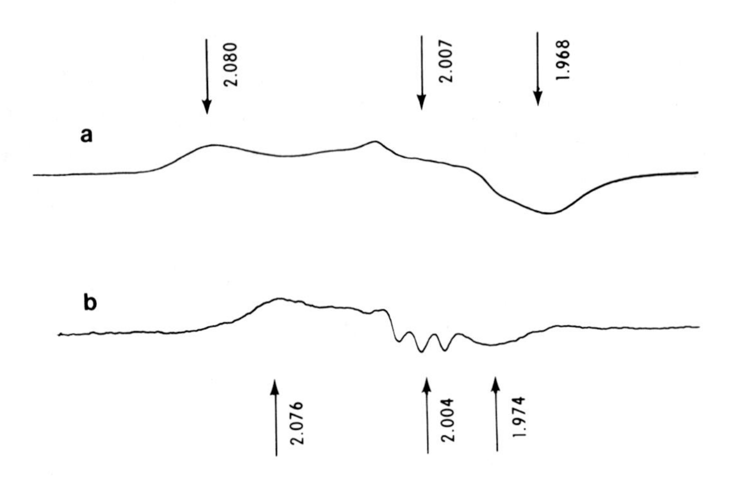

Fig. 1. EPR spectra of nitrosyl derivatives of hemoglobin A (0.2 mM in heme) recorded at 1.6 K. (a) at pH = 7.00, no phosphates present, (b) at pH = 6.95 in the presence of inositol hexaphosphate (0.1 mM). Arrows indicate the apparent g values based on the turning points of the absorption spectra.

addition of the molecular effector, inositol hexaphosphate. Also, it has been shown that a similar change both in quarternary structure and affinity for NO could be brought about by a change in pH in the absence of a molecular effector for the allosteric transition. As these findings now provide the opportunity to study the electronic structure at the hemes of nitrosyl hemoglobin when the protein is either in a high- or a low-affinity form, we decided to compare a few human hemoglobin variants under conditions which can be related to other measurements involving protein structure and ligand affinity.

The low temperature (1.6 K in our studies) EPR spectrum [7, 8] of nitrosyl hemoglobin A at pH 7 is a typical rhombic spectrum (Figure 1a). Here, the SHF structure is poorly resolved at g_z (g = 2.007) but by analogy with other heme proteins [2, 3, 9], a nine-line pattern is suggested. If now IHP is added, the spectrum is altered to a second type (Figure 1b) confirming the previous studies of Rein et al. [10]. Here, a three-line SHF pattern is clearly resolved.

Hemoglobin$_{\text{Kansas}}$, a mutant hemoglobin $[\alpha_2\beta_2^{102\ \text{Asn} \rightarrow \text{Thr}}]$ having an amino acid substitution in the β chains far removed from the heme [11] has lower affinity for oxygen, a reduced degree of coorperativity, a greater tendency to dissociate into dimers and a normal alkaline Bohr effect when compared to hemoglobin A. When nitrosyl Hemoglobin$_{\text{Kansas}}$ stripped of phosphates is studied, the two types of EPR spectra observed with hemoglobin A are observed here as well. The first type, with the three-line SHF pattern (Figure 2b) is observed in the pH range below 8.2, while the other type with higher multiplicity in the g_z region (Figure 2a) is observed above pH 8.2. In the case of Hemoglobin$_{\text{Kansas}}$ in the presence of IHP a similar behavior is observed. However, the

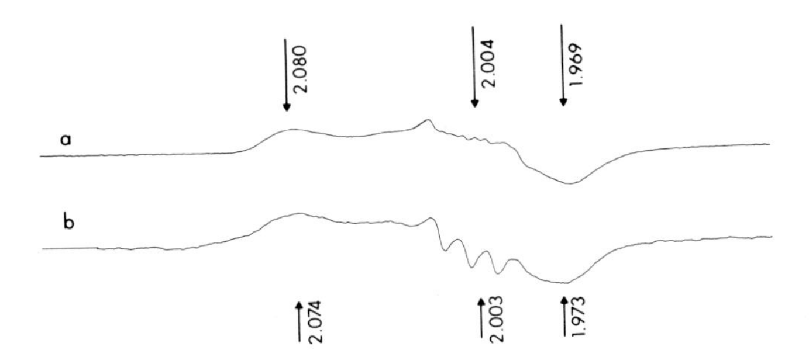

Fig. 2. EPR spectra of nitrosyl derivatives of Hemoglobin $_{\text{Kansas}}$ (0.2 mM in heme) recorded at 1.6 K. (a) at pH = 8.50, no phosphates present, (b) at pH = 7.95 in the presence of inositol hexaphosphate (0.1 mM). Arrows indicate the apparent *g* values.

transition between the two types of spectra is shifted up to pH 8.5. Good agreement is shown on comparison of these transitions with the transitions between the two quaternary structures as studied by NMR (Figure 3).

Although in all of these studies with hemoglobin, it has been assumed that the superhyperfine pattern with a higher multiplicity contained nine lines, analogous to that which is found for myoglobin, this had never been confirmed. In order to do so, we introduced a new technique into EPR spectroscopy. In this technique, which is currently used in a variety of other spectroscopies [12 – 19], one records the third harmonic component of the spectrum. This is achieved by applying a fundamental modulation frequency to the sample and tuning the phase-sensitive detector to the third harmonic frequency. With this technique a significant increase in the signal-to-noise ratio is achieved by the obligatory condition of using relatively high fundamental modulation amplitudes. The amplitude used is of the same magnitude as the signal under study. Figure 4 is an example in which the first and the third harmonics of the same sample were recorded. The third harmonic component clearly reveals nine lines. These are analyzed as a triplet of triplets with splittings of 6.5 – 7.0 G and 23 – 24 G, typical values for a nitrosyl hemoglobin in a high affinity form.

Careful examination of the EPR spectra of either hemoglobin A or Hemoglobin$_{\text{Kansas}}$ by the third harmonic technique, revealed that on no occasion a spectrum with a single three-line SHF pattern and with no contribution from a species giving rise to a nine-line SHF pattern was ever seen, even at low pH and in the presence of IHP. There was always an underlying nine-line pattern. As the amplitude of the nine-line SHF pattern is much smaller than that associated with the three-line pattern, it would be difficult to observe the nine-line component under the predominant three-line feature using conventional EPR spectroscopy. Quantitative comparison of the third harmonic EPR spectrum with the same

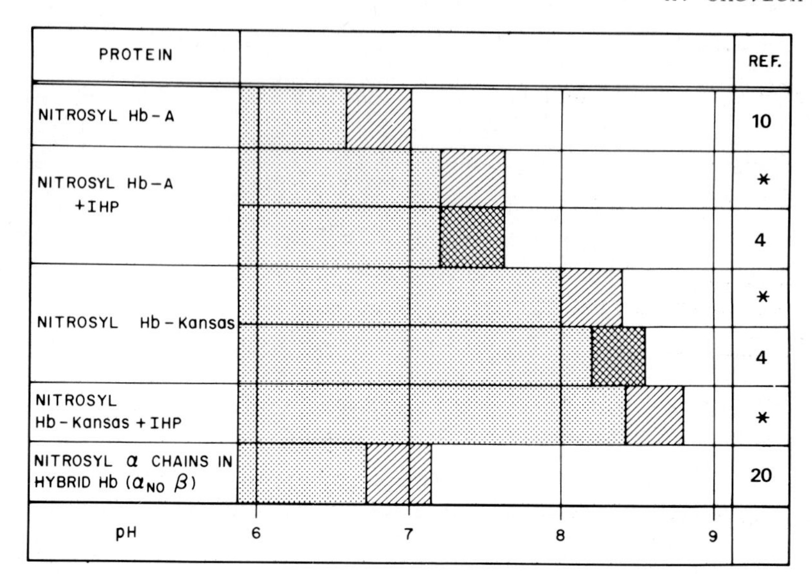

Fig. 3. The transition from low affinity (, dotted area) to high affinity form (, blank area) of nitrosyl hemoglobins. The striped ( ) and cross-hatched ( ) regions show the pH of the transition as determined by EPR and proton NMR, respectively. The width of these regions represents the uncertainty of each of these determinations. * indicates transitions based on data collected in this work.

hemoglobin in the high-affinity form shows that about 50% of the nine-line component persists even when the three-line feature predominates. This is taken as direct evidence that the EPR of α and β chains behave differently within tetrameric nitrosyl hemoglobin. This evidence, suggesting chain inequivalence, is in agreement with the findings by Henry and Banerjee [20] for the mixed hybrids $\alpha_{NO}\beta_{deoxy}$ and $\alpha_{deoxy}\beta_{NO}$. In both cases, only the nitrosylated chains contributes to the EPR. These authors found that the former $\alpha_{NO}\beta_{deoxy}$, exhibits either a three- or a nine-line SHF spectrum depending upon the state of oxygenation of the non-nitrosylated chain, while the latter, $\alpha_{deoxy}\beta_{NO}$, shows *only* the nine-line type spectrum independent of the state of oxygenation [2]

[2] From optical studies of Sugita [21], it was suggested that the spectral change seen in the transition from a high- to a low-affinity form of hemoglobin A arises almost exclusively from the α chains, even though both α and β chains alter their affinity to nearly the same extent

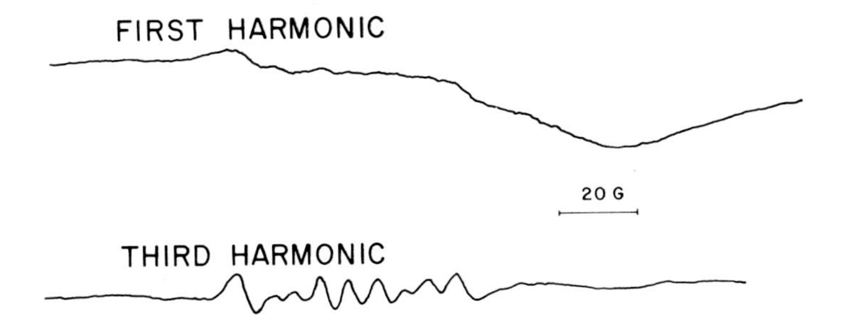

Fig. 4. A first and a third harmonic low temperature EPR spectrum of nitrosyl derivative of fetal hemoglobin (0.5 mM in heme) at pH = 7.2 in the absence of organic phosphate. (a) first derivative recorded with modulation amplitude of 1.5 G. (b) third harmonic recorded with modulation amplitude of 6.0 G. Analysis of the nine-line superhyperfine pattern into a triplet of triplets reveals that one triplet has a splitting of 23–24 G and the other triplet arises from an additional interaction with a splitting of 6.5–7 G.

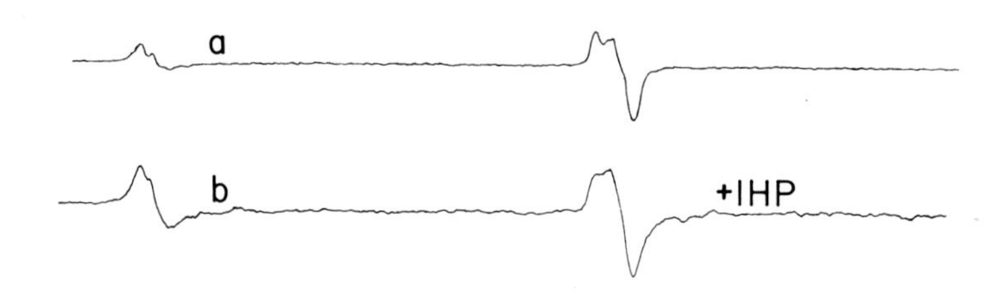

Fig. 5. EPR spectra of nitrosyl derivatives of Hemoglobin M_{Iwate} (500-5500 G) (0.9 mM in heme) recorded at 1.6 K. (a) at pH = 7.10, no phosphate present (g = 6.20, 5.86, 2.00). (b) at pH 7.10 in the presence of inositol hexaphosphate (1.1 mM) recorded with lower gain than a, (g = 6.20, 5.83, 2.00). The apparent g values for the high spin ferric iron are based on the turning points of the absorption derivative spectra.

On the basis of this study we propose that the α-chains can exhibit either a three- or nine-line SHF pattern while the β-chains can exhibit only a nine-line pattern in the tetramer.

In order to test this proposal, we examined the EPR of Hemoglobin M_{Iwate}, a variant hemoglobin in which the covalently bound proximal histidine imidazole of the α chain is substituted by a tyrosyl group. In this protein only the β chains can bind NO, while the α chains reside in the ferric form. The EPR spectrum of the nitrosyl derivative of Hemoglobin M_{Iwate} stripped of phosphates at pH 7 shows the presence of two different paramagnetic species (Figure 5a), the high-spin ferric

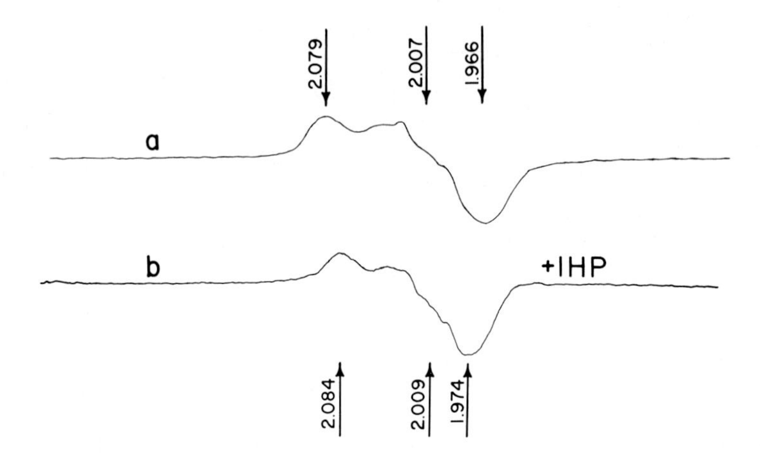

Fig. 6. EPR spectra of nitrosyl derivatives of Hemoglobin M_{Iwate} (0.9 mM in heme) recorded at 1.6 K. (a) at pH = 7.10, no phosphates present. (b) at pH = 7.10 in the presence of inositol hexaphosphate (1.1 mM). Arrows indicate the apparent *g* values.

iron (S = 5/2) of the α^{Mmet} chains (g = 6.20, 5.86 and 2.00) [22] and the unpaired electron originating in the NO (g = 2.079, 2.007 and 1.966) bound to the beta chains. The part of the spectrum around g = 2 is similar to Figures 1a and 2a and the SHF pattern consists of a multiplet of nine-lines (Figure 6a). The addition of IHP does not change the g values nor the SHF pattern (Figure 6b). Hemoglobin M_{Iwate} at pH 7 is believed to be in a low-affinity form [23] and the addition of IHP would favor such a quaternary state even more. Yet, under no experimental condition are we ever able to demonstrate a three-line SHF pattern as is observed with hemoglobin A or $Hemoglobin_{Kansas}$ under conditions favoring a low-affinity form.

In another EPR study carried out with human fetal hemoglobin, driving the protein to a low-affinity form with IHP or lowering pH caused the spectrum to change from one with a nine-line SHF pattern to one with a predominantly three-line pattern and having an underlying nine-line pattern which is well-resolved by the third harmonic technique As in the case of hemoglobin A, it is suggested that the spectrum of α chains in tetrameric nitrosyl fetal hemoglobin is altered when the protein changes both its quaternary state and ligand affinity while the γ chains, structurally similar to β chains in hemoglobin A and $Hemoglobi_{Kansas}$ exhibit only a nine-line SHF pattern.

Since EPR is a local probe of paramagnetic centers, what structural parameters in the hemoglobin molecule can be considered responsible for the changes observed in the SHF pattern of nitrosyl hemoglobin? Kon [24] has recently suggested that the three-line spectrum may be taken as a diagnostic for pentacoordinated heme iron. He states that whenever this

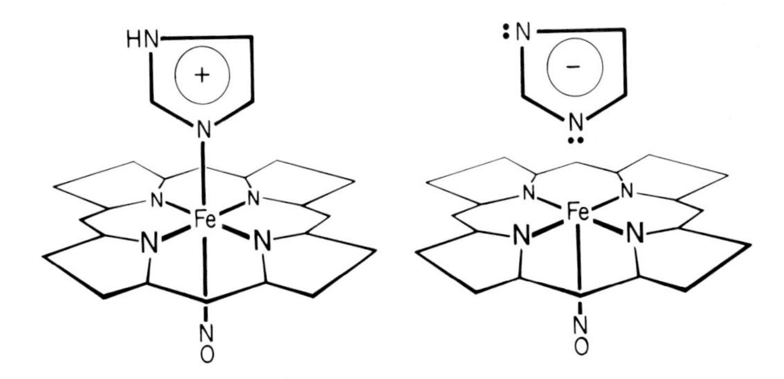

Fig. 7. The two aromatic chemical structures of imidazole-NO heme complexes. In the structure on the left the imidazole molecule binds in the neutral form, the nitrogen atom (N-1) not bound to the iron remaining protonated. The electronic structure is formally that of the imidazolium cation. In the structure on the right the imidazole has lost the proton, and its electronic structure is formally that of the imidazolide anion. The σ and π bonds to the iron atom are different in the two cases.

type of spectrum is observed for nitrosyl hemoglobin the covalent bond of the iron to the protein via the proximal histidine is dissociated. Although the EPR spectrum of pentacoordinated heme-NO studied by Kon had a three-line SHF, the shape of the pattern and the g anisotropy were different from hexacoordinated cases as are found for nitrosyl hemeproteins which also exhibit the three-line SHF pattern, and thus another explanation for the three-line spectrum must be sought.

Studies on model compounds by Henry et al. [25] have shown that for a series of heme-NO complexes having various ring substituted nitrogenous bases at the trans axial ligand, the EPR spectrum observed could have SHF structure consisting of either nine or three lines depending on whether the substitution was electron-supplying or electron-withdrawing, respectively. As the g anisotropy does not vary much in both instances, these authors conclude that the iron remains hexacoordinated in both cases. By analogy, in hemoglobin it is suggested that the imidazole remains bound to the heme when either a three-line or a nine-line spectrum is elicited, but the bonding of the imidazole to iron in those species represented by the nine-line spectrum is different from that in those species giving rise to the three-line spectrum.

What structural changes can effect the bonding of the imidazole to the heme? It has been demonstrated that for *bis* imidazole heme complexes, imidazole can exist in two alternative structures, one of which has the proton on N-1 and the other of which has the proton removed [26, 27] (Figure 7). Similarly, the differences in the EPR characteristics of the low-affinity form of nitrosyl hemoglobin compared to the high-affinity form may be a result of an alteration of the structure of the imidazole ligated to the heme, giving rise to two different SHF patterns.

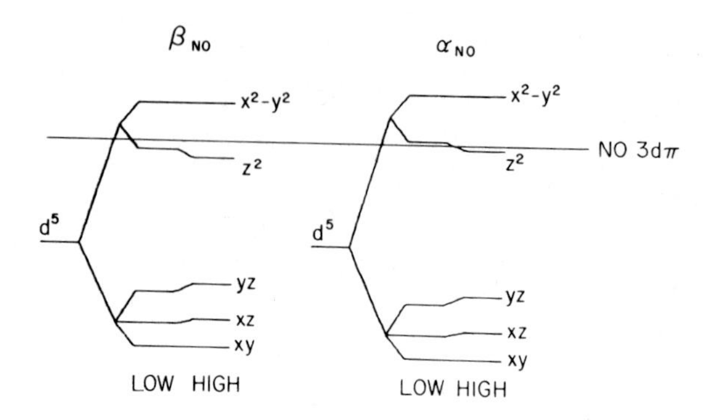

Fig. 8. Schematic arrangement of the iron 3d orbitals in NO-ligated α and β chains of hemoglobin A and Hemoglobin$_{Kansas}$. In the case of α chains, a change in the affinity is accompanied by sufficient shift in the energy levels that the d_{z^2} orbital of the iron atom and the $3d\pi$ orbital of the NO exchange order. Although the shift in the energy levels of the β chains is approximately the same under the same conditions, the levels are arranged so that no exchange of order takes place.

It is suggested that when the hemoglobin is in a low-affinity form, the proximal imidazole has a proton on N-1, while in a high-affinity form, the proton is removed.

Energetic considerations can explain how these alternative structures of imidazole can so radically alter the EPR spectrum of nitrosyl hemoglobin. The interaction of heme with NO and other ligands removes the five-fold degeneracy of the d orbitals of the iron. The octahedral coordination of the iron in heme compounds splits the e_g and t_{2g} orbitals leaving the former at higher energy. Lower symmetry fields completely remove the remaining degeneracy. These levels are shown schematically (but not to scale) in Figure 8. The hybridization demonstrated by the spectra of Figures 1a and 2a indicate that the energy level of the unpaired electron in the NO is comparable to the energy of the d_{z^2} orbital of the iron is ferrous hemoglobin. Considering the principal resonance forms of the imidazole ring that are shown in Figure 7, we suggest that when the ring has the proton, the d_{z^2} orbital participates in an σ -antibonding molecular orbital with the nitrogen atom of the ring, raising its energy. In addition, π electron donation from the metal to the ring will lower the energies of the d_{xz} and d_{yz} orbitals relative to the d_{xy} orbitals providing a smaller axial crystal field splitting within the three t_{2g} levels. When the ring is deprotonated, there is little σ bonding interaction, but there is a large π electron donation from the ring to the metal atom, increasing the splitting of the t_{2g} levels. This change of crystal field of the t_{2g} levels has been demonstrated [9] in the case of ferric low-spin heme compounds. The change in EPR spectra of the NO adducts shows that the 3dπ orbital [28] of the NO α chains and the d_{z^2} orbital of the iron can invert their order, thus changing the orbital in which the unpaired spin resides. When the energy level of the NO is lower than the d_{z^2}, as is the case when the proximal imidazole has a proton on N-1, the electron interacts only with one nitrogen nucleus. that of the NO, and the spectrum will show three lines (Figures 1b, 2b). In the case of deprotonated imidazole, on the other hand, there would be weaker interaction and weaker cubic crystal field resulting in smaller splitting of the d-orbitals so that the energy of the iron d_{z^2} is lower than that of the NO and the electron will have interaction with both nitrogen nuclei but in an inequivalent manner, giving rise to a nine-line spectrum (Figures 1a, 2a).

In the case of β or γ chains, on the other hand, a weaker crystal field in these chains causes the e_g orbitals of the iron to be lower in energy. Thus when the protein is either in a high- or a low-affinity form, there is no crossing of the d_{z^2} and the 3dπ orbitals (Figure 8). Indeed the β or γ chains would exhibit nine-line SHF spectrum even when in low-affinity form as the unpaired spin resides continuously in the d_{z^2} orbital of the iron, and interacts with the two nitrogen nuclei in each affinity state.

In summary, we suggest that the SHF pattern observed in the low temperature EPR spectrum of nitrosyl hemoglobins can serve as a marker for the quaternary structure and ligand affinity. There is good agreement between markers of *globin* characteristics, such as the 14 ppm line downfield from DSS in the NMR or the 285 nm negative peak in the CD, and *heme* characteristics as studied by EPR. Evidence is presented here indicating subunit inequivalence in the hemoglobin tetramer. Also, we present a theory that relates SHF patterns to a molecular orbital picture of the nitrosyl heme complex.

REFERENCES

1. Griffith, J.S.: (1956), Proc. Roy. Soc. (London) A235, 23.
2. Chien, J.C.W.: (1969), J. Chem. Phys. 51, 4220.
3. Dickinson, L.C. and Cien, J.C.W.: (1971), J. Amer. Chem. Soc. 93, 5036.
4. Salhany, J.M., Ogawa, S., and Shulman, R.G.: (1975), Biochemistry 14, 2180.
5. Moore, E.G. and Gibson, Q.H.: (1976), J. Biol. Chem. 251, 2788.
6. Salhany, J.M.: (1974), FEBS Letters 49, 84.
7. Feher, G.: (1957), Bell System Technical J. 26, 449.
8. Berzofsky, J.A., Peisach, J., and Blumberg, W.E.: (1971), J. Biol. Chem. 246, 3367.
9. Yonetani, T., Yamamoto, H., Erman, J.E., Leigh, J.S., and Reed, G.H;: (1972) J. Biol. Chem. 247, 2447.
10. Rein, H., Ristau, O., and Scheler, W.: (1972), FEBS Letters 24, 24.
11. Bonaventura, J. and Riggs, A.: (1968), J. Biol. Chem. 243, 980.
12. Wilson, G.V.H.: (1963) J. Appl. Phys. 34, 3276.

13. Allen, L.C., Gladney, H.M., and Glarum, S.H.: (1964), J. Chem. Phys. 40, 3135.
14. Glarum, S.H.: (1965), Rev. Sci. Instrum. 36, 771.
15. Beynon, J.H., Clough, S., and Williams, A.E.: (1958), J. Sci. Instrum. 35, 164.
16. Butler, W.L. and Hopkins, D.W.: (1970), Photochem. and Photobiol. 12, 439.
17. Harris, L.A.: (1968), J. Appl. Phys. 39, 1419.
18. Smith, N.V. and Traum, M.M.: (1970), Phys. Rev. Lett. 25, 1017.
19. Rowe, J.E. and Ibach, H.: (1973), Phys. Rev. Lett. 31, 102.
20. Henry, Y. and Banerjee, R.: (1973), J. Mol. Biol. 73, 469.
21. Sugita, Y.: (1975), J. Biol. Chem. 250, 1251.
22. Watari, E., Hayashi, A., Morimoto, H., and Kotani, M.: (1969), in S. Fujiwara and L.H. Piette (eds.), Recent Developments of Magnetic Resonance in Biological Systems, Hirokawa Publishing Company, Tokyo, p. 128.
23. Mayer, A., Ogawa, S., Shulman, R.G., and Gersonde, K.: (1973), J. Mol. Biol. 81, 187.
24. Kon. H.: (1975), Biochim. Biophys. Acta 379, 103.
25. Henry, Y., Peisach, J., and Blumberg, W.E.: (1975), Abstracts of the 19th Annual Meeting of the Biophysical Society, Philadelphia, Biophys. J. 15, 286a.
26. Peisach, J., Blumberg, W.E., and Adler, A.: (1973) Ann. N.Y. Acad. Sci. 206, 310.
27. Peisach, J. and Mims, W.B.: submitted for publication.
28. McMillan, J.A.: (1968), in Electron Paramagnetism, Reinhold, New York, p. 28.

DISCUSSION

CAUGHEY: Recently we reported (Maxwell and Caughey, Biochemistry (1976) infrared data on the N—O stretching frequencies as well as Soret-visible and EPR spectral data on five and six-coordinate protein-free hemes and on hemoglobin A nitrosyls with and without IHP present. We interpreted the effects of IHP on Hb A NO as consistent with two of the four subunits experiencing cleavage of the proximal histidine to iron bond upon IHP binding. Thus IHP caused two hemes to become five coordinate. Do your EPR results reported here appear consistent with such an interpretation?

CHEVION: Five coordinated heme-NO complexes as reported by Kon [7] are characterized by two features: (a) A very sharp and narrow three spikes and (b) g anisotropy that is quite different from that of hexa-coordinated heme-NO complexes and is usually located at the g_{min} region.

In our studies a similar pattern to that was observed with denatured hemoglobin. This is in contrast to the pattern I showed here for both hemoglobin A and hemoglobin Kansas that is characteristic of native fully ligated protein in the low affinity state. Thus we do not take the view that in the presence of IHP two subunits of each tetramer are penta-coordinated. We do suggest that coupled with the transition between the two quaternary states, an alteration in the bonding of the NO to the heme takes place in both α and β subunits. As these chains are not identical the alteration is expressed differently for the two subunits.

CARBON-13 NMR STUDIES OF HUMAN HEMOGLOBIN AND SOME OTHER HEMOPROTEINS

R. BANERJEE*, J.M. LHOSTE**, and F. STETZKOWSKI*

1. INTRODUCTION

The ligand of primary interest for hemoglobin and myoglobin is oxygen, the physiological role of these proteins being precisely the transport and storage of this gas for metabolic use. Carbon monoxide, a redoubtable competitive inhibitor of oxygen has nevertheless occupied an important place in hemoglobin research. Reasons for this preference are manifold; it has been realized very early that the properties of carbonyl hemoglobins are in most respects closely similar to those of oxyhemoglobin. More important, protein structural changes [1, 2] which give rise to the physiologically important property of cooperative binding and release of oxygen are the same for the two ligands, oxygen and carbon monoxide. Similar considerations apply to the Bohr effect which is the dependence of apparent ligand affinity on pH. From a practical point of view, carbon monoxide offers the advantage of being more convenient for kinetic studies, either by rapid mixing or by flash photolysis, the carbonyl derivative being highly photodissociable. Moreover, the advent of ^{13}C NMR spectroscopy permits to dispose of a ligand, ^{13}CO which, at the same time 'reports' on the environment perceived in its vicinity.

Initiated several years ago (3-5), the ^{13}C NMR of human and other mammalian $Hb^{13}CO$ have yielded an interesting result; in general, the carboxyhemoglobins show two distinct resonances of equal intensity which have been shown to arise from ^{13}CO bound either to the α or the β chain. The environment experienced by the carbonyl ligand in the two types of chain inside the hemoglobin tetramer is thus different. This result which confirms other evidence of chain heterogeneity also suggests a direct method for the identification and quantitative estimation of the two kinds of chains in systems where they may not be present in equivalent amounts.

The study described in this report was undertaken with a view to gain further insight into the underlying causes of different chemical shifts shown by the ^{13}C of carbonyl ligand in different hemoproteins. Measurements have been performed on several hemoproteins with widely different CO-affinities and the chemical shifts considered in relation to available knowledge on the stability of the carbonyl complexes.

---

* Institut de Biologie Physico-chimique, 13 rue Pierre et Marie Curie, 75005 Paris
** Fondation Curie - Institut du Radium, Orsay, France

B. Pullman and N. Goldblum (eds.), Metal-Ligand Interactions in Organic Chemistry and Biochemistry, part 2, 163-171.

Also, the possibility of chain quantitation by integration procedures were made use of in studying chain heterogeneity for an interesting reaction, namely oxidation-reduction equilibrium.

2. MATERIALS AND METHODS

The hemoproteins used in this study were: human hemoglobin and its isolated α and β chains; valency hybrids derived from human hemoglobin (see below); leghemoglobin isolated from Soya bean root nodules; horse myoglobin and its sulf-derivative. The experimental material were prepared according to standard published methods cited elsewhere [6-7]. When obtained in the ferrous oxygenated from, the proteins were converted into the carboxy derivatives by allowing to equilibrate with a slight positive pressure of ^{13}CO (90.5 % ^{13}C, Merck, Sharp & Dohme, Canada). Leghemoglobin and myoglobin which are generally isolated in the ferri form were converted into the ferrous carboxy derivatives by reduction with a 20-fold excess of sodium dithionite under 1 atm. ^{13}CO.

NMR samples tube was of 10 mm o.d. A 3 mm tube containing D_2O (to serve as a field frequency lock) and TMS (in a sealed capillary, to serve as an external reference) was positioned concentrically inside the sample tube. ^{13}C NMR spectra were recorded at 25.2 MHz and 34° using a Varian XL-100 spectrometer operated in the Fourier Transform mode with proton noise decoupling. Spectral widths of 5 500 Hz (220 ppm) were used for shift measurements ($\pm$ 0.03 ppm) with respect to the primary external TMS reference and of 200 Hz for measurement of integrated intensities. Data accumulation were carried out for 4 to 12 h using 90° r.f. pulses (45 μsec) and 0.7 or 1 s acquisition times respectively. Under these conditions neither the acquisition rate ($\geqslant 3\ \underline{T_1}$) nor the nuclear Overhauser effect due to proton decoupling have an influence upon the relative integrated intensities of the ^{13}C resonances corresponding to ^{13}CO bound respectively to the α and β chains.

The correction of the shifts for macroscopic susceptibility effects of the hemoglobin solution (valency hybrids) upon the external reference were measured by comparison of the proton resonance of the external TMS, with that of internal TTS. A low field shift of 0.065 ppm for the external reference was observed upon complete oxidation to high spin metHb of a 4 mM diamagnetic HbO_2 solution

3. RESULTS AND DISCUSSION

3.1. *Chemical shifts of multiple chain and single chain hemoglobin carbonyls*

For $Hb^{13}CO$, the chemical shifts do not depend on subunit structure since the α and β resonances of the tetrameric protein are in the same positions as for the isolated monomeric α and tetrameric β chains [3, 6]. Nor are they sensitive to allosteric effectors (inositol hexaphosphate [IHP], H^+) which change the *apparent* overall binding parameters of the carbonyl ligand. Clearly, the ^{13}C of the carbonyl complex 'feels' only its local environment and ignores all about the previous 'history' of the reaction, i.e. changes of protein structure or of ionizations associated with ligand binding, changes which can have different free energy contributions depending on pH or on the presence of effectors. Such subsidiary free energy contributions are appreciable for the normal hemoglobin but are believed to be small for the isolated chains as well as for the single chain proteins examined. The chemical shifts cover a range of several ppm; when they are considered in relation to the stability of the respective carbonyl complexes, an approximate relationship appears, larger stability of the carbonyl complex corresponding to a larger shielding of the carbon nucleus of the ligand (Table I).

TABLE I
^{13}C chemical shifts of ^{13}CO in some carbonyl hemoglobins and myoglobins together with reported values of the free energy of formation of the carbonyl derivatives.

Protein	Chemical shift[a)]	$-\Delta G°$ (Kcal)
Leghemoglobin a	207.33	12.0
Human β chain	206.69	11.7
Human α chain	207.14	11.4
Horse myoglobin	208.29	10.0
Horse sulfmyoglobin	211.17	6.0

a) in ppm (± 0.03) downfield from external TMS

Such shielding may arise from charge transfer from the haem group as well as from polarization of the CO-ligand [8, 9]. Lb_a does not seem to follow this alignment. Other proteins examined all belong to or are derived from those for which the haem environment as seen from X-ray studies seem to be similar [1, 10], notable features being a hydrophobic cavity and the presence of an imidazole group on the distal side; the structure around the hæm in leghemoglobin is not known in detail and may be different from that of others. However, a definitive correlation on the lines suggested above should await further results on other hemoproteins and model complexes.

3.2. *Human hemoglobin valency hybrids:* $\alpha^{II}\beta^{III}$ *&* $\alpha^{III}\beta^{II}$

Of particular interest is the shift of resonances, when they occur for the same protein as a result of perturbations exercised through various interactions with external agents: solutes or the partner subunit in an oligomer. A notable example is the effect of changed tertiary structure of partner ferri-chains in tetrameric valency hybrids [11]. These are molecules in which the haem groups belonging either to the α or to the β chains have been selectively oxidized to the ferri state and consequently cannot bind oxygen or carbon monoxide. Half of the sites being thus 'freezed' in a ligated state, these molecules have been considered to resemble in some way the real short--lived intermediates formed in the course of saturation of hemoglobin by gaseous ligands. Various studies have already been performed on these compounds with a view to understand the molecular events associated with the change of the tetramer from half-ligated to the fully ligated form [12-14], sometimes parallely with change of quaternary structure.

The purpose of the present ^{13}C NMR experiments with the hybrids was to know if the carbonyl ligand bound to the respective ferrous chains would sense a change of tertiary structure of the partner ferri chain. This structure change can be brought about by replacing the weakfield water molecule by cyanide for the sixth ligand of the hexa-coordinated ferric ion. This ligand replacement reaction ($H_2O \rightarrow CN^-$) results in a change of spin state of the iron from $\underline{S} = 5/2$ to $\underline{S} = 1/2$ and is known to shorten the Fe-N bond to the proximal histidine by about o.3 Å [2]. It appears from the results (Table II) that the two hybrids behave differently.

TABLE II
Effect of exchanging the sisth ligand ($H_2O \rightarrow CN$) of the ferri-chain haem on the carbonyl ^{13}C resonance of ferrous-chain for valency hybrids.

Hybrid	Chemical shift a)	
$(\alpha^{II}CO)_2\ (\beta^{III}H_2O)_2$	207.28	
$(\alpha^{II}CO)_2\ (\beta^{III}CN)_2$	207.12	
$(\alpha^{III}H_2O)_2\ (\beta^{II}CO)_2$		206.65
$(\alpha^{III}CN)_2\ (\beta^{II}CO)_2$		206.61

a) in ppm (± 0.03) downfield from external TMS. Corrected for macroscopic paramagnetic susceptibility.

The ^{13}C resonance of the carbonyl ligand bound to the ferrous α chain of $\alpha^{II}\beta^{III}$ moves significantly upfield, but no change is observed for the β chain carbonyl for the other hybrid, $\alpha^{III}\beta^{II}$. This finding, together with other observations [14-17] suggest that the haem environment of the α chain may be more sensitive to certain protein structural changes even when caused by agents which act preferentially on the β chain.

3.3. *Study of chain heterogeneity in oxidation-reduction equilibrium*

The possibilities offered by ^{13}C NMR of $Hb^{13}CO$ for the identification and quantitation of individual chains can be applied to cases where α and β chains of intact hemoglobin may possess different reaction parameters and, as a result, be present in non equivalent amounts in course of a reaction or at equilibrium. An example is the oxidation--reduction equilibrium of human hemoglobin which differs from the oxygenation equilibrium in an important aspect. For the two reactions, the Hill coefficient, $\underline{n}$, is affected differently by a change of pH; the value of $\underline{n}$ for the oxidation-reduction equilibrium changes from 1.2 at pH 6 to 2.5 at pH 9 [18] whereas the oxygenation equilibrium remains equally cooperative ($\underline{n}$ = 2.7) throughout this region. A proposal to explain the properties of the oxidation-reduction system is based on *pH-dependent* chain heterogeneity [19], i.e. a difference of intrinsic $\underline{E}_{1/2}$ values of the α and β chains in neutral solution but not at high pH, in about the same way as the $\underline{E}_{1/2}$ values of the isolated chain have been found to differ [20]. This interpretation has been contested recently [21] on the basis of kinetic studies which tended to show a constant difference at pH 6 and at pH 8.7 [22], i.e. absence of pH-dependence. Since the kinetic evidence was indirect, we have undertaken a reinvestigation of the problem by using the straightforward method offered by ^{13}C NMR for identifying and estimating the ferrous chains of both types.

A 1:1 mixture ferrous deoxyhemoglobin and ferri-hemoglobin in dye-mediated oxidation-reduction equilibrium is characterized by an overall potential at a given pH. This half reduction potential corresponds to [Ox] = [Red], but since each of the above species includes in its turn two subspecies, namely α^{III}, β^{III} and α^{II}, β^{II} the condition $\alpha^{III}/\alpha^{II} = \beta^{III}/\beta^{II}$ at equilibrium may be satisfied only if the intrinsic potentials of the α and β chains are equal. Conversely, if these ratios are unequal but known, they can give the differences of the individual $\underline{E}_{1/2}$ values. In practice, it is sufficient to measure α^{II}/β^{II}, since the systems considered contain as many total ferrous (α + β) as total ferric chain (α + β). The ratio α^{II}/β^{II} at equilibrium can be measured by 'freezing' the ferrous components by reacting with ^{13}CO and by integrating the ^{13}C NMR spectra. The dye is present only in catalytic amounts and the high combination rate of ^{13}CO ensures that the proportions of $\alpha^{13}CO$ and $\beta^{13}CO$ remain the same as those of α^{II} and β^{II} at equilibrium.

Many measurements were done on such equilibrium mixtures at different pH values; some others were carried out in the presence of the polyanions, inositol hexaphosphate (IHP) and benzene hexacarboxylate (BHC) which are known to significantly raise the overall oxidation-reduction potential. It is clear from the values reported in Table III that the ratio α^{II}/β^{II} is pH-dependent; β^{II} is predominant in neutral solution [$\underline{E}_{1/2}(\beta) > \underline{E}_{1/2}(\alpha)$], but not at pH 9 and above where the ratio α^{II}/β^{II} is nearly equal to unity [$\underline{E}_{1/2}(\beta) \approx \underline{E}_{1/2}(\alpha)$] (Figure 1).

TABLE III
The proportions [a] of ferrous α and β chains in 1:1 mixtures of ferrihemoglobin and deoxy ferrous hemoglobin in dye-mediated oxidation-reduction equilibrium, at different pH values.

	pH	α^{II}	β^{II}
	6.0	28	72
	6.3	30	70
	6.5	35	75
	7.1	30	70
	7.4	35	65
	7.8	35	65
	8.0	40	60
	8.2	38	62
	8.4	39	61
	8.6	41	59
	8.9	43	57
	9.2	49	51
	9.3	50	50
	9.7	50	50
+ IHP	6.1	30	70
	6.6	32	68
	7.0	30	70
	7.2	31	69
+ BHC	6.9	29	71

a) The error of the integration procedure is normally within ± 3%. For samples having a β:α ratio of 2 or more, the error may be higher, up to ± 5%.

The values of $\Delta E_{1/2}$ $[E_{1/2}(\beta) - E_{1/2}(\alpha)]$ as calculated at different pH values by assuming the involvement of a single electron for each couple α^{III}/α^{II} and β^{III}/β^{II} clearly show the existence of a pH- dependent difference of intrinsic potential of the chains (Figure 2). Moreover, it appears that the increase of the overall potential by IHP and BHC (inset Figure 2) is not contributed specifically by any one type of chain. If the individual values were to change at all, they should do so in a manner as to keep the difference, $\Delta E_{1/2}$, about the same. More likely, the observed changes of overall $E_{1/2}$ could take place without any modification of the intrinsic potentials. The effect of the polyanions could possibly arise solely from their actions on protein structure remote from the haem group, the only thermodynamic requirement being that they act differently on the oxidized and reduced form.

The above results may be safely be taken as confirming *pH-dependent* differences of intrinsic oxidation-reduction potentials of the chains in the hemoglobin tetramer. In an allosteric scheme. Perutz et al. [21] have totally disregarded the possible effect of such hetero-

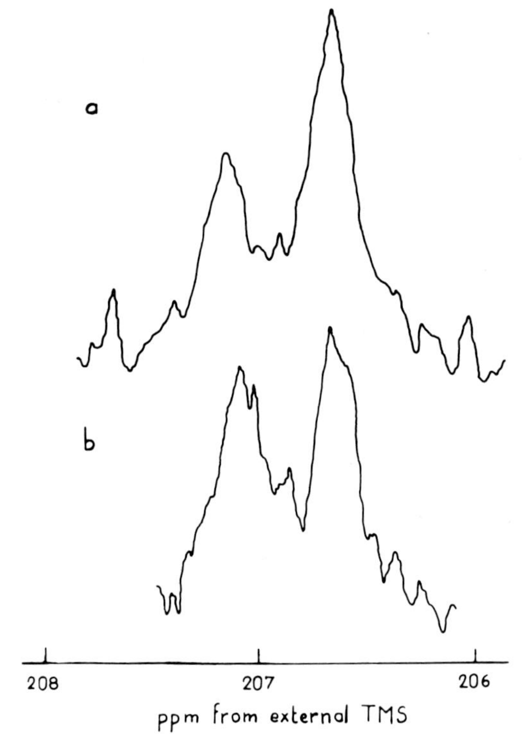

Fig. 1. Proton noise decoupled ^{13}C NMR spectra in the ^{13}CO region of carbonylated samples derived from oxidation-experiments at pH 6.5 (a) and pH 9.5 (b).

geneities on the Hill coefficient. It is obvious, however, that pH-dependent difference of $\underline{E}_{1/2}(\beta)$ and $\underline{E}_{1/2}(\alpha)$ must show up for the overall equilibrium and make the Hill coefficient to change as a function of pH. However, one cannot say at present whether the observed change of $\underline{n}$ with pH can be fully accounted for by chain heterogeneity or requires supplementary contributions arising from allosteric effects.

ACKNOWLEDGEMENTS
This work was supported by grants from the Centre National de la Recherche Scientifique, Délégation à la Recherche Scientifique et Technique and the Institut National de la Santé et de la Recherche Médicale.

Abbreviations. $\underline{E}$: potential referred to the standard hydrogen electrode for [Ox] = [Red]; $\underline{E}_{1/2}(\alpha)$, $\underline{E}_{1/2}(\beta)$ are the $\underline{E}_{1/2}$ values for the respective chains; α^{III}, α^{II}, β^{III}, β^{II}: the roman numerals refer to the valency states; IHP: inositol hexaphosphate; BHC: benzene hexacarboxylate; TMS: tetramethyl silane; TTS: tetradeuterotrimethylsilyl propionate.

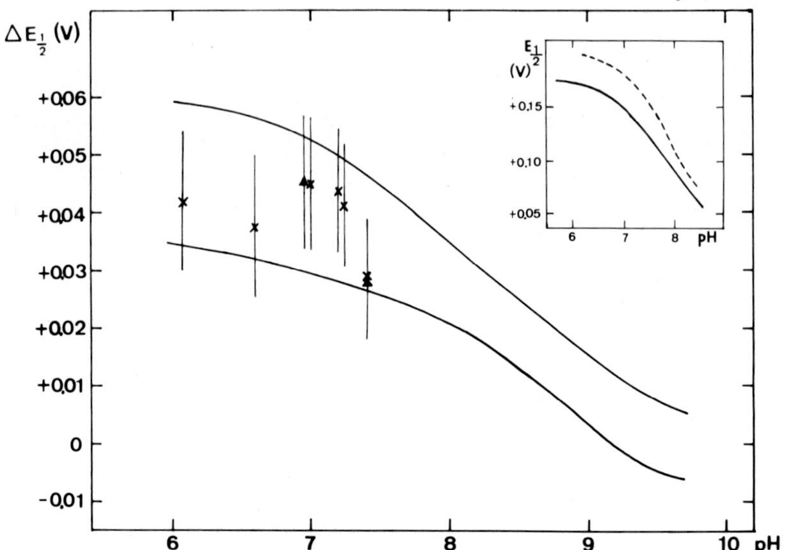

Fig. 2. Estimated values of $\Delta \underline{E}_{1/2}$, the difference between the oxidation-reduction potentials of chains in intact hemoglobin, at different pH values. $\Delta \underline{E}_{1/2} = [\underline{E}_{1/2}(\beta) - \underline{E}_{1/2}(\alpha)]$, as calculated from ratios of β^{II}/α^{II} obtained by integration of the NMR spectra. The $\Delta \underline{E}_{1/2}$ values, with error, lie between the two curved lines. The points with vertical error bars represent the same, in experiments where IHP (✕) and BHC (▲) were present.

Inset: the effect of IHP (dashed line) on the overall $\underline{E}_{1/2}$. BHC exerces a similar effect.

REFERENCES

1. Perutz, M.F., Muirhead, H., Cox, J.M. and Goaman: 1968, Nature 219 131-139.
2. Perutz, M.F.: 1970, Nature 228, 726-739.
3. Moon, R.B. and Richards, J.H.: 1972, J. Amer. Chem. Soc. 94, 5093-5095.
4. Vergamini, P.J., Matwiyoff, N.A., Wohl, R.C. and Bradley, T.: 1973, Biochem. Biophys. Res. Comm. 55, 453-461.
5. Antonini, E., Brunori, M., Conti, F. and Geraci, G.: 1973, FEBS Letters, 34, 69-70.
6. Banerjee, R., Stetzkowski, F. and Lhoste, J.M.: 1976, FEBS Lett (in press).
7. Banerjee, R. and Lhoste, J.M.: 1976, Eur. J. Biochem. 67, 349-356.
8. Weiss, J.J.: 1964, Nature 202, 83-84.
9. Pauling, L.: 1964, Nature 203, 182-183.
10. Kendrew, J.C., Watson, H.C., Strandberg, B.E., Dickerson, R.E., Phillips, D.C. and Shore, V.C.: 1961, Nature 190, 666-670.
11. Banerjee, R. and Cassoly, R.: 1969, J. Mol. Biol. 42, 351-361.
12. Ogawa, S. and Shulman, R.G.: 1972, J. Mol. Biol. 70, 315-336.

13. Cassoly, R. and Gibson, Q.H.: 1972, J. Biol. Chem. 247, 7332-7341.
14. Banerjee, R., Stetzkowski, F. and Henry, Y.: 1973, J. Mol. Biol. 73, 455-467.
15. Henry, Y. and Banerjee, R.: 1973, J. Mol. Biol. 73, 469-482.
16. Cassoly, R.: 1975, J. Mol. Biol. 73, 581-595.
17. Sugita, Y.: 1975, J. Biol. Chem. 250, 1251-1256.
18. Antonini, E., Wyman, J., Brunori, M., Taylor, J.F., Rossi-Fanelli, A.,and Caputo, A.: 1964, J. Biol. Chem. 239, 907-912.
19. Brunori, M., Alfsen, A. Saggese, U., Antonini, E. and Wyman, J.: 1968, J. Biol. Chem. 243, 2950-2954.
20. Banerjee, R. and Cassoly, R.: 1969, J. Mol. Biol. 42, 337-349.
21. Perutz, M.F., Fersht, A.R., Simon, S.R. and Roberts, G.C.K.: 1974, Biochemistry 13, 2174-2186.
22. Mac Quarrie, R.A. and Gibson, Q.H.: 1971, J. Biol. Chem. 246, 517-522.

STRUCTURES OF THE NON-HEME IRON PROTEINS MYOHEMERYTHRIN AND HEMERYTHRIN

KEITH B. WARD and WAYNE A. HENDRICKSON
Laboratory for the Structure of Matter, U.S. Naval Research Laboratory, Washington D.C. 20375, U.S.A.

1. OXYGEN TRANSPORT PROTEINS

Oxygen transport plays a vital role in the physiology of multicellular animals. Although several species of Antartic fish and certain invertebrates can survive solely on oxygen dissolved in plasma, all other species depend upon respiratory transport pigments to enhance the oxygen capacity of their circulatory or coelomic fluids. In every instance these pigments are metalloproteins capable of reversibly binding oxygen. Knowledge of the metal-ligand interactions in these proteins is essential for understanding the mechanisms by which they function. On the basis of fundamental differences in the active site, these oxygen transport pigments may be divided conveniently into three classes - hemoglobin, hemocyanin, and hemerythrin (Table I).

Since it occurs in humans, hemoglobin is the most familiar of these pigments. Hemoglobins, including the related proteins, myoglobin, erythrocruorin, and chlorocruorin, are also the most prevalent of the three classes. Distributed throughout many different phyla, they vary considerably in molecular weight and subunit constitution. The distinguishing feature they share is the presence of heme, a prostheric group comprising a porphyrin moiety and an iron ion, to which oxygen binds in the ratio 1Fe:O_2. These molecules have been investigated extensively, and a variety of techniques have led successfully to detailed knowledge about metal-ligand interactions in hemoglobins (Antonini and Brunori, 1971). Indeed, a number of current results bearing on this subject will be presented in this Symposium.

About hemocyanins, the more rare pigments found only in certain molluscs and crustaceans, much less is known (Van Holde and Van Bruggen, 1971). These blue proteins have high molecular weights and contain neither iron nor porphyrin, but copper instead, to which oxygen binds in the ratio 2Cu:O_2. Electron microscopy has revealed the quaternary arrangement of a gastropod hemocyanin that contains 120 oxygen binding subunits (Mellema and Klug, 1972; Siezer and van Bruggen, 1974). In the absence of more detailed structural information, Gray (1971) has presented several models of the active site based on spectral evidence. In addition, Loehr et al. (1974) used resonance Raman spectroscopy to show that the bound oxygen is a superoxide and is

B. Pullman and N. Goldblum (eds.), Metal-Ligand Interactions in Organic Chemistry and Biochemistry, part 2, 173-187.

TABLE I
Comparison of oxygen-transport proteins

	Hemoglobin	Hemerythrin	Hemocyanin
Metal	Fe	Fe	Cu
Oxidation state of metal in deoxy protein	II	II	I
Metal: O_2	Fe: O_2	2Fe: O_2	2Cu: O_2
Color oxygenated	Red	Violet-pink	Blue
Color deoxygenated	Red-purple	Colorless	Colorless
Coordination of Fe	Porphyrin & protein side chain	Protein side chains	Protein side chains
Molecular Weight	65 000	108 000	400 000-20 000 000
Number of subunits	4[a]	8[b]	Many

[a] In some species (e.g. Glycera) hemoglobins are monomeric. In other species (e.g. Arenicola) hemoglobins contain many subunits and have molecular weights $>10^6$.

[b] Myohemerythrin from T. dyscritum is monomeric and the hemerythrin of Phascolosoma agassizii is a trimer.

probably not in a π-bonded complex with copper. A recent report (Kuiper et al., 1975) of x-ray diffraction studies on crystals of a hexameric hemocyanin from the spiny lobster offers hope that more detailed information about the structure of the active site will soon be forthcoming.

Hemerythrins represent the third class of transport proteins and they occur even less frequently than the hemocyanins. They are restricted mainly to members of the phylum Sipuncula, the marine peanut worms, although they are also found in isolated species of Annelida, Brachiopoda, and Priapulida. Members of this class have active sites containing two iron atoms which bind oxygen in the ratio $2Fe:O_2$.

Numerous biochemical, spectroscopic, and magnetic investigations of hemerythrin have been reviewed by Klotz (1971) and Klotz et al. (1976). These studies have revealed much about the structure of the molecule in general and have led to interesting deductions concerning the structure of the active site in particular. Crystal structure analyses currently under way in our laboratory and others should increase our knowledge of the metal-ligand interactions in hemerythrin.

2. STRUCTURE OF MYOHEMERYTHRIN

Although hemerythrin usually occurs as an octamer consisting of eight 13 500 molecular weight subunits, Klippenstein et al. (1972) isolated a monomeric form, myohemerythrin, from the retractor muscle of the sipunculan *Themiste pyriodes*. Numerous physical measurements suggested it was quite similar in structure to the subunits of the octameric form, a relationship reminiscent of the structural homology between vertebrate myoglobin and hemoglobin. The tertiary structure of myohemerythrin was revealed by a single crystal structure analysis using the multiple isomorphous replacement technique (Hendrickson et al., 1975).

The initial 5.5 Å resolution electron density map was readily interpretable because of its extraordinary quality, the high helix content of the molecule (75%), and the extensive chemical data available for hemerythrin. A model representing the tertiary structure of myohemerythrin is shown in Figure 1. The two iron atoms are located close

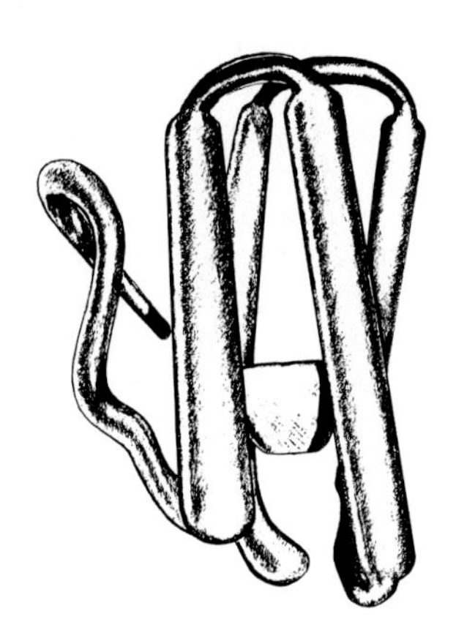

Fig. 1. A model of the tertiary structure of myohemerythrin. The dimeric iron center lies between four helical segments.

to one another near the center of the molecule in an active site embraced by four interconnected quasi-parallel helix segments represented by the long straight cylindrical structures. The polypeptide chain runs through the amino terminal arm from the molecular bottom, first forth and then back along the left side of the molecule. At the very top, it turns acutely and goes through the first helix (A) to the far end of the molecule, makes a short corner and returns via a second helix (B). Thereupon, the chain moves downward in a U-turn, courses through a third helix (C) and returns through the fourth helix (D). The chain then turns sharply upward and to the right, ending in a short helical (E) carboxyl terminal stub.

The model shown in Figure 1 shows the presence of a pseudodiad axis running roughly parallel to the major helices and passing through the iron center. The pseudosymmetry relates helices A and B to the C and D helices, and is correlated with an apparent repeat in the amino acid sequence. Similar pseudosymmetry has been detected in the structure of carp muscle calcium binding protein (Kretsinger, 1972) and bacterial ferredoxin (Adman et al., 1973).

3. STRUCTURE OF OCTAMERIC HEMERYTHRIN

The first x-ray diffraction study of crystals of octameric hemerythrin was made by Love (1957) in order to determine the molecular weight. North and Stubbs (1974) studied another crystal form of hemerythrin and were able to show that the molecule has a 4-fold axis of symmetry and that it quite likely possesses 422 (D_4) symmetry. Further analysis was hampered in part by the large size of the crystallographic asymmetric unit which contained an entire octamer. A preliminary analysis of crystals of another octameric hemerythrin also indicated that the molecule might possess a 4-fold axis (Loehr *et al.*, 1975).

Klippenstein (1972) found that octameric hemerythrin from *Phascolopsis gouldii* contains subunits of two types, A and B, which differ slightly in amino acid sequence. Octameric hemerythrin B (HrB) was reconstituted from only B type subunits in the hope that this new material would yield crystals more suitable for structure analysis. Tetragonal crystals were obtained having space group symmetry P422 and with a = b = 104.82 Å and c = 54.08 Å. The lattice constants and knowledge that the molecule contained a 4-fold axis limited the possible molecular packing arrangements. These constraints, coupled with the pseudosymmetry apparent in precession photographs, indicated that the octamers almost certainly were centered about the two spe ial positions in the unit cell having 422 (D_4) symmetry. Thus, HrB itself must exhibit 422 symmetry, and the crystallographic asymmetric unit contains only two subunits, one from each octamer.

This arrangement was confirmed by a subsequent HrB crystal structure analysis (Ward *et al.*, 1975). Sin e hemerythrin subunits were thought to be isostructural with myohemerythrin, the HrB structure was determined by molecular search techniques using the low resolution image of myohemerythrin. The major difference in structure appears in the C-D corner region and corresponds to a five residue deletion present in hemerythrin in this area.

The quaternary structure of octameric HrB is presented in Figure 2. The molecule is arranged in two layers with approximate dimensions 75 x 75Å, each comprising four subunits related in an end-to-side fashion by the molecular tetrad, R. The layers are related by four diads lying normal to R, (P1, Q1, P2, Q2), to form a molecule approximately 50 Å thick. A channel along R, some 20 Å square, is formed by B and C helices and leads to a central chamber bounded by helices A, B, and E which is 30 Å in diameter and 15 Å high. A tentative model of hemerythrin (Hendrickson and Ward, 1975) indicate that the number of contacts between subunits across the P, Q, and R interfaces is in the ratio 22:71:42.

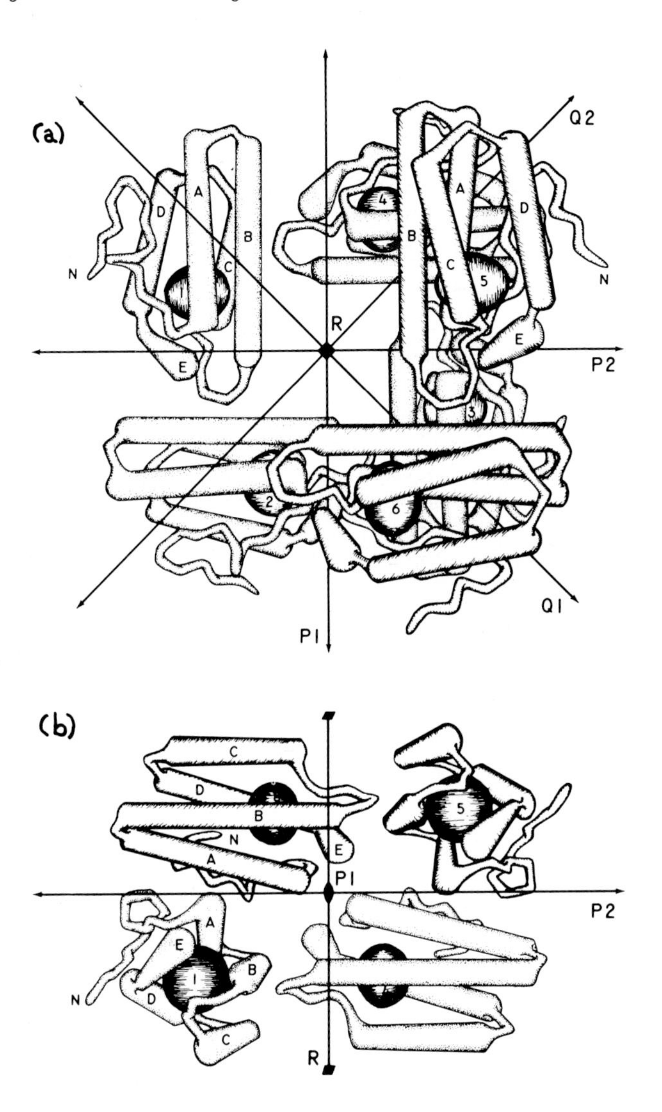

Fig. 2. Quaternary structure of octameric hemerythrin B. Subunits were drawn by Diane Ward from computer-generated schematic representations of the polypeptide chains. The amino terminal arm is labeled N, and the five helices are marked A-E. The oblate spheroids nestled among the helices and upon which subunit numbers appear, represent the dimeric iron centers. The octamer exhibits 422 (D_4) symmetry, and the subunits are interrelated by the tetrad R and the diads P1, Q1, P2, and Q2. View (a) is along R and shows four subunits (stippled) in the lower layer and two of four upper layer subunits (hatched) related by diads P and Q to the former. View (b) shows four subunits, two from each layer, as seen from within the octamer along the diad P1.

Stenkamp *et al.* (1976a) have determined the structure of octameric hemerythrin from the sipunculan *Themiste dyscritum* by a single isomorphous replacement. The space group is P4, and in these crystals the molecular tetrad coincides with the crystallographic 4-fold axis so that the asymmetric unit contains four subunits, two from each octamer. The tertiary and quaternary structures revealed by this analysis are essentially identical to those obtained for HrB.

4. ACTIVE SITE

These low resolution images of myohemerythrin and hemerythrin show that both irons are located in an active site near the center of the molecule. Of major interest is the manner in which the irons interact with ligands supplied by the protein and with exogenous ligands such as O_2, N_3^-, OH^-, etc. Hemerythrin exists in three distinct ligand states - deoxyhemerythrin, oxyhemerythrin, and liganded methehemerythrin. The iron-protein coordination, to be discussed presently, is probably quite similar in all states. However, a variety of studies (reviewed by Klotz, 1971, and Klotz et al., 1976) show that other features of the active site differ between ligand states. Figure 3 schematically illustrates several of these differences.

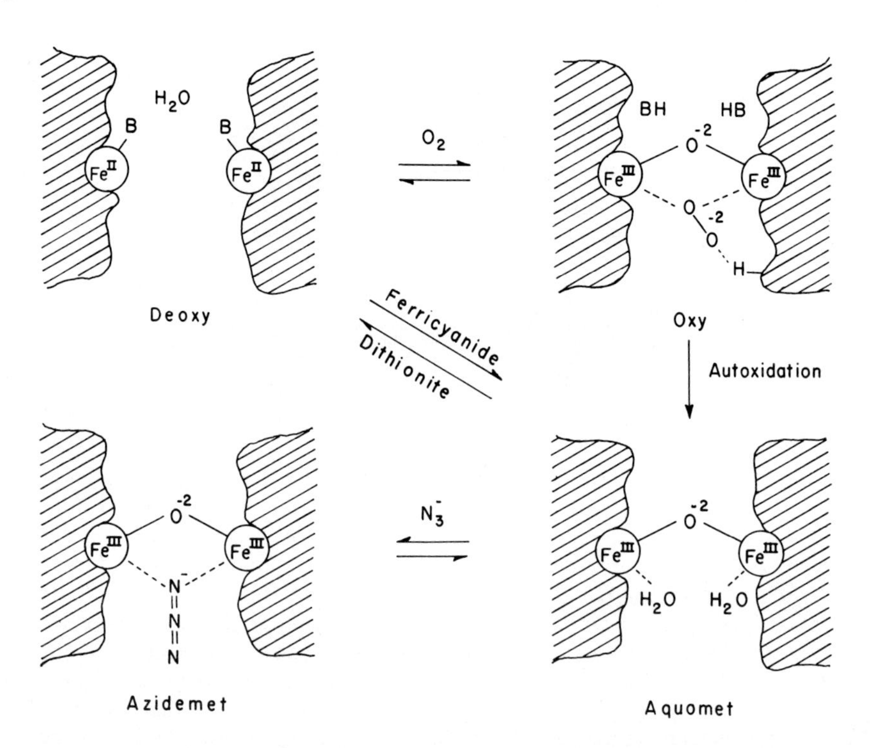

Fig. 3. Models proposed for the active site of hemerythrin in various ligand states. Adapted from Garbett *et al.* (1971).

In deoxyhemerythrin both irons are high spin Fe(II) as suggested by chemical studies (Klotz and Klotz, 1955) and magnetic susceptibility measurements (Kubo, 1953). This assignment is confirmed by Mossbauer results (Okamura et al., 1969; York and Bearden, 1970; Garbett et al., 1971) which also indicate that both irons are in equivalent chemical environments.

When oxygen reacts with deoxyhemerythrin to form oxyhemerythrin, there is a change in the optical absorption spectrum and a decrease in magnetic susceptibility. The Mossbauer investigations mentioned above show that both iron atoms are high spin Fe (III), but that they are located in different chemical environments. Resonance Raman spectra (Dunn et al., 1975) of oxyhemerythrin indicate that the bound oxygen is a peroxide, O_2^{-2}. The low apparent moment is due to antiferromagnetic coupling between the irons, most likely through a μ-oxo, O^{-2}, bridge. (Dawson et al., 1972). The two irons are oxidized to Fe (III) and the O_2 is reduced to form peroxide, O_2^{-2}. The source of the bridging O^{-2} is less certain, but it might result from the reduction of H_2O by basic groups, B, present in deoxyhemerythrin (Garbett et al., 1971). The unsymmetrical binding of O_2^{-2} shown in Figure 3 is typical of models proposed to explain the evidence for two iron environments. Dawson et al., 1972, measured the exchange coupling constant, $J = -77cm^{-1}$, for oxyhemerythrin. They argue that its large magnitude implies there is but a single bridge, and present several models in which the peroxide is bound to one iron only. A model is which the oxygens are equidistant from the metal is consistent with resonance Raman spectra of oxyhemerythrin (Dunn et al., 1975), which show frequenc es typical of such dioxygen complexes (Valentine, 1971).

Changes in the coordination and electronic structure of the dimeric iron complex upon oxygenation might be expected to effect local conformation changes in the protein. Bossa et al. (1970) report circular dichroic spectral evidence for such changes. Similar changes might lead to the cooperative oxygen binding observed in brachiopod hemerythrin (Manwell, 1960), although sipunculan hemerythrins, upon which most studies have been conducted, show no cooperativity in vitro. However, recent measurements of deoxygenation of sipunculan coelomic cells indicate that fairly high cooperativity exists in vivo (Mangum and Kondon, 1975).

Methemerythrin is produced by autoxidation of oxyhemerythrin, or by reacting deoxhemerythrin with an oxidizing agent such as ferricyanide. The latter reaction is highly cooperative, and Brunori et al. (1971) were unable to detect molecular species containing one Fe (II) and one Fe (III). Methemerythrin exhibits Mossbauer, electronic absorption, and circular dichroic spectra similar to those of model binuclear oxo-bridged Fe (III) complexes (Murray, 1974). Two high spin Fe (III) iron atoms are antiferromagnetically coupled via an O^{-2} bridge, but, in contrast with oxyhemerythrin, Mossbauer data indicate identical chemical environments for both iron atoms (Okamura et al., 1969; York and Bearden, 1970; Garbett et al., 1971).

Many small anions, including N^-, NCS^-, NCO^-, Br^-, Cl, CN^-, F^-, and OH^-, serve as ligands for methemerythrin (Keresztes-Nagy and Klotz, 1965). Below pH 8, the two hydroxyl ligands are protonated to form aquomethemerythrin (Figure 3). Azidemethemerythrin contains a single N_3^- per active site. The dichroism exhibited by crystals of azidemetmyohemerythrin (Hendrickson *et al.*, 1975) suggests the N_3^- axis lies roughly perpendicular to the Fe-Fe axis. Despite evidence for equivalent iron environments, the azide may be bound to one iron only since measured values of the N=N stretching frequency, 2046 cm^{-1} and 2049 cm^{-1} (York and Bearden, 1970; Dunn et al., 1975), are virtually identical to the value (2045 cm^{-1}) measured for the high spin Fe (III) azide complex in azidemetmyoglobin (McCoy and Caughey, 1970).

5. IRON-PROTEIN COORDINATION

In every ligand state, both iron atoms are held in position by direct coordination to amino acid side chains. Klotz et al. (1976) discuss the extensive biochemical studies that limit which of the 113 amino in *P. gouldii* hemerythrin might participate in metal coordination. Seventy may be eliminated because their side groups either are apolar or have such extreme pK values as to make them unlikely candidates. Various group-specific reagents have been used to block certain side groups without disturbing the spectral properties of the protein. Such experiments eliminate from consideration the single cysteine (Cys 50) and methionine (Met 62) as well as all eleven lysines, both the amino and carboxyl termini, and all seventeen side chains containing carboxyl groups.

There then remain only seven histidines and five tyrosines as potential iron-protein ligands. His 34 and His 82 can be eliminated since they are replaced by Cys 34 and Glu 82 in myohemerythrin (Klippenstein et al., 1976). In addition, experiments by Morrisey (1971) suggest that these two histidines and either His 73 or His 77 may be reacted with iodoacetamide without causing changes in the fraction of α-helix, iron content, or absorbtion spectrum due to the iron chromophore. Rill and Klotz (1971) used the group specific reagent, tetranitromethane, to show that Tyr 18 and Tyr 70 are not involved in iron coordination, and the work of Fan and York (1972) suggests that not more than two tyrosines may be involved. After these chemical modification criteria are applied, the amino acid residues potentially available for coordination are His 25, His 54, either His 73 or His 77, His 101, Tyr 8, Tyr 67, and Tyr 109.

6. INTERPRETATION OF MYOHEMERYTHRIN ELECTRON DENSITY MAP

Besides revealing the image of myohemerythrin presented in Figure 1, the initial study of azidemetmyohemerythrin disclosed certain details about the active site (Hendrickson et al., 1975). An anomolous scattering analysis using native data collected to 2.8 Å resolution showed the iron-iron distance to be 3.44 Å $\pm$ 0.05 Å. This distance is comparable to iron-iron distances found in binuclear oxo-bridged iron (III) complexes (Murray, 1974) and, if the iron-oxygen distance is taken as 1.80 Å, corresponds to a reasonable Fe-O-Fe bridging angle of 145°.

The 5.5 Å resolution electron density map contained six apparent density connections between the protein chain and iron center, one from each of the four helical regions, one from the carboxyl terminal stub, and one from the BC corner.

These connections were correlated with the amino acid sequence by an unusually detailed analysis of the map made possible by its exceptional quality and by the extensive biochemical data available for hemerythrin, including the amino acid sequence of myohemerythrin. It was reasonably assumed that (1) the connections from each helix might correspond to four putative histidine ligands and (2) the mercury sites found during isomorphous substitution experiments marked the positions of the two lone cysteinyl residues. When these benchmarks were used, the length of the continuous tube of density corresponding to polypeptide chain was adequately explained by assuming an axial translation of 1.5 Å/residue along helical protions and 3.3 to 3.8 Å/ residue along regions of apparently extended chain. This procedure allowed particular segments of the sequence to be assigned to individual stretches of electron density. The three connections between the iron center and helices A and B and the carboxyl terminal stub corre spond in hemerythrin to the locations of His 25, His 54, and Tyr 109 respectively. Those from the BC corner; and from the C and D helices were assigned to Tyr 67, His 73, and His 101 respectively.

This identification of iron-protein ligands is shown in Figure 4.

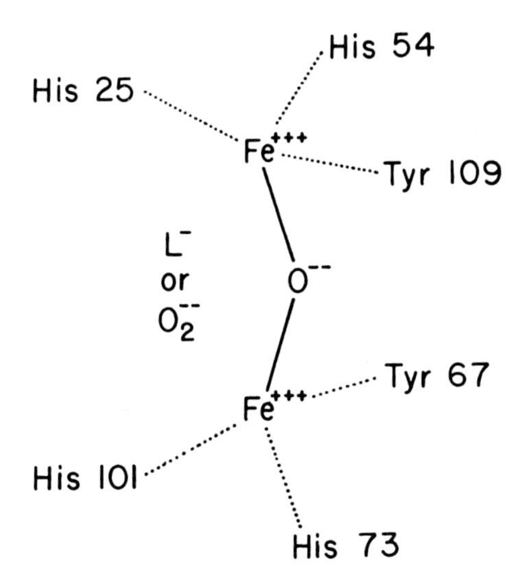

Fig. 4. Amino acid residues believed to serve as iron ligands in hemerythrin. Note that due to a five residue insertion, His 101 and Tyr 109 correspond to His 105 and Tyr 114, respectively, in myohemerythrin.

The structure analysis of myohemerythrin is currently being extended to higher resolution in order to better describe the structure of the active site.

Atomic models of the polypeptide backbone of both myohemerythrin and hemerythrin were built by computer using the low resolution electron density maps (Hendrickson and Ward, 1975). The construction process achieved a weighted optimization of the following five criteria: (1) the preservation of the standard bond length and angles, (2) the maintenance of standard conformations in helical segment, (3) the maximization of electron density at atomic positions, (4) the minimization of energy due to repulsive interactions between nonbonded atoms, and (5) the constraint of presumed iron ligand residues by attachment to the iron positions. The model so derived for myohemerythrin is presented in Figure 5.

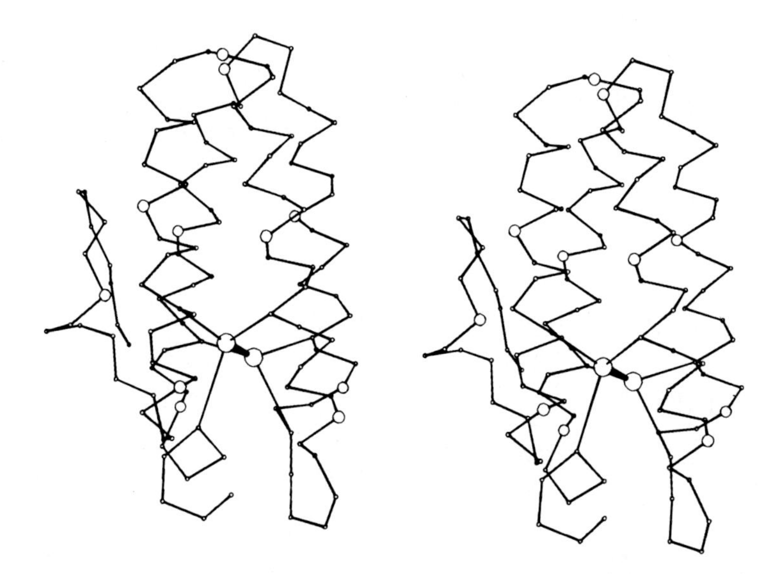

Fig. 5. Stereo view of a model of the alpha-carbon backbone of myohemerythrin with the same orientation as in Figure 1. Every tenth alpha-carbon is shown enlarged. Lines are drawn between iron atoms and the alpha carbon atoms of the residues having side chains that serve as iron ligands. See figure 4.

Beginning with this tentative model of myohemerythrin, phase information was extended to higher resolution data by means of a constrained least squares structure factor refinement procedure (Konnert, 1976). As the structure refinement proceeded, side chain atoms were added to the model whenever their positions were apparent in Fourier maps synthesized using calculated phases for those data of greater than 5.5 Å resolution. At the present time a total of 835 of the 979 non-hydrogen atoms are included in the model, and the conventional residual, R = 0.305 based on all data to 2.8 Å resolution.

The accuracy of the atomic model is checked by viewing computer-

-drawn stereo images which show the model superposed with the electron density. At the present state of refinement, it is apparent the model still has certain deficiencies. Several portions of model, for example the B helix shown in Figure 6, fit the electron density extremely well.

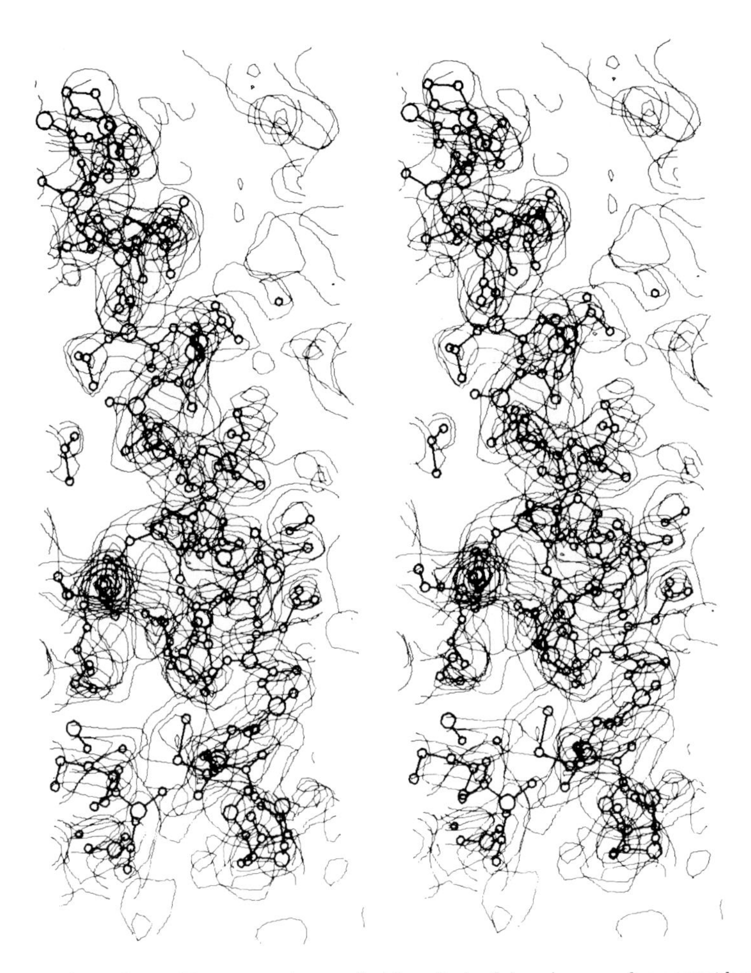

Fig. 6. Stereo view of the B helix in myohemerythrin. The orientation is the same as in Figure 5. The electron density was synthesized using the observed diffraction amplitudes and the phases calculated from the current atomic model. The circles of larger size represent alpha-carbon atom positions. His 54 appears above the iron center near the middle of the left side of the figure.

In other regions, especially in some parts of the extended chain connecting helical segments, the fit is less accurate. In view of the disparate quality of the map, the current structural interpretations must be considered quite tentative.

The region around the active site as viewed from the AB corner is presented in Figure 7. The electron density representing the μ-oxo

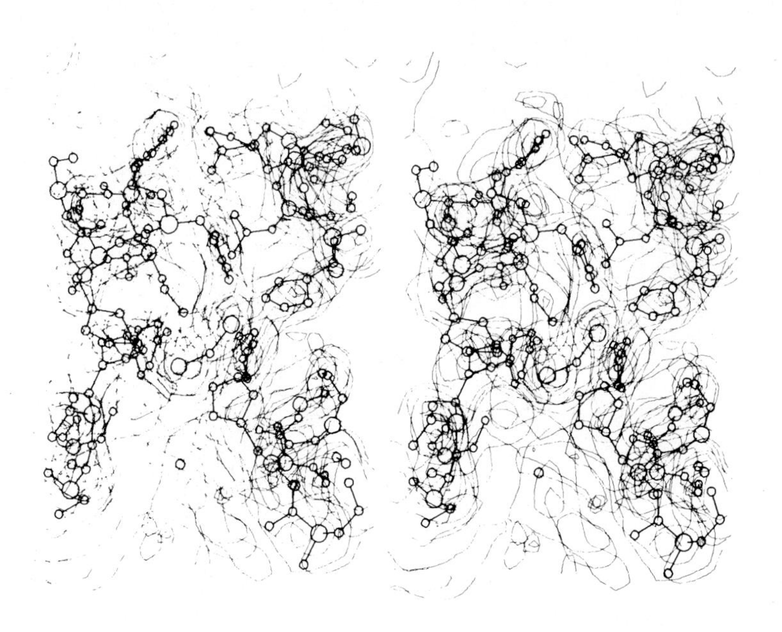

Fig. 7. Stereo view of the active center in myohemerythrin as seen from the A-B corner of the molecule. Calculation methods were the same as for Figure 6.

bridged iron dimer, located at the center, is contiguous with six electron density features representing the iron ligands.

The features behind, above, and to the right of the upper iron atom correspond to the side groups of Tyr 109, His 54, and His 25, respectively, in hemerythrin. Likewise, the regions behind, to the left of and to the right of the lower iron atom contain the side chains of Tyr 67, His 73, and His 101, respectively. In addition, the side group of Glu 58 readily fits into the density above and slightly behind the lower iron atom, indicating that this residue is in close contact with the iron center. The position of the azide ion is uncertain. The explanation of other electron density features in this region must await further model refinement. The structural interpretation might be biased because atoms in the iron complex and in the side groups involved in iron coordination were included in the phase calculation. To test this possibility, new phases were calculated after excluding these atoms, but the new electron density distribution was almost identical to that shown in Figure 7.

Stenkamp *et al.* (1976b) have recently presented a model of the iron complex based on their inspection of a 2.8 A resolution electron density map of aquomethemerythrin from the sipunculid *T. dyscritum*. They assign His 25, His 54 and Tyr 109 as ligands for one iron atom, in agreement with the myohemerythrin interpretation. However, they believe that His 73, His 77, and His 101, rather than Tyr 67, His 73, and His 101, respectively, serve as ligands for the second iron atom. In addition, they find both Gln 58 and Asp 106 in position to form bridges between the iron atoms. These apparent differences in iron--protein coordination may reflect true structural differences between the two molecules, or they may represent different coordination present in azide - as opposed to aquomethemerythrin. There is little chemical or biochemical evidence to support these possibilities, however. It is more likely that the differences appear because both interpretations were made at intermediate resolution, and that the discrepancy will be resolved as the model refinement continues.

ACKNOWLEDGEMENT

Keith B. Ward is supported by a National Research Council Resident Research Associateship.

REFERENCES

Adman, E.T., Sieker, L.C., and Jensen, L.H.: 1973, J. Biol. Chem. 248, 3987.
Antonini, F. and Brunori, M.: 1971, Hemoglobin and Myoglobin and Their Reactions with Ligands, North-Holland Publ. Co., Amsterdam.
Bossa, F., Brunori, M., Bates, G.W., Antonini, E., and Fasella, P.: 1970, Biochem. Biophys. Acta 207, 41.
Brunori, M., Rotilio, G.C., Rotilio, G., and Antonini, E.: 1971, FEBS Letters 16, 89.
Dawson, J.W., Gray, H.B., Hoenig, H.E., Rossman, G.R., Schredder, J.M., and Wang, R.: 1972, Biochemistry 11, 461.
Dunn, J.B.R., Schriver, D.F., and Klotz, I.M.: 1975, Biochemistry 14, 2689.
Fan, C.C. and York, J.L.: 1969, Biochem. Biophys. Res. Comm. 36, 365.
Garbett, K., Johnson, C.E., Klotz, I.M., Okamura, M.Y., and Williams, R.J.P.: 1971, Arch Biochem. Biophys. 142, 574.
Gray, H.B.: 1971, Adv. Chem. Ser. 100, 365.
Hendrickson, W.A., Klippenstein, G.L., and Ward, K.B.: 1975, Proc. Natl. Acad. Sci. U.S.A. 72, 2160.
Hendrickson, W.A. and Ward, K.B.: 1975, Biochem. Biophys. Res. Comm. 66, 1349.
Keresztes-Nagy, S. and Klotz, I.M.: 1965, Biochemistry 4, 919.
Klippenstein, G.L., Van Riper, D.A., and Oosterom, E.A.: 1972, J. Biol. Chem. 247, 5959.
Klippenstein, G.L.:1972, Biochemistry 11, 372.
Klippenstein, G.L., Cote, J.L., and Ludlam, S.E.: 1976, Biochemistry 15, 1128.
Kretsinger, R.H.: 1972, Nature New Biol. 240, 85.
Klotz, I.M.: 1971, in Timasheff, S.N. and Fasman, G.D. (eds.), Subunits in Biological Systems, Marcel Dekker, New York, pp. 55-103.

Klotz, I.M., Klippenstein, G.L., and Hendrickson, W.A.: 1976, Science 192, 335-344.
Klotz, I.M. and Klotz, T.A.: 1955, Science 121, 477.
Konnert, J.H.: 1976, Acta Cryst. A32, 614-617.
Kubo, M.: 1953, Bull. Chem. Soc. Jap. 26, 244.
Kuiper, H.A., Gaastra, W., Beintema, J.J., Van Bruggen, E.F.J., Schepman, A.M.H., and Drenth, J.: 1975, J. Mol. Biol. 99, 619.
Loehr, J.S., Freedman, T.B., and Loehr, T.M.: 1974, Biochem. Biophys. Res. Comm. 56, 510.
Loehr, J.S. Meyerhoff, K.N., Sieker, L.C., and Jensen, L.H.: 1975, J. Mol. Biol. 91, 521.
Love, W.E.: 1957, Biochim. Biophys. Acta 23, 465.
Mangum C.P. and Kondon, M.: 1975, Comp. Biochem. Physiol. 50A, 777.
Manwell, C.: 1963, Science 139, 175.
McCoy, S. and Caughey, W.S.: 1970, Biochemistry 9, 2387.
Mellema, J.E. and Klug. A.: 1972, Nature 239, 146.
Morrisey, J.A.: 1971, Ph.D. Dissertation, University of New Hampshire.
Murray, K.S.: 1974, Coord. Chem. Rev. 12, 1.
North, A.C.T. and Stubbs, G.J.: 1974, J. Mol. Biol. 88, 125
Okamura, M.Y., Klotz, I.M., Johnson, C.E., Winter, M.R.C., and Williams, R.J.P.: 1969, Biochemistry 8, 1951.
Rill, R.L. and Klotz, I.M.: 1971, Arch. Biochem. Biophys. 147, 226.
Siezer, R.J. and Van Bruggen, E.F.J.: 1974, J. Mol. Biol. 90, 77.
Stenkamp, R.E., Sieker, L.C., Jensen, L.H., and Loehr, J.S.: 1976a, J. Mol. Biol. 100, 23.
Stenkamp, R.E., Sieker, L.C., and Jensen, L.H.: 1976b, Proc. Natl. Acad. Sci. U.S.A. 73, 349.
Valentine, J.S.: 1973, Chem. Revs. 73, 235.
Van Holde, K.E. and Van Bruggen, E.F.J.: 1971, in Timasheff, S.N. and Fasman, G.D. (eds.), Biological Macromolecules Series, 5A, pp. 1-55.
Ward, K.B., Hendrickson, W.A., and Klippenstein, G.L.: 1975, Nature 257, 818.
York, J.L. and Bearden, A.J.: 1970, Biochemistry 9, 4549.

DISCUSSION

Sarkar: I would like to make a similar comment which I made in Prof. Jensen's lecture about the trend that I have observed in the metal-binding sites of metalloenzymes and metalloproteins concerning the natural occurence of one to three amino acid residues in between two ligands for the metal-binding while the other ligands originate from a distant part of the polypeptide chain. Your hemerythrin structure is the first one I have known now which does not follow the above trend. However, the hemerythrin structure determined by Prof. Jensen does show the trend I have observed earlier: His (25) [His (54), Gln (58)], [His (73), His (77)], His (101), [Asp (106), Tyr (109)] (Stenkamp, Sieker and Jensen, Proc. Natl. Acad. Sci. U.S.A. 73, 349-351 (1969).

Ward: The trend you mention is an interesting one. It appears to hold in the active site interpretation presented by Stenkamp et al. I might mention that in addition to the six putative iron ligands, our current model places Glu 58 near the binuclear iron complex, and this would be consistant with the trend you have observed.

Caughey: Do the X-ray data make unlikely bridging between irons by a tyrosine hydroxyl oxygen ?

Ward: With our present model, we think it unlikely that a tyrosine hydroxyl oxygen is serving as a bridge. I wouldn't rule it out completely, however, since we are still refining our model, and a new interpretation might place one of the tyrosines in a bridging position

Eichhorn: I just want to clear up a point about myohemerythrin. Myoglobin and hemeglobin are two different substances. I assume that myohemerythrin is not a different substance from hemerythrin, but only the monomeric form ? Is that correct ?

Ward: No. Myohemerythrin is as distinct from hemerytrhin as is myoglobin from hemoglobin. At this time, myohemerythrin has been isolated from only one species, the sipunculan worm, *Themiste pyroides*. Myohemerythrin and the subunits of octameric hemerythrin are, of course, very similar in structure. Our X-ray analysis indicates major structural differences only in the region between the C and D helices, where myohemerythrin contains a five residue insertion.

Werber: I wonder about the nature of the μ-oxo bridge in hemerytrhin. What are the B ligand groups that are replaced by the μ-oxo bridge ? Is there a pH dependence of the formation of this μ-oxo bridge ?

Ward: The B ligands are just meant to represent two potentially basic groups which might serve to remove the protons from water when the μ-oxo bridge is formed. There is no evidence I know of which suggests what these postulated ligands might be.

The oxygen equilibrium has been shown by Bates et al. (Biochem. 7, 3016 (1968) to be insensitive to changes in pH and ionic strength.

DIVALENT CATION-LIGAND INTERACTIONS OF PHOSPHOLIPID MEMBRANES: EQUILIBRIA AND KINETICS

DUNCAN H. HAYNES
Dept. of Pharmacology, University of Miami Medical School, P.O. Box 520875, Biscayne Annex, Miami, Fla. 33152, U.S.A.

Summary. The properties of the negatively-charged lipid phosphatidic acid (PA^-) as a phosphate ligand for divalent cations are discussed. Within phospholipid membranes, PA^- binds divalent cations with binding constants and ion specificity ratios comparable to those of $HPO_4{}^{2-}$. These binding reactions are influenced by the electrostatic effect of surface charge. The apparent binding constant has been shown to be directly proportional to an exponential function of the membrane surface potential calculated by the Gouy-Chapman theory in modified form. Divalent cation binding has been shown to cause a redistribution of PA^- on the surface of mixed membranes of PA^- and phosphatidyl choline (PC), such that the average distance between occupied and unoccupied PA^- molecules becomes smaller.

The kinetics of binding were studied using the temperature jump technique with murexide as a spectroscopic indicator of Ca^{2+} and PA^- in dimyristoyl PC/PA^- (mole ratio 1:1) is 1.9 x 10^7 M^{-1} s^{-1} which is two orders of magnitude lower than the corresponding value for phosphate and other simple ligands in solution. Various explanations for this discrepancy are considered, and it is concluded that the reduction in rate is due to steric blockage of the ligand.

Binding of Ca^{2+} and other divalent cations to phospholipid vesicles containing PA^- or phosphatidyl serine (PS^-) at high mole fractions approaching 1.0 gives rise to vesicle aggregation. The reaction shows threshold behavior with respect to divalent cation concentration which indicates that it is necessary to achieve 0.5 Ca^{2+}/PA^- binding stoichiometry and full neutralization of the membrane surface charge for aggregation to occur. The kinetics of the aggregation reaction were analyzed and the rate constant for dimer formation for two vesicles was found to be in the range of 10^7 M^{-1} s^{-1}, or two orders of magnitude lower than that for diffusion control. The explanation arrived at for this is that a stable aggregate must form salt bridges between the phosphate ligands of one vesicle and those of another. The time for correlation between occupied and unoccupied ligands on the surfaces of two apposed membranes is 1msec and this represents the rate--limiting step in the aggregation reaction.

B. Pullman and N. Goldblum (eds.), Metal-Ligand Interactions in Organic Chemistry and Biochemistry, part 2 ,189-212.

1. INTRODUCTION

The essential properties of phospholipid membranes and phospholipid regions of natural membranes as hydrophobic barriers can be explained by a minimum number of concepts. Listed in approximate decreasing order of importance they include (1) *hydrophobic interaction* of non-polar portions of the phospholipids, (2) *hydrophilic contact* of the polar head groups with water and (3) *dipolar interactions* between them, and (4) *electrostatic* interaction of polar head groups bearing net negative charge. The hydrophobic interaction energy of the alkane chains of the phospholipids is necessary for the integrity of the bilayer structure, and this energy increases with increasing chain length and with increasing opportunity for contact between the chains. Orientation of the lipids within the bilayer is maintained by hydrophilic contact and hydration of the lipid polar head groups at the water--membrane interface. Dipolar steric interactions between polar head groups and surface charge and surface potential effects on the membrane can modulate the degree of hydrophobic interaction, and thus the degree of membrane rigidity, in addition to affecting properties of the membrane surface. Such changes can also be produced by the binding of divalent cations to the membrane surface.

The discussion above, which represents a gross oversimplification of the physical state of affairs in membranes, should be considered as an attempt to put into perspective the studies of the cation-ligand interactions in membranes reported below. Cation-ligand interactions are involved in all major processes under study by membrane biologists. A few examples are the role of Ca^{2+} in the nerve action potential, the release and uptake of Ca^{2+} as a trigger for muscle contraction, and the role of Ca^{2+} in the fusion of synaptic vesicles with the presynaptic membrane to release neural transmitter substances. The half-times of the above reactions can be as low as 1 msec and have extremely high specificity for Ca^{2+} over other divalent cations. The studies reported here on simple systems were carried out in order to provide a basis for interpretation and prediction of behavior in these more specific and complicated systems.

The present communication deals with the effects of ion binding on negatively-charged phospholipid membranes. In keeping with the spirit of this symposium, we will start with the question of whether the major thermodynamic and kinetic features of divalent cation (M^{2+}) binding and interaction with membranes composed of negatively-charged phospholipids can be explained by the above principles coupled with a knowledge of the aqueous phase complexation behavior of the liganding groups. This question will be discussed in detail for the binding of Ca^{2+} and other divalent cations to membranes containing the phosphate--bearing lipid phosphatidic acid (PA). My studies have been carried out on 'monolayer' (Träuble and Grell, 1971) and bilayer vesicles (Haung, 1969) which have the structures and gross dimensions given in Figure 1.

Figure 2 compares the ligand phosphate with phosphatidyl choline (PC) and a number of negatively-charged phospholipids. Phosphate in solution and PA^- in membranes have moderate Ca^{2+} binding affinity,

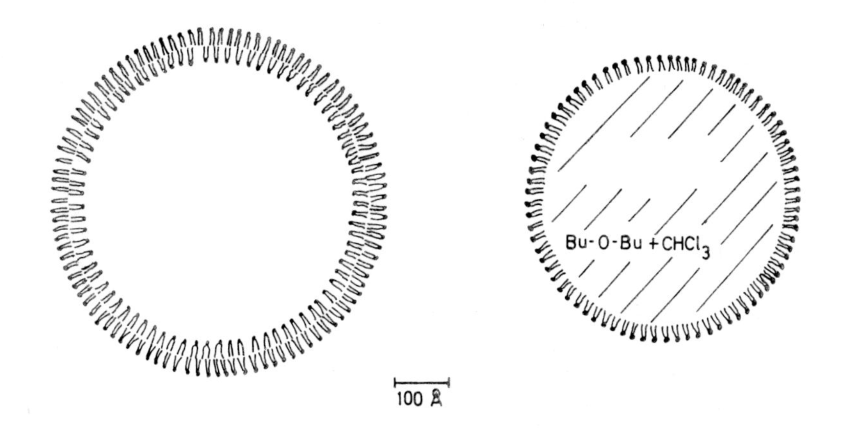

Fig. 1. Monolayer and bilayer vesicles prepared by sonication of phospholipids.

Phosphate		$O{=}P(OH)(O^-)\text{-}O\text{-}H$
Phosphatidic Acid	(PA)	$R\text{-}C(=O)\text{-}O\text{-}CH_2$ / $R\text{-}C(=O)\text{-}O\text{-}CH\text{-}CH_2\text{-}O\text{-}P(=O)(O^-)\text{-}O\text{-}H$
Phosphatidyl Choline	(PC)	$R\text{-}C(=O)\text{-}O\text{-}CH_2$ / $R\text{-}C(=O)\text{-}O\text{-}CH\text{-}CH_2\text{-}O\text{-}P(=O)(O^-)\text{-}O\text{-}CH_2\text{-}CH_2\text{-}\overset{+}{N}(CH_3)_3$
Phosphatidyl Glycerol	(PG)	$R\text{-}C(=O)\text{-}O\text{-}CH_2$ / $R\text{-}C(=O)\text{-}O\text{-}CH\text{-}CH_2\text{-}O\text{-}P(=O)(O^-)\text{-}O\text{-}CH_2\text{-}CH(OH)\text{-}CH_2(OH)$
Phosphatidyl Serine		$R\text{-}C(=O)\text{-}O\text{-}CH_2$ / $R\text{-}C(=O)\text{-}O\text{-}CH\text{-}CH_2\text{-}O\text{-}P(=O)(O^-)\text{-}O\text{-}CH_2\text{-}CH(NH_3^+)\text{-}C(=O)\text{-}O^-$

Fig. 2. Structures of phosphatidyl choline and acidic phospholipids. The R's represent n-alkane chains of 13 or 15 carbon atoms length (dimyristoyl or dipalmitoyl). These correspond to the average length of the hydrocarbon chains of natural lipids. Inorganic phosphate was included in the figure for the sake of comparison.

whereas the zwitterionic PC is essentially devoid of divalent cation binding activity (Haynes, 1974). This result can be rationalized by a consideration of the structures attached to the phosphate ligand in Figure 2. Not only does the positively-charged choline of PC effectively neutralize the negative charge of the phosphate, but it also can serve as a steric factor of the phosphate environment which could hinder coordination. In PA^- these factors are absent and essentially normal complexation can occur.

2. INFLUENCE OF MEMBRANE SURFACE POTENTIAL ON DIVALENT CATION BINDING

Ion binding to membranes is expected to influence and be influenced by the membrane surface potential, and quantitative evaluation of the effects of added cations requires the application of a model which gives the correct relationship between surface charge, ionic strength, and surface potential. Such a model is provided by the Gouy-Chapman theory (cf. Overbeek, 1949) which predicts that the distribution of ionic species at the membrane surface will be governed by a Boltzmann distribution of the membrane surface potential ψ_0. Figure 3 shows how ψ_0 and ψ_0 (the potential at the distance r from the membrane) fall off with increasing ionic strength. A practical consequence of the model

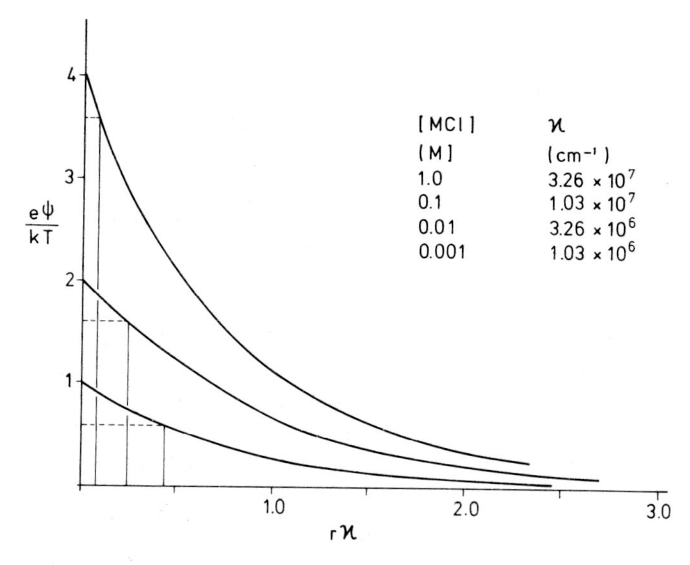

Fig. 3. Potential distribution as a function of ionic strength as predicted by the Gouy-Chapman theory.

is that membrane negative surface charge will concentrate cations near the surface in an electric double layer. It is important to distinguish between true binding (involving inner sphere coordination) and entrapment in the double layer. The latter case gives only the appearance of binding. The two cases can be distinguished by studying the binding with indicators which are not affected by potential (Haynes, 1974).

Divalent cation binding to PA^- containing membranes was studied by use of 1-anilino-8-naphthalenesulfonate (ANS^-) as fluorescent indicator of membrane surface potential, and by use of murexide (Schwarzenbach and Gysling, 1949) as an indicator of free Ca^{2+} in the aqueous phase (Haynes, 1974). The observed binding was fitted according to the equation

$$Ca^{2+}_{aq} + PA^-_{membrane} \underset{}{\overset{K_2}{\rightleftharpoons}} Ca^{2+}\text{-}PA^-_{membrane} \tag{1}$$

The equilibrium constant K_2, which uses the bulk aqueous phase as the reference state, depends upon membrane surface potential according to

$$K_2 = K_{20} \exp(-2e\psi_0/kT) \tag{2}$$

where e is the electronic charge, k is the Boltzmann constant, T the absolute temperature and K_{20} is the 'chemical' or non-electrostatic binding constant (for $\psi_0=0$). The value of ψ_0 could be calculated from the degree of ANS^- binding. This was done by the apparent and 'chemical' binding constants of ANS^-. The latter was measured at very high ionic strength where $\psi_0=0$. For ANS^- binding the ratio of the two constants is equal to $\exp(e\psi_0/kT)$. The competence of ANS^- to measure surface potentials was determined in experiments in which surface charge and ionic strength were varied in the absence of Ca^{2+}. The degree of Ca^{2+} binding could then be determined as the difference between the initial surface charge given by the PA^- concentration within the membrane and the surface charge in the presence of Ca^{2+} calculated from ψ_0 as experimentally measured by the ANS^- method. The accuracy of this procedure for the determination of Ca^{2+} binding was confirmed by binding studies using murexide as the indicator.

3. EFFECTS OF DIVALENT CATION BINDING ON LIPID DISTRIBUTION

Table I gives the values for K_{20} obtained for membranes made from PA^- and PA^-/PC mixtures. This 'chemical' or non-electrostatic binding constant is also the receprocal Ca^{2+} concentration at which one Ca^{2+} is bound per two PA^- and at which the membrane surface charge is fully neutralized. In a later section this will also be shown to be the threshold concentration for divalent cation-induced aggregation of the vesicles. The value of $K_{app(\frac{1}{4})}$, the reciprocal concentration for $\frac{1}{4}$ saturation of the PA^- shows an expected ionic strength dependence.

Use of the ANS^- fluorescent indicator technique to study divalent cation binding in PC/PA mixtures showed that M^{2+} binding results in an increase in the number of binding sites for the probe. As illustrated in Figure 4, these binding sites are made up of four neighboring PC polar head groups (cf. Haynes and Staerk, 1974). When ANS^- is bound to these binding sites it is protected from quenching processes involving water, and it exhibits a fluorescence increase due to its high quantum yield in this protected state. Lipids which lack a polar head group

TABLE I
The influence of ionic strength on the apparent binding constant for Ca^{2+} binding at pH 7.3

Mole fraction PA	1:1 Electrolyte (M)	K_{20} (M^{-1})	$K_{app(1/4)}$ (M^{-1})
0.1	0.0039	176 ± 30	5.9×10^3
0.1	0.0136	255 ± 60	2.4×10^3
0.1	0.0506	311 ± 45	1.6×10^3
0.1	0.2529	120 ± 40	4.4×10^2
0.2	0.0039	100 ± 50	8.3×10^3
0.2	0.0136	292 ± 30	5.3×10^3
0.2	0.0506	175 ± 40	1.1×10^3
0.2	0.2529	280 ± 50	1.0×10^3
0.3	0.0039	100 ± 30	1.6×10^4
0.3	0.0136	249 ± 30	6.7×10^3
0.3	0.0506	355 ± 70	2.8×10^3
0.3	0.2529	240 ± 60	1.0×10^3
0.5	0.0039	50 ± 20	1.4×10^4
0.5	0.0136	167 ± 30	7.7×10^3
0.5	0.0506	221 ± 50	2.2×10^3
0.5	0.2529	170 ± 50	7.7×10^2
1.0	0.019	(∿280) [a]	$2.8 \pm 0.4 \times 10^3$ [b]
1.0	0.319	(∿500) [a]	$1.6 \pm 0.4 \times 10^3$ [b]

[a] Estimated from the value of $K_{app(1/4)}$.

[b] Determined at pH 8.50 using the murexide technique, with 19 mM Tris buffer and 0 or 0.3 M NaCl.

Experimental Conditions: Dimyristoyl PC/PA monolayer vesicles, 3.9 mM imidazole buffer, pH 7.3, 1.24×10^{-4}M ANS, 30 °C.

The value of $K_{app(1/4)}$ is defined as the Ca^{2+} necessary for 1/4 saturation of the PA binding sites and half-neutralization of the membrane surface charge.

capable of supporting this hydrophobic association with the probe (eg. PA^-) do not react with ANS^- to produce a fluorescent species with those which do not, the number of binding sites is reduced. These data were compared with two models, the first one assuming that the PC and PA form separate phases in the membrane and another assuming random mixing of the two species. The second model is illustrated in Figure 5. Inclusion of the PA^- within the four-membered sites destroys them as ANS^- binding sites. Table II presents experimental data on the dependence of the number of sites on the mole fraction of PA and shows good

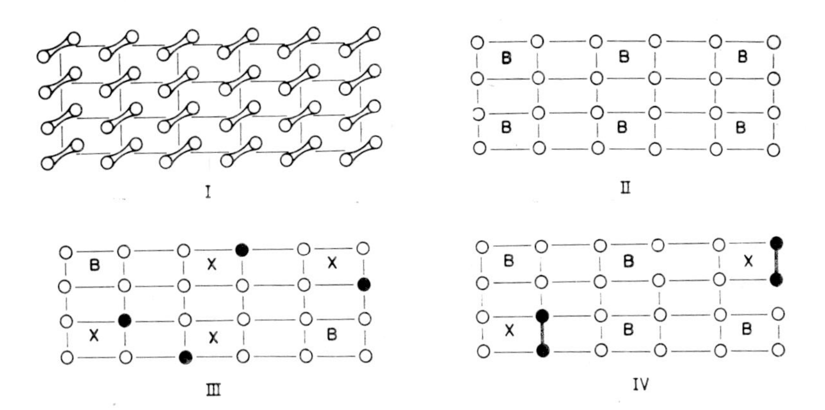

Fig. 4. Model for the effect of Ca^{2+} binding on the distribution of PA^- and PC within the membrane: (I) The approximate hexagonal packing of the fatty acid chains (circles) of the phospholipid molecules (dumbbells) gives rise to a 'quadratic' array of polar head groups. (II) Binding within one quadratic array precludes binding in the four neighboring arrays. The binding site is denoted with B. (III) Insertion of PA, precludes binding within this site, (IV) Association of Ca^{2+} -PA^- with PA^- or their mutual attraction, ●–●, liberates binding sites.

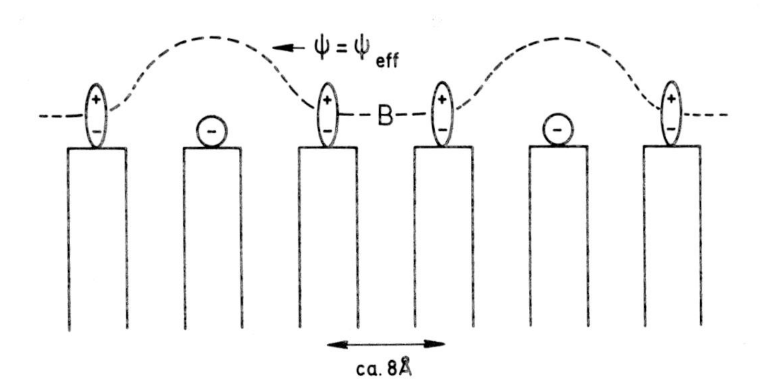

Fig. 5. Schematic representation of potential isotherm in the case of discreteness of charge effects. Citations of discussion of these properties are found in an earlier paper (Haynes, 1974).

TABLE II
Dependence of the number of ANS^- binding sites and K_3 values on PA content of dimyristoyl PC/PA monolayer vesicles

Mole fraction PA	$[B]_t$ experimental	$[B]_t$ ideal mixing	$[B]_t$ no mixing	K_3
0	0.25	-	-	-
0.1	0.13 ±0.03	0.163	0.225	2.67±0.50
0.2	0.07 ±0.02	0.102	0.200	1.00±0.30
0.3	0.07 ±0.02	0.060	0.175	0.92±0.08
0.5	0.016±0.002	0.0154	0.125	0.54±0.20

Experimental conditions: Dimyristoyl PC/PA monolayer vesicles 3.9 mM imidazole buffer, pH 7.3, 1.24×10^{-4} M ANS^-, 30 °C. $[B]_t$ values were determined in 1.5 M KCl and the K_3 values given were determined for 0.249M KCl.

$[B]_t$ are the numbers of binding sites per phospholipid in the membrane. Values for ideal mixing were calculated as $[B]_t$ (mixture) = $[B]_{t\ (PC)} \times (f_{PC})^4$ where $[B]_{t(PC)}$ is the number of binding sites for membranes of pure PC and where f_{PC} is the mole fraction of PC in the mixed membrane. The values for no mixing were calculated as $[B]_t$ (no mixing) = $[B]_{t(PC)} \times f_{PC}$. The value of K_3 is an index for the tendency of Ca^{2+} binding to separate PA molecules from PC molecules on the microscopic scale. It was calculated from experimental data on the effect of Ca^{2+} to increase the number of binding sites of PC/PA mixtures according to:

$$[B]_t(\text{mixture}) = [B]_{t(PC)} \times (1-[PA]_t+K_3 \times [Ca^{2+}-PA^-] \times [PA^-]/[PA]_t)^4$$

A value of K_3=4 means that each Ca^{2+} bound removes one PA from an ANS^- binding site and indicates the formation of separate PA and PC microphases within the membrane.

agreement with the second model and poor agreement with the first. PC and PA are thus randomly mixed within the membrane on the microscopic scale.

Addition of Ca^{2+} to these mixed vesicles results in an increase in the number of binding sites. This can be explained as the tendency of $Ca^{2+}-PA^-$ to attract PA^- such that the approach of these two species effectively liberates an ANS^- binding site. This is illustrated in configurations III and IV of Figure 4. A qualitative index of the tendency for association is given by the K_3 values of Table II. A value of 4 indicates 100% probability of an association of PA^- with $Ca^{2+}-PA^-$. Although the values obtained were considerably smaller than 4, they provide a clear indication that either association or electrostatic effects between PA^- and $Ca^{2+}-PA^-$ can produce deviations from randomness of their distribution in the plane of the membrane.

4. DIVALENT CATION SPECIFICITY OF BINDING

Table III shows that the ion specificity for binding of PA within a dimyristoyl PC Matrix is $La^{3+} > Mn^{2+} \sim Mg^{2+} > Ca^{2+} > Ba^{2+}$. These data are in

TABLE III
Comparison of specificity of divalent cation binding to PA in dimyristoyl PC/PA membranes with HPO_4^{2-} binding specificity

Ligand	Ion specificity[a] normalized to Ca^{2+}				
	Ba^{2+}	Ca^{2+}	Mg^{2+}	Mn^{2+}	La^{3+}
PA in PC/PA membranes	0.42±0.07	1.00	1.46±0.18	1.62±0.31	10-100
HPO_4^{2-} in Water[b]	-	1.00	1.52	7.62	-

[a] $K_{20}(M^{2+})/K_{20}(Ca^{2+})$

[b] Data from Smith and Alberty (1956) obtained for a medium containing 0.2 M $(n\text{-}C_3H_7)_4NCl$. The Ca^{2+} binding constant was reported as 50±2 M^{-1}. Extrapolated to zero ionic strength, the binding constant becomes 500 M^{-1}.

qualitative agreement with those of Smith and Alberty (1956) for complexation of divalent cations with HPO_4^{2-} in water. Thus the ion specificity of phosphate as a complexing ligand is essentially unmodified by its covalent linkage to the lipid at the water-membrane interface. The binding constants for Ca^{2+} can be compared directly. The binding constant for Ca^{2+} with $H_2PO_4^-$ extrapolated to zero ionic strength is 12 M^{-1} (Davies and Hoyle, 1953). The corresponding value for HPO_4^{2-} is 500 M^{-1}. The former value is one order of magnitude lower than the K_{20} values given in Table I. A partial explanation for this may be that the electrostatic self-energy of the phosphate at the water-membrane interface is greater than for the corresponding ion in water. However, considering the degree of accuracy of electrostatic calculations for the process of cation-anion association, a one order of magnitude difference in complexation constants in small. Some problems involved with these calculations will be discussed below.

5. SEPARATION OF 'ELECTROSTATIC' AND 'CHEMICAL' EFFECTS ON BINDING

The binding energy of a cation-anion pair can be thought of as the sum of chemical and electrostatic contributions. The electrostatic term would be the product of the cationic charge and the electrostatic potential at that distance from the center of the ligand which characterizes the distance between cation-anion centers in the stable complex. (cf. Robinson and Stokes, 1956; Gurney, 1943). In order to compare 'chemical' binding constants for a ligand in an aqueous and in a membrane environment, it will be necessary to calculate the electrostatic interaction energy.

For the case of the ligand in an aqueous phase this could be calculated simply as $(Z_{cation} Z_{anion} e^2)/\varepsilon R$ where Z represents the absolute value of the charge, e is the electronic charge, ε is the dielectric constant of water and R is the distance of separation. This calculated energy would be attenuated by contributions from the Debye-Hückel effects of the ionic strength on the activity coefficients of the separate species. Applying this to the Ca^{2+}-$H_2PO_4^-$ case with R = 2.3 Å (the sum of Ca^{2+} and O^{2-} crystal radii) we would arrive at an energy of 12 kT. Not only is our estimate very sensitive to assumptions such as the value of the local dielectric constant, but it is extremely sensitive to the distance of complexation, due to its R^{-1} dependence. It should be further pointed out the the Debye-Hückel theory can not correctly predict the effect of ionic strength on the potential drop in the immediate vicinity of the anion because it has been derived using a linearized form of the exponential term in the Boltzmann equation. It would thus seem that a better procedure for estimating 'chemical' binding constants is to extrapolate the measured binding constant to infinite ionic strength under conditions where the supporting electrolyte does not itself bind the anion.

The Gouy-Chapman theory can be used to calculate the electrostatic energy of bringing a charge from a point infinitely removed up to the membrane surface. However, the theory has some serious drawbacks which indicate that it will not account for the total electrostatic energy of the ion binding process in the same sense as it was defined in the preceding paragraph. The theory treats the charges of the ligands in the membrane as being 'smeared out' but considers the counter ions and binding cations as point charges. This model becomes physically unreasonable for calculating the electrostatic energy involved in the last ca. 10 Å of approach of Ca^{2+} to its phosphate ligand in PA^- membranes. There have been a large number of calculations which have estimated sizable discreteness-of-charge effects and deviations from Gouy-Chapman behavior (cf. Haynes, 1974). My findings show a lack of agreement between the experimentally determined ionic strength dependencies of Ca^{2+}, ANS^- and H^+ binding and the dependencies predicted by the Gouy-Chapman equations and indicate that discreteness-of-charge effects are playing a role.

It was found that the Gouy-Chapman equations could be empirically modified to give potentials which are smaller than those predicted by the theory, and which give the experimentally-measured ionic strength dependence of the binding of the fluorescent anion ANS^- to PC and PC/PA mixed membranes and for H^+ and M^{2+} binding to PA^- containing membranes. This potential corresponds to ψ_{eff} illustrated in Figure 5. In contrast to the case with the Gouy-Chapman theory, the isopotentials are not considered to run strictly parallel to the membrane, but can rather approach the membrane more closely at points where there is no fixed charge. A procedure for calculating these isopotentials has been published recently (Nelson and McQuarrie, 1975). The dependence of ψ_{eff} on surface charge and ionic strength from the experiments reviewed here is given in Figure 6. It should be pointed out that meth-

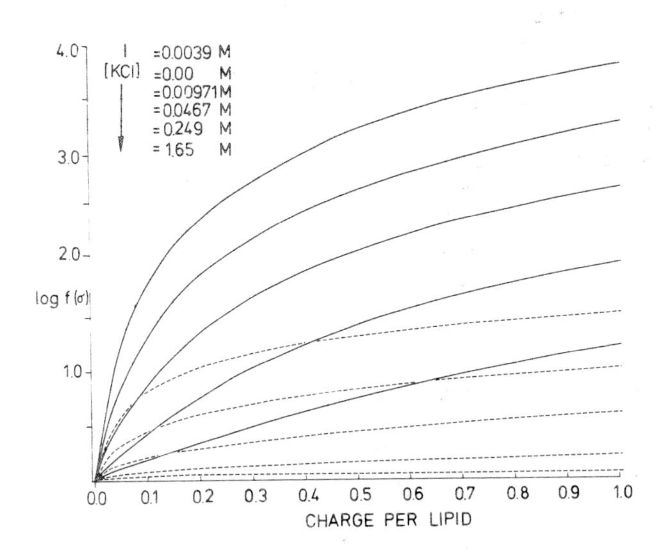

Fig. 6. The dependence of the membrane surface potential on surface charge per lipid area (per 58 Å^2) and on ionic strength. Each vertical unit corresponds to 59 mV. The solid lines and the broken lines indicate the dependenceis for the Gouy-Chapman equation and for the equation as modified to fit the ion binding data in the previous study (Haynes, 1974).

ods which sense the potential averaged over the whole membrane surface (electrophoresis and ionizing electrodes, cf. Hayden and Meyers, 1973) or methods which sense the potential in the center of the hydrocarbon chain region of the membrane (ionophore-induced conductivity cf. McLaughlin et al., 1971) may be less sensitive to discreteness-of-charge effects and might be expected to give higher estimates of the surface potential.

Since these potentials derived from electrostatic models for both solutions and for membranes are very sensitive to starting assumptions, it would seem advisable to determine K_{20} for both cases by extrapolating the apparent binding constants to infinite ionic strength. This procedure has left us with a more than one order of magnitude difference in K_{20} for $H_2PO_4^-$ and PA^-. At least part of this difference must be related to the electrostatic self-energy of the PA^- molecules in their ordered array. Finally, we note that the K_{20} values for both ligands contain a sizable but essentially constant (ionic strength independent) electrostatic component.

6. KINETICS OF Ca^{2+} BINDING

Studies of the rate of Ca^{2+} binding were carried out not only for the sake of learning more about the fundamental interactions of the cation with the membrane, but also to provide a model for the specific Ca^{2+} triggered reactions of membranes. A particularly relevant biological example is the exocytotic release of acetylcholine at the nerve terminal occurring in times of less than 200 μsec. The theoretical maximum for the bimolecular rate constant for the reaction of two spherical electrically neutral and like-sized species in aqueous medium can be calculated from the theory of Smoluchowski (1916 and 1917) as 6.8 x 10^9 M^{-1} s^{-1}. This value depends only upon the absolute temperature and the viscosity of the medium. Subsequent formulations by Debye (1942) showed that this value is increased for the reaction of oppositely--charged species and is decreased for the reaction of like-charged species.

The studies of Eigen and coworkers (Eigen, 1963) have shown that the rates of reaction of Ca^{2+} with simple ligands are almost in the range of diffusion control. Examples are the bimolecular rate constant of 7 x 10^8 M^{-1} s^{-1} for the reaction of Ca^{2+} a 'metal-phthalein' indicator dye (Czerlinski et al., 1959) and the lower limit of 10^9 M^{-1} s^{-1} for the reaction of Ca^{2+} with ATP^{4-} (Diebler et al., 1960). The corresponding rate constants for Mg^{2+} complexation were 1.6 x 10^6 M^{-1} s^{-1} and 1.2 x 10^7 M^{-1} s^{-1}, respectively. These and other comparative studies showed that the reason for this two order of magnitude difference was that inner sphere complexation of Mg^{2+} has as a rate-limiting step the dissociation of a water of hydration from its inner coordination sphere. For Ca^{2+}, this step constitutes no substantial rate limitation. It was of interest to see whether Ca^{2+} could bind to PA^- within the membrane as rapidly as it combines with these model ligands in an aqueous milieu.

Rapid kinetic experiments were carried out using the temperature--jump method (Eigen and de Maeyer, 1963) using murexide as an indicator of free Ca^{2+} in the aqueous phase (Haynes, 1974). The method consists of raising the temperature in the reaction cell rapidly within 3 to 10 μsec to alter the equilibrium constant of the reaction. Information about the rate constants of the reaction can be obtained by studying the dependence of the time constant for exponential approach τ to the new equilibrium condition upon equilibrium concentrations of reactants and products. The method is applicable if (a) the reaction to be studied has a sufficiently large ΔH value, (b) if τ is greater than 3-10 sec and (c) if the reaction is attended by or can be coupled to spectral change such as absorbance, fluorescence or light scattering.

In the temperature-jump experiments reported here, murexide was used as a Ca^{2+} indicator. The dye has moderate binding affinity (log (K_s)=2.8 at pH 7.85; Schwarzenbach and Gysling, 1949) which is well matched to the binding constant of Ca^{2+} on PA^-. Our temperature-jump experiments showed that the complexation reaction is faster than the ca. 10 μsec RC time constant of the electric discharge of the instrument. The dye is thus competent as an indicator of Ca^{2+} binding kinetics.

Figure 7 shows ca. 75 μsec ($t_{1/2}$) relaxation obtained upon T-jump perturbation of the Ca^{2+} binding equilibrium of dimyristoyl PC/PA 5/5

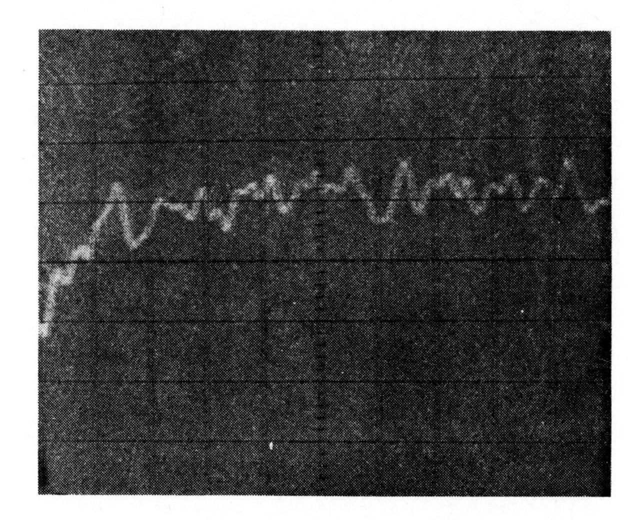

Fig. 7. Oscillograph trace of T-jump Relaxation of Ca^{2+} binding to PA^-. Murexide was used as an indicator. The horizontal axis represents 100 μsec per scale division and the vertical axis displays transmission of 550 nm light through the sample, with one scale division representing a 4.0% change in transmission. Increasing light intensity is plotted downward. The discharge voltage was 30 kV and the capacitor was 0.05 uF. The medium contained 5×10^{-4}M Ca^{2+}. Other experimental details are given in Figure 8.

vesicles. The rising trace corresponds to increased optical absorbance at 550 nm (the absorbance maximum of free murexide), indicating that the temperature rise has resulted in an increase in binding of Ca^{2+} to the indicator which in turn was the result of a decrease in binding to the PA in the membranes. The amplitude of this relaxation showed isobestic behavior at 505 nm proving that it is indeed Ca^{2+} complexation that was being followed. The direction of the change indicates a positive enthalpy of reaction. The reaction was analyzed according to the reaction mechanism.

$$Ca^{2+}_{aq} + PA^-_{membrane} \underset{k_{-1}}{\overset{k_1}{\rightleftharpoons}} Ca^{2+}\text{-}PA^-_{membrane} \tag{3}$$

If it is assumed that the reaction has the kinetic properties of a simple A + B → C reaction, the expected dependence of τ on the equilibrium concentrations of reactants, $[Ca^{2+}]_{eq}$ and $[PA^-]_{eq}$, is given by:

$$1/\tau = k_1 \left([Ca^{2+}]_{eq} + [PA^-]_{eq}\right) + k_{-1} \tag{4}$$

For small PA^-concentrations, this predicts a linear dependence of the rate of relaxation on the Ca^{2+} concentration. Figure 8 is a plot of rate data obtained at different Ca^{2+} concentrations. From the slope

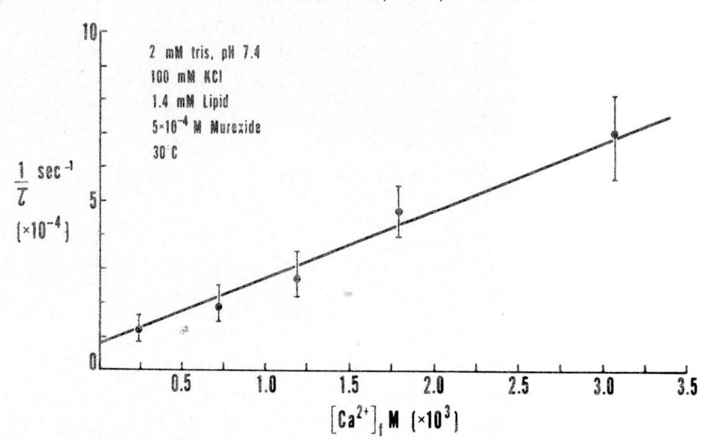

Fig. 8. Dependence of the relaxation rate on Ca^{2+} concentration. The experimental conditions are given in the figure. Other experimental parameters are identical to those of Figure 7.

and intercept, apparent* values of ca. $1.9 \times 10^7 M^{-1} s^{-1}$ and ca. $7 \times 10^3 s^{-1}$ are obtained for k_1 and k_{-1} respectively. The k_1 value is only 0.027 as large as the rate constant for Ca^{2+} complexation with the 'metal phthalein' indicator or with ATP^{4-}, indicating that the phosphate in PA^- is somehow much less kinetically accessible to Ca^{2+} than are the complexing centers of simple ligands. It should be pointed out these are the kinetics of surface binding and not those of Ca^{2+} permeation which has a half-time of hours.

Possible explanations for the low value of k_1 could include (a) the effect of water structure to produce a barrier to diffusion at the membrane surface, (b) blockage of the approach of the cation by the PC polar head groups and (c) structural rearrangements within the membrane as an obligitory step in the binding mechanism. Comparison of the Ca^{2+} rate constants with those for ANS^- binding to sites in PC membranes indicates that explanation (a) is unlikely. The bimolecular rate constant for ANS^- combining with its binding site composed of four PC polar head groups is $3 \times 10^8 M^{-1} s^{-1}$ (Haynes, 1972; Haynes,

---

*These values must be considered as apparent since they would be expected to depend upon the surface potential, which becomes smaller with increasing $[Ca^{2+}]$. The values are correct for the experimental conditions at which they were determined. Thus the apparent binding constant from the kinetic experiments $k_1/k_{-1} = 2.7 \times 10^3 M^{-1}$ is within the range of K_2 the values determined at the Ca^{2+} concentrations of Figure 8, but is 10 times as large as K_{20} (cf. Equation (2) and Table I)

1974) is an order of magnitude higher than the corresponding rate constant for Ca^{2+} on PA^-. The rate constant has been shown to be inversely proportional to the solution viscosity which was adjusted by sucrose addition (D. Haynes, unpublished), indicating that it is a reaction which is essentially under diffusion control.

Blockage of access to the PA^- phosphate by the intervention of the $choline^+$ head groups of PC offers a more likely explanation. Experiments with molecular models show that it is sterically possible for the quaternary nitrogen to approach the phosphate group closely. The most obvious test of this explanation could be provided by a study of the binding rate for pure PA membranes. Unfortunately this experiment is rendered difficult by the vesicle aggregation reactions promoted by divalent cation binding to membranes whose mole fraction of acidic phospholipids approach 1.0. This phenomenon will be discussed in the next section.

The third possibility is that structural alterations are an obligatory step to the Ca^{2+} binding reaction. A molecular interpretation of this would suggest that binding requires a redistribution of lipid within the membrane or a structural change within the membrane. However, the equilibrium studies of the Ca^{2+} binding phenomenon do not give evidence that this is necessary. The finite values of K_3 given in Table II indicate that there is some effect of Ca^{2+} binding on the distribution of uncomplexed PA in the plane of the membrane. However, the values are generally much smaller than 4, indicating that complexation with the formation of a separate PA phase is not necessary for Ca^{2+} binding to occur. Furthermore it can be calculated that the time necessary for PA^- to diffuse in the plane of the membrane to combine with Ca^{2+}-PA^- is smaller than the measured τ values or the time for complex dissociation ($1/k_{-1}$) implying that lipid rearrangements could not be the rate limiting step. The diffusion constants of these membrane lipids are estimated as ca. 1×10^{-8} cm^2 s^{-1} by comparison with the diffusion constant of spin-labelled dipalmitoyl PC (1.8×10^{-8} cm^2 s^{-1}, Devaux and Mc Connell, 1972) and with the corresponding constant for an andrestan spin label in PC membranes (1×10^{-8} cm^2 s^{-1}, Sackmann and Träuble, 1972). Using the diffusion equation for root mean square distance $\overline{r}$ in time t,

$$\overline{r}^2 = 2\,Dt \qquad (5)$$

and using 24 Å (two lipid cross sections) for $\overline{r}$ we calculate t = 2.9 µsec. This would be the time necessary for the formation of a 2:1 complex from a 1:1 complex on a membrane compound of PC/PA 5/5. It would not be possible to resolve this process with the present instrumentation.

Another molecular interpretation would suggest that the rate of the net binding reaction is limited by the rate of membrane structural changes attendant to binding, such as a compression of the membrane upon Ca^{2+} binding. Evidence for such changes in fluorescent lifetime studies of ANS^--B in PC membranes (Haynes and Staark, 1974). Membrane surface charge and potential were perturbed by the incorporation of PA and the variation of monovalent and divalent electrolyte. There is some evidence that the hydrophobic association of the probe was weakened by the inclusion of PA. However, these electrostatic effects are anticooperative rather than cooperative in nature, implying that these changes are not necessary for Ca^{2+} binding. Furthermore, temperature-jump relaxation experiments (D. Haynes, unpublished) monitoring light scattering and ANS^- fluorescence indicate that these relaxation processes are an order of magnitude slower that the dissociation rate for Ca^{2+} determined here.

7. VESICLE AGGREGATION PROMOTED BY DIVALENT CATIONS

In a recent study (Lansman and Haynes, 1975) we showed that divalent cations can produce a rapid aggregation of PA^- and PS^--containing vesicles. This reaction was shown to proceed with 'polymerization' kinetics, such that after several minutes the vesicles were in such a high stage of aggregation that they flocculated out of suspension. It was possible to resolve kinetically the aggregation reaction of the vesicles to form 'dimers'. This was analyzed for Ca^{2+} and vesicle concentration dependence and was shown to proceed according to a two-step mechanism in which the vesicles first bound Ca^{2+}

$$n\ Ca^{2+} + \text{vesicle} \underset{k_{-0}}{\overset{k_0}{\rightleftharpoons}} (\text{vesicle*n } Ca^{2+}) \quad (6)$$

Where n represents a number of Ca^{2+} per vesicle sufficient to promote aggregation. The dimerization of the vesicles then occurs in a slower step, with an apparent rate constant k_{app}

$$(\text{vesicle*n } Ca^{2+}) + (\text{vesicle*n } Ca^{2+}) \underset{k_{-app}}{\overset{k_{app}}{\rightleftharpoons}} (\text{vesicle*n } Ca^{2+})_2 \quad (7)$$

From the data of the previous section we can calculate that the first reaction (Equation 6) will be essentially complete within 100 μsec for the case of Ca^{2+} binding. Stopped-flow experiments with Mg^{2+}, Mn^{2+}, Ba^{2+} and La^{2+} indicate that the half time for binding of these cations are less than 3 msec, the time resolution of the method. The vesicle aggregation represents the slow step of the reaction.

Figures 9 and 10 show that the extent and rate of the aggregation reaction had a very strong dependence on the Ca^{2+} concentration. The dashed lines in these figures indicate the concentration of Ca^{2+} necessary for achievement of a one Ca^{2+} per two PA^- binding stoichiometry and full neutralization of the membrane surface charge. Thus the number n seems to be 0.5 Ca^{2+}/PA^-. Log-log plots of data of the type shown in Figures 9 and 10 give slopes as high as 3, indicating a fair degree of cooperativity in the aggregation process. This is not surprising since the vesicle contains ca. 10^4 negative charges which must be neutralized for a vesicle-vesicle approach to be effective.

Table IV shows that the values of $t_{1/2}$ and extent of reaction are relatively independent upon the choice of acidic lipid (PA^- or PS^-) or divalent cation.

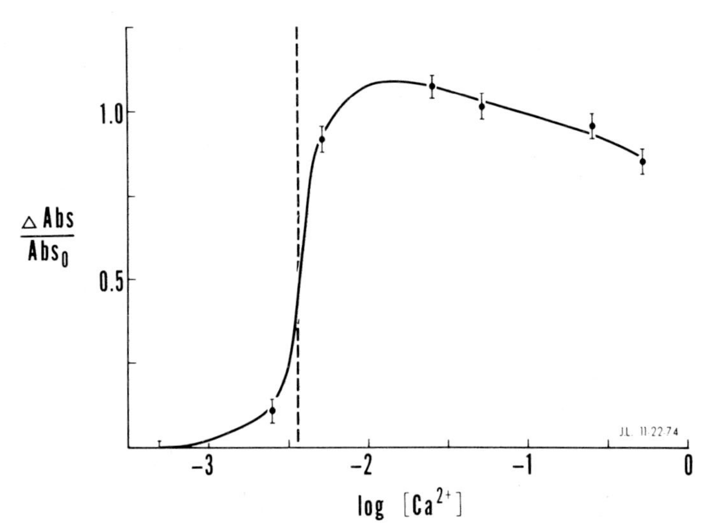

Fig. 9. The dependence of the reaction amplitude on Ca^{2+} concentration. The data were obtained from a stopped-flow experiment monitoring optical transmittance at 360 nm. Syringe A contained phosphatidic acid vesicles (0.7 mg/ml) in 10 mM NaCl, 10 mM Tris-HCl, pH 7.4; syringe B contained $CaCl_2$ at twice the concentration given in the figure. The mixing volumes were equal. A value of 1.0 on the vertical scale represents a doubling of turbidity and light scattering and a doubling of the weight average molecular weight of the aggregates.

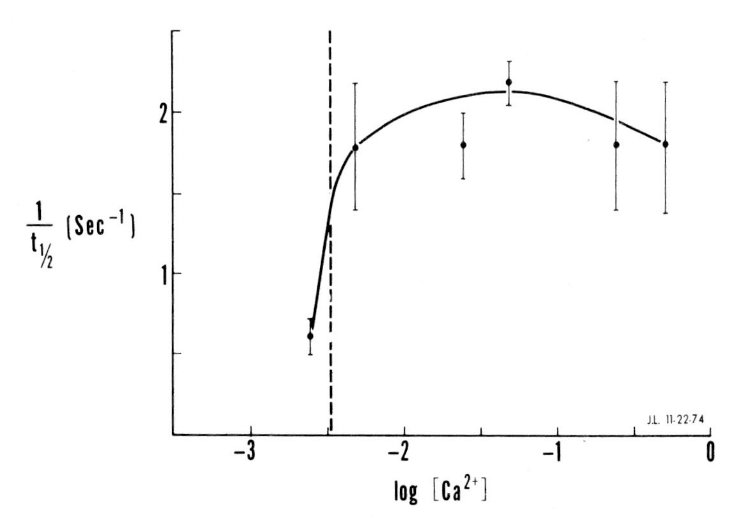

Fig. 10. The dependence of the reaction rate on Ca^{2+} concentration. Conditions were identical to those of Figure 9.

TABLE IV
The effect of Lipid and Cation on $t_{1/2}$ and Reaction Amplitude
The relative amplitude was measured at 340 nm. The cation concentration was 50 mM, unless otherwise indicated T=21 °C.

Lipid	Lipid (mg/ml)	Cation	$t_{1/2}$	$\Delta abs/abs_0$
Phosphatidic acid	0.5	Ca^{2+}	0.43±0.10	1.16 ±0.02
Phosphatidic acid/lecithin (1:1)	0.5	Ca^{2+}	1.00±0.20 [a]	0.044±0.022
Phosphatidic acid	0.5	Mg^{2+}	0.46±0.10	1.30 ±0.2
Phosphatidic acid	0.5	Mn^{2+} (5.0 mM)	0.45±0.07	1.3 ±0.1
Phosphatidic acid	0.5	Ba^{2+}	0.38±0.10	1.1 ±0.1
Phosphatidic acid	0.5	La^{3+} (0.5 mM)	0.56±0.02	1.2 ±0.5
Phosphatidylserine	0.3	Ca^{2+}	0.50±0.1	1.0 ±0.1
Phosphatidylserine	0.3	Mg^{2+}	0.67±0.1	1.0 ±0.1
Phosphatidylserine	0.3	Mn^{2+}	0.42±0.1	0.6 ±0.2
Phosphatidylserine	0.3	Sr^{2+}	0.35±0.1	0.6 ±0.2
Phosphatidylserine	0.3	Ba^{2+}	0.40±0.1	0.6 ±0.2
Phosphatidylserine	0.3	La^{3+} (0.5 mM)	0.35±0.1	0.6 ±0.2

[a] To be interpreted as time necessary for half-maximal change in α

The reciprocal half-time for the process of Equation (10) is roughly proportional to the vesicle concentration (Figure 11) in

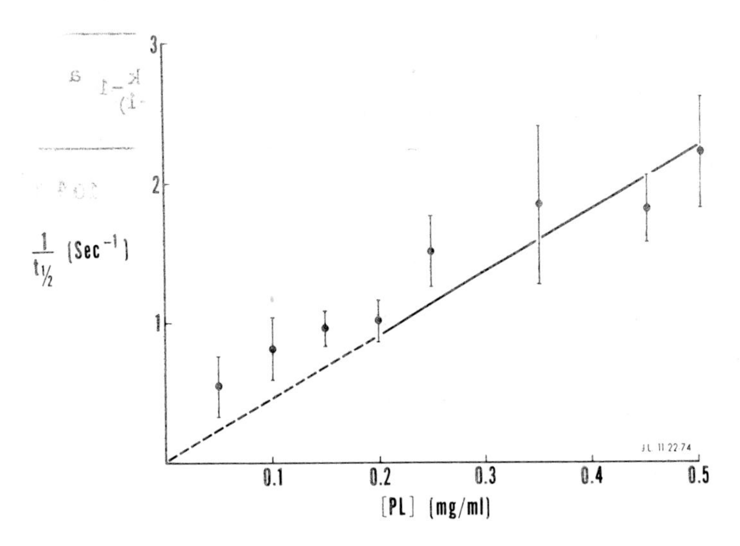

Fig. 11. Dependence of $1/t_{\frac{1}{2}}$ on phosphatidic acid vesicle concentration. The experimental conditions were identical to those of Figure 9 and 10 except that syringe B contained 100 mM $CaCl_2$.

keeping with the expectations of a dimerization reaction. The rate constant for the reaction of Equation (7) could be determined from the slope according to

$$k_{app} = \frac{\alpha}{2(1-\alpha)} \frac{N_{PL}}{[PL]\ t} \tag{8}$$

where α is the extent of the reaction as measured by the light scattering change [PL] is the concentration of lipid and N_{PL} is the number of lipid molecules per vesicle. The values of k_{app} for phosphatidic acid and phosphatidyl serine vesicles with Ca^{2+} are given in Table V. These numbers are two orders of magnitude below the value calculated for diffusion control ($6.8 \times 10^9\ M^{-1}\ s^{-1}$) calculated from the Smoluchowski theory (Smoluchowski, 1916, 1917). It thus seems that processes occurring at the membrane surface represent that rate-limiting step in the aggregation reaction. The foregoing discussion indicates that the reaction of Equation (7) is an oversimplification and that the minimal mechanism would be:

$$\begin{aligned}(\text{vesicle}^{*}\text{n } Ca^{2+}) + (\text{vesicle}^{*}\text{n } Ca^{2+}) &\underset{k_{-1}}{\overset{k_1}{\rightleftharpoons}} (\text{encounter complex}) \\ &\underset{k_{-2}}{\overset{k_2}{\rightleftharpoons}} (\text{stable complex})\end{aligned} \tag{9}$$

TABLE V
Calculation of kinetic constants for the aggregation reaction

Lipid	k_{app} ($M^{-1}s^{-1}$)	k_1/k_{-1} [a] (M^{-1})	k_2 [b] (sec^{-1})
Phosphatidic acid	3.9×10^7 c	3.9×10^4 c	1.0×10^3 c
	9.9×10^6 d	5.1×10^3 d	1.9×10^3 d
Phosphatidylserine	6.9×10^7 c	3.9×10^4 c	1.7×10^3 c
	1.7×10^7 d	5.1×10^3 d	3.4×10^3 d

[a] Calculated using $k_1/k_{-1} = (4/3\pi R^3 N \times 10^{-3}$ (M^{-1}).
[b] Calculated according to Equation (12).
[c] Calculated using R = 250 Å; $N_{pl} = 2.66 \times 10^4$
[d] Calculated using R = 125 Å; $N_{pl} = 6.67 \times 10^3$

where k_1 and k_{-1} are determined by the rate of diffusion according to the Smoluchowski treatment and where k_2 is determined by process occurring at the membrane surface. These are related to k_{app} by

$$k_{app} = (k_1/k_{-1})k_2 \tag{10}$$

where $k_1 = 6.8 \times 10^9\ M^{-1}\ s^{-1}$ and where $k_{-1} = (1.7\text{-}7.9) \times 10^5\ s^{-1}$. The lower and upper limits for k_{-1} are defined by the upper and lower estimates of the vesicle radius (R = 250 Å and 125 Å, respectively). The value of k_2, the rate constant for the transformation of the encounter complex of PA^- vesicles to a stable complex, is thus $(1.0\text{-}1.9) \times 10^3$ s^{-1}. The fraction of the total number of collisions which are effective in producing a stable complex is thus only 0.0014-0.0059 ($=k_2/(k_{-1} \times k_2)$). A second physical interpretation of this result is that two membranes which have bound Ca^{2+}, which have ca. 500 Å cross sections and which are apposed within reaction distance of each other will require 0.5-1.0 msec ($1/k_2$) to aggregate. Tables IV and V show a similar situation for PS vesicles and for the other divalent cations.

The identity between the divalent cation concentrations necessary for binding one Ca^{2+} per two PA^- and the treshold concentration for aggregation, taken together with the observation that the monovalent cations do not promote aggregation, is good evidence that the aggregation reaction is the result of a M^{2+} - mediated salt bridging interaction. The following discussion is taken directly from the previous report (Lansman and Haynes, 1975).

Figures 12A and 12B illustrate the two possible modes of this salt bridging interaction. Figure 12A treats the interaction as the transformation of a 1:1 Ca^{2+}-(polar head group) complex into a 1:2 intermembrane complex. Evidence for the 1:1 interaction was given in

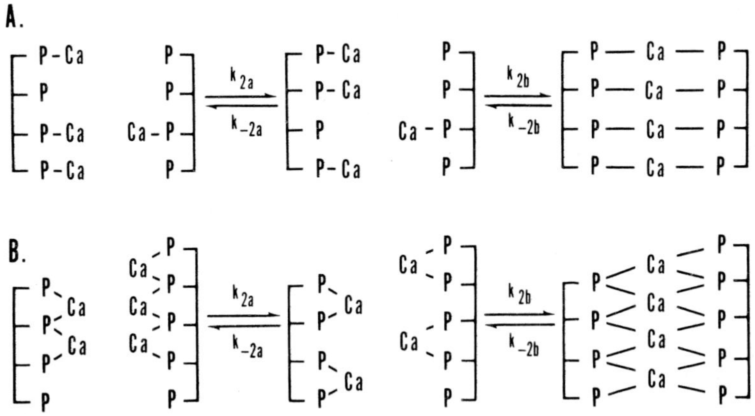

Fig. 12. Schematic model for the vesicle aggregation reaction. The membrane surfaces of two vesicles are depicted with the symbol P denoting the charged polar head groups. The coordination of the cation is indicated by the solid lines. For visual clarity, the charges of the cation Ca^{2+} and the polar head group P^- are not given in the figure. (A) The bound cation is singly coordinated in the non-aggregated state and doubly coordinated in the aggregated state. (B) The bound cation is subject to higher degrees of intramembrane and transmembrane coordination.

a study of Ca^{2+} binding to membranes whose phosphatidic acid/lecithin content was varied (Haynes, 1974). Figure 4B describes the membrane-membrane interaction as the conversion of a 1:2 Ca^{2+}-(polar head group) complex into a 1:4 intermembrane complex. Since both of the models share the same essential feature, that aggregation requires binding of divalent cations to the extent of one cation per two polar head groups, the above data do not allow us to distinguish between these two models.

Figures 12A and 12B show why Ca^{2+} binding and apposition of the two membranes are not in themselves sufficient to cause the membranes to aggregate. Salt bridging requires the positions of the Ca^{2+}-occupied and unoccupied polar head groups to be correlated, but such correlation of occupation of a large number of polar head groups is unlikely if their disposition is statistical. Aggregation thus requires rearrangement of the Ca^{2+} on the membrane surface. For phosphatidic acid and phosphatidylserine vesicles, this step (k_{2a}) and the subsequent aggregation of the membranes (K_{ab}) must require 0.3-1.0 ms.

Our finding that the threshold Ca^{2+} concentration for aggregation coincides with the concentration for 1:2 Ca^{2+}: PA^- binding stoichiometries can be readily understood in terms of Figure 12. Low degrees of occupation of the phosphatidic acid binding sites would not result in sufficient probability of correlation of the positions of the occupied phosphatidic acid molecules for stable aggregation.

The extent of the aggregation reaction is reduced drastically by addition of lecithin to the phosphatidic acid. Aggregation of vesicles composed of lecithin is probably suppressed by short-range repulsive interactions resulting from the perpendicular orientation and dipolar character of the polar head groups. We would thus expect the formation of stable junctions between phosphatidic acid. For statistical distribution of phosphatidic acid and lecithin on the membrane surfaces of the encounter complex, the probability of occurrence of two apposed microregions containing only phosphatidic acid molecules would be $P = (f_{PA})^q$, where f_{PA} is the mole fraction of phosphatidic acid within the membrane and where q is the number of phosphatidic acid molecules within the two microregions. A value of q between 4 and 6 would be sufficient to explain the observed 25-fold reduction in k_2/k_{-2} for the 1:1 phosphatidic acid/lecithin mixture (cf. Table IV).

The irreversible step in the aggregation reaction requires 0.5-1.0 ms and involves the rearrangement of Ca^{2+} on the apposed surfaces of the two membranes. This rearrangement can occur by mechanisms involving movement of the vesicles, Ca^{2+} or phospholipids. (a) An unsuitable encounter complex can dissociate and then reassociate to form an encounter complex in which other areas of the membranes are brought into contact. The probability that this new complex is stable is $P = k_2/(k_2+k_{-1})$. This process would be repeated until a stable complex is formed. (b) Rearrangement of Ca^{2+} on the membrane surface can occur in a process in which Ca^{2+} dissociates from one polar head group and associates with a second one. Our temperature-jump experiments described in the previous section indicate that any single PA^- binding site might be expected to undergo this reaction about 10 times during the 1 msec required for the encounter complex to transform into the stable complex. (c) A phospholipid molecule whose polar head group is occupied by Ca^{2+} can change places with an unoccupied neighbor. The calculation in the previous section shows that this time will be of the order of 3 μsec, so a given lipid should be able to make ca. 300 changes in place during the 1 msec required for the transformation in Equation (9)

We have shown that k_{app} decreases with increasing viscosity of the solutions. This is predicted by mechanisms a, b and possibly c. The lack of dependence of the rate constant on the type of divalent cation and lipid polar head group might favor models a and c. It is concluded that all three mechanisms probably contribute to the overall process and that the present data do not allow us to determine their relative importance.

8. CONCLUSIONS

The material discussed above indicates that the simple interactions between divalent cations and phosphate can result in fairly complex behavior when the phosphate ligand is introduced into the membrane as PA^-. Our major conclusions can be summarized as follows: The equilibrium properties of complexation are essentially those of $H_2PO_4^-$. Divalent cation binding can result in a change in the distribution of lipid constituents of the membrane on the microscopic scale. The rate constants for complexation of PA^- in PC/PA membranes are two orders of magnitude lower than for simple ligands, indicating that the phosphate group of PA^- is sterically hindered. Finally, divalent cation binding of PA^- membranes can result in aggregation. The kinetics, cation concentration dependence and influence of lipid composition on this process indicate that divalent cation-mediated salt-bridging between the PA^- molecules is responsible for aggregation.

ACKNOWLEDGEMENTS

This work was supported by Grant 1 PO1 HL 16117-01 from the National Institutes of Health. I thank Joan Clark for her help with the preparation of this manuscript.

REFERENCES

Czerlinski, G., Diebler, H., and Eigen, M.: 1959, 'Relaxationsuntersuchungen zur Kinetik der Metallkomplexbildung', Z. Physik. Chem. 19, 246-249.

Davies, C.W. and Hoyle, B.E.: 1953, 'The Interaction of Calcium Ions with Some Phosphate and Citrate Buffers', J. Chem. Soc., 4134.

Debye, P.: 1942, 'Reaction Rates in Ionic Solutions', Trans. Electrochem. Soc. 82, 265-272.

Devaux, P. and McConnell, H.M.: 1972, 'Lateral Diffusion in Spin-Labeled Phosphatidylcholine Multilayers', J. Amer. Chem. Soc. 94, 4475-4481.

Diebler, H., Eigen, M., and Hammes, G.G.: 1960, 'Relaxationsspektrometrische Untersuchungen schneller Reaktionen von ATP in wässeriger Lösung', Z. Naturforschung 15b, 554-560.

Eigen. M.: 1963, 'Fast Elementary Steps in Chemical Reaction Mechanisms', Pure Appl. Chem. 6, 97-115.

Eigen, M. and de Maeyer, L.: 1963, 'Relaxation Methods', in A. Weissberger (ed.), Techniques of Organic Chemistry, Interscience, New Yrok, vol. VII/2, p. 895.

Gurney, R.: 1943, Ionic Processes in Solution, McGraw-Hill, New York

Haung, L.: 1969, 'Studies on Phosphatodylcholine Vesicles. Formation and Physical Characteristics', Biochem. 8, 344-352.

Haydon, D.A. and Meyers, V.B.: 1973, 'Surface Charge, Surface Dipoles and Membrane Conductance', Biochim. Biophys. Acta. 307, 429.

Haynes, D.H.: 1972, 'Studien der Bindung und des Transportes von Ionen und Molekülen an Phospholipid-Membranen', in Dechema Monographien, Tutzing Symposium der DECHEMA, Vol. 71, p. 119.

Haynes, D.H. and Staerk, H.: 1974, '1-Anilino-8-Naphthalenesulfonate: A fluorescent Probe of Membrane Surface Structure, Composition and Mobility', J. Membrane Biol. 17, 313-340.

Haynes, D.H.: 1974, '1'Anilino-8-Naphthalenesulfonate: A Fluorescent Indicator of Ion Binding and Electrostatic Potential on the Membrane Surface', J. Membrane Biol. 17, 341-366.

Lansman, J. and Haynes, D.H.: 1975, 'Kinetics of a Ca^{2+}-Triggered Mem-

brane Aggregation Reaction of Phospholipid Membranes', Biochim. Biophys. Acta 394, 335-347.
McLaughlin, S.G.A., Szabo, G., Eisenman, G.: 1971, 'Divalent Ions and Surface Potential of Charged Phospholipid Membranes', J. Gen. Physiol. 58, 667.
Nelson, A.P. and McQuarrie, D.A.: 1975, 'The Effect of Discrete Charges on the Electrical Properties of a Membrane. I. J. Theor. Biol. 55, 13-27.
Overbeek, J.Th.G.: 1949, 'The Gouy-Chapman Theory' in H.R. Kruyt, (ed.), Colloid Science, Elsevier Publishing Co., Amsterdam, vol. I, pp. 128-132.
Schwarzenbach, G. and Gysling, H.: 1949, 'Metallindikatoren I. Murexid als Indikator auf Calcium- und andere Metall-Ionen. Komplexbildung und Lichtabsorption', Helv. Chim. Acta 32, 1314-1325.
Smith, R.M. and Alberty, R.A.: 1956, 'The Apparent Stability Constants of Ionic Complexes of Various Adenosine Phosphates with Divalent Cations', J. Amer. Chem. Soc. 78, 2376.
Smoluchowski, M.: 1916, 'Drei Vorträge über Diffusion, Brownsche Molekularbewegung und Koagulation von Kolloidteilchen', Physik. Zeitschr. 17, 557-599.
Smoluchowski, M.: 1917, 'Versuch Einer mathematischen Theorie der Koagulationskinetik kolloider Lösungen', Z. Physik Chem. 92, 129-168.
Träuble, H. and Grell, E.: 1971, 'The Formation of Asymmetrical Spherical Lecithin Vesicles', Neurosci. Res. Prog. Bull. 9(3), 373.
Träuble, H. and Sackmann, E.: 1972, 'Studies of the Crystalline-Liquid Crystalline Phase Transition of Lipid Model Membranes. III. Structure of a Steroid-Lecithin System below and above the Lipid-Phase Transition', J. Amer. Chem. Soc. 94, 4499-4510.

DISCUSSION

Jagur-Grodzinski: Your computation of the diffusion controlled reaction rates is based on the assumption that Ca^{2+} must diffuse, through aqueous phase only, to reach the phosphate group. However, if the latter is only partially burried within a membrane phase, a diffusion through the latter would be also required. The rules of diffusion in this case would be of course by several orders of magnitude slower. Could such an assumption account for the experimentally observed reaction rules? This would actually represent the hindrance you mentioned in your talk as explanation of the phenomenon.

Haynes: Our work on the kinetics of the binding of the fluorescent anion 1-aniline-8-naphthalenesulfonate to lecithin membranes indicates a bimolecular rate constant of 2×10^8 M^{-1} s^{-1} for the process of reaction with a binding site composed of four lecithin polar head groups. This process would be also represent 'diffusion' in the polar head group region of the membrane. Since the rate constant for this process is close to that expected for a diffusion-controlled reaction in aqueous media (ca. 1×10^9 M^{-1} s^{-1}), we concluded that changes in the polar head group conformation do not represent a kinetic limitation for the binding of the probe. In the case of Ca^{2+} binding in lecithin-phosphatidic acid mixtures, however, they apparently do.

^{31}P NUCLEAR MAGNETIC RESONANCE STUDIES OF METAL-LIGAND INTERACTIONS IN HUMAN BLOOD COMPONENTS

THOMAS O. HENDERSON
Dept. of Biological Chemistry, University of Illinois at the Medical Center, Chicago, Ill. 60612, U.S.A.

1. INTRODUCTION

Since this Symposium is concerned with metal-ligand interactions in organic and biological chemistry, I would like to describe some of our recent work on the use of metal ions in ^{31}P nuclear magnetic resonance (^{31}P NMR) studies of human blood components. Although we have been investigating several phenomena related to human erythrocytes with this technique (Henderson et al., 1974a; Costello et al., 1976) this report will focus on our efforts to determine the structural arrangement of the phospholipids in human serum lipoproteins.

2. ^{31}P NMR OF PHOSPHOLIPIDS

Over the last several years, we have been applying ^{31}P NMR to the study of biological systems. These studies have ranged from qualitative and quantitative analysis of biological samples (Glonek et al., 1970; Hilderbrand et al., 1971; Henderson et al., 1972) to the determination of the structural arrangement of phospholipids in human serum lipoproteins (Glonek et al., 1974; Henderson et al., 1974b; Henderson et al., 1975; Brasure et al., 1976). Early in our studies, we found that crude phospholipid extracts in organic solvents gave rise to two resonance bands in the orthophosphate region of the ^{31}P NMR spectrum (Figure 1). We were curious as to whether the difference in the two peaks derived from (1) differences in the nature of the polar head groups of the phospholipids; (2) differences in the fatty acyl substituents present; or (3) other differences which might affect the nature of the magnetic microenvironment of the phospholipid phosphates.

In order to gain insight into this, we determined the ^{31}P chemical shifts (δ) of a number of phospholipids and related phosphorylated compounds (Henderson et al., 1974c). (Structures of these compounds are shown below). The effect of the simple alkyl esterification on the chemical shift of orthophosphate at pH 7 is shown by lines 2-4 of Table I, where the chemical shift of the diester is upfield from the monoester. (The δs of non-esterified and mono-esterified orthophosphates are markedly pH-dependent, with increasing protonation producing up-

B. Pullman and N. Goldblum (eds.), Metal-Ligand Interactions in Organic Chemistry and Biochemistry, part 2, 213-230.

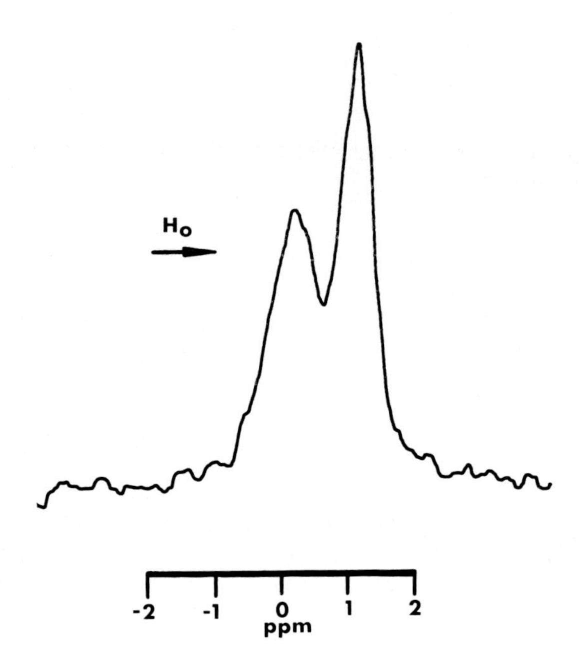

Fig. 1. ^{31}P NMR spectrum of total bovine liver lipids in chloroform-methanol (2:1, v/v). (Adapted from Henderson et al., 1974c.)

CH_2-O-C(=O)-$(CH_2)_n$-CH_3
CH_3-$(CH_2)_n$-C(=O)-O-CH
CH_2-O-P(=O)(O^-)-O-X

X = HO-CH_2-CH_2-$\overset{+}{N}H_3$
Phosphatidylethanolamine (PE)
X = HO-CH_2-CH_2-$\overset{+}{N}(CH_3)_3$
Phosphatidylcholine (PC)
X = HO-CH_2-CH($\overset{+}{N}H_3$)-COO^-
Phosphatidylserine (PS)

$^-$O-P(=O)(O^-)-O-CH_2-CH_2-$\overset{+}{N}H_3$, Phosphorylethanolamine (EP)

$^-$O-P(=O)(O^-)-O-CH_2-CH_2-$\overset{+}{N}(CH_3)_3$, Phosphorylcholine (CP)

$^-$O-P(=O)(O^-)-O-CH_2-CH($\overset{+}{N}H_3$)-COO^- , Phosphorylserine

field displacement of δ(Crutchfield et al., 1967)). Water soluble compounds of interest in the study of phospholipids are found in lines 5-7 (monoesters) and 8-10 (diesters). In general, compounds 5-7 have chemical shifts where one might expect monoesters of orthophosphate to come into resonance at pH 7 (-3.2 to -3.8 ppm), whereas the glycerylphosphoryl esters of choline, ethanolamine, and serine behave as one might expect for simple diesterified orthophosphate (-0.4 to +0.1 ppm). It should be noted that in both lines 5-7 and 8-10, the choline derivatives are to higher field than are the ethanolamine or serine compounds by 10 to 18 Hz, perhaps reflecting a greater shielding of the phosphorus nuclide by the quarternary nitrogen group.

TABLE 1
^{31}P NMR chemical shifts of selected water-soluble orthophosphate esters related to phospholipids[a]

#	Compound[b]	Chemical Shift[c]	
		Hz	ppm
1	Orthophosphoric acid (85%)	0	0.0
2	Orthophosphate	-98	-2.7
3	Monoethylphosphate	-138	-3.8
4	Diethylphosphate	-25	-0.7
5	Phosphorylcholine	-116	-3.2
6	Phosphorylethanolamine	-133	-3.7
7	Phosphorylserine	-134	-3.7
8	Glycerophosphorylcholine	+5	+0.1
9	Glycerophosphorylethanolamine	-13	-0.4
10	Glycerophosphorylserine	-5	-0.1

[a] Adapted from Henderson et al. (1974c).

[b] All compounds (except 85% orthophosphoric acid) were converted to the Na^+ salt and measured at a phosphorus concentration of 0.02 M in aqueous 0.2 M EDTA, Na^+ ion, pH7.0.

[c] Chemical shifts are given relative to an external standard of 85% H_3PO_4 (Compound 1) in all cases in this report.

We next determined the chemical shifts of a number of natural and synthetic phospholipids. We found that phosphatidylserine (PS), phosphatidylthanolamine (PE), N-monomethyl-PE, and N,N-dimethyl-PE all came into resonance at 0 to -0.3 ppm, whereas egg phosphatidylcholine (PC) had a chemical shift of +0.9 ppm (Table II). These resonances corresponded to the low field and high field signals, respectively, observed with phospholipid extracts of biological extracts (cf. Figure 1).

From these observations, it would appear that choline-lipids come into resonance at a position upfield from PE and PS, perhaps as a result of increased shielding of the phosphorus produced by the quarternary nitrogen. However, when we examined two other naturally occurring lipids which are esters of phosphorylcholine, 1-monoacyl glycerylphosphorylcholine (lyso-PC) and sphingomyelin (SPH), we found that these

TABLE II
^{31}P NMR chemical shifts of phosphatidylethanolamine and related phospholipids[a]

Compound[b]	Chemical shift	
	Hz	ppm
PC (egg)	+34	+0.9
N-dimethyl-PE	-11	-0.3
N-monomethyl-PE	-1	0.0
PE	-8	-0.2
PS	+2	+0.0

[a] Adapted from Henderson et al. (1974c).

[b] All compounds were measured at a phosphorus concentration of 0.02 M in chloroform-methanol (2:1, v/v). The abbreviations used are: PC, phosphatidylcholine; phosphatidylethanolamine; PS, phosphatidylserine.

two compounds had low field chemical shifts of +0.2 and 0 ppm, respectively, i.e., much closer to PE and PS than to PC (Table III). The lyso-derivatives of PE and PS were essentially the same as the parent compounds.

TABLE III
^{31}P NMR chemical shifts of choline-containing phospholipids and related lysophosphatides[a]

Compound[b]	Chemical shift	
	Hz	ppm
SPH	-1	0.0
Lyso-PC	+6	+0.2
PC (avg.)	+31	+0.8
Lyso-PE	-9	-0.2
PE	-8	-0.2
Lyso-PS	-6	-0.2
PS	+2	0.0

[a] Adapted from Henderson et al. (1974c).

[b] All compounds were measured at a phosphorus concentration of 0.02 M in chloroform-methanol (2:1, v/v), except lyso-PE (ca. 0.007 M) and lyso-PS (ca. 0.005 M). The abbreviation SPH represents sphingomyelin.

We wished to insure that we were not dealing with anomolous behavior of the PC produced by the fatty acyl esters present in egg PC. Therefore, we examined several synthetic phospholipids of defined fatty acid composition. From the chemical shift data presented in Table IV, it is obvious that the fatty acid composition plays no major role in determining the shift positions of the phospholipids.

TABLE IV
Influence of fatty acid composition on the ^{31}P NMR chemical shift of phosphatidylcholine[a]

Compound[b]	Chemical shift	
	Hz	ppm
$(16:0_2$-PC	+30	+0.8
$(18:0_2$-PC	+28	+0.8
$(18:1_2$-PC	+31	+0.8
PC (egg)	+34	+0.9

[a] Adapted from Henderson et al. (1974c).

[b] All compounds were measured at a phosphorus concentration of 0.02 M in chloroform-methanol (2:1, v/v). The numbers 16:0, 18:0, and 18:1 refer to the fatty acid substituents palmitic acid, stearic acid, and oleic acid, respectively.

3. CONFORMATION OF PHOSPHOLIPIDS

Figure 2 summarizes the salient features of Tables I through IV and illustrates the significant chemical shift differences between PC and the other phospholipids examined. The quarternary nitrogen of PC quite likely contributes to the upfield shift of the compound. We estimate this contribution to be ca. 15 Hz, based on the chemical shifts of the water-soluble orthophosphate esters (Table I) and from the difference in shift between lyso-PC and lyso-PE (Table II). This is too small to account for the difference between PC and the other phospholipids of 25 to 45 Hz. Neither does it account for the observation that lyso-PC and SPH, both of which contain phosphorylcholine, give rise to signals in the same region of the ^{31}P spectrum as do PE, PS, lyso-PE, etc.

We observed that there was one property which was possible related to these ^{31}P NMR chemical shift measurements. All of the phospholipid examined which contain phosphate diesters and have downfield chemical shifts potentially can form 7-element hydrogen-bonded ring structures involving the anionic phosphate oxygen and a proton on an amine, amide, or hydroxyl group of the phospholipid in organic solvents. This includes, of course, lyso-PC and SPH, as well as cardiolipin and phosphatidylglycerol (PG) (Henderson et al., 1974c). PC does not contain a similar dissociable proton and, therefore, could not

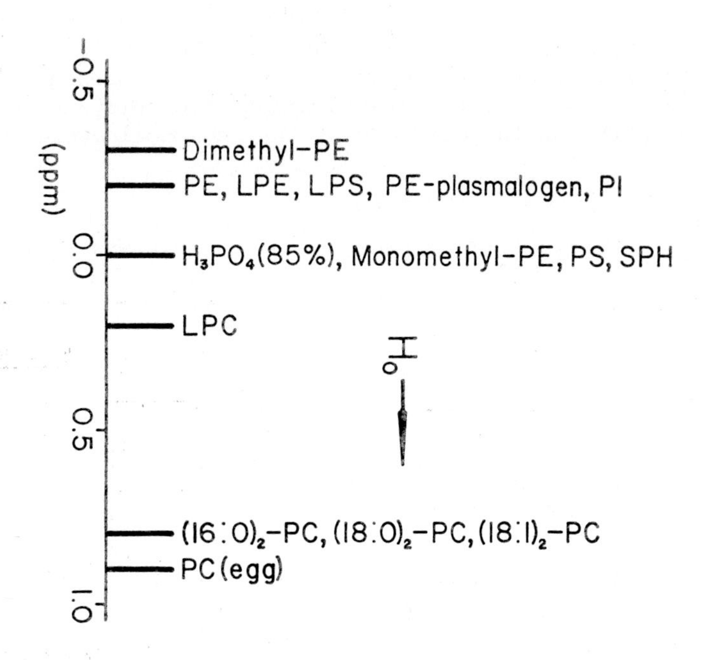

Fig. 2. Correlation chart of phospholipid ^{31}P NMR chemical shifts (see Tables II-IV for experimental conditions). (Adapted from Henderson et al., 1974c.)

form such hydrogen-bonded conformers. (This would not preclude, however, possible electrostatic interactions between the positively-charged quarternary nitrogen group and the anionic phosphate group (cf., Pullman and Berthod, 1974; and Pullman et al., 1975)). Schematic structures consistent with our interpretation are shown for: I, PE; II, lyso-PC; III, SPH; and IV, PG.

Phosphatidylethanolamine I

Lysophosphatidylcholine II

Sphingomyelin III

Phosphatidylglycerol IV

B. Pullman and associates have recently carried out quantum-mechanical studies on the conformational properties of the polar head groups of phospholipids by both the SCF ab initio and PCILO methods (Pullman and Berthod, 1974; Pullman et al., 1975). The specific compounds they studied were phosphorylethanolamine (EP), phosphorylcholine (CP), glycerylphosphorylethanolamine (GPE), and glycerylphosphorylcholine (GPC). Their computations of global energy minima indicated that, indeed, the seven-membered intramolecularly hydrogen-bonded conformation for EP (analogous to PE, structure I, above) is the intrinsically most stable conformation for that compound and is ca. 60 kcal mol^{-1} more stable than the extended conformer predicted from crystallographic studies of others (Kraut, 1961).

4. EFFECT OF METAL CONTAMINATION ON PHOSPHOLIPID SPECTRA

During the course of the above studies, we were constantly plagued with the problem of polyvalent cation contamination of the phospholipids and related compounds, and the resulting deleterious effects on their ^{31}P NMR spectra. An example of this is shown in Figure 3. In this experiment, the bovine liver lipids (spectrum shown in Figure 1) were fractionated by silicic acid chromatography into PC, PE, and neutral lipids (cf. Henderson et al., 1974c).

The PC and PE were examined by ^{31}P NMR and then recombined. The initial spectrum obtained after recombination was markedly broad and the signals from PC and PE were not resolved (Spectrum A, Figure 3). When this sample was equilibrated against a solution of EDTA (Na^+, pH 7.0), Spectrum B of Figure 3 was obtained in which the PC and PE were well resolved and, in fact, were narrower than they were in the original extract (Figure 1). However, when the neutral lipid fraction was combined with the EDTA-treated PC + PE sample, the spectrum obtained (Spectrum C, Figure 3) was nearly identical to that obtained originally with the crude extract. Upon washing with EDTA, this sample gave rise to a well-defined spectrum similar to Spectrum B, Figure 3. These results demonstrate that trace polyvalent cation contamination markedly alters the ^{31}P NMR spectrum of phospholipids. They further indicate that different phospholipid species differ in their affinities for these cations. Specifically, it would appear that PE has a higher affinity for these trace contaminants than does PC, since the PE signal undergoes greater peak broadening and concommitant decreased signal amplitude (compare Spectra B and C, Figure 3).

5. ^{31}P NMR OF HUMAN SERUM LIPOPROTEINS

We recognized the possible great potential of ^{31}P NMR in the study of phospholipid interactions in biological systems. For example, we knew from the work described above that PC could be differentiated from other phospholipids present in most biological systems by virtue of its upfield chemical shifts. Furthermore, we also had found that metal ions affected the ^{31}P spectrum of phospholipids with some degree of

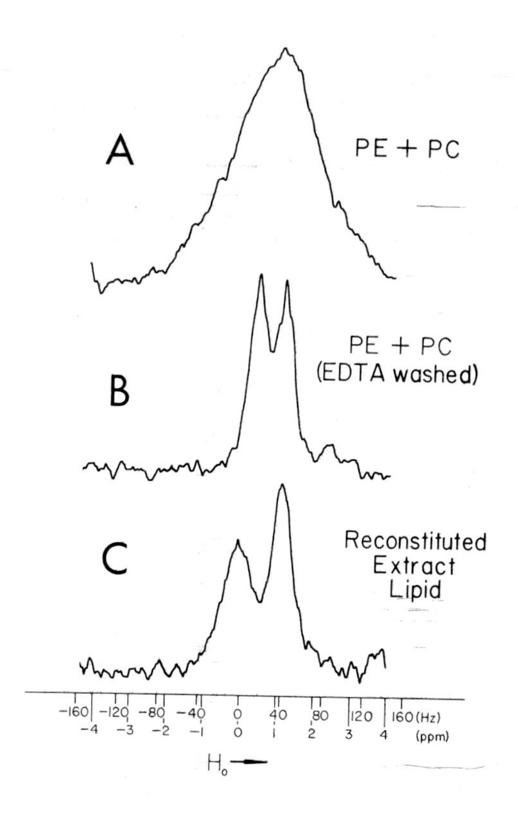

Fig. 3. The bovine liver lipids used to obtain the spectra shown in Figure 1 were fractionated by silicic acid column chromatography into PE, PC, and neutral lipid fractions. When the PE and PC fraction were combined and examined by ^{31}P NMR, Spectrum A was obtained. Washing this sample with 0.2 M sodium EDTA (pH 7.0) gave rise to Spectrum B. Addition of the neutral lipid fraction for the sample used for B produced the results shown in Spectrum C. (Adapted from Henderson et al., 1974c)

selectivity. With this background, we set about determining the feasibility of studying human serum lipoproteins with ^{31}P NMR.

Human serum lipoproteins are macromolecular complexes composed of varying proportions of proteins, neutral lipids (triglyceride, cholesterol, cholesterol esters), and phospholipids. These complexes are involved in the inter-organ transport of fat. The lipoproteins can be separated into different classes by flotational ultracentrifugation, with those having the greater proportion of protein having higher hydrated densities than those with lower relative protein content. The classes we will be most concerned with here are the high density lipoproteins (HDL) and low density lipoproteins (LDL).

Human HDL contain, by weight, about 50% protein and 25% phospholipid, with the remainder composed of 14% cholesterol esters, 4% cholesterol, and 4% triglycerides. The phospholipids consist of ca. 75% PC, 14% SPH, and small amounts of PE and other phospholipids (Skipski et al., 1967). In contrast, LDL contain 21% protein and 79% lipid, with the lipids being 47% cholesterol esters, 10% cholesterol, 28% phospholipids, and 14% triglycerides. PC is the most predominant phospholipid in LDL, as in HDL (Scanu and Kruski, 1973). The molecular weight of LDL is 2.3 x 10^6, while HDL subclass-3 (HDL_3) is 175,000 (Scanu and Kruski, 1973).

We determined that we could obtain usable ^{31}P NMR spectra from intact native HDL (Glonek et al., 1973; Glonek et al., 1974) and that, indeed, the PC signal was upfield from the signal of the other phospholipids. (Typical LDL and HDL_3 spectra are shown in Figure 4 and 5 (upper spectrum).) We then set about carrying out a series of experiments designed to assess the geometric arrangement of the phospholipids in these complexes, using the rationale developed below.

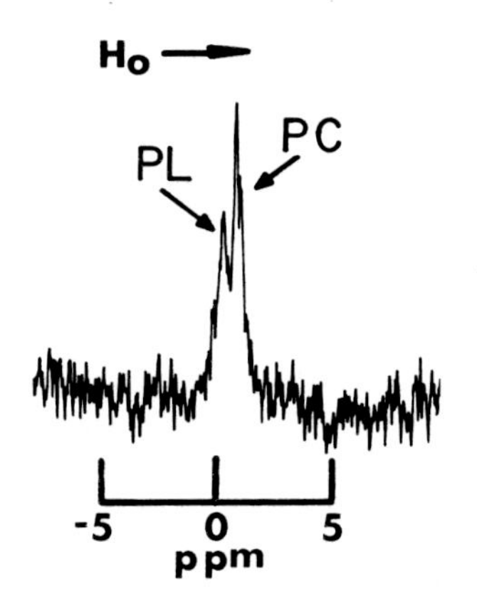

Fig. 4. ^{31}P NMR spectrum of native human serum low density lipoprotein (LDL). The signals, proceeding upfield, are: PL, the signal from phospholipids other than PC, principally sphingomyelin; and PC, phosphatidylcholine (0.9 ppm). Chemical shifts are given relative to 85% H_3PO_4 (external) as the zero (0) ppm reference. The spectrum was obtained on a Bruker HFX-5 90 MHz spectrometer operating at 36.43 MHz for ^{31}P. Fourier transform techniques were employed, and proton broadband decoupling was used to eliminate H-P coupling. (Adapted from Henderson et al., 1976.)

6. EFFECT OF PARAMAGNETIC IONS ON THE ^{31}P NMR SPECTRA OF SERUM LIPOPROTEINS

Bystrov et al. (1971), studying sonicated aqueous dispersions of egg yolk PC by ^1H NMR, were able to estimate the distribution of PC between the inside and outside layers of the bilayered vesicles. This was done by measuring the selective broadening or upfield shift of a portion of the ^1H signal of the N-methyl protons of PC when low concentrations of the paramagnetic ions Mn^{++} or Eu^{+++} were added to the vesicles. The premise of their experiment was that those ^{1}H resonances which were broadened by Mn^{++} or shifted upfield by Eu^{+++}, arose from N-methyl groups of PC molecules that were in the outside layer of the vesicles and therefore exposed to the Mn^{++} or Eu^{+++} added into the bulk water; the ^{1}H signals arising from the N-methyls of the inside PC molecules would not be exposed to the bulk solvent or paramagnetic ions therein and therefore would not be broadened or shifted. The ratio of perturbed to unperturbed N.methyl ^{1}H resonances could therefore be used to estimate the relative distribution of PC in the outside (perturbed by Mn^{++} or Eu^{+++}) or inside (unperturbed) layers of the bilayered vesicles.

More recently, Michaelson et al. (1973) carried out similar studies with vesicles prepared by cosonication of equimolar amounts of PG and PC as well as with vesicles of pure PC and pure PG separately. By the use of both ^{1}H and ^{31}P NMR and Mn^{++} and Eu^{+++}, those workers confirmed and extended the findings of Bystrov et al. (1971). In the PC vesicles studied by Michaelson et al. (1973), the ratio of phospholipid inside to that outside was found to be 37%:63%. In the cosonicated PG:PC vesicles, they concluded that the PG and PC were not equally distributed between the inside and outside layers of the vesicles, but rather, there was an assymetric distribution, with there being twice as much PG as PC in the outer layer of the vesicles. Conversely, the inner layer is enriched in PC.

The rationale behind the experimental approach of Michaelson et al. (1973) involved the selective broadening of the N-methyl proton resonances in the presence of Mn^{++} or Eu^{+++} as described by Bystrov et al. (1971), and the observation that the ^{31}P resonance of PG was downfield from PC in the PG:PC vesicles. This latter observation allowed them to determine the relative effects of paramagnetic ions on the two different phospholipids in the two different layers of the vesicles.

The rational behind our approach to strucural studies on LDL and HDL, then, was derived from the approach of Bystrov et al. (1971) and Michaelson et al. (1973) described above. We conducted preliminary studies on HDL and LDL with the lanthamide paramagnetic ions Eu^{+++}, Pr^{+++}, and Sm^{+++} (all of which have been employed as NMR shift reagents) and with Mn^{++} (which has been demonstrated to produce line broadening and decreased amplitude of the NMR signal). The hydrated ions and the EDTA- and NTA-chelated ions were tested for their ability to produce changes in the ^{31}P NMR spectrum of a given lipoprotein without altering the physical properties of the lipoproteins (as determined by CD-ORD and ultracentrifugation). These studies eventually led us to select Mn^{++}/EDTA (1/2.2) as the paramagnetic titration reagent for the investigation. The concentration of Mn^{++} in the titrant used was 0.1 M.

The experimental procedure followed after the preliminary studies was as follows. Samples of HDL_2, HDL_3, or LDL of known lipid composition, content, and concentration were placed in 10 mm tubes and a seale melting-point capillary tube containing ca. 0.1 M sodium pyrophosphate in water (pH = 6.8) was mounted coaxially in the NMR sample tube as an intensity reference. The ^{31}P NMR spectrum was obtained, and the ratio of the peak height of the phospholipid-phosphate signal to the peak height of the signal from the pyrophosphate in the sealed capillary was determined. The ratio obtained with no addition of Mn^{++}/EDTA (1/2.2) was assigned a value of 100%.

A typical ^{31}P spectrum of HDL_3 is shown in the upper spectrum of Figure 5. When aliquots of the Mn^{++}/EDTA reagent were added to the HDL_3 sample to give a Mn^{++} to phospholipid-phosphorus ratio of 1:4, the bottom spectrum of Figure 5 was obtained. Note the decrease in the PC signal amplitude in the treated HDL_3 compared to the untreated HDL_3.

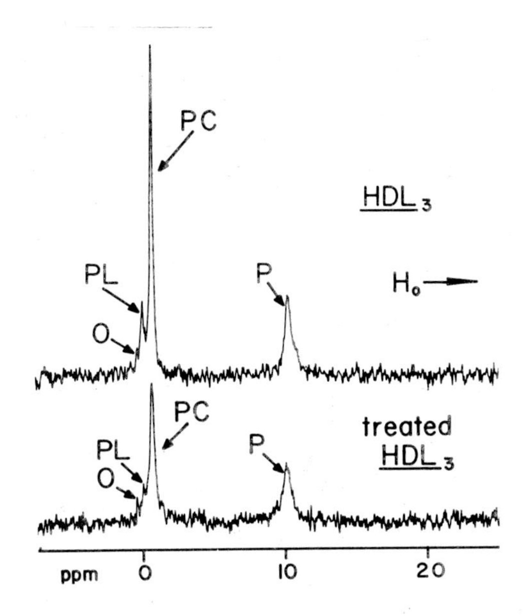

Fig. 5. ^{31}P NMR spectra of normal human high density lipoprotein-3 (HDL_3) (upper spectrum); and (lower spectrum) HDL_3 to which the Mn^{++}/EDTA (1:2.2) reagent was added (Mn^{++}/P= 0.25). Instrumental conditions were as described in Figure 4. The symbol O at -0.3 ppm indicates the signal from the inorganic phosphate impurity in the pyrophosphate intensity reference capillary; the symbol P (10.4 ppm) represents the signal from the inorganic pyrophosphate intensity reference capillary (sodium pyrophosphate, pH 6.8). (Adapted from Henderson et al., 1975).

The results of these titration studies are depicted graphically in Figure 6, where the percent of original peak height remaining (normalized against the sealed pyrophosphate reference signal) is plotted against the mole ratio of Mn^{++} to phospholipid-phosphate in the sample tube.

Appropriate model systems were also titrated with the Mn^{++}/EDTA reagent in order to aid the interpretation of the results of the studies of LDL, HDL_2, and HDL_3. These model systems were: (1) diethyl phosphate in H_2O; (2) PC in anhydrous MeOH; (3) SPH in anhydrous MeOH; and (4) bilayered egg PC vesicles in water. Systems (1)-(3) represent conditions in which these compounds exist as monodisperse species in which all of the phosphate groups should be freely accessible to the Mn^{++}/EDTA reagent introduced into the solvent; system (4) should have only ca. 60% of the phospholipid-phosphate groups accessible to the Mn^{++}/EDTA reagent cf., Michaelson et al., 1973).

The results of the Mn^{++} titration of the model systems are depicted graphically in Figure 6 and are in accord with the predictions mentioned above; i.e., the amplitude of the ^{31}P signals from diethyl phosphate, PC in MeOH, or SPH in MeOH were diminished upon Mn^{++} titration to the point where they could no longer be differentiated from the baseline noise level, whereas the ^{31}P signal from the PC vesicles reached a plateau at a value of ca. 40% of the original peak height remaining. This indicates that ca. 60% of the PC-phosphates in the vesicles were accessible to the Mn^{++}/EDTA and coincides with outside:inside ratios obtained for these vesicles by Bystrov et al. (1971) and Michaelson et al. (1973).

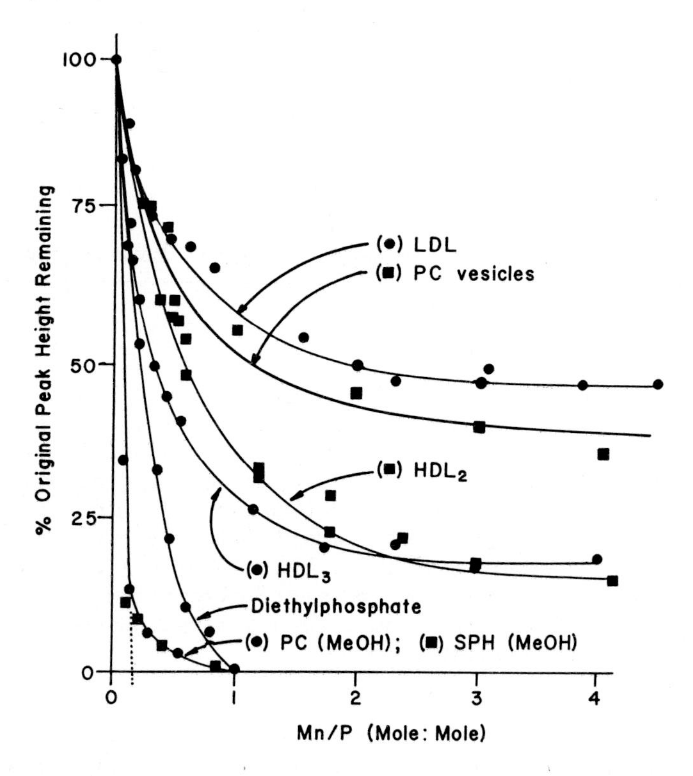

Fig. 6. ^{31}P NMR paramagnetic titration curves of HDL_2, HDL_3, LDL, PC vesicles and model compounds (see text for details). The titrant was 0.1 M Mn^{++} as Mn^{++}/EDTA (1:2.2). (Adapted from Henderson et al., 1975.)

We concluded from these model studies that the extent of ^{31}P line broadening and diminution in signal amplitude produced by the Mn^{++}/EDTA reagent were a function of the accessibility of the phosphate groups in question and that the Mn^{++}/EDTA reagent could be used to assess the relative accessibility of the PL polar head groups in LDL, HDL_2, and HDL_3 (Henderson et al., 1975).

Titrations of LDL and HDL with the Mn^{++}/EDTA reagent produced two fundamentally different curves (Figure 6). In the case of LDL, as the concentration of Mn^{++}/EDTA was increased, the decrease in PC-peak height reached a plateau corresponding to 50% of the original peak height. In contrast, the curve obtained upon titration of HDL_3 reached a plateau at 20% of the original peak height. The curve for HDL_2 was similar to that for HDL_3, except that the plateau occurred at 17% of the original peak height.

We interpret these data as indicating that only 50% of the phospholipid-phosphates of LDL are accessible to the paramagnetic reagent, whereas in HDL , about 80% are accessible. Conversely, in HDL_3, 20% of these groups are *not* accessible. This latter point differs from the results of Assman et al. (1974) who interpreted their Eu^{+++} titration data as indicating that all of the phospholipid-phosphates of HDL are exposed to the external aqueous environment.

The effects of the Mn^{++}/EDTA on the ^{31}P spectra could be reversed by exhaustive dialysis against the original buffer, indicating no permanent change in the microenvironment of the phospholipid polar head group. If the PC vesicles or lipoproteins were disrupted by sonication in the presence of the reagent, the signals were broadened to such an extent as to no longer be observable.

If the extent of quenching produced by the paramagnetic reagent Mn^{++}/EDTA accurately reflects the differences in the degree of accessibility of the phospholipid-phosphate groups, our data indicate that there is a marked difference in the surface arrangement between HDL and LDL (Figure 6). X-ray diffraction studies on human LDL (Mateu et al., 1972) and ^{1}H NMR studies on pig lipoproteins (Finer et al., 1975) have indicated that, in LDL, the phospholipids exist in two layers, one in which the polar head groups are oriented on the exterior surface, and the other sequestered away from the exterior aqueous environment. As can be seen from Figure 6, our results agree with this arrangement; indeed, the response of LDL and the bilayered PC-vesicles to the Mn^{++}/EDTA are nearly identical.

On the other hand, there is no evidence that HDL contains a phospholipid bilayer. Small angle X-ray diffraction studies on HDL_2 and HDL_3 (Shipley et al., 1972) indicate, in fact, that the phospholipids are likely to be oriented in such a way as to be exposed to the exterior and, therefore, would be accessible to Mn^{++}/EDTA in the aqueous buffer. Indeed, Assman et al. (1974) concluded from ^{31}P NMR studies in which Eu^{+++} was employed as a paramagnetic shift reagent that all of the phospholipid-phosphates were accessible. In general, our results agree with this interpretation (Henderson et al., 1975), except that only ca. 80% of the phosphates were accessible to the Mn^{++}/EDTA reagent, with a consistent, significant number of the polar head groups (20%) which do not respond to the reagent. It may be that this 20%

represents those phospholipids which are undergoing strong electrostatic or hydrophilic interactions with proteins in the complexes and are, therefore, prevented from interacting with the chelated Mn^{++} ion.

7. PHOSPHOLIPASE A_2 TREATMENT OF HDL_3

Another approach we have recently used in our efforts to understand the structure of serum lipoproteins has been to determine the effect of α-phospholipase A_2 on HDL_3, specifically with respect to the effect on the ^{31}P NMR resonances from the affected phospholipids (Brasure et al., 1976). The α-phospholipase A_2 used in these studies was prepared from the venom of the rattlesnake *Crotalus adamanteus* and was a gift of Dr Ferenc Kézdy of the University of Chicago. Experimental details are given in the captions to Figure 7 and 8. The enzyme catalyzes the conversion of PC to lyso-PC, with a fatty acid being removed at the 2-OH of PC:

$$\underset{\text{(PC)}}{\begin{array}{l} \qquad\quad CH_2\text{-O-}\overset{O}{\overset{\|}{C}}\text{-}R_1 \\ R_2\text{-}\overset{O}{\overset{\|}{C}}\text{-O-}C\text{-H} \\ \qquad\quad CH_2\text{-O}\overset{O}{\overset{\uparrow}{P}}\underset{O^-}{}\text{O-}CH_2CH_2\overset{+}{N}(CH_3)_3 \end{array}} \xrightarrow{\text{phospholipase } A_2,\ Ca^{++}} \underset{\text{(lyso-PC)}}{\begin{array}{l} \quad CH_2\text{-O-}\overset{O}{\overset{\|}{C}}\text{-}R_1 \\ \text{HO-}C\text{-H} \\ \quad CH_2\text{-O}\overset{O}{\overset{\uparrow}{P}}\underset{O^-}{}\text{O-}CH_2CH_2\overset{+}{N}(CH_3)_3 \end{array}} + R_2\text{-}\overset{O}{\overset{\|}{C}}\text{-}O^-$$

Figure 7 shows the ^{31}P NMR spectra of HDL_3 after 16, 220, and 840 min of enzymatic treatment. The predominant signal at ca. 1.1 ppm in the top spectrum arises from PC, whereas the lowfield signal at ca. 0.5 ppm represents mainly SPH since, at 16 min, only a small amount of lyso-PC is present. As the length of incubation proceeded, the relative area of the PC decreased and the lowfield signal (ca. 0.6 ppm), arising from lyso-PC, increased to the extent that, at 840 min, 75% of the PC originally present had been converted to lyso-PC. Of importance here is the observation that the proton-decoupled linewidths of the PC present initially and the lyso-PC subsequently produced were identical (7.5 Hz). In general, NMR linewidth at a given magnetic field strenth is related to the mobility of the nuclide, with increasing linewidth reflecting decreased mobility (cf., Finer et al., 1975). Furthermore, it has been observed that, in the cases of phospholipid vesicles, increases in the ^{31}P linewidth are associated with increases in the vesicle size (Gent and Prestegard, 1974; Berden et al., 1975). We can conclude, then, that the mobility of the phospholipids in the phos-

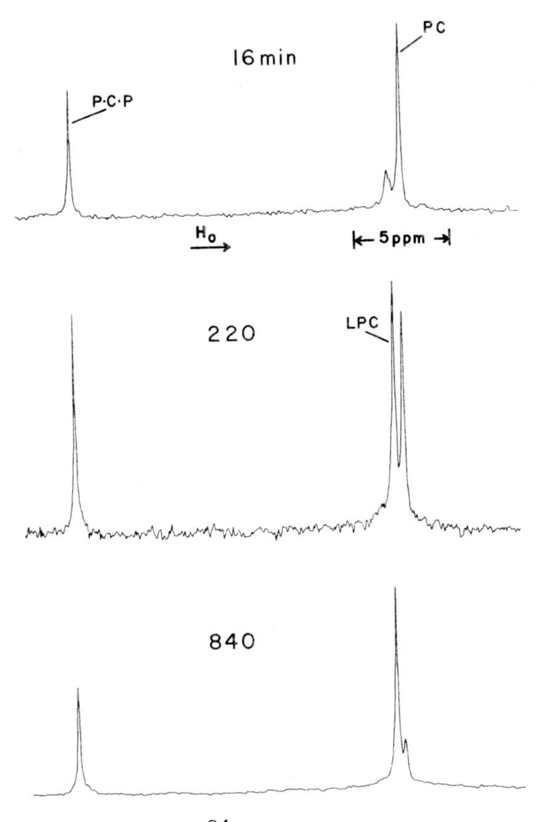

Fig. 7. Changes in the ^{31}P NMR spectrum of HDL_3 produced by digestion with α-phospholipase A_2 from *C. adamanteus*. PC, phosphatidylcholine; LPC, lysophosphatidylcholine; and P-C-P, the signal from a methylene diphosphonic acid intensity reference capillary. The upper spectrum was obtained after 16 min, the middle spectrum after 220 min, and the bottom spectrum after 840 min of incubation at 28 °C (the temperature of the probe). Conditions are given in the caption to Figure 4. The enzyme reaction mixture contained, in 4 ml: HDL_3, 9.5 x 10^{-3} M with respect to PC; $CaCl_2$, 7 x 10^{-3} M; and α-phospholipase A_2, 5.9 x 10^{-9} M; in 0.15 M NaCl containing 0.05% EDTA, pH 7.2. No bovine serum albumin was added. (Adapted from Brasure et al., 1976.)

pholipase A_2-digested HDL_3 is the same as in the original HDL_3 complexes. We can further conclude that there is essentially no change in particle size (i.e., they did not coalesce into larger particles) as a result of the enzymatic treatment. These results substantiate and extend the kinetic observations of Pattnaik et al., (1976) who found that α-phospholipase A_2 treatment of HDL_3 had little, if any, effect on the immunological and physical properties of HDL_3.

The time course of a typical α-phospholipase A_2 experiment is shown in Figure 8A, in which the percent PC remaining is plotted against length of incubation. The reaction generally slowed or stopped after ca. 500 min, most likely due to increasing enzyme denaturation and/or end-product inhibition of the enzyme by the free fatty acid produced. When more enzyme was added (indicated by the arrow in Figure 8A, the reaction resumed until all of the PC had been converted to lyso-PC, as reported by Pattnaik et al. (1976)). These results would indicate that all of the PC present in the HDL_3 is exposed to (or in equilibrium with) the surface of the complexes. However, when fully-digested HDL_3 were titrated with the Mn^{++}/EDTA reagent described above, there were still 20% of the phosphate groups which were not affected as determined by ^{31}P NMR (E. Brasure, unpublished observations; data not shown).

Finally, in Figure 8B, four experiments like those described in Figure 7 and 8A are presented. In this figure, the ^{31}P NMR area measurements have been converted to molarity of PC remaining, and the $\log_{10}$ molarity PC plotted against the length of incubation. These plots show that the α-phospholipase A_2-catalyzed reactions follow first order kinetics through at least 500 min under the conditions of the experiments, and have an average measured rate constant (K_{exp}) of 3.35 ($\pm$ 0.45) x 10^{-5} s^{-1} (n=4). This value compares closely with values for K_{exp} determined by Pattnaik et al (1976) when differences in the concentration of enzyme and substrate are taken into account.

In conclusion, we have demonstrated that ^{31}P NMR can be used in assessing the structural arrangement of phospholipid polar head groups by determining the relative apparent accessibility of these groups to metal ions. In addition, we have shown that the NMR technique can be used not only to follow the course of an enzyme-catalyzed reaction, but perhaps more importantly, to determine the changes, if any, which occur in the microenvironment of the substrate in these biological complexes.

ACKNOWLEDGEMENTS

The author would like to acknowledge his colleagues at the University of Illinois at the Medical Center, Chicago, Illinois, who were collaborators on aspects of this work: Professor Terrell C. Myers, Dr Thomas Glonek, Mr Leonard G. Davis, and Ms Elizabeth Brasure. He wishes also to acknowledge the collaboration of Professor Angelo M. Scanu, Dr Arthur Kruski, and Dr Nikhil Pattnaik, of the University of Chicago on aspects of the serum lipoprotein studies. The author's contribution to the work described here was supported in part by United States Public Health Service Grants NS-9354 and GM-20127, Chicago Heart Association Grant C75-37, American Heart Association Grant 74-1011, funds from the University of Illinois College of Medicine General Research Support Grant, and funds from the University of Illinois Graduate College Research Board. The Bruker 90 MHz spectrometer used for these studies is operated by the NMR Facility of the Research Resources Laboratory in the Graduate College of the University of Illinois at the Medical Center.

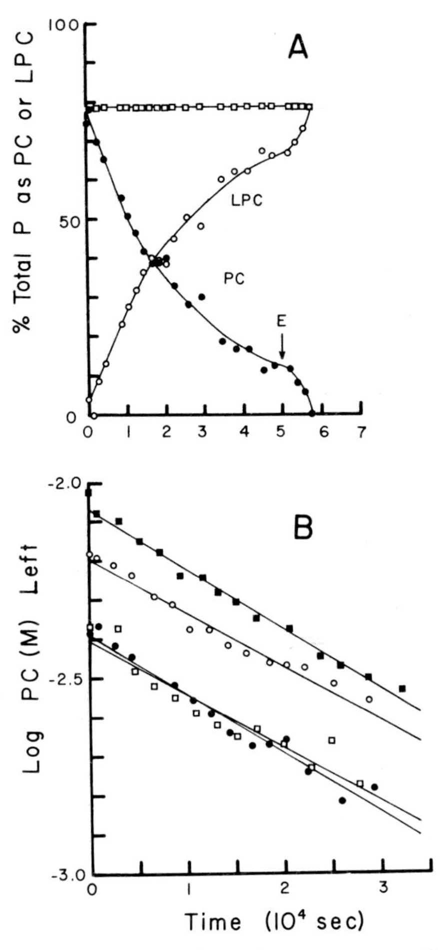

Fig. 8. (A) represents the time course of a typical experiment in which HDL_3 were digested with α-phospholipase A_2. The percent PC and LPC were taken from the relative areas of the respective peaks on ^{31}P NMR spectra such as those shown in Figure 7. (B) represents a plot of $\log_{10}$ [PC] (in M) versus time calculated from the ^{31}P NMR spectra of four different digestion experiments. The experimental conditions were as described in the caption to Figure 7. The average rate constant as determined from the slopes of the lines is 3.35 ($\pm$ 0.45) x 10^{-5} s^{-1}. The symbols indicate the experimental data; the lines represent the best fit for the experimental data as determined by linear regression analysis ($P \geqq 0.95$). The enzyme and PC concentrations for the individual experiments were: (■) PC, 9.5 x 10^{-3} M; enzyme, 8 x 10^{-9} M; (○) PC, 6.5 x 10^{-3} M; enzyme, 5.9 x 10^{-9} M; (●) PC, 4.1 x 10^{-3} M; enzyme, 5.9 x 10^{-9} M; and (□) PC, 4.3 x 10^{-9} M; enzyme, 6.25 x 10^{-9} M. (Adapted from Brasure et al., 1976.)

REFERENCES

Assman, G., Sokoloski, E.A., and Brewer, H.B., Jr.: 1974, Proc. Natl. Acad. Sci. U.S.A. 71, 549.

Brasure, E.B., Henderson, T.O., Glonek, T., Pattnaik, N.M., and Scanu, A.M.: 1976, Fed. Proc. 35, 1678 (Abs. No. 1634).

Bystrov, V.F., Dubrovina, N.I., Barsukov, L.I., and Bergelson, L.D.: 1971, Chem. Phys. Lipids 6, 343.

Costello, A.J., Marshall, W.E., Omachi, A., and Henderson, T.O.: 1976, Biochim. Biophys. Acta 427, 481.

Crutchfield, M.M., Dungan, C.H., Letcher, J.H., Mark, V., and Van Wazer, J.R.: 1967, Top. Phosphorus Chem. 5, 173.

Finer, E.G., Henry, R., Leslie, R.B., and Robertson, R.N.: 1975, Biochim. Biophys. Acta 380, 320.

Glonek, T., Henderson, T.O., Hilderbrand, R.L., and Myers, T.C.: 1970, Science 169, 192.

Glonek, T., Henderson, T.O., Kruski, A.W., and Scanu, A.M.: 1973, Abstracts of the 166th National Meeting of the American Chemical Society, Chicago, Ill., Biol-209.

Glonek, T., Henderson, T.O., Kruski, A.W., and Scanu, A.M.: 1974, Biochim. Biophys. Acta 348, 115.

Henderson, T.O., Costello, A., and Omachi, A.: 1974a, Proc. Natl. Acad. Sci. U.S.A. 71, 2487.

Henderson, T.O., Davis, L., Glonek, T., Kruski, A.W., and Scanu, A.M.: 1974, Circulation, Supplement to vol. 49 and 50, p. III-265 (Abs. No. 1025).

Henderson, T.O., Davis, L., Glonek, T., Kruski, A.W., and Scanu, A.M.: 1975, Biochemistry 14, 1915.

Henderson, T.O., Glonek, T., Hilderbrand, R.L., and Myers, T.C.: 1972, Arch. Biochem. Biophys. 149, 484.

Henderson, T.O., Glonek, T., and Myers, T.C.: 1974c, Biochemistry 13, 623.

Hilderbrand, R.L., Henderson, T.O., Glonek, T., and Myers, T.C.: 1971 Fed. Proc. 30, 1072 (Abs. No. 116).

Kraut, J.: 1961, Acta Crystalog. 14, 1146.

Mateu, L., Tardieu, A., Luzzati, V., Aggerbeck, L., and Scanu, A.M.: 1972, J. Mol. Biol. 70, 105.

Michaelson, D.M., Horwitz, A.F., and Klein, M.P.: 1973, Biochemistry 12, 2637.

Pattnaik, N.M., Kezdy, F., Scanu, A.M.: 1976, J. Biol. Chem. (in press).

Pullman, B., and Berthod, H.: 1974, FEBS Lett. 44, 266.

Pullman, B., Berthod, H., and Gresh, N.: 1975, FEBS Lett. 53, 199.

Scanu, A.M., and Kruski, A.W.: 1973, in International Encyclopedia of Pharmacology and Therapeutics, Masoro, E., Ed., New York, N.Y., Pergamon Press, Section 24, pp. 21.

Shipley, G.V., Atkinson, D., and Scanu, A.M.: 1972, J. Supramol. Struct. 1, 98.

Skipski, V.P., Barclay, J., Barclay, R.K., Fetzer, V.A., Good, J.J., and Archibald, F.M.: 1967, Biochem. J. 104, 340.

STRUCTURES OF CALCIUM-CARBOHYDRATE COMPLEXES

WILLIAM J. COOK
Institute of Dental Research, Department of Pathology

and

CHARLES E. BUGG
Institute of Dental Research, Department of Biochemistry, Cancer Research and Training Center, University of Alabama in Birmingham University Station Birmingham, Ala. 35294, U.S.A.

1. INTRODUCTION

Calcium ions and carbohydrates are found in close proximity at many places within biological systems, and there is indirect evidence that calcium-carbohydrate interactions may be involved in several processes including calcium transport [10, 37], stabilization of membrane structures [9, 32], calcification [7, 24, 48], calcium storage [25], cell-cell adhesion [30, 33, 35, 50], binding of glycoproteins to cell surfaces [22, 33, 41], and agglutination of polysaccharides [38, 47]. Numerous studies have demonstrated that calcium ions bind to ionic and uncharged carbohydrates in aqueous solution [2-4, 36, 42], thus supporting the possible importance of calcium-carbohydrate complexes in biological systems. About twenty calcium-carbohydrate complexes and salts have been crystallized from aqueous solution and studied by X-ray diffraction methods. In this paper we review these crystal structures, emphasize those structural features that we feel may be of general importance, and speculate about the possible roles that calcium-carbohydrate complexes might play in biological systems.

2. HYDRATED CALCIUM IONS

Normally, calcium ions in aqueous solution are highly hydrated. Therefore, to form stable calcium complexes in aqueous environments, carbohydrates must be able to compete with water-calcium interactions. Unfortunately, little is known about the detailed properties of hydrated calcium ions or about the specific processes by which various ligands bind to calcium ions in aqueous solution. For the purpose of this review, we are particularly concerned with the geometrics of the inner spheres of water molecules found around calcium ions and with possible

B. Pullman and N. Goldblum (eds.), Metal-Ligand Interactions in Organic Chemistry and Biochemistry, part 2, 231-256 .

relationships between the structures of hydrated calcium ions and of hydrated calcium-carbohydrate complexes. In aqueous solution, water molecules are probably entering and leaving the calcium hydration shell at rapid rates, so that a specific structure cannot really be ascribed to hydrated calcium ions. However, we shall assume that certain geometrical arrangements of water molecules result in particularly stable interactions with the calcium ions, and that these stable arrangements are the ones found most frequently in aqueous solutions of calcium salts. We shall then try to relate the observed crystal structures of calcium-carbohydrate complexes to the postulated structures of hydrated calcium ions.

Models for the preferred geometries of calcium ions in aqueous solution have come from recent crystallographic analyses of structures that contain completely hydrated calcium ions. One of these crystal structures (calcium potassium arsenate octahydrate) possesses a calcium ion that is coordinated to eight water molecules [23], and another (calcium dichromate bis (hexamethylenetetramine) heptahydrate) displays a calcium ion that is coordinated to seven water molecules [20]. The geometries of the hydration shells around these calcium ions are depicted in Figure 1. In both structures the water molecules are positioned with their oxygen atoms directed toward the calcium ions. In

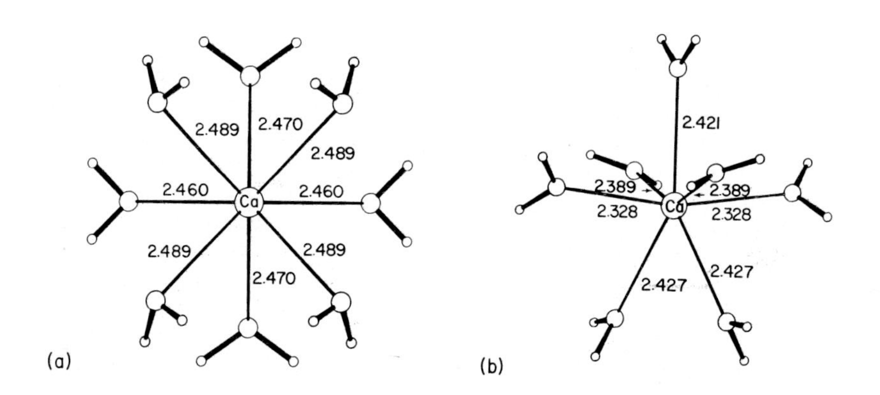

Fig. 1. Stereochemistry of the calcium ion hydration shell in (a) $CaKAsO_4 \cdot 8H_2O$ and (b) $CaCr_2O_7 \cdot [(CH_2)_6N_4] \cdot 7H_2O$.

the arsenate structure (Figure 1a), the eight oxygen atoms assume a square-antiprism arrangement, and the Ca-O distances range from 2.46 to 2.49 Å with an average value of 2.48 Å. In the dichromate structure (Figure 1b) the seven oxygen atoms assume a pentagonal bipyramidal arrangement, and the Ca-O distances range from 2.33 to 2.43 Å with an average value of 2.39 Å. Within both types of coordination polyhedra the neighboring water molecules form close-packed arrangements; the distances between neighboring oxygen atoms range from 2.7 to 3.1 Å with an average separation of 3 Å. Neither structure displays hydrogen bonding between water molecules within the hydration shell.

Although additional coordination numbers have been found for calcium complexes that involve ligands other than water, we feel that the sevenfold (pentagonal bipyramidal) and eightfold (square-antiprismatic) coordination polyhedra are probably the most stable for hydrated calcium ions. Eight water molecules can effectively pack around the calcium ion, while forming stable calcium-water contacts, without forcing the water molecules significantly closer than normal van der Waals separations. By removing one water molecule from the hydration shell and altering the geometry of the coordination polyhedron, it is possible to form even shorter (more stable) calcium-water contacts (Ca-O distances decreased by about 0.1 Å). Thus the transition from eightfold coordination involves the loss of one calcium-water contact, with an offsetting gain in the strengths of the seven remaining calcium-water contacts. Since ninefold and higher coordination would require the formation of Ca-O contacts considerably longer than normal, coordination numbers greater than eight are probably relatively unstable. It is also unlikely that calcium-water contacts that are significantly shorter (stronger) than those in the sevenfold hydration state could be formed; therefore, hydration numbers below seven are probably also of relatively minor importance.

Considering these factors, we believe that the geometries depicted in Figure 1 are typical of the more stable states for the inner-sphere hydration shells around calcium ions in aqueous media. These two geometrical arrangements are also typical of the coordination patterns found in crystal structures of calcium-carbohydrate complexes and salts.

3. CALCIUM INTERACTIONS WITH UNCHARGED CARBOHYDRATES

Many investigations have demonstrated that calcium ions can bind to uncharged carbohydrates in aqueous solution [2-4, 36, 42], and recent NMR studies of these solution complexes suggest that the complexes are stabilized by interactions between the calcium ion and hydroxyl groups of the carbohydrates [2-4]. The crystal structures of eleven hydrated calcium halide complexes of uncharged carbohydrates have been reported, involving a total of eight different sugars [8, 11-16, 18, 19, 43, 49]. Pertinent data relating to these crystalline complexes are compiled in Table I, and the structural formulas and numbering schemes of the sugars are depicted in Figure 2.

These eleven complexes have several important features in common. Most of these features are illustrated by the lactose-calcium halide complexes. Hydrated calcium complexes of lactose are easily crystallized, by simply dissolving approximately equimolar amounts of lactose and either calcium bromide or calcium chloride in water, followed by gradual evaporation of the aqueous solution. The resultant crystalline complexes have the stoichiometry lactose·calcium halide·$7H_2O$, and they are quite suitable for high-resolution crystallographic analyses. The calcium bromide and calcium chloride complexes are isostructural; i.e., except for minor perturbations attributable to differences in the ionic radii of the chloride and bromide ions, the two lactose-calcium halide complexes have identical crystal structures. These lactose complexes are heavily hydrated, containing seven water molecules per calcium ion.

TABLE I
Chelation data for hydrated calcium halide complexes of uncharged carbohydrates.

Carbohydrate	Space Group	Stoichiometry	Number of carbohydrate molecules in Ca shell	Number of water molecules in Ca shell	Calcium coordination number	Ca-O distances
α,α-Trehalose	$C222_1$	$C_{12}H_{22}O_{11} \cdot CaBr_2 \cdot H_2O$	4	1	7	2.32-2.47 Å
β-D-Fructose	$P2_1$	$C_6H_{12}O_6 \cdot CaCl_2 \cdot 2H_2O$	3	2	7	2.33-2.46 Å
β-D-Fructose	$P2_1$	$C_6H_{12}O_6 \cdot CaBr_2 \cdot 2H_2O$	3	2	7	2.32-2.47 Å
β-D-Fructose	C2	$(C_6H_{12}O_6)_2 \cdot CaCl_2 \cdot 3H_2O$	4	2	8	2.45-2.50 Å
α-D-Fucose	$P2_12_12_1$	$C_6H_{12}O_5 \cdot CaBr_2 \cdot 3H_2O$	2	3	7	2.32-2.44 Å
α-D-Xylose	$P2_1$	$C_5H_{10}O_5 \cdot CaCl_2 \cdot 3H_2O$	2	3	7	2.32-2.51 Å
α-D-Galactose	$P2_12_12_1$	$C_6H_{12}O_6 \cdot CaBr_2 \cdot 3H_2O$	3	3	8	2.35-2.55 Å
β-L-Arabinose	C2	$C_5H_{10}O_5 \cdot CaCl_2 \cdot 4H_2O$	2	4	8	2.33-2.70 Å
D-Lactose	$P2_12_12_1$	$C_{12}H_{22}O_{11} \cdot CaCl_2 \cdot 7H_2O$	2	4	8	2.39-2.54 Å
D-Lactose	$P2_12_12_1$	$C_{12}H_{22}O_{11} \cdot CaBr_2 \cdot 7H_2O$	2	4	8	2.38-2.54 Å
myo-Inositol	$P\bar{1}$	$C_6H_{12}O_6 \cdot CaBr_2 \cdot 5H_2O$	2	4	8	2.37-2.52 Å

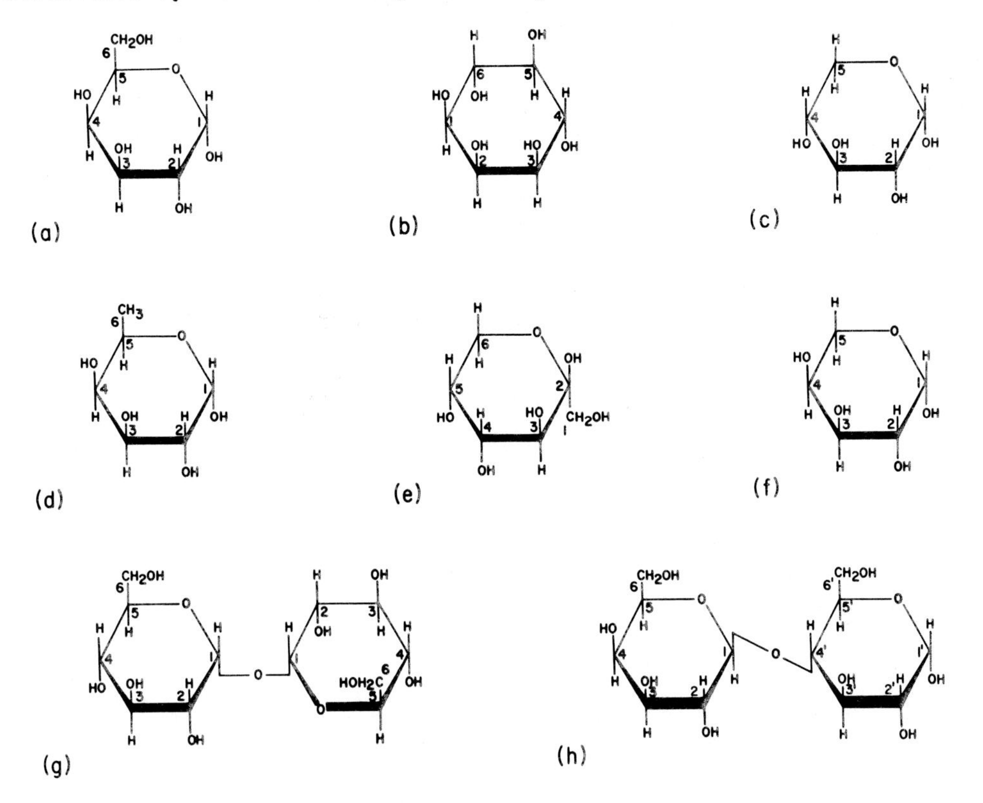

Fig. 2. Structural formulas and numbering schemes for the uncharged carbohydrates included in this review:
(a) α-D-galactose [14]; (b) myo-inositol [16];
(c) α-D-xylose [43]; (d) α-D-fucose [12];
(e) β-D-fructose [11, 18, 19]; (f) β-L-arabinose (49);
(g) α,α-trehalose [13]; (h) D-lactose [8, 15].

The calcium ion environment in these lactose complexes is depicted in Figure 3. Each calcium ion is coordinated to two symmetry-related lactose molecules. One lactose molecule chelates the calcium ion through the O(2')-O(3') pair of hydroxyl groups from its glucose moiety, and the second molecule chelates the calcium ion through the O(3)-O(4) pair of hydroxyl groups from its galactose moiety. The coordination polyhedron around the calcium ion is completed by four water molecules. Therefore, the calcium ion is coordinated to eight oxygen atoms: four from hydroxyl groups and four from water molecules.

Since the detailed structural properties of these lactose complexes have been discussed in earlier publications [8, 15], we shall not attempt to describe the complete crystal structures. However, it is worthwhile to summarize features that are typical of the whole group of calcium-carbohydrate complexes. As can be seen from Figure 3, the coordination polyhedron around the calcium ion looks much like the structure of a hydrated calcium ion (Figure 1a), except that four of

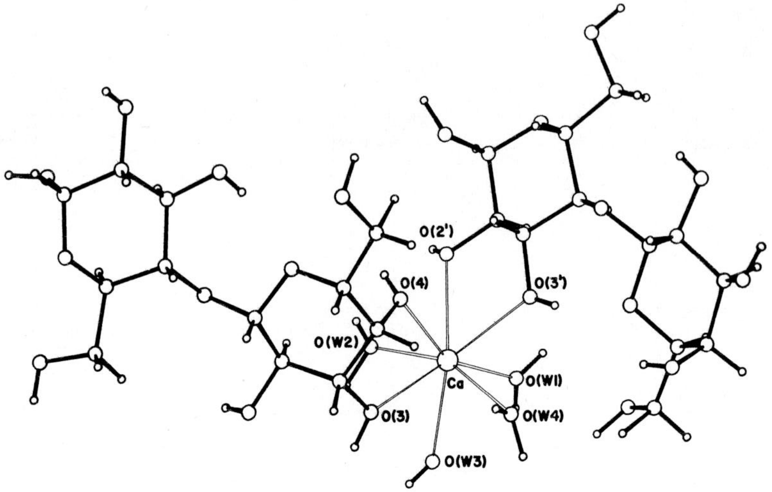

Fig. 3. Environment of the calcium ion in the lactose $-CaBr_2 \cdot 7H_2O$ and lactose-$CaCl_2 \cdot 7H_2O$ complexes.

the water sites have been replaced by hydroxyl groups. The geometry around the calcium ion is much like that shown in Figure 1a: the Ca-O distances range from about 2.4 to 2.5 Å, and the eight oxygen atoms form a distorted square-antiprism. The calcium-carbohydrate interaction is best described as a simple substitution reaction, involving the replacement of water molecules by lactose hydroxyl groups.

An interesting feature of this complex is that neither the chloride nor the bromide anion is directly coordinated to the calcium cation. This feature is also characteristic of the other carbohydrate--calcium halide complexes that have been examined: in no case are the halide anions involved in the inner-sphere coordination shell around the calcium ion. Therefore, it appears that the carbohydrate-calcium interactions are sufficiently strong to compete with the calcium cation-halide anion coulombic attraction, at least in these solid state environments.

Another pertinent feature of the lactose-calcium complex is the effect exerted by the calcium ion on the conformation of the sugar residues. The hydroxyl groups appear to be drawn toward the calcium ion, thus perturbing the carbohydrate conformation at the calcium-binding sites. The distortions are such that, at both chelation sites, the two calcium-binding hydroxyl groups are drawn about 0.2 Å closer together than those in the crystal structure of lactose monohydrate [26]. These decreases in the spacings between chelating oxygen atoms are accomplished by altering both the bond angles and the torsion angles in the neighborhood of the calcium-binding sites. Relative to the conformation found in the crystal structure of lactose monohydrate, calcium binding to the glucose moiety is accompanied by a decrease of 7° in the O(2')-C(2')-C(3')-O(3') torsion angle, and a decrease of 5° in the C(3')-C(2')-O(2') bond angle. Calcium binding to the galactose moiety of lactose is accompanied by decreases of 9° in the O(3)-C(3)-C(4)-O(4) torsion angle and 5° in the C(4)-C(3)-O(3) bond angle. Detailed discussions of conformational changes found in other calcium-carbohydrate complexes have been published [11, 16]; the general finding is that calcium binding to adjacent hydroxyl groups is responsible for decreases of about 0.2 Å between the hydroxyl oxygen atoms, and concomitant perturbations ranging up to 7° for bond angles and up to 15° for torsion angles.

The lactose-calcium complex displays another feature that we consider particularly interesting and significant. As shown in Figure 3, the calcium ion acts as a bridge between the two lactose molecules, thereby linking them together. In the crystal structure, these bridges continue indefinitely, cross-linking the lactose molecules end-to-end throughout the structure. We feel that hydrated carbohydrate-calcium-carbohydrate bridges similar to that shown in Figure 3 may be involved in a variety of calcium-dependent agglutination processes, a point that we shall discuss later in this paper.

Most of the structural features found for the lactose-calcium complex are typical of those in other calcium complexes of uncharged carbohydrates. The only major exception is that several structures display sevenfold pentagonal bipyramidal coordination geometry to the calcium ions. A typical example of this alternate pattern of coordination is depicted in Figure 4, which displays the calcium environment in the fucose-calcium bromide trihydrate complex [12]. As in the lactose complex, the calcium ion is chelated by pairs of hydroxyl groups from two symmetry-related fucose molecules. In this case, however, the coordination polyhedron is completed by only three water molecules, resulting in a final inner-sphere complex that contains four hydroxyl groups and

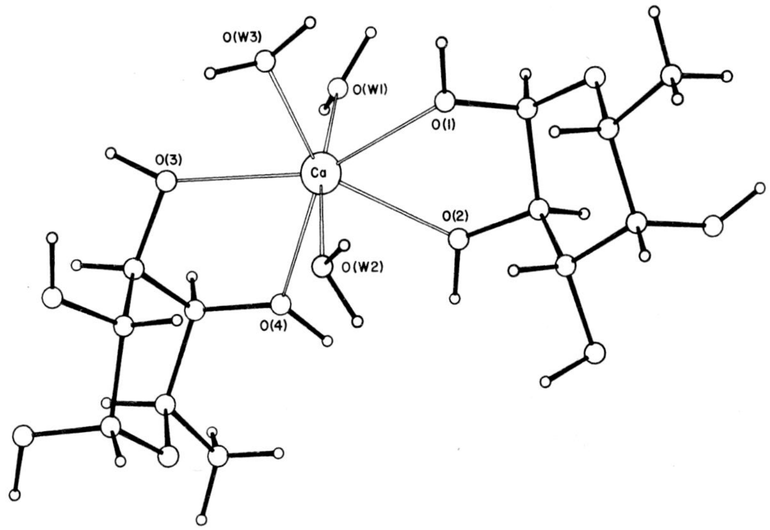

Fig. 4. Environment of the calcium ion the the fucose-$CaBr_2$ $3H_2O$ complex.

three water molecules. The geometry of this shell is almost identical to that of the sevenfold hydrated calcium ion (Figure 1b).

Table I summarizes the major structural features of the other calcium-carbohydrate complexes. The general features typified by the lactose and fucose complexes are found in all these structures: (1) only sevenfold and eightfold coordination patterns are seen; (2) the calcium shells contain from one to four water molecules; (3) the halide anions are never involved in the coordination polyhedra; (4) the calcium ion is always coordinated to two or more sugars, thus cross-linking the carbohydrate molecules; and (5) the calcium-oxygen contacts fall within the range 2.3 to 2.6 Å. The geometries of the calcium-binding sites in these complexes are depicted in Figures 5 and 6.

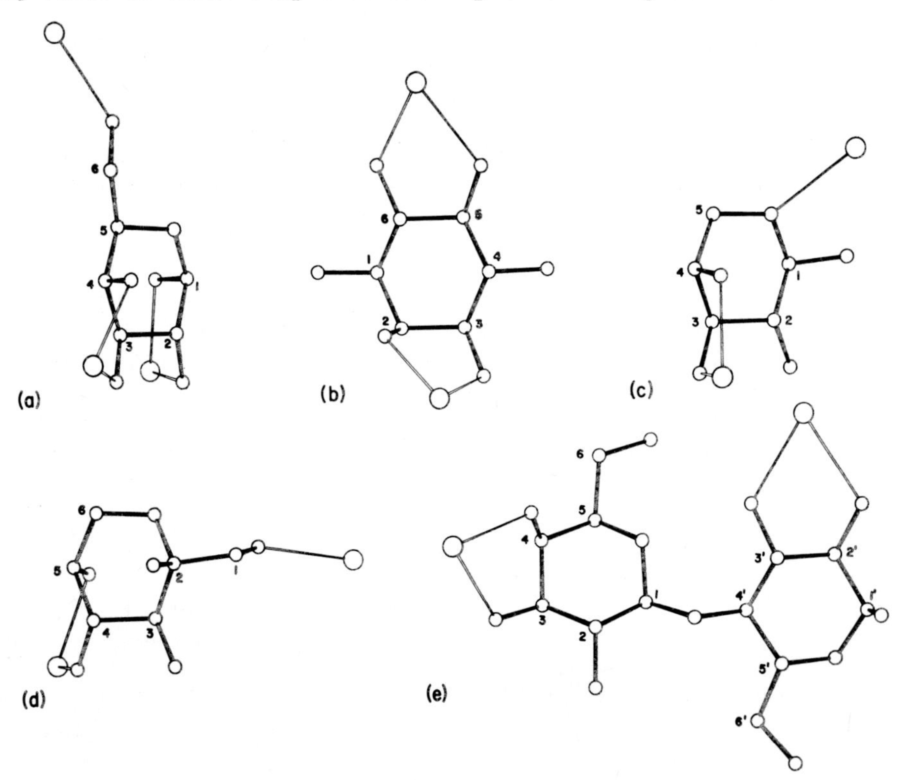

Fig. 5. Calcium-binding sites on several uncharged carbohydrates:
(a) α-D-galactose; (b) myo-inositol; (c) β-L-arabinose; (d) β-D-fructose (8-fold coordination); (e) D-lactose.

The most common binding sites are pairs of adjacent hydroxyl groups, which are either in axial-equatorial or equatorial-equatorial arrangements. Four crystal structures also display interactions between hydroxymethyl groups and calcium ions. One crystal structure - arabinose-calcium chloride tetrahydrate - involves a calcium interaction with the ring oxygen atom.

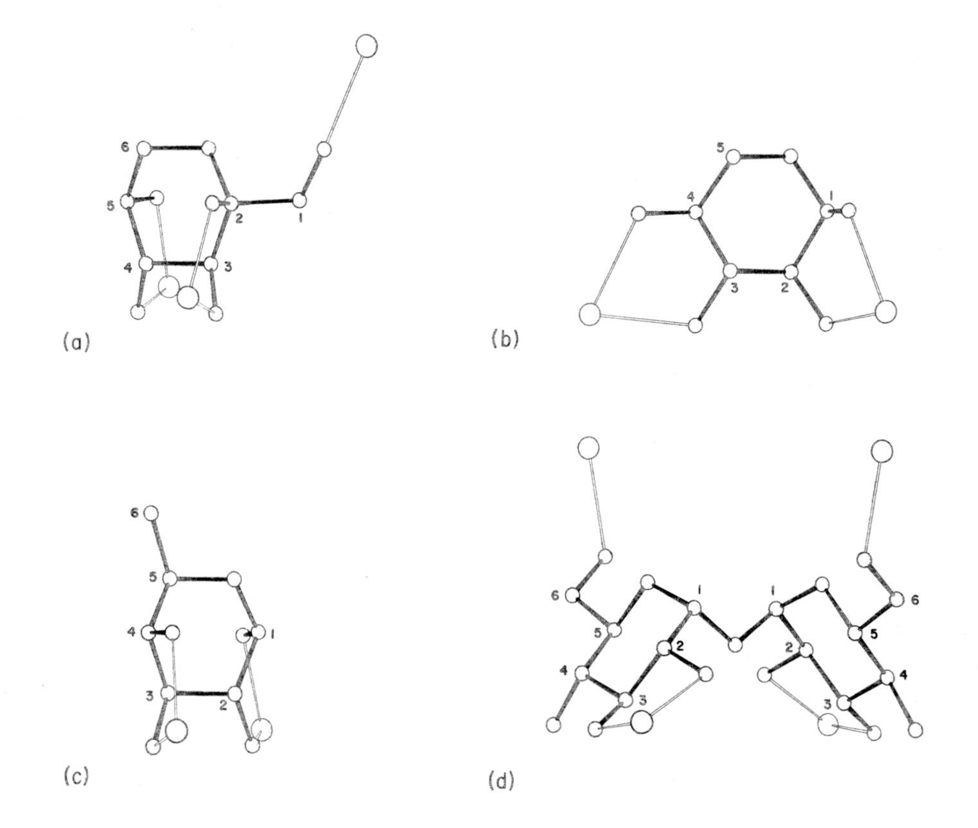

Fig. 6. Calcium-binding sites on several uncharged carbohydrates:
(a) β-D-fructose (7-fold coordination); (b) α-D-xylose;
(c) α-D-fucose; (d) α,α-trehalose.

These eleven crystal structures indicate that uncharged carbohydrates contain a variety of different sites that are suitable for binding calcium ions. The most important chelation sites are provided by pairs of hydroxyl groups. However, almost any other oxygen atom from the carbohydrates is also capable of participating in the interactions, if it is in an orientation that permits substitution in the calcium coordination polyhedron.

4. CALCIUM INTERACTIONS WITH ANIONIC CARBOHYDRATES

Several crystal structures of calcium salts of anionic carbohydrates and related compounds have been determined [1, 6, 17, 21, 27, 29, 21, 34, 39]. Table II lists those included in this review, and the structural formulas and numbering schemes for these compounds are given in Figures 7 and 8

TABLE II
Chelation data for hydrated calcium complexes of anionic carbohydrates

Carbohydrate	Space Group	Stoichiometry	Number of carbohydrate anions in Ca shell	Number of water molecules in Ca shell	Calcium coordination number	Ca-O distances
α-D-Glucoisosaccharinic acid	$P2_12_12$	$Ca(C_6H_{11}O_6)_2$	2	0	8	2.38-2.51 Å
L-Tartaric acid	$P2_12_12_1$	$CaC_4H_4O_6 \cdot 4H_2O$	4	2	8	2.39-2.54 Å
5-keto-D-Gluconic acid	A2	$Ca(C_6H_9O_7)_2 \cdot 2H_2O$	2	2	8	2.39-2.47 Å
D-Arabonic acid	C2	$Ca(C_5H_9O_6)_2 \cdot 5H_2O$	2	2	8	2.44-2.52 Å
L-ascorbic acid	$P2_1$	$Ca(C_6H_7O_6)_2 \cdot 2H_2O$	3	2	8	2.41-2.53 Å
α-D-Glucuronic acid	$P2_1$	$C_6H_9O_7$ $CaBr \cdot 3H_2O$	3	2	8	2.38-2.57 Å
Lactobionic acid	$P2_12_12_1$	$C_{12}H_{21}O_{12} \cdot CaBr_2 \cdot 4H_2O$	2	3	8	2.37-2.52 Å
α-D-Galacturonic acid	$P6_3$	$NaCa$ $(C_6H_9O_7)_3 \cdot 6H_2O$	3	3	9	2.52-2.87 Å

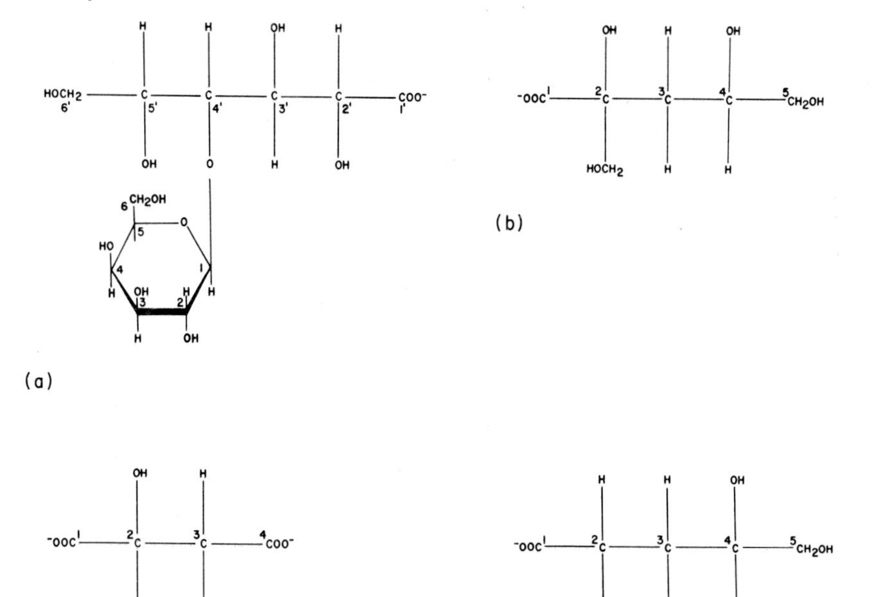

Fig. 7. Structural formulas and numbering schemes for some of the anionic carbohydrates included in this review: (a) lactobionate [17]; (b) α-D-glucoisosaccharate [39]; (c) L-tartrate [1]; (d) D-arabonate [27].

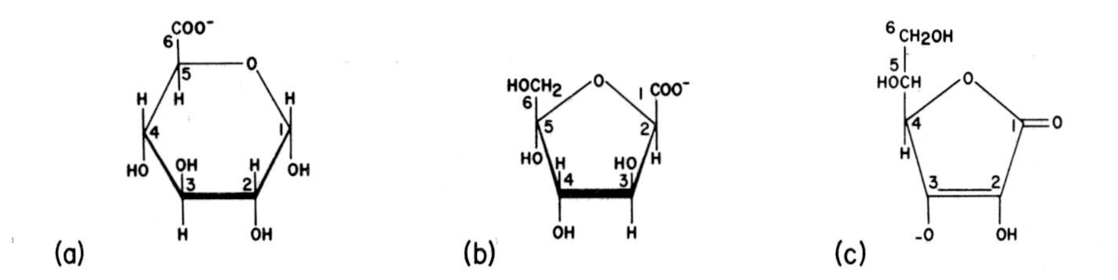

Fig. 8. Structural formulas and numbering schemes for some of the anionic carbohydrates included in this review: (a) α-D-glucuronate [21]; (b) 5-keto-D-gluconate [6]; (c) L-ascorbate [31, 34].

Most of the structural features of calcium-carbohydrate salts can be seen from an examination of the hydrated calcium bromide salt of lactobionic acid (Figure 7a) [17]. Lactobionic acid is an oxidation product of lactose, wherein the glucose residue is converted to a gluconate anion and the galactose moiety is unaltered. The lactobionate ion contains many of the functional groups present on lactose as well as an additional carboxyl group. Crystals of a calcium bromide salt of lactobionic acid are readily grown by dissolving approximately equimolar quantities of calcium bromide and calcium lactobionate in water and slowly evaporating the aqueous solution. The resultant crystals contain four molecules of water per calcium ion.

The environment of the calcium ion in this crystal structure is shown in Figure 9. In contrast to the crystalline calcium-lactose complex, the galactose moiety is not involved in calcium interactions.

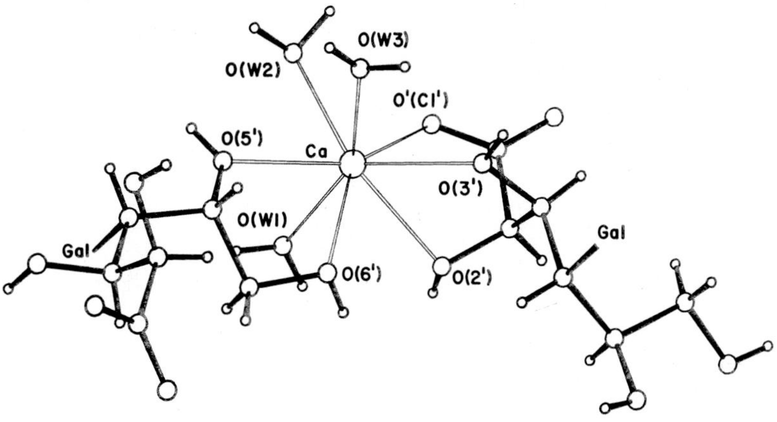

Fig. 9. Environment of the calcium ion in the lactobionate-$CaBr \cdot 4H_2O$ complex. The galactose moieties are omitted for clarity.

Two symmetry-related lactobionate anions are bound to the calcium ion through their gluconate residues. One lactobionate ion chelates the calcium ion through a tridentate site that involves an oxygen atom from the carboxyl group acting in concert with O(2') and O(3'), the oxygen atoms from the adjacent hydroxyl groups that are in the α and β positions relative to the carboxyl group. The second lactobionate ion chelates the calcium ion through the O(5')-O(6') pair of hydroxyl groups, which are located at the terminal end of the gluconate moiety. Three water molecules complete the coordination shell. The resultant structure contains eight oxygen atoms that are coordinated to calcium: three from water molecules, one from the carboxyl group, and four from hydroxyl groups. These eight oxygen atoms form a square antiprism that looks much like that shown in Figure 1a for the hydrated calcium ion. As in the calcium halide complexes of uncharged carbohydrates, the bromide ion does not participate in the calcium coordination polyhedron. Again, the calcium ion forms a hydrated bridge between two carbohydrate residues.

Essential details concerning the other calcium-carbohydrate salts are given in Table II, and the calcium-binding sites used by these carbohydrate anions are shown in Figures 10 and 11. In all these crystal structures, the anionic substituents participate in the calcium interactions. However, they never act alone, but are combined with neighboring hydroxyl groups and ring oxygen atoms to produce bidentate

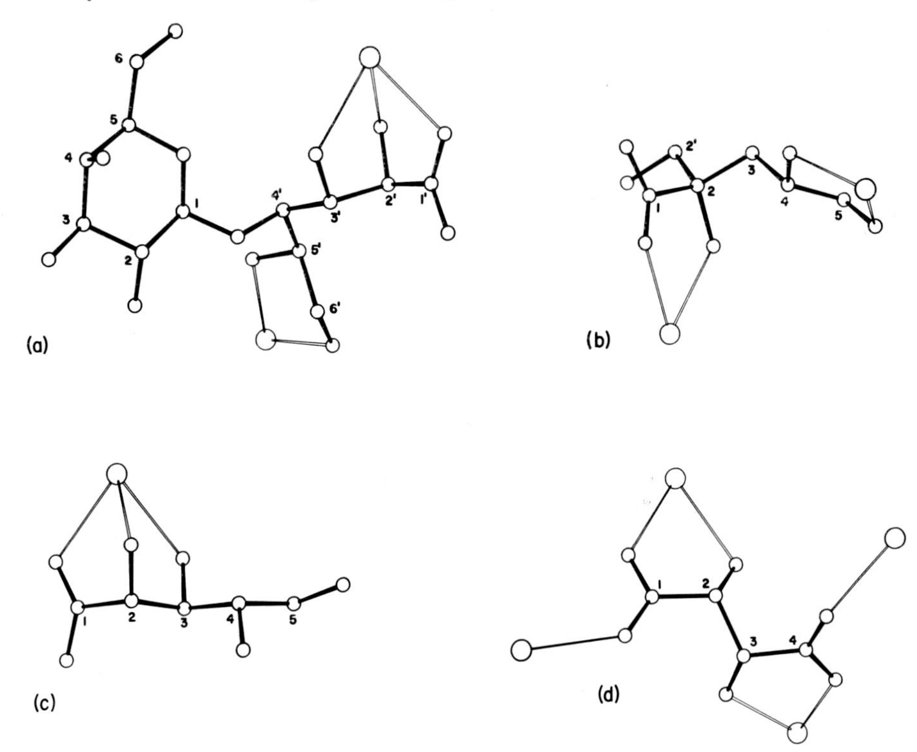

Fig. 10. Calcium-binding sites on several anionic carbohydrates: (a) lactobionate; (b) α-D-glucoisosaccharate; (c) D-arabonate; (d) L-tartrate.

or tridentate chelation sites. Sets of hydroxyl groups that are distant from the anionic sites may also chelate the calcium ion, as exemplified by the O(5')-O(6') pair of hydroxyl groups in the lactobionate structure. We consider it particularly important that, even when anionic sites are available, the calcium interactions are not determined simply by cation-anion coulombic factors, but are dependent upon the availability of sets of ligands that combine to form suitable chelation sites.

Eightfold coordination geometry is found in all the crystal structures of calcium-carbohydrate salts, except that of the sodium-calcium double salt of galacturonic acid [29]. This single exception displays ninefold coordination and involves some calcium-oxygen distances much longer than normal. In all these structures, except the calcium salt of glucoisosaccharinic acid, either two or three water molecules are coordinated to the calcium ion. As for the uncharged carbohydrates, the calcium ion is always coordinated to two or more sugar residues, thus forming bridges between carbohydrate molecules.

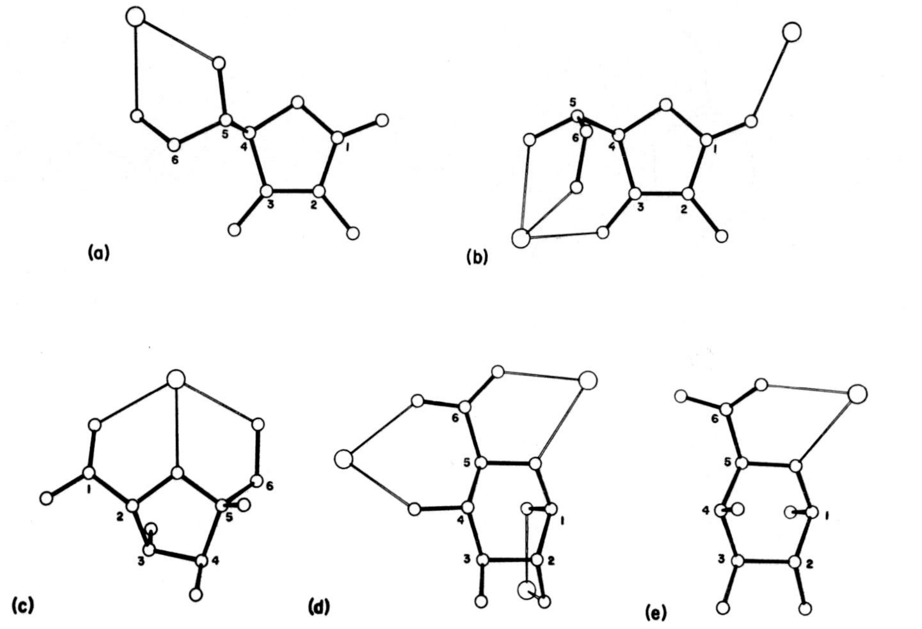

Fig. 11. Calcium-binding sites on several anionic carbohydrates: (a) L-ascorbate; (b) L-ascorbate; (c) 5-keto-D-gluconate; (d) α-D-glucoronate; (e) α-D-galacturonate [29].

5. GEOMETRY OF THE CALCIUM COORDINATION SHELLS

Figures 12-14 show examples of the three fundamental types of calcium coordination polyhedra in these crystal structures of calcium-carbohydrate salts and complexes. Among the 19 structures examined are five examples of sevenfold coordination, thirteen examples of eightfold coordination and one example of a crystal structure with ninefold coordination. Based upon this set of structures, it appears that the eightfold mode of coordination is preferred, but the sevenfold pattern is also of major importance. In aqueous solution, one might resonably expect that calcium-carbohydrate complexes would exist primarily in these two coordination states, much like the situation postulated earlier for hydrated calcium ions. With regard to this possibility, it is worth noting that the fructose-calcium chloride complex crystallizes from aqueous solution in two distinct forms: one that displays sevenfold (pentagonal bipyramidal) coordination [19] and one that shows eightfold (square-antiprismatic) coordination [18].

Examples of the coordination polyhedra in these calcium-carbohydrate structures are shown in Figures 12 and 13. Figure 12 shows the sevenfold coordination pattern in the crystal structure of the fructose-calcium bromide trihydrate complex. Five ligands are nearly coplanar with the calcium ion, and are approximately related by a fivefold axis that lies perpendicular to the plane and passes through the

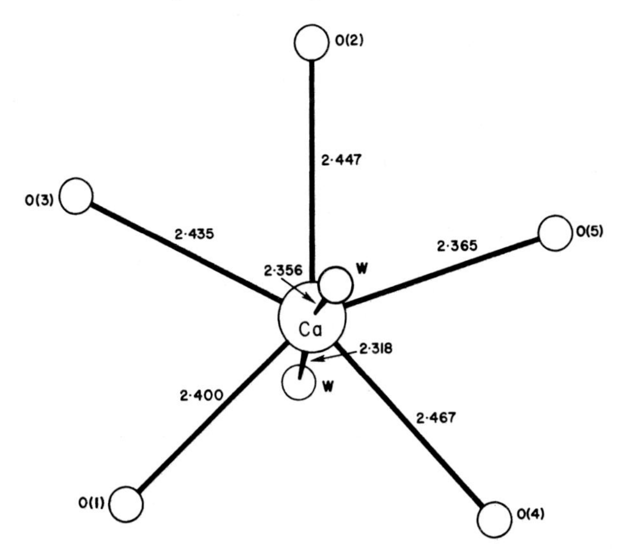

Fig. 12. Geometry of the calcium ion coordination shell in the fructose-$CaBr_2 \cdot 3H_2O$ complex.

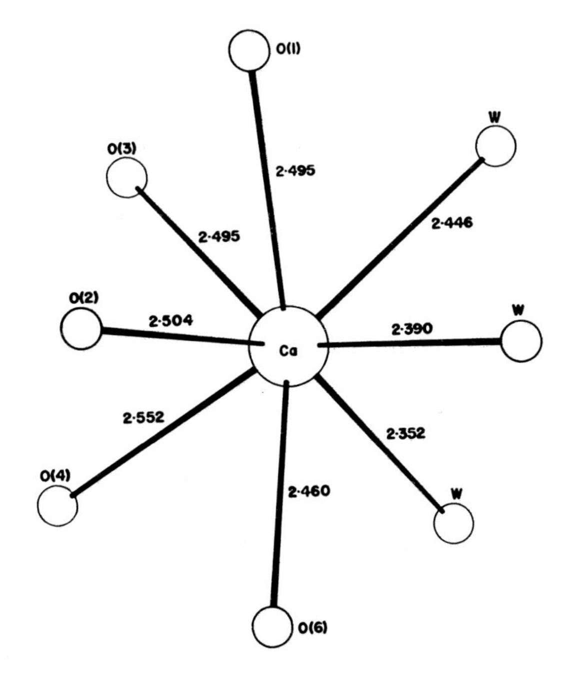

Fig. 13. Geometry of the calcium ion coordination shell in the galactose-$CaBr_2 \cdot 3H_2O$ complex.

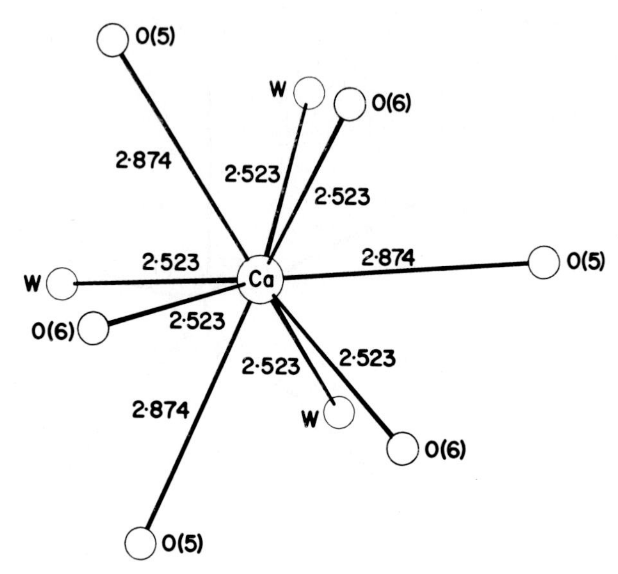

Fig. 14. Geometry of the calcium ion coordination shell in the galacturonate-CaNa·$2H_2O$ complex.

calcium ion. The final two ligands, which are water molecules, are situated above and below this plane, very near the fivefold axis. The calcium-oxygen distances range from 2.32 to 2.47 Å, with a mean value of 2.40 Å. Within the coordination shell, the oxygen atoms are closely packed, with no hydrogen bonding between the ligands.

Figure 13 depicts the eightfold, square-antiprismatic coordination polyhedron in the galactose-calcium bromide trihydrate complex [14]. In this geometrical arrangement there are four ligands lying close to a plane which is above the calcium ion and four lying near a parallel plane below the calcium ion. There is a pseudo-fourfold axis passing through the calcium ion and situated perpendicular to these two planes, and the planes are rotated about 45° relative to each other about the fourfold axis. The calcium-oxygen contacts range from 2.35 to 2.55 Å, with an average value of 2.46 Å. Again, the oxygen atoms within the coordination shell are in a close-packed arrangement, and there is no hydrogen-bonding between the ligands.

The ninefold coordination polyhedron in the hydrated sodium calcium galacturonate complex is shown in Figure 14. Six ligands form a trigonal prism, and three additional oxygen atoms lie in the faces of the prism. There is a crystallographic threefold axis passing through the calcium ion and perpendicular to the ends of the prism (approximately perpendicular to the plane of the paper in Figure 14). The calcium contacts with ligands of the prism are all 2.52 Å, and the calcium distances to the three additional ligands are 2.87 Å. The average calcium-oxygen distance for these nine ligands is 2.64 Å, which is considerably longer than normal calcium-oxygen contacts. Within this coordination shell the oxygen-oxygen contacts are somewhat longer than in the sevenfold and eightfold coordination shells.

Figure 15 shows a series of histograms that summarize the calcium-oxygen distances in the sevenfold and eightfold coordination shells observed in calcium-carbohydrate crystal structures. The calcium-hydroxyl contacts in the eightfold shells range from 2.41 to 2.70 Å, with a mean value of 2.50 Å; for the sevenfold shells, the calcium-hydroxyl range is 2.36 to 2.51 Å, with a mean value of 2.43 Å. For the eightfold shells, the calcium-water contacts lie between 2.33 and 2.57 Å, with a mean value of 2.41 Å; in the sevenfold shells, the corresponding range is 2.32 to 2.36 Å, and the mean distance is 2.34 Å. The calcium-carboxyl contacts are found only in eightfold coordination shells; for these contacts the calcium-oxygen distances lie between 2.38 and 2.52 Å, with a mean value of 2.45 Å.

The overall pattern that emerges from examination of the coordination shells in these crystal structures is as follows: (1) the preferred coordination geometries are sevenfold pentagonal-bipyramids and eightfold square-antiprisms; (2) coordination polyhedra that contain carboxyl groups tend to display eightfold geometry; (3) in both sevenfold and eightfold coordination shells, the calcium-water contacts tend to be about 0.1 Å shorter than the calcium-hydroxyl contacts; (4) the distances for calcium contacts with carboxyl groups are intermediate between those for hydroxyl groups and those for water molecules; (5) overall, the calcium-oxygen distances increase in the following order: sevenfold calcium-water<eightfold calcium-water<sevenfold calcium-hydroxyl<eightfold calcium-carboxyl<eightfold calcium-hydroxyl. Presumably, the average strengths of calcium-oxygen contacts follow the reverse order.

6. CALCIUM-CARBOHYDRATE INTERACTIONS IN BIOLOGICAL SYSTEMS

The crystallographic results indicate that carbohydrates can provide a variety of suitable sites for interaction with calcium ions, at least within hydrated, solid-state environments. Considering the extensive evidence that carbohydrates also bind calcium ions in aqueous solution, we feel that the crystallographic results may be extrapolated to biological systems to yield tentative models for the types of functions that calcium-carbohydrate interactions might serve. Unfortunately, most of the evidence that calcium-carbohydrate interactions are even involved in biological processes is indirect and hypothetical, so we are faced with the uncomfortable situation of attempting to postulate models for hypothetical biological processes. Despite this, we shall attempt to summarize those structural features that we believe to be of likely importance.

At many biological sites calcium salts are bound to organic matrices that contain polysaccharides and glycoproteins. A prime example is bone, which contains a preponderance of calcium phosphate minerals embedded in a matrix that is rich in proteins and carbohydrates as well as other organic components [45]. Collagen is the principal component of the organic matrix [45], and much work has been directed toward an understanding of interactions that might occur between bone collagen and calcium salts [5, 28, 44]. Collagens from bone and uncalcified

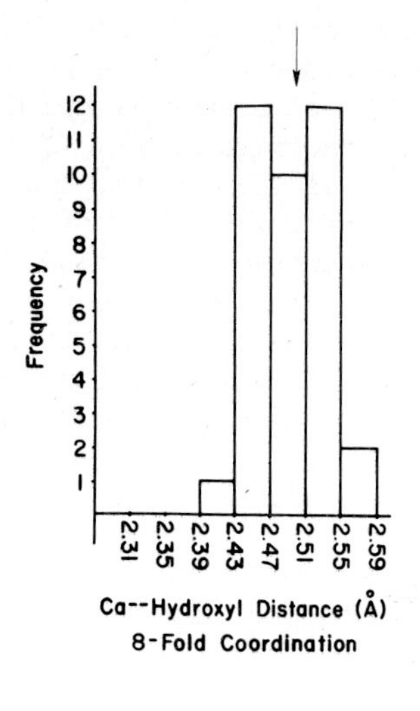

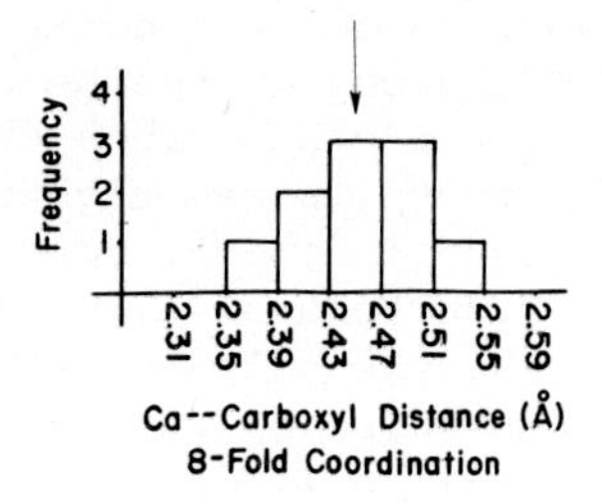

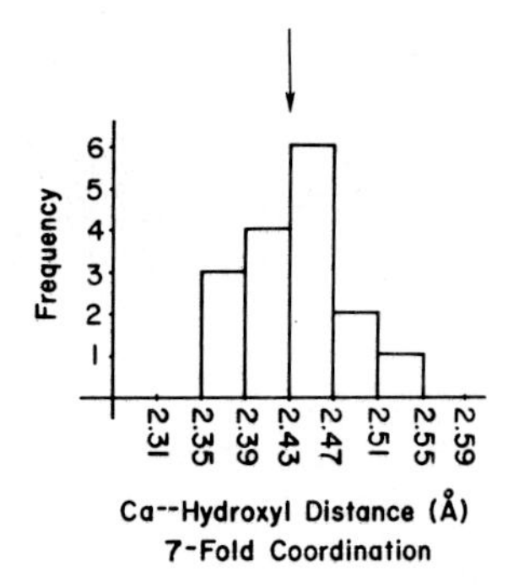

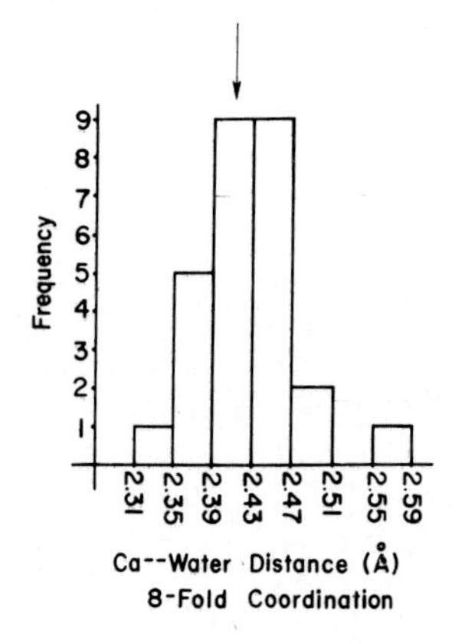

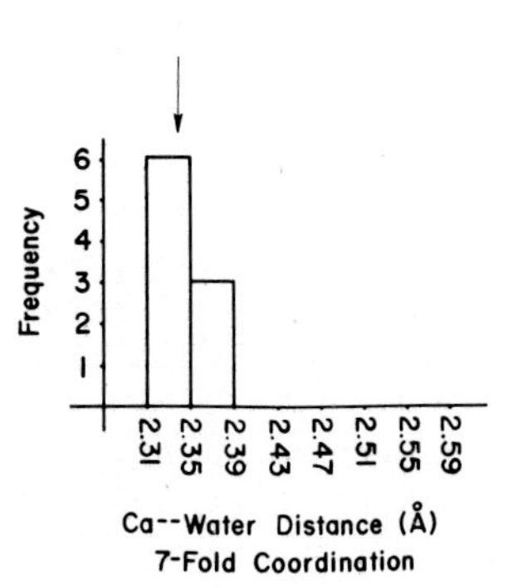

Fig. 15. Calcium-oxygen distances in the crystal structures of calcium-carbohydrate salts and complexes. Arrows mark the average calcium-oxygen distances for the various types of contacts.

tissues display only slight chemical differences; among these differences is the pattern of glycosylation. Bone collagen contains a preponderance of galactose, whereas soft tissue collagens contain a glucosyl-galactosyl disaccharide as the major carbohydrate residue [40, 46]. Based on our crystallographic study of a galactose-calcium complex, we have suggested that these galactose residues on collagen might provide sites for interactions with calcium salts [14].

Figure 16 depicts the hypothetical mechanism that we envision for calcium binding to galactose residues of collagen. The crystal struc-

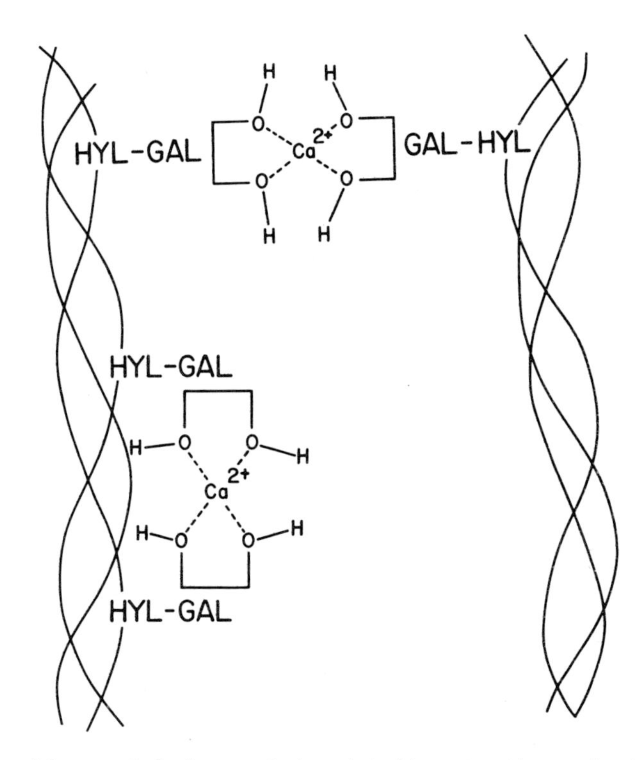

Fig. 16. Model for calcium-binding to the galactose moieties of bone collagen. HYL corresponds to hydroxylysine residues, and GAL denotes the galactose moieties. Only single pairs of hydroxyl groups from the galactose moieties are shown.

ture of the calcium-galactose complex demonstrates that galactose residues possess two sites at which pairs of hydroxyl groups can chelate calcium ions. One of these sites, which includes the O(3)-O(4) pair of hydroxyls, is also involved in calcium binding to the galactose residues in the crystal structure of the calcium-lactose complex, and it would be available for calcium binding to the O(1)-linked galactose residues of bone collagen. NMR studies show that uncharged carbohydrates bind calcium ions strongly only if they can provide sites with three or more hydroxyl groups in a geometrical arrangement that is suitable for calcium chelation [2-4]. An individual galactose residue can provide only a single hydroxyl group or pairs of hydroxyl groups for calcium binding. Consequently, individual, isolated galactose moieties of bone collagen are not likely to have a high affinity for calcium ions. However, if two or more galactose moieties were in close proximity on the collagen matrix, and were geometrically oriented so that each could contribute a pair of hydroxyl groups to the calcium binding site, then strong interactions with calcium could occur. Such neighboring galactose residues might be found within a triple helical collagen molecule, or might be contributed by neighboring collagen molecules within the organic matrix. If two or more galactose moieties with the proper spatial arrangement are required for strong calcium binding, then the availability of such sites would be dependent upon the primary structure and conformation of the collagen, the sites of glycosylation, and the packing pattern of collagen molecules. All these properties are probably subject to biological controls.

Though speculative, this model for calcium-galactose interactions exemplifies the major features we believe to be important at biological sites that may involve calcium-carbohydrate interactions. Most of the sugars that are naturally occurring components of glycoproteins and polysaccharides seem to be capable of providing chelation sites that involve pairs of hydroxyl groups, or sites where charged groups can combine with adjacent hydroxyl groups and/or ring oxygen atoms. The evidence from solution studies indicates that, in aqueous media, such bidentate sites do not bind calcium ions strongly enough to be of much importance in most biological processes. However, sugar residues seldom occur as isolated units at those biological sites where calcium interactions take place; they are usually incorporated into polysaccharides or into oligosaccharide chains on the surface of glycoproteins. At many biological sites, such as cell surfaces, intercellular matrices, organic matrices of mineralized tissues, etc., the carbohydrates are found in condensed, hydrated phases, where multiple intermolecular interactions can occur. As depicted for the collagen matrix, adjacent carbohydrate residues might form multidentate chelation sites, if they are in geometrical arrangements that place suitable ligands in the orientations required for calcium interactions.

The geometrical arrangements that are likely to be required at biological sites can be deduced from the crystallographic studies of calcium-carbohydrate complexes and salts. The complete set of ligands must be able to form calcium contacts that lie somewhere near the range of 2.3 to 2.6 Å, and the resultant geometry of the coordination polyhedra is likely to be either sevenfold (pentagonal bipyramidal) or eightfold (square antiprismatic). Assuming that three of more ligands are required and that these geometrical restraints must be met, it is reasonable to assume that calcium binding sites would be highly stereospecific and subject to all of the biological controls that regulate conformations, glycosylation patterns, and packing arrangements of proteins and polysaccharides. Since such interactions depend more on geometry than on the specific presence of any given functional group, they should be readily reversible, a desirable feature for many of the biological processes that involve calcium ions.

The crystallographic results also suggest that calcium-carbohydrate interactions may contribute to those numerous biological agglutination and adhesion processes that involve both calcium ions and carbohydrates [30, 33, 35, 38, 47, 50]. Carbohydrate-calcium-carbohydrate bridges are found in all crystal structures included in this review. Similar bridges would provide a logical mechanism for the formation of intermolecular cross-links, and might thereby contribute to various calcium-dependent aggregation processes. The general mechanism suggested by the crystallographic results is depicted schematically in Figures 17 and 18. Figure 17 shows the types of hydrated calcium-carbohydrate bridges typically

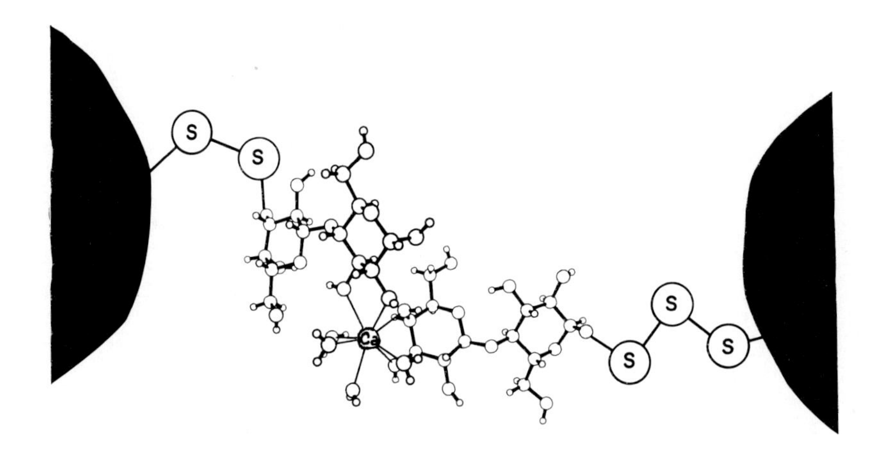

Fig. 17. Model for calcium-dependent cross-linking between two surfaces, utilizing one hydrated calcium-carbohydrate bridge. Sugar residues that are linked to the terminal disaccharides are designated by large circles.

found in the crystal structures. If the sugars bridged by the calcium ion were components of oliosaccharide chains attached to macromolecular surfaces, then the interactions would serve to cross-link the surfaces. Since the cross-links would be dependent upon the ability of the bridging sugar-moieties to chelate the calcium ions, the bridges would be subject to all of the structural restrictions discussed earlier. Hydrated carbohydrate-calcium-carbohydrate bridges might thereby provide a selective, stereospecific mechanism for cross-linking glycoproteins, polysaccharides, glycolipids, etc.

As hypothesized in Figure 18, surfaces that have complementary sets of carbohydrate chains could interact through calcium ions in a concerted fashion, thus forming multiple cross-links. Stringent geometrical constraints would have to be satisfied to establish multiple cross-links between carbohydrate chains; the successful formation of complete sets of calcium-carbohydrate bridges would be possible only when the lengths, compositions, and conformations of the carbohydrate chains simultaneously satisfied the coordination requirements of the calcium ions.

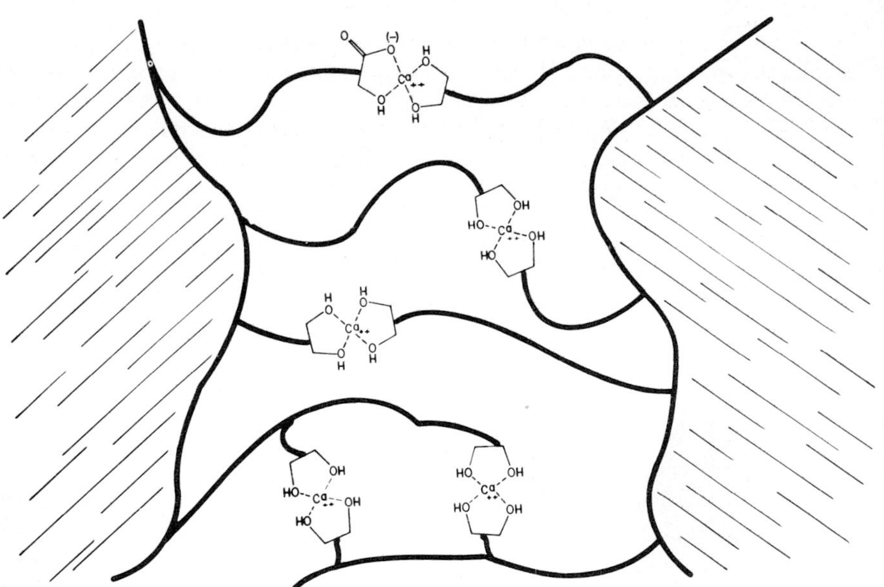

Fig. 18. Model for calcium-dependent cross-linking of two surfaces through multiple calcium-carbohydrate bridges. Oligosaccharide chains are depicted as curved lines protruding from the surfaces. Bidentate calcium-binding sites from individual sugar moieties are shown.

7. CONCLUSIONS

The crystal structures of nineteen calcium-carbohydrate salts and complexes have been examined. They exhibit the following general structural features:

(1) Carbohydrates bind calcium ions through sets of hydroxyl groups or through hydroxyl groups that act in concert with carboxyl groups and ring oxygen atoms. Most carbohydrates possess several sites at which adjacent ligands are suitably arranged for chelating calcium ions.

(2) Calcium-carbohydrate interactions are subject to considerable geometrical constraints. The calcium-ligand distances generally fall within the range 2.3-2.6 Å, and the calcium ions usually display either sevenfold (pentagonal bipyramidal) or eightfold (square-antiprismatic) coordination.

(3) The calcium-carbohydrate interactions are sufficiently strong to produce significant changes in carbohydrate conformations at the calcium-binding sites, and to compete successfully with calcium cation--halide anion coulombic attractions.

(4) The calcium ions are always coordinated to two or more sugar moieties, thus linking carbohydrates together.

(5) Calcium-carbohydrate complexes usually contain varying amounts of water, which serve to complete the coordination polyhedra of the calcium ions.

Considering the ubiquity of these structural features in a variety of crystalline environments, the general roles that water plays in the crystal structures, and the extensive evidence that carbohydrates also interact with calcium ions in aqueous solutions, we believe that the crystallographic results may be relevant to calcium-carbohydrate complexes in biological systems. If so, carbohydrates might provide strong, biological calcium-binding sites. The crystallographic results further suggest that carbohydrate-calcium-carbohydrate bridges may provide an effective, stereospecific mechanism for cross-linking carbohydrate chains in biological systems, and might thereby contribute to calcium-dependent agglutination and adhesion processes.

ACKNOWLEDGEMENTS

We thank Miss Catherine Sims and Miss Mary Ann Comer for assistance with the preparation of this manuscript. This work was supported by U.S.P.H.S. grant number DE-02670, CA-12159 and CA-13148. We also wish to thank the Department of Pathology for travel support.

REFERENCES

1. Ambady, G.K.: 1968, 'The Crystal and Molecular Structures of Strontium Tartrate Trihydrate and Calcium Tartrate Tetrahydrate', Acta Cryst. B24, 1548-1557.
2. Angyal, S.J.: 1972, 'Complexes of Carbohydrates with Metal Cations', Aust. J. Chem. 25, 1957-1966.
3. Angyal, S.J. and Davies, K.P.: 1971, 'Complexing of Sugars with Metal Ions', J. Chem. Soc. D. 500-501.
4. Angyal, S.J. and Hickman, R.J.: 1975, 'Complexes of Carbohydrates with Metal Cations. IV Cyclitols', Aust. J. Chem. 28, 1279-1287.
5. Bachra, B.N. and Fischer, H.R.A.: 1968, 'Mineral Deposition in Collagen In vitro', Calcif. Tissue Res. 2, 343-352.
6. Balchin, A.A. and Carlisle, C.H.: 1965, 'The Crystal Structure of Calcium 5-Keto-D-Gluconate (Calcium D-Xylo-5-Hexulosonate)', Acta Cryst. 19, 103-111.
7. Bowness, J.M.: 1968, 'Present Concepts of the Role of Ground Substance in Calcification', Clin. Orthop. Rel. Res. 59, 233-247.
8. Bugg, C.E.: 1973, 'Calcium Binding to Carbohydrates. Crystal Structure of a Hydrated Calcium Bromide Complex of Lactose', J. Am. Chem. Soc. 95, 908-913.
9. Cameron, L.E. and Le John, H.B.: 1972, 'On the Involvement of Calcium in Amino Acid Transport and Growth of the Fungus Achlya', J. Biol. Chem. 247, 4729-4739.
10. Charley, P. and Saltman, P. 1963, 'Chelation of Calcium by Lactose: Its Role in Transport Mechanisms', Science 139, 1205-1206.
11. Cook, W.J. and Bugg, C.E.: 1976, 'Effects of Calcium Interactions on the Conformations of Sugars: Crystal Structure of β-D-Fructose-Calcium Bromide Dihydrate, Acta Cryst. B32, 656-659.

12. Cook, W.J. and Bugg, C.E.: 1975, 'Calcium-Carbohydrate Bridges Composed of Uncharged Sugars: Structure of a Hydrated Calcium Bromide Complex of α-Fucose', Biochim. Biophys. Acta 389, 428-435.
13. Cook, W.J. and Bugg, C.E.: 1973, 'Calcium Interactions with D-glucans: Crystal Structure of α,α-Trehalose-Calcium Bromide Monohydrate', Carbohydr. Res. 31, 265-275.
14. Cook, W.J. and Bugg, C.E.: 1973, 'Calcium Binding to Galactose. Crystal Structure of a Hydrated α-Galactose-Calcium Bromide Complex', J. Am. Chem. Soc. 95, 6442-6446.
15. Cook, W.J. and Bugg, C.E.: 1973, 'A Lactose-Calcium Chloride Heptahydrate Complex', Acta Cryst. B29, 907-909.
16. Cook, W.J. and Bugg, C.E.: 1973, 'Effects of Calcium Interactions on the Conformations of Sugars: Crystal Structure of Myo-Inositol-Calcium Bromide Pentahydrate', Acta Cryst. B29, 2404-2411.
17. Cook, W.J. and Bugg, C.E.: 1973, 'Calcium Binding to Carbohydrates: Crystal Structure of a Hydrated Calcium Bromide Salt of Lactobionic Acid', Acta Cryst. B29, 215-222.
18. Craig, D.C., Stephenson, N.C., and Stevens, J.D.: 1974, 'Bis-(β-D-Fructopyranose) Calcium Chloride Trihydrate', Cryst. Struct. Comm. 3, 195-199.
19. Craig, D.C., Stephenson, N.C., and Stevens, J.D.: 1974, 'β-D-Fructopyranose Calcium Chloride Dihydrate', Cryst. Struct. Comm. 3, 277-281.
20. Dahan, F.: 1975, 'The Crystal Structure of Calcium Dichromate Bis (hexamethylenetetramine) Heptahydrate', Acta Cryst. B31, 423-426.
21. Delucas, L., Bugg, C.E., Terzis, A., and Rivest, R.: 1975, 'Calcium Binding to D-Glucuronate Residues: Crystal Structure of a Hydrated Calcium Bromide Salt of D-Glucuronic Acid', Carbohydr. Res. 41, 19-29.
22. Deman, J., Mareel, M., and Bruyneel, E.: 1973, 'Effects of Calcium and Bound Sialic Acid on the Viscosity of Mucin', Biochim. Biophys. Acta 297, 486-490.
23. Dickens, B. and Brown, W.E.: 1972, 'The Crystal Structure of $CaKAsO_4 \cdot 8H_2O$', Acta Cryst. B28, 3056-3065.
24. Dunstone, J.R.: 1962, 'Ion Exchange Reactions Between Acid Mucopolysaccharides and Various Cations', Biochem. J. 85, 336-351.
25. Farber, S.J., Schubert, M., and Schuster, N.: 1957, 'The Binding of Cations by Chondroitin Sulfate', J. Clin. Invest. 36, 1715-1722.
26. Fries, D.C., Rao, S.T., and Sundaralingam, M.: 1971, 'Structural Chemistry of Carbohydrates. III. Crystal and Molecular Structure of 4-O-β-D-Galactopyranosyl-α-D-Glucopyranose Monohydrate (α-Lactose Monohydrate)' Acta Cryst. B27, 994-1005.
27. Furberg, S. and Hellend, S.: 1962, 'The Crystal Structure of the Calcium and Strontium Salts of Arabonic Acid', Acta Chem. Scand. 16, 2372-2383.
28. Glimcher, M.J., Hodge, A.J., and Schmitt, F.O.: 1957, 'Macromolecular Aggregation States in Relation to Mineralization: The Collagen-Hydroxyapatite System as Studied in vitro', Proc. Natl. Acad. Sci. U.S.A. 43, 860-867.
29. Gould, S.E.B., Gould, R.O., Rees, D.A. and Scott, W.E.: 1975, 'Interactions of Cations with Sugar Anions. Part I. Crystal Structures of Calcium Sodium Galacturonate Hexahydrate and Strontium Sodium Galacturonate Hexahydrate, J. Chem. Soc., Perkin Trans. II, 237-242.

30. Hays, R.M., Singer, B., and Malamed, S.: 1965, 'The Effect of Calcium Withdrawal on the Structure and Function of the Toad Bladder', J. Cell. Biol. 25, 195-208.
31. Hearn, R.A., and Bugg, C.E.: 1974, 'Calcium Binding to Carbohydrates: Crystal Structure of Calcium Ascorbate Dihydrate', Acta Cryst. B30, 2705-2711.
32. Howath, G., and Sovak, M.: 1973, 'Membrane Coarctation by Calcium as a Regulator for Bound Enzymes', Biochim. Biophys. Acta 298, 850-860.
33. Humphreys, T.: 1967, 'The Cell Surface and Special Cell Aggregation', in B.D. Davis and L. Warren (eds.), The Specificity of Cell Surfaces, Prentice-Hall, Inc., Englewood Cliffs, N.Y., pp. 195-210.
34. Hvoslef, J. and Kjellevold, K.E.: 1974, 'The Crystal Structure of Calcium Ascorbate Dihydrate', Acta Cryst. B30, 2711-2716.
35. Manery, J.F.: 1966, 'Effects of Ca Ions on Membranes', Fed. Proc., Fed. Am. Soc. Exp. Biol. 25, 1804-1810.
36. Mills, J.A.: 1961, 'Association of Polyhydroxy Compounds with Cations in Solution', Biochem. Biophys. Res. Commun. 6, 418-421.
37. Moore, C.L.: 1971, 'Specific Inhibition of Mitochondrial Ca^{2+} Transport by Ruthenium Red', Biochem. Biophys. Res. Commun. 42, 298-305.
38. Morris, E.R., Rees, D.A., and Thom, D.: 1973, 'Character of Polysaccharide Structure and Interactions via Circular Dichroism: Order-Disorder in a Calcium Alginate System', J. Chem. Soc., Chem. Commun., 245-246.
39. Norrestam, R., Werner, P.E., and Von Glehn, M. 1968, 'The Crystal Structure of Calcium-α-Glucoisosaccharate and Some Extended Hückel Calculations on α-D-Glucoisosaccharinic Acid', Acta Chem. Scand. 22, 1395-1403.
40. Pinnel, S.R., Fox, R., and Krane, S.M.: 1971, 'Human Collagens: Differences in Glycosylated Hydroxylsine in Skin and Bone', Biochim. Biophys. Acta 229, 119-122.
41. Pricer, W.C., Jr., and Ashwell, G.: 1971, 'The Binding of Desialylated Glycoproteins by Plasma Membranes of Rat Liver', J. Biol. Chem. 246, 4825-4833.
42. Rendleman, J.A., Jr.: 1966, 'Complexes of Alkali Metals and Alkaline Earth Metals with Carbohydrates', Adv. Carbohydr. Chem. 21, 209-271.
43. Richards, G.F., 1973, 'An X-Ray Crystallographic Study of a Metal-Carbohydrate Complex: α-D-Xylose $CaCl_2.3H_2O$, Carbohydr. Res. 26, 448-449.
44. Santanam, M.S.: 1959, 'Calcification of Collagen', J. Mol. Biol. 1, 65-68.
45. Schiffman, E., Martin, G.R., and Miller, E.J.: 1970, 'Matrices That Calcify', in H. Schraer (ed.), Biological Calcification: Cellular and Molecular Aspects, Appleton-Century-Crofts, New-York, N.Y., pp. 27-67.
46. Segrest, J.P. and Cunningham, J.W.: 1970, 'Variations in Human Urinary O-Hydroxylsyl Glycoside Levels and Their Relationship to Collagen Metabolism', J. Clin. Invest. 49, 1497-1509.
47. Smidsrod, O. and Haug, A.: 1972, 'Dependence Upon the Gel-Sol State of the Ion-Exchange Properties of Alginates', Acta Chem. Scand. 26, 2063-2074.

48. Smith, Q.T. and Lindenbaum, A.: 1971, 'Composition and Calcium Binding of Proteinpolysaccharides of Calf Nasal Septum and Scapula', Calc. Tissue Res. 7, 290-298.
49. Terzis, A.: 1975, 'Calcium Arabinose Complex', ACA Newsletter 6, 12.
50. Zeidman, I.: 1947, 'Chemical Factors in the Mutual Adhesiveness of Epithelial Cells', Cancer Res. 7, 386-389.

DISCUSSION

Haynes: Can you comment on differences between Ca^{2+} and Mg^{2+} in any of the complexation properties which you have described ?

Cook: As measured by nuclear magnetic resonance studies, calcium binding to carbohydrates in aqueous solution is considerably stronger than magnesium binding. The difference in the strength of complex formation is reflected also in certain biological systems. For example, when magnesium is substituted for calcium in cell cultures, cell-cell adhesion is markedly decreased.

Concerning crystal structures of carbohydrate complexes with magnesium and calcium, the principal difference is the coordination number. Whereas calcium coordination is generally eightfold or sevenfold, magnesium coordination is sixfold, with a corresponding decrease in the magnesium-oxygen distances. This is in agreement with the difference in ionic radii (Ca^{2+} = 0.99 Å; Mg^{2+} = 0.65 Å).

A. Pullman: I do not think that it is surprising that the binding constant of Ca^{++} to these molecules reveals small binding in water, because the intrinsic binding energy to sugar hydroxyl oxygen is very little different from the corresponding binding energy to water.

WHY DOES CALCIUM PLAY AN INFORMATIONAL ROLE UNIQUE IN BIOLOGICAL SYSTEMS?

R.H. KRETSINGER
Dept. of Biology, University of Virginia, Charlottesville, Va. 22901, U.S.A.

I recently had the opportunity to prepare a review on 'Calcium Binding Proteins' (1976) and with Don Nelson another on 'Calcium in Biological Systems' (1976). Most biochemists and physiologists appreciate the importance of calcium; only recently have I realized the uniqueness of its role. First I want to describe briefly the informational role played by calcium. Most of the documentation is in these review articles. Then I want to consider why calcium plays this role.

Twenty years ago Heilbrunn (1956) summarized and described a variety of seemingly unrelated phenomena - muscle contraction, ameboid movement, secretion, cell adhesion etc. - that require calcium. Douglas (1968) correctly suggested that calcium links stimulus to secretion. Ebashi (1963) identified troponin as the calcium binding protein that couples excitation to muscular contraction. Rasmussen et al. (1972) reviewed the evidence and clearly stated the hypothesis that calcium functions as a second messenger. One of the most universal characteristics of eukaryotic cells is that they maintain a free Ca^{2+} concentration in the cytoplasm of 10^{-7} to 10^{-8} M. These cells normally have extracellular environments of free Ca^{2+} 10^{-3} M. This 10^4 gradient is used as an information potential. An external stimulus - chemical, electrical or mechanical - touches the surface of the cell. A pulse of Ca^{2+} enters the cell with the ultimate result of a cellular response. The original stimulus never touches the responding molecules: calcium conveys the signal as a second messenger.

Why calcium?

It seems reasonable, with the power of hindsight, that cells should employ such second messengers. Cyclic adenosine monophosphate (cAMP) is also a very general second messenger, often functioning coordinately with Ca^{2+}. Note that in the more slowly responding cAMP systems the second messenger is synthesized following the stimulus. My first assumption is that for some unknown reason, possibly related to metabolic economy, an inorganic ion was chosen as one of the second messengers.

It seems reasonable that a cation be used. Biological molecules have many more electronegative ligands and hence have a greater range of cation affinity and specificity.

The ocean provided the original extracellular environment for the first prokaryotic cells (over 3.5×10^9 yr ago) as well as for the first

B. Pullman and N. Goldblum (eds.), Metal-Ligand Interactions in Organic Chemistry and Biochemistry, part 2, 257-263.

eukaryotic cells (about 2 x 10 yr ago). Its composition has changed little since that time, Table I. If we assume that the information

TABLE I

	Ionic	Radii[a]	Electro-[a]	Total	Concentration[b]
	Pauling	Goldschmidt	Negativity	Ocean	Water
Na^+	0.95 A	0.98 A	1.01	470	mM
K^+	1.33	1.33	0.91	10.0	
Mg^{2+}	0.65	0.65	1.23	53.6	
Ca^{2+}	0.99	0.94	1.04	10.2	

[a] Cotton and Wilkinson (1962)

[b] Sillén (1961)

system should function by allowing the cations to pass down a preformed gradient, then we find that only Na^+, K^+, Mg^{2+} and Ca^{2+} are present in adequate concentrations in the external reservoir. Also both iron and manganese are subject to oxidation and reduction and hence might appear less desirable messengers.

Let me outline several considerations, no single one of which really seems satisfactory, that might have favored the use of Ca^{2+} in preference to Na^+, K^+ or Mg^{2+}.

The calcium carbonates, calcite and aragonite, are abundant in the Echinodermata, Mollusca and other simple marine organisms such as corals. Hydroxyapatite, $Ca_{10}(PO_4)_6(OH)_2$, is a constituent of the endo and exoskeleton of Arthropoda and Chordata. These are good building materials. Perhaps cells developed the ability to extrude and to concentrate Ca^{2+} to form their skeletons. Secondarily this Ca^{2+} pumping ability was utilized in an informational system.

Very few enzymes require Na^+ or K^+. Of those that require Ca^{2+} only the nuclease of Staphylococcus is known to bind Ca^{2+} near the active site. For all the remainder, most of which are extracellular, Ca^{2+} plays a stabilizing, not catalytic role. In contrast many intracellular enzymes, particularly those binding phosphate groups, use Mg^{2+}, apparently at their active sites. Lohrmann and Orgel (1973) suggested that polymetaphosphates could provide energy for amino acid and for nucleoside polymerization in prebiotic conditions. In these systems Mg^{2+} plays a key catalytic function. Perhaps Ca^{2+} (and Na^+) pumps were evolved to protect Mg^{2+} enzyme systems.

A similar interpretation was recently suggested to me by Weber (1976). Much of the cell's metabolism is centered about phosphate compounds. Intracellular phosphate concentrations range from 0.005 to 0.02 M

Sillén (1961) summarized the following solubility products:

	Ca^{2+}	+	HPO_4^{2-}	⇄	$CaHPO_4$	log K = - 7.0
3 x	Ca^{2+}	+	2 x PO_4^{3-}	⇄	$Ca_3(PO_4)_2$	-26.0
5 x	Ca^{2+}	+	3 x PO_4^{3-} + OH^-	⇄	$Ca_5(PO_4)_3OH$	-55.9
5 x	Ca^{2+}	+	3 x HPO_4^{2-} + H_2O	⇄	$Ca_5(PO_4)_3OH + 4H^+$	- 5.0

If intracellular Ca^{2+} were at ocean (or plasma) levels, 10^{-2} M, then HPO_4^{2-} concentrations could not exceed $10^{-7.7}$ M ($[HPO_4^{2-}]^3 = 10^{-5} \times (10^{-7})^4 \times (10^{-2})^{-5}$). Or assuming an intracellular HPO_4^{2-} concentration 10^{-2} M, Ca^{2+} could not exceed $10^{-5.2}$ M. As in the previous argument one postulates that calcium is actively extruded from the cell since it would interfere with normal metabolism. However once the cell had expended the energy to create this Ca^{2+} gradient it 'decided' to use it to carry information.

In contrast mitochondria do accumulate Ca^{2+} from the cytoplasm and under normal physiological circumstances apatite precipitates are formed. There are some suggestions that mitochondria evolved from parasitic or symbiotic bacteria. It is not yet known whether prokaryotes actively secrete or accumulate calcium.

Margulis (1976) noted that microtubules form the spindles found in the mitotic apparatus common to all eukaryotes. Ca^{2+} at 10^{-4} M induces depolymerization of the microtubules to the dimeric tubulin. She too suggested calcium removal to allow stabilization of an intracellular function.

These suggestions of calcium extrusion imply the prior or concomitant evolution of calcium pumping mechanisms. These may be related to sodium pumps. Axonal membranes contain a sodium pump with ATP'ase (Na^+) activity. Calcium is then removed by coupling a Ca^{2+} (outward) with a Na^+ (inward) exchange (Blaustein & Russell, 1975). Most other eukaryotic cells contain an ATP'ase (Ca^{2+}) calcium pump. Apparently prokaryotes contain a sodium pump. The original calcium pump was possibly a sodium, calcium exchange mechanism; alternatively the present calcium pump(s) and sodium pump(s) evolved from a common precursor.

The effects of these cations on phospholipid membranes is another consideration. Exocytosis is one of the most important results of calcium's functioning as a second messenger. One interpretation suggests that Ca^{2+} neutralizes mutually repulsive changes on the surface of the secretory vesicle and on the surface of the plasma membrane thereby allowing their fusion with the subsequent loss of the contents of the vesicle (Muller and Finkelstein, 1974). Other membrane related phenomena such as cell fusion, adhesion, and growth (Rubin and Koide, 1976) are affected by calcium. Calcium does reduce fluidity and induce phase separations in synthetic membranes. (Ohnishi and Ito, 1974). However it is not clear how the quasi random distribution of oxygen ligands on the membrane surfaces might distinguish between Ca^{2+} and Mg^{2+}. The ability of chelators to distinguish between Ca^{2+} and Mg^{2+} seems to depend primarily on the difference in ionic radii, Table I.

Last year, I presented the hypothesis that 'Calcium modulated proteins contain EF hands' (Kretsinger, 1975). The EF hand is a protein conformation consisting of an α-helix (the E helix forefinger) a loop containing six oxygen ligands coordinating Ca^{2+} and another α-helix (the F helix thumb), Figure 1. This conformation was first seen in the

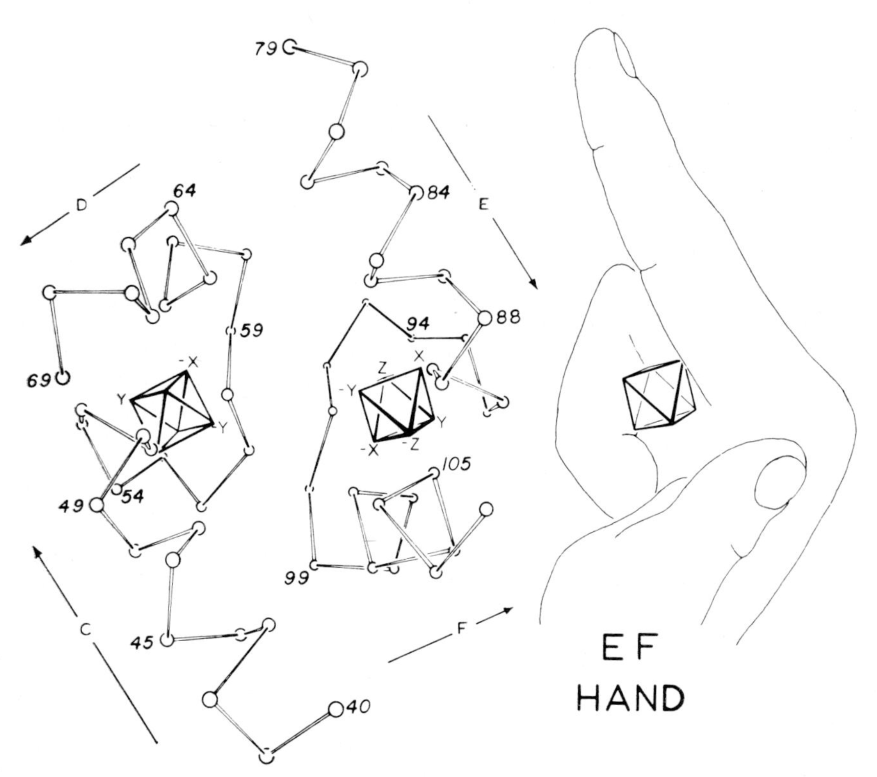

Fig. 1. The EF Hand. The EF hand as seen in the crystal structure of carp muscle calcium binding parvalbumin consists of an α-helix, residues 80 through 90, a calcium coordinating loop and a second α-helix, residues 99 through 108. The calcium ion is six coordinate with oxygen ligands supplied by amino acid residues 90, 92, 92, 96, 98 and 101.

crystal structure of muscle calcium binding parvalbumin. By considering the critical amino acid residues of the E and the F helices and particularly those of the calcium coordination octahedron, other investigators recognized homologous EF hands in the calcium binding component of troponin (Collins et al., 1973 and van Eerd and Takahaski, 1975), alkali extractable myosin light chains (Weeds and McLachlan, 1974 and Tufty and Kretsinger, 1975), -SH containing myosin light chains (Collins, 1976), vitamin D induced calcium binding protein (Huang et al., 1975) and phosphodiesterase activator protein (Watterson et al., 1976).

There is an important distinction between extracellular proteins that bind calcium, usually for stabilization, and those intracellular proteins that are modulated by calcium. None of these extracellular proteins contain EF hands. The calcium modulated proteins are the targets of calcium functioning as a second messenger. They release their calcium during rest (10^{-7} M free Ca^{2+}) and bind calcium following a stimulus (10^{-4} M free Ca^{2+}). All of the calcium modulated proteins appear to share a common evolutionary origin. Is it possible that the original EF hand protein preceeded the utilization of calcium as an informational ion? By chance did Nature find a protein conformation both specific for calcium and amenable to calcium modulation? No corresponding magnesium specific conformation has been found.

Crick (1968) suggested that the genetic code might have resulted from a 'frozen accident' L-amino acids may also have been chosen by chance. Analagously one might suggest that the informational role of calcium is a historical accident dependent upon the appearance of an EF hand protein.

This suggestion, that calcium evolved as a second messenger because a well suited target existed, begs the question of how the calcium gradient was established. This line of reasoning would lead one to suggest that an EF hand might be part of the membrane ATP'ase (Ca^{2+}) Then both the generator and the target of the Ca^{2+} gradient share a common evolutionary origin.

As an experimentalist I find none of these explanations - calcium carbonate or phosphate shell formation, selective magnesium-phosphate compound chemistry, specific membrane effects, common evolution of sodium and/or calcium pumps, or accidental origin of EF hand proteins - very convincing. Nonetheless I do find it rewarding to view biology in an evolutionary context. These evolutionary considerations should focus our research on several questions:

(1) To what extent are those proteins and processes involved in calcification and in transcellular calcium transport related to their informational counterparts?

(2) What are the chemical reasons for the selective involvement of magnesium in phosphate biochemistry?

(3) What physiological effects do Ca^{2+} and Mg^{2+} have on synthetic and natural phospholipid bilayers?

(4) What are the intracellular concentrations of free Ca^{2+}, Mg^{2+}, Na^{+} and K^{+} in prokaryotes as well as in a broad range of eukaryotes?

(5) Why do eukaryotes pump out sodium and is the sodium pump homologous to the calcium pump?

(6) How widely distributed are EF hand proteins and in particular is the membrane ATP'ase (Ca^{2+}) one of them?

REFERENCES

Blaustein, M.P. and Russell, J.M.: 1975, 'Sodium-Calcium Exchange and Calcium-Calcium Exchange in Internally Dialyzed Squid Giant Axons', J. Memb. Biol. 22, 285-312.

Collins, J.H.: 1976, Homology of Myosin DTNB Light Chain with Alkali Light Chains, Troponin C and Parvalbumin' Nature 259, 699-700.

Collins, J.H., Potter, J.D., Horn, M.J., Wilshire, G., and Jackman, N.: 1973, 'The Amino Acid Sequence of Rabbit Skeletal Muscle Troponin C:

Gene Replication and Homology with Calcium Binding Proteins from Carp and Hake Muscle' FEBS Lett. 36, 268-272.
Cotton, F.A. and Wilkinson, G.: 1962, Advanced Inorganic Chemistry, Interscience Publishers, New York, pp. 1136.
Crick, F.H.C.: 1968, 'The Origin of the Genetic Code' J. Mol. Biol. 38, 367-379.
Ebashi, S.: 1963, 'Third Component Participating in the Superprecipitation of Natural Actomyosin', Nature 200, 1010.
Heilbrunn, L.V.: 1956, The Dyamics of Living Protoplasm, Academic Press (New York), 327 pp.
Huang, W.-Y., Cohn, D.V., Hamilton, J.W., Fullmer, C., and Wasserman, R.H.: 1975, 'Calcium-Binding Protein of Bovine Intestine: The Complete Amino Acid Sequence', J. Biol. Chem. 250, 7647-7655.
Kretsinger, R.H.: 1975, 'Hypothesis: Calcium Modulated Proteins Contain EF Hands', in E. Carafoli, F. Clementi, W. Drabikowski, and A. Margreth (eds.), Calcium Transport in Contraction and Secretion, North-Holland Publishing Co., Amsterdam, pp. 469-378.
Kretsinger, R.H.: 1976, 'Evolution and Function of Calcium Binding Proteins', Int. Rev. Cytol. 45, 323-393.
Kretsinger, R.H. and Nelson, D.J.: 1976, 'Calcium in Biological Systems' Coord. Chem. Rev. 18, 29-124.
Lohrmann, R. and Orgel, L.EE.: 1973,'Prebiotic Activation Processes', Nature 244, 418-420.
Margulis, L.: 1976, 'Evolution of Mitosis and the Late Appearance of Metazoa, Metaphyta and Fungi', to be published. Proceedings of College Park Colloquia on Chemical Evolution, Academic Press, New York.
Onishi, S. and Ito, T.: 1974, 'Calcium-Induced Phase Separation in Phosphotidylserien-Phosphotidylcholine Membranes', Biochem. 13, 881-887.
Rasmussen, H., Goodman, D.B.P., and Tenenhouse, A.: 1972, 'The Role of cyclic AMP and Calcium in Cell Activation', C.R.C. Crit. Rev. Biochem. 1, 95-148.
Rubin, H. and Koide, T.: 1976, 'Mutual Potentiation by Magnesium and Calcium of Growth in Animal Cells', Proc. Natl. Acad. Sci. 73, 168-172.
Sillén, L.G.: 1961, 'The Physical Chemistry of Sea Water' in M. Sears (ed.), Oceanography, pp. 549-581.
Tufty, R.M. and Kretsinger, R.H.: 1975, 'Troponin and Parvalbumin Calcium Binding Regions Predicted in Myosin Light Chain and T-4 Lysozyme', Science 187, 167-169.
van Eerd, J.-P. and Takahashi, K.: 1975, 'The Amino Acid Sequence of Bovine Cardiac Troponin-C. Comparison with Rabbit Skeletal Troponin C', Biochem. Biophys. Res. Com. 64, 122-127.
Watterson, D.M., Harrelson, W.G., Keller, P.M., Sharief, F., and Vanaman, T.C.: 1976, 'Structural Similarities Between the Ca^{2+}-Dependent Regulatory Proteins of 3':5'-Cycle Nucleotide Phosphodiesterase and Actomyosin ATP'ase', J. Biol. Chem., submitted.
Weber, A.: 1976, Symp. Soc. Exp. Biol., in press.
Weeds, A.G. and McLachlan, A.D.: 1974, Structural Homology of Myosin Alkali Light Chains, Troponin C and Carp Calcium Binding Protein', Nature 252, 646-649.

DISCUSSION

Werber: (1) With regard to your idea of Ca^{++} must be expelled from cells because in 10^{-3}M phosphate there could not be more than 10^{-5} M Ca^{++}, don't you think that Ca^{++} could have been retained in the cells by binding to nucleotides and unspecific protein sites?

(2) Would it not be possible that Ca^{++} could act at two concentration levels: one of 10^{-6} M, where very specific sites would be modulated, and a second concentration of about 10^{-3} M which could be non-specific metal ion sites?

Eichhorn: You have presented an impressive list of properties of calcium that could account for its biological activities. I should like to add another property to this last: of all the physiologically important divalent metal ions calcium is the one that binds most weakly to ligands generally. Therefore,it is possible that calcium is useful in processes for which a weakly binding metal is required, but in which stronger binding might cause conformational changes in proteins, nucleic acids, etc., that could be deleterious.

Kretsinger: Answers to:

Werber 1: Certainly high concentrations of complexed Ca^{2+} are found within eukaryotic cells, for example inside the mitochondrion or bound to calsequestrin in the sarcoplasmic reticulum. I wish to emphasize that my discussion refers to free Ca^{2+} in the cytoplasm.

Werber 2: Under physiological conditions free Ca^{2+} levels as high as 10^{-3} M do not occur in the cytoplasm of eukaryotic cells, even following a stimulus.

Eichhorn: This may, in general, be true, but for the EF hand proteins like TN-C, Ca^{2+} is bound a thousand times more strongly than is Mg^{2+}.

INFLUENCE OF TRANSITION ELEMENT IONS ON THE HYDROGEN BONDS FORMED BY THEIR FIRST HYDRATION SHELL

GEORG ZUNDEL and ALFRED MURR
Physikalisch-Chemisches Institut der Universität München, Theresienstrasse 41, D 8000 München 2, West Germany

ABSTRACT.

The enhancement of the hydrogen bond donor property of the OH groups of water molecules due to cations is studied by IR spectroscopy using salts of polyelectrolytes. The band of the OH stretching vibration of the hydrogen bonded groups shifts with decreasing cation radius, i.e., increasing electrostatic field strength, toward smaller wave numbers. Thus the increase of the hydrogen bond donor property is strongly related to the ion-induced dipole interaction cation - water. Besides this electrostatic effect with transition element ions, an additional effect is observed: The covalent interaction between the lone pairs of the water molecules and the d orbits increases the hydrogen bond donor property, too, as indicated by additional band shifts. The enhancement of the hydrogen bond donor property shows the same dependence on the position of the ion in the periodic table as the stability of the complexes formed. It is shown that the number of water molecules attached under comparable hydration conditions is not only determined by the strength of the interaction but also by the coordination number. Finally, the nature of complexes formed by Ni^{2+}, Cu^{2+} and Zn^{2+} ions with water molecules is discussed.

We studied the influence of cations on the hydrogen bond donor property of the OH groups of water molecules by IR spectroscopy. We performed these investigations using 5μ thick membranes of polyelectrolytes, especially of salts of polystyrene sulfonic acid [1]. With these systems, the degree of hydration can be varied by the humidity of the air at the membranes. Of special importance is the fact that with these systems the number of water molecules per ion may be reduced to any required degree without disturbance of the amorphous character of the hydrate structure due to crystallization.

As proved in ref. [1] p. 66 ff. and illustrated in Figure 1, at low degrees of hydration the water molecules in these polyelectrolytes are attached to the cations with their lone electron pairs and bound via hydrogen bonds to O atoms of neighboring $-SO_3^-$ ions. The OH stretching vibration of these water molecules is observed as a broad band with maximum in the region 3470-3180 cm^{-1}. Several examples are shown in Figure 2. Sometimes - as, for instance, with the Y^{3+} and the La^{3+}

B. Pullman and N. Goldblum (eds.), Metal-Ligand Interactions in Organic Chemistry and Biochemistry, part 2, 265-271.

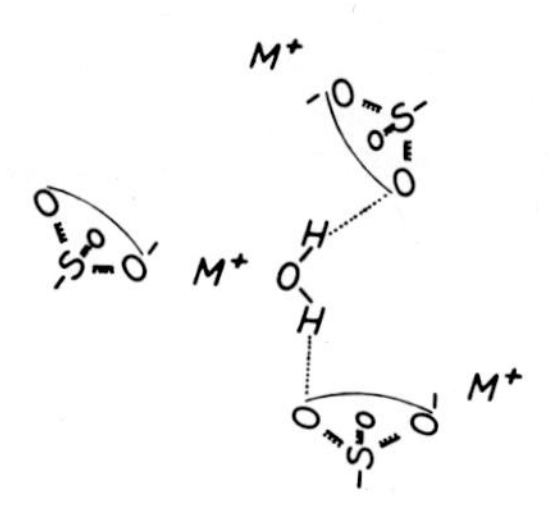

Fig. 1. The attachment of hydration water molecules at low degree of hydration.

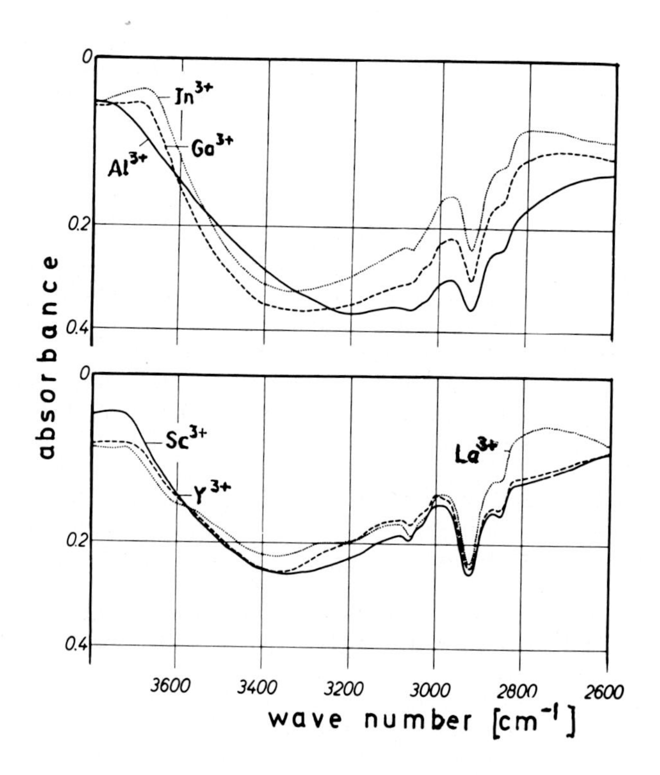

Fig. 2. IR spectra of membranes of In^{3+}, Ga^{3+}, Al^{3+} and Y^{3+}, Sc^{3+}, La^{3+} salts of polystyrene sulfonic acid.

sample in Figure 2 - the stretching vibration of non-hydrogen-bonded, i.e., free OH groups, is found as a shoulder at 3615 cm^{-1} (ref. [1] p. 83 ff.).

1. ELECTROSTATIC CATION FIELD, INDUCED DIPOLE INTERACTION AND HYDROGEN BOND DONOR PROPERTY OF HYDRATION WATER MOLECULES

In Table I the maximum of the broad band of the OH stretching vibration of the hydrogen bonded water molecules is given for various cations, together with their radii and their electrostatic fields at the water molecules of the first hydration shell. The latter values are relative

TABLE I
Position of the band maximum of the OH stretching vibration in hydrogen bonds of hydration water molecules in salts of polystyrene sulfonic acid at low degrees of hydration

Ion	Maximum of the band of the OH stretching vibration in cm^{-1}	Ionic radius Å	Electric field $[ESU \times cm^{-2}] \times 10^6$
Ba^{2+}	3440	1.34	1.51
Sr^{2+}	3412	1.12	1.85
Ca^{2+}	3406	0.99	2.11
Mg^{2+}	3394	0.66	3.07
Be^{2+}	3201	0.35	4.46
In^{3+}	3338	0.81	3.83
Ga^{3+}	3315	0.62	4.78
Al^{3+}	3195	0.51	5.53
La^{3+}	3373	1.14	2.73
Y^{3+}	3364	0.93	3.40
Sc^{3+}	3350	0.81	3.83

values, taking into account only the charge of the cations and the geometry (see ref. [1] p. 70 ff.). Let us consider the position of the maximum of the broad band of the hydrogen bonded groups dependent on the type of cations present. This band shifts strongly toward smaller wave numbers with decreasing radius of the cations, i.e., with increasing relative electrical field strength at the OH groups of the water molecules. This is true with bivalent cations: The band shifts from Ba^{2+} to Be^{2+}; with trivalent cations, the band shifts from In^{3+} to Al^{3+} as well as from La^{3+} to Sc^{3+}.

The reason of the shift of the OH stretching vibration caused by these cations is as follows: The OH bonds of the water molecules are stretched and loosened due to the electrostatic fields. These cation fields, on the one hand, repell the H nuclei of the water molecules and, on the other hand, remove the electrons slightly from the OH bonds, i.e., they polarize these bonds more or less strongly [2]. Hence the hydrogen bond donor property increases with increasing cation field. The hydrogen bonds become stronger. The well of the potential energy surface becomes broadened and hence the energy levels become narrower. This is observed as a shift of the OH stretching vibration toward smaller wave numbers.

Thus the observed shift is strongly connected with the ion-induced dipole interaction between the cations and induced dipoles in the water molecules induced due to the cation fields. Hence this cation-dependent band shift directly indicates this type of interaction.

2. INTERACTION OF TRANSITION ELEMENT IONS WITH WATER MOLECULES

Figure 3 shows spectra of Cu^{2+} and Zn^{2+} salts of polystyrene sulfonic acid and, for comparison, the spectrum of a Mg^{2+} sample. With the Cu^{2+}

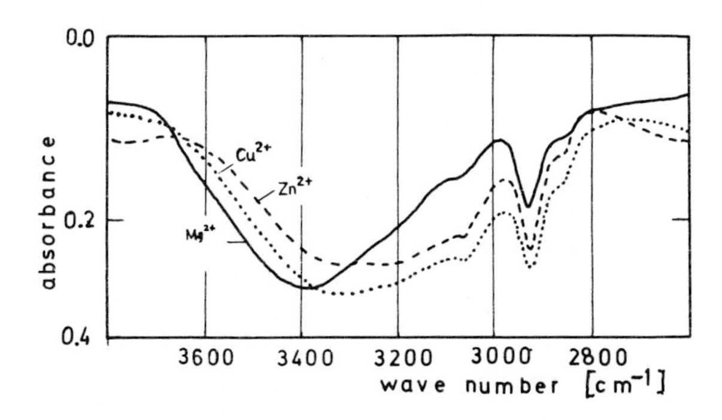

Fig. 3. IR spectra of membranes of the Cu^{2+} and the Zn^{2+} and, for comparison, of the Mg^{2+} salt of polystyrene sulfonic acid.

and Zn^{2+} samples, the shift of the band toward smaller wave numbers is much stronger than in the case of the Mg^{2+} sample. Nevertheless, the radii of these ions and hence also the electrostatic fields at the water molecules are similar to that of the Mg^{2+} sample. This is shown by a number of examples in Table II. In the case of all these cations - the only exception is Mn^{2+} - the band of the OH stretching vibration of the hydration water molecules is more strongly shifted toward smaller wave numbers than is the case with Mg^{2+} ions, although the electrostatic field is larger in the case of Mg^{2+} than in the case of any investigated transition element ions.

The question arises how this additional band shift with the transition element ions may be explained. With regard to the interaction of transition element ions with ligands, the following is known: The electrons of these cations are rearranged by the dipole fields of the water molecules. The bonding orbits of these cations, which are occupied by the d electrons, interact with the lone electron pairs of the water molecules, whereby a covalent bonding component appears. Thus

TABLE II
Position of the band maximum of the OH stretching vibration in hydrogen bonds of hydration water molecules with transition element ions in salts of polystryrene sulfonic acid. The band position with the Mg^{2+} sample for comparison.

Ion	Maximum of the band of the OH stretching vibration in cm^{-1}	Ionic radius Å	Electric field [ESU x cm^{-2}] x 10^6
Mg^{2+}	3394	0.66	3.07
Mn^{2+}	3406	0.80	2.59
Co^{2+}	3381	0.72	2.87
Ni^{2+}	3373	0.69	2.94
Cu^{2+}	3314	0.72	2.87
Zn^{2+}	3317	0.74	2.80

the observed additional band shifts demonstrate the following: The interaction of the lone pairs of the water molecules with the d holes of the transition element ions leads to an additional polarization of the OH groups of the water molecules and hence to an additional enhancement of the hydrogen bond donor property of these groups.

At first it seems that this explanation cannot be applied to Zn^{2+} ions, since non-hydrated Zn^{2+} ions have 10 d electrons, i.e., completely occupied orbits. Nevertheless, the observed band shift is very large. The energy required to lift one of the d electrons into the next-higher shell is, however, only 9.7 eV [3]. Hence, due to the interaction with the water molecules, hybrids between 3 d orbits and orbits of the fourth shell may be formed, interacting with the lone pairs of the water molecules.

2.1. *Nature of Transition Element Ion-Hydration Water Complexes and Strength of the Interaction*

Let us now consider the strength of this interaction effect. The maximum of the band shows the following shift as a function of the position of the cation in the periodic table:

Ion	Mn^{2+} <	Co^{2+} <	Ni^{2+} <	Cu^{2+} >	Zn^{2+}	
Band position	3406	3381	3373	3314	3317	cm^{-1}

The shift increases from Mn^{2+} to Cu^{2+} and decreases slightly from Cu^{2+} to Zn^{2+}. Irving and Williams [4] studied the stability of complexes formed by these ions with the result that the stability of complexes formed by these ions increases from Mn^{2+} to Cu^{2+} and decreases slightly from Cu^{2+} to Zn^{2+}. Hence the effect of transition element ions on the hydrogen bond donor properties of hydration water molecules shows the same dependence on the nature of the cation as the stability of complexes formed by these cations.

Figure 4 shows spectra of Ni^{2+}, Cu^{2+} and Zn^{2+} samples at very low and at 7% atmospheric humidity. Next I wish to discuss the nature of

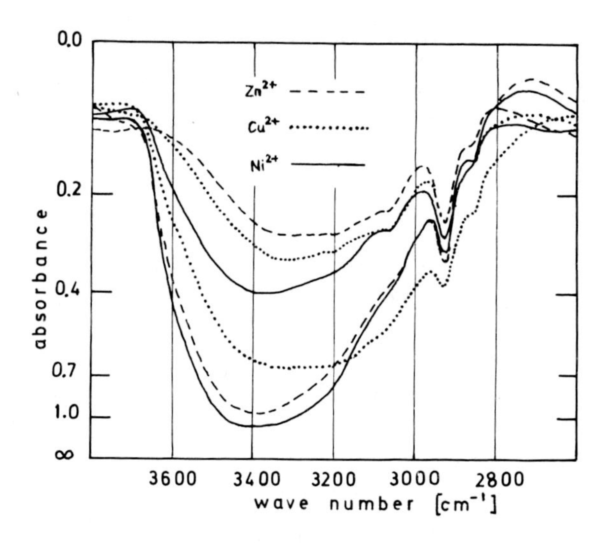

Fig. 4. IR spectra of membranes of Ni^{2+}, Cu^{2+} and Zn^{2+} salts of polystyrene sulfonic acid. Upper set of curves: membranes thoroughly dried; lower set: membranes at 7% relative atmospheric humidity.

the complexes formed by these ions with hydration water molecules. With the same atmospheric humidity at the samples, i.e., under comparable hydration conditions, the intensity of the band is stronger with Ni^{2+} than with Cu^{2+} and Zn^{2+} ions. This is true with lower as well as with higher humidity. Hence, under comparable conditions, more water is attached to Ni^{2+} than to Cu^{2+} and Zn^{2+} ions. Nevertheless, the interaction between Cu^{2+} and Zn^{2+} ions and water is stronger as indicated by the shift of the band toward smaller wave numbers. This can easily be understood, as it is well known that Ni^{2+} forms octahedral complexes whereas Cu^{2+} as well as Zn^{2+} form at low degrees of hydration only complexes with four ligands. Thus, under comparable atmospheric humidity at the sample, the number of water molecules attached per cation is determined not only by the strength of the cation-water bond but, to an equal degree, also by the coordination number of the respective cation. When the water molecules present are not sufficient to occupy all coordination positions, O atoms of $-SO_3^-$ anions are bound to these positions, as indicated by the bands of the $-SO_3^-$ groups (ref. [1] p. 80

The comparison of the Cu^{2+} and Zn^{2+} samples in Figure 4 shows the following: At low degrees of hydration, the OH stretching vibration band is more intense with Cu^{2+} than with Zn^{2+} ions. The opposite is tr at larger degrees of hydration. Furthermore, in the case of Zn^{2+} ions the OH stretching vibration shifts strongly toward higher wave numbers with increasing degree of hydration, whereas with the Cu^{2+} ions hardly any shift is observed. Hence, in the case of Zn^{2+} ions the strength of the cation water bond decreases strongly with increasing degree of hydration.

The different behaviour of Cu^{2+} and Zn^{2+} ions may easily be understood when considering the nature of the complexes formed. Cu^{2+} ions form planar complexes as a result of the Jahn-Teller Effect. Hence these ions may attach four water molecules, quadratically arranged in a plane, with relatively strong bonds. In the case of the Zn^{2+} ions, the result can also be understood when considering the complex formation properties of these ions. Orgel writes ([5] p. 83) on the hydration complex of Zn^{2+}: The Zn^{2+} ion, more than any other of the group, is stable in tetrahedral environment. It seems, however, that in aqueous solutions $[Zn(H_2O)_6]^{2+}$ is the predominant species (perhaps due to the enormous excess of water). From this statement we can draw the conclusion that, at low degree of hydration, tetrahedral complexes are formed with strongly bound water molecules. With increasing degree of hydration, these complexes change into octahedral complexes. More water molecules are bound with the Zn^{2+} complexes, since at larger humidities the OH band is more intense with Zn^{2+} than with Cu^{2+}. The binding of the water molecules is, however, less strong, since the OH band is not as strongly shifted toward smaller wave numbers.

Our thanks are due to the Deutsche Forschungsgemeinschaft and the Fonds der Chemischen Industrie for providing the facilities for this work.

REFERENCES

1. Zundel, G.: 1969, 'Hydration and Intermolecular Interaction - Infrared Investigations with Polyelectrolyte Membranes', Academic Press, New York; 1972, Mir, Moscow.
2. Zundel, G. and Weidemann, E.G.: 1973, Z. Physik. Chem. 83, 327. Zundel, G.: 1976, in Schuster, P., Zundel, G. and Sandorfy, C. (eds.), 'The Hydrogen Bond', North-Holland Publ. Co., Amsterdam, ch. 15, p. 748 ff.
3. Hartmann, H.: 1954, Theorie der Chemischen Bindung, Springer, Berlin, p. 167.
4. Irving, H. and Williams, R.J.R.: 1948, Nature 162, 746.
5. Orgel, L.E.: 1963, An Introduction to Transition-Metal Chemistry, Methuen, London, and Wiley, New York.

THE NUCLEOPHILICITY OF METAL BOUND HYDROXIDE, MECHANISMS OF DISPLACEMENT ON ESTERS, AND TRANSESTERIFICATION INVOLVING THE METAL-ACYL ANHYDRIDE BOND *#

MICHAEL A. WELLS** and THOMAS C. BRUICE##

The role of the metal ion in metallohydrolases continues to receive much attention. Most of the information from model studies on the possible role of the metal come from studies on stable complexes of Co(III). In these studies it has been shown that both carbonyl activation by the metal ion to intermolecular attack of hydroxide and intramolecular attack of a metal bound hydroxide can occur. In the case of amides (Buckingham et al., 1970) the latter pathway proceeds about 10^3 faster than the former. The advantage of the Co(III) system lies in the very slow rate of ligand exchange, which means that the nature of the complex under study can be readily determined. Scheme I represents a typical Co(III) system where the relative importance of the inter- and intramolecular processes might be evaluated (Buckingham et al., 1969). However, the rate determining step is often found to be partitioning of the five coordinate intermediate (A) between B. and C. Since the breakdown of B is rapid, most studies are limited to the breakdown of C. Thus the very feature which makes the stable complexes so attractive also means that certain aspects of metal ion reactions are not amenable to study.

In the case of the labile complexes of, for example, Co(II), one might expect that competition between the inter- and intramolecular reactions might be more easily studied, since the equilibrium between the intermediates corresponding to B and C should be rapid. Of course the problem lies in the labile nature of the complexes, which means that thermodynamic as well as kinetic data must be collected. Even then, the nature of the reactive intermediate can only be determined under special circumstances. In this paper we report a successful study of a Co(II) system, which provides some new insights into the scope of metal ion reactivity in ester hydrolysis.

---

* This research was supported by a grant from the National Institute of Health
# Contribution from the Department of Chemistry, University of California at Santa Barbara, California 93106.
** Macy Faculty Scholar 1975-76: Permanent address: Department of Biochemistry, College of Medicine, University of Arizona, Tuscon, Arizona.
To whom correspondence should be directed.

B. Pullman and N. Goldblum (eds.), Metal-Ligand Interactions in Organic Chemistry and Biochemistry, part 2 , 273-284.

Scheme I

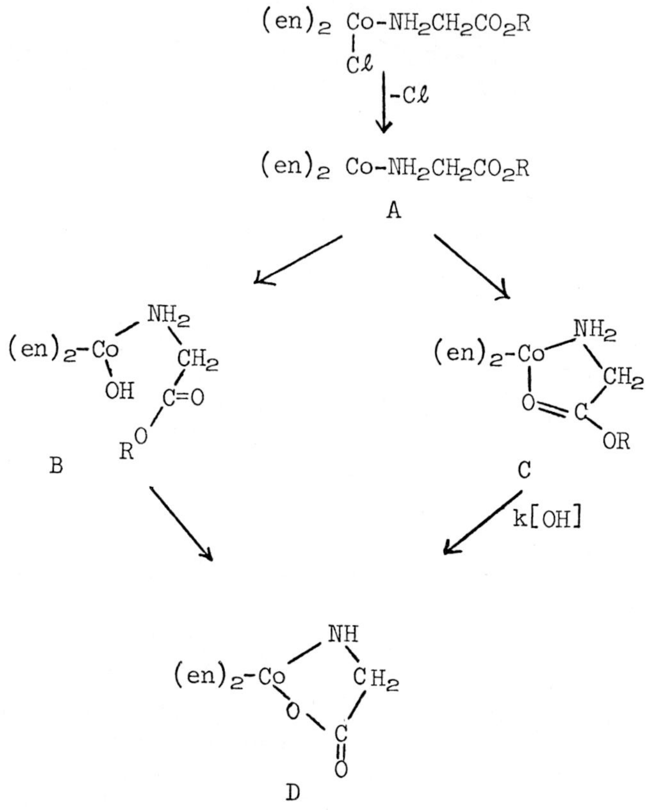

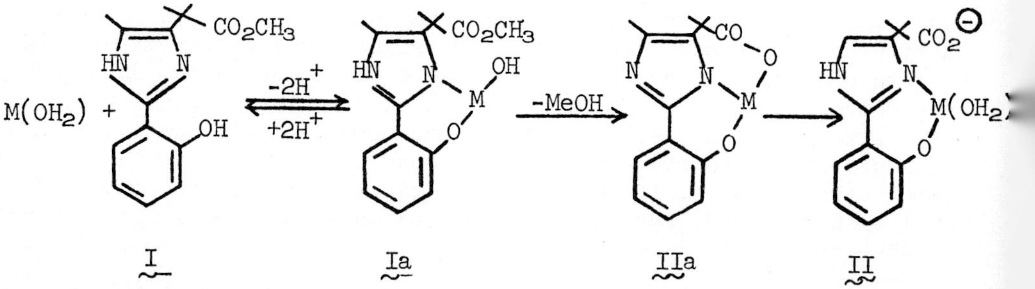

$(en)_2\,Co{-}NH_2CH_2CO_2R$ (with Cl)

$\downarrow -Cl$

$(en)_2\,Co{-}NH_2CH_2CO_2R$

A

B: $(en)_2{-}Co(OH)$–NH_2–CH_2–$C{=}O$–OR

C: $(en)_2{-}Co$–NH_2–CH_2–$C(OR){=}O$–Co

k[OH]

D: $(en)_2{-}Co$–NH–CH_2–C(=O)–O–Co

$M(OH_2)$ + I $\underset{+2H^+}{\overset{-2H^+}{\rightleftharpoons}}$ Ia $\xrightarrow{-MeOH}$ IIa $\rightarrow$ II

I: CO_2CH_3, HN, N, OH

Ia: CO_2CH_3, HN, N, OH, M, O

IIa: CO, O, N, N, M, O

II: $CO_2^{\ominus}$, HN, N, $M(OH_2)$, O

$pK_1\ (ImH^+)\ 6.1$

$pK_2\ (-OH)\ 9.6$

$pK_a(CO_2H)\ \ 3.2$

$pK_b(ImH^+)\ \ 6.9$

$pK_c(-OH)\ \ 9.8$

(Equation (1))

Hydrolysis of I (Rodgers and Bruice, 1974) (Equation (1)) has been investigated as a function of pH and metal ion concentration. Our experimental results find quantitative expression in the reaction of Scheme II. The values of the determined constants at 30° in H_2O at μ = 0.2 ($NaClO_4$) are: $pK_x(Co^{+2}) = 9.4$, $pK_x(Ni^{+2}) = 9.4$; $\log K_m(Co^{+2}) = 5.42$, $\log K_m(Ni^{+2}) = 4.85$; $pK_3(Co^{+2}) = 9.6$; $pK_3(Ni^{+2}) = 9.6$; $K_{eq}(Co^{+2}) = 14.0$, $K_{eq}(Ni^{+2}) = 2.0$; $k_r(Co^{+2}) = 1.05\ min^{-1}$, $k_r(Ni^{+2}) = 0.60\ min^{-1}$.

Scheme II

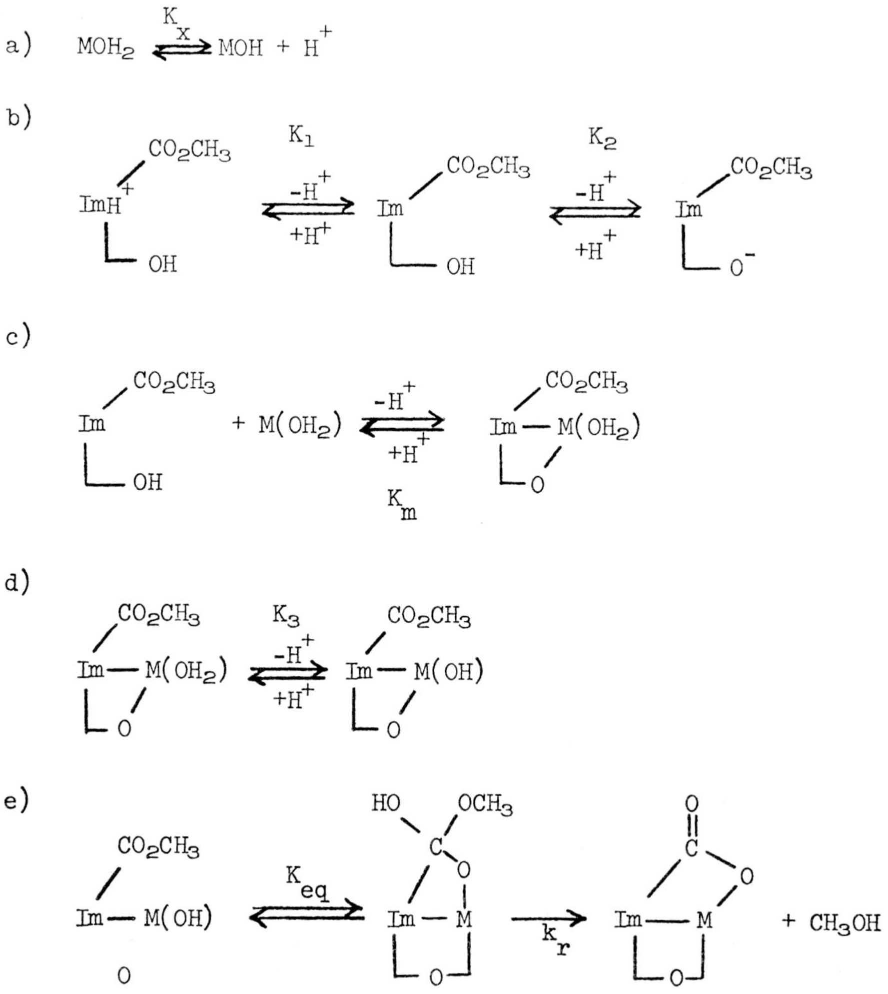

a) $$MOH_2 \underset{}{\overset{K_x}{\rightleftharpoons}} MOH + H^+$$

b) $$ImH^+(CO_2CH_3)(OH) \underset{+H^+}{\overset{K_1,\ -H^+}{\rightleftharpoons}} Im(CO_2CH_3)(OH) \underset{+H^+}{\overset{K_2,\ -H^+}{\rightleftharpoons}} Im(CO_2CH_3)(O^-)$$

c) $$Im(CO_2CH_3)(OH) + M(OH_2) \underset{+H^+}{\overset{-H^+}{\rightleftharpoons}} Im(CO_2CH_3)\text{—}M(OH_2)\text{—}O \qquad K_m$$

d) $$Im(CO_2CH_3)\text{—}M(OH_2)\text{—}O \underset{+H^+}{\overset{K_3,\ -H^+}{\rightleftharpoons}} Im(CO_2CH_3)\text{—}M(OH)\text{—}O$$

e) $$Im(CO_2CH_3)\text{—}M(OH)\ O \overset{K_{eq}}{\rightleftharpoons} Im\text{—}M\text{—}O\ [C(HO)(OCH_3)\text{—}O\text{—}M] \xrightarrow{k_r} Im\text{—}M\text{—}O\ [C(=O)\text{—}O\text{—}M] + CH_3OH$$

The thermodynamic constants of Equations (a) to (d) (Scheme I) can be determined titrimetrically* since proton transport and metal ion complexation are rapid compared to the hydrolytic steps of reaction (e). The complexation of I with metal ions is evidenced by a lowering of the pK of the phenolic hydroxyl which may be quantitatively monitored at 292 nm (an isosbestic point for ionization of the imidazolyl group). The pH dependence of the metal-substrate complex dissociation (K_{app}) is shown in Figure 1. The theoretical curve in Figure 1 was calculated

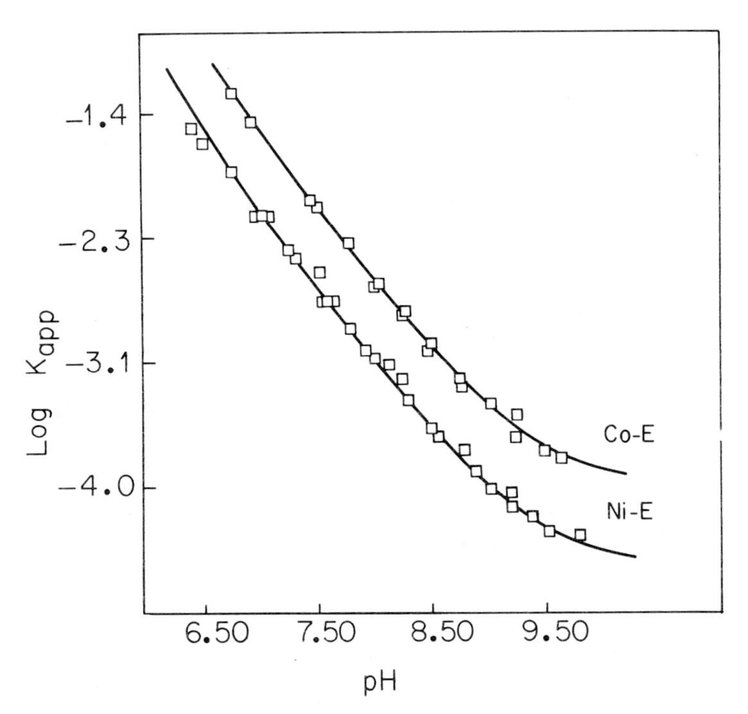

Fig. 1. pH-Dependence of the apparent dissociation constant (K_{app}) of metal complexes of ester I. The solid curves represent the best ift of Equation (2) to the experimental data

according to Equation (2). The fit was obtained

$$\log K_{app} = \log K_m - pH + \log(1+ \frac{a_H}{K_1} + \frac{K_2}{a_H}) + \log(1+ \frac{K_X}{a_H}) - \log(1+ \frac{K_3}{a_H}) \quad (2)$$

---

* All studies were carried out in the Cary-pH-stat (Bruice and Maley, 1970) under N_2. A modified burrett assembly prevented precipitation of $M(OH)_2$ and the calomel electrode was separated from the reaction by a liquid junction through which 0.2 M $NaClO_4$ was passed. This apparatus prevents leakage of Cl^- into the reaction.

by iteration in K_m and K_3, which are the only constants which cannot be determined independently. It is important to note that the value of K_3, the pK of the metal-OH_2 moiety of the complex controls the fit at high pH.

Conversion of I → IIa was followed spectrophotometrically (258.5 nm) at various pH's and metal ion concentrations ($[M_T]$); first-order kinetics were observed for at least 2-3 half-lives. The rate constant (k_{app}) increased with increasing metal concentration until saturation of substrate by metal ion. Plots of $1/k_{app}$ vs $1/[M_T]$ at each pH investigated were linear and gave $1/k_{obs}$ as the intercept at $1/[M_T] = 0$, while the intercept on the $-1/[M_T]$ axis provided the kinetically determined metal substrate dissociation constant (K_{app}). Plots of log k_{obs} vs pH in the presence of Co^{+2} or Ni^{+2} and for the OH^- catalyzed reaction in the absence of metal ($k_{OH} = 1.63$ min^{-1} M^{-1}) are shown in Figure 2. The metal

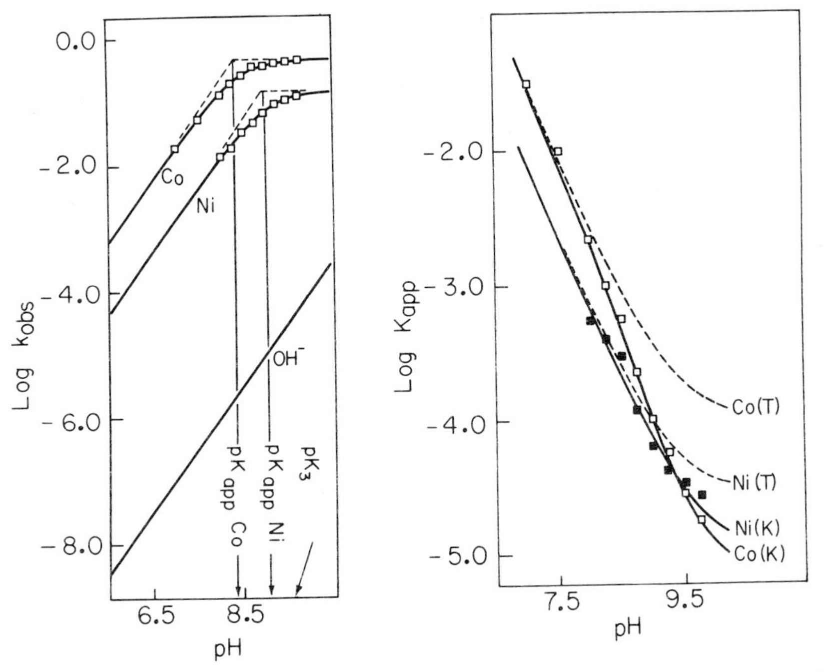

Fig. 2. Left Panel: pH-rate profile for the hydrolysis of ester I in the presence of Co^{+2}, Ni^{+2}, and the OH^- catalyzed reaction. The arrows on the pH ordinate show the kinetically determined pK for $Co(H_2O)$ and $Ni(H_2O)$, as well as the thermodynamic pK_3. Right Panel: pH-dependence of the kinetically determined dissociation constant of metal complexes of ester I. (□) Co^{2+}, (■) Ni^{2+}. Solid lines labelled Co(K) and Ni(K) represent the fit of Equation (5) to the data. Dotted lines labelled Co(T) and Ni(T) represent the best fit of the data in Figure 1 to Equation (2).

ion data were fit to eq (3) using $k = 0.98\ min^{-1}$, $pK'_3 = 8.4$ for Co^{+2}, and $k = 0.4\ min^{-1}$, $pK'_3 = 9.1$ for Ni^{+2}. The fact that pK'_3 is less than

$$k_{obs} = k \frac{K'_3}{K'_3 + a_H} \qquad (3)$$

pK_3 allows distinction to be made between attack of the substrate bound metal-hydroxide (Scheme II) and the apparently identical kinetic mechanism which involves attack of external OH^- on the carbonyl group activated by the bound metal.

Thus, mechanisms A and B (where T = tetrahedral intermediate) both

$$\begin{array}{ccc} \text{Ester-MOH}_2 & \underset{+H^+}{\overset{K_3,\ -H^+}{\rightleftharpoons}} & \text{Ester-MOH} \\ K_{eq} \updownarrow & & \\ \text{Ester-M} & \underset{k_{-2}}{\overset{k_2[OH]}{\rightleftharpoons}} \text{T} \xrightarrow{k_3} \text{Products} & \end{array} \qquad A$$

$$\text{Ester-MOH}_2 \underset{+H^+}{\overset{K_3,\ -H^+}{\rightleftharpoons}} \text{Ester-MOH} \overset{K_{eq}}{\rightleftharpoons} \text{T} \xrightarrow{k_3} \text{Products} \qquad B$$

provide the kinetic expression of Equation (3). For A, $K'_3 = K_3/(1 + K_{eq})$, whereas for B, $K'_3 = K_3(1 + K_{eq})$. Therefore, only mechanism B can account for the observation that pK'_3 is less than pK_3. In addition only mechanism B can account for the observation that the kinetically determined metal substrate dissociation constant is less than the thermodynamically determined constant (see Equation (5)). Assumption of steady state in T for both A and B show that $K'_3 = K_3$. For a fuller discussion of the influence of preequilibria on kinetically determined pK_a' see Bruice and Schmir (1959).

We propose the formation of a stable tetrahedral intermediate along the reaction path (Scheme II) in order to account for the fact that pK'_3 determined kinetically is significantly different from the thermodynamic contant pK_3. The non-rate determining formation of an additional intermediate after ionization of the $M(OH_2)$ moiety would result in the apparent ionization constant (pK'_3) of the $M\text{-}OH_2$ moiety of the complex being less than pK_3 (Bruice and Schmir, 1959). On the basis of Scheme II k_{obs} is given by Equation (4) it can be seen that K'_3 of the complex $M(OH_2)$ moiety is provided by $K_3(1 + K_{eq})$ and that

$$k_{obs} = \frac{k_r K_{eq}}{(1 + K_{eq})}\left[\frac{K_3(1 + K_{eq})}{a_H + K_3(1 + K_{eq})}\right] \qquad (4)$$

$k = k_rK_{eq}/(1 + K_{eq})$. The data in Figure 2 can be fitted to eq (4) using the thermodynamic pK_3 and $k_r = 1.05\ min^{-1}$, $K_{eq} = 14.5$ for Co^{+2}, and $k_r = 0.6\ min^{-1}$, $K_{eq} = 2.5$ for Ni^{+2}.

According to Scheme II the value of K_{app} determined kinetically is given by Equation (5) which may be compared to Equation (2) which provides the value of K_{app} determined from the titrimetric data. The fit of the experimental data to Equation (5) is shown in Figure 2; K_{eq} =

$$\log K_{app} = \log K_m - pH + \log(1+\frac{a_H}{K_1} + \frac{K_2}{a_H}) + \log(1+\frac{K_x}{a_H}) - \log\left[1+\frac{K_3(1 + K_{eq})}{a_H}\right] \quad (5)$$

13.5 for Co^{+2} and 1.5 for Ni^{+2}. The agreement between K_{eq} determined from Equations (4) and (5) is reasonable. The combination of the pH-rate profile data and the difference between the thermodynamic and kinetic metal-substrate complex dissociation constants strongly support the suggestion that intramolecular attack of metal bound hydroxide and not metal ion activation of the carbonyl group to intermolecular attack by hydroxide is the mechanism operative in this model system.

The presence of a kinetically important tetrahedral intermediate formed in the metal-OH catalyzed hydrolysis of ester I was shown in a study of the effect of the pK_a of the leaving alcohol on the rate of hydrolysis. In Figure 3 are compared the rate of OH^- catalyzed hydrolysis in the absence of metal ion to the reaction catalyzed by the Co(II) complex at pH 7. For the OH^- catalyzed reaction (lower panel) a linear Brønsted relationship was found with $\beta = 0.5$. For the metal catalyzed reaction (upper panel) a break in the Brønsted plot occurs. For alcohols with a high pK and poor leaving tendency (methanol and chloroethanol) there was little dependence of rate on pK, whereas, for alcohols with lower pK's and good leaving tendency (propargyl and trifluoroethyl alcohols) the slope was 1.1. Such a discontinuous Brønsted plot has been noted in studies on neighboring carboxyl attack in mono phthalate esters of the same alcohols (Thanassi and Bruice, 1966). The explanation offered in this case as well as in the similar case of maleate half ester hydrolysis (Aldersley et al., 1974) is shown in Scheme III. For poor leaving groups a proton must be transferred to the alkyl oxygen (path B) and the alcohol leaves as ROH. Path B should be relative insensitive to the pK of ROH. For good leaving groups direct expulsion of RO^- occurs (path A), and the rate of expulsion of RO^- depends directly upon the pK of ROH.

For the case under study here we suggest that the tetrahedral intermediate formed by attack of the metal-OH can breakdown as shown in Scheme IV. For the better leaving groups path A, which depends on the pK of the alcohol predominates, whereas, for poor leaving groups path B, which does not depend on the alcohol pK predominates.

Thus we arrive at the unexpected and intriguing suggesting that a metal bound OH behaves as though it were a neighboring carboxyl group. The higher pK of $M\text{-}OH_2$ (9.6) compared to COOH (4) would of course suggest that M-OH would be a much better nucleophile. In the intermolecular hydrolysis of propinoic anhydride by stable complexes of the form $L_5M\text{-}OH$ it was found that those complexes with higher pK's were the better nucleophiles (Buckingham and Engelhardt, 1975).

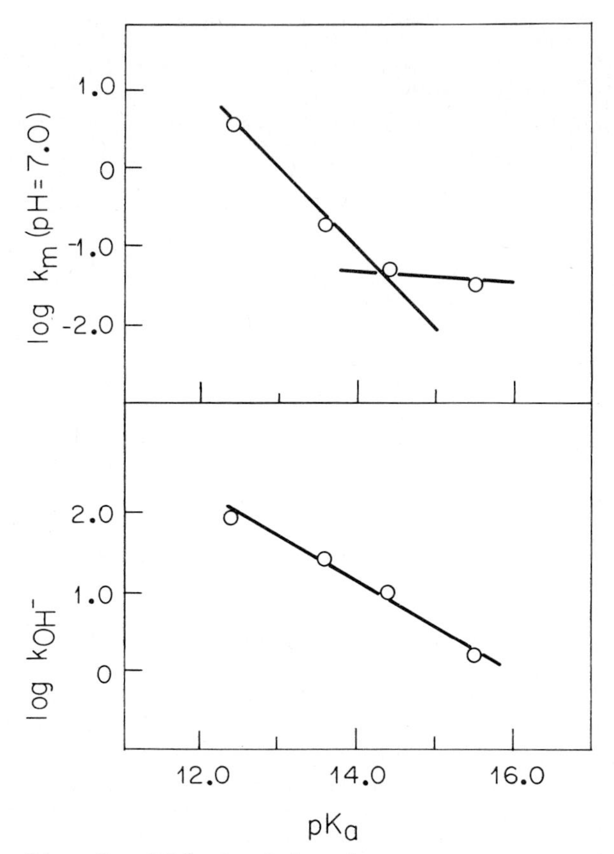

Fig. 3. Effect of leaving group pK_a on the rate of hydrolysis of various esters of I. Lower panel: hydroxide catalyzed reaction in the absence of metal ion. Uppel panel: hydrolysis of the Co(II) complex at pH = 7.0. Esters were methyl (pK_a = 15.49); 2-chloroethyl (pK_a = 14.31); propargyl (pK_a = 13.55); trifluoroethyl (pK_a = 12.36).

Another unusual and highly unique observation in this system relates to evidence for the existence of a highly reactive intermediate formed after expulsion of the alcohol group and before the formation of the ultimate metal-carboxylic acid product. We have termed this intermediate an acylmetal anhydride by analogy to the anhydride intermediate formed during neighboring carboxyl hydrolysis of half ester of maleate and phthalate. The existence of this intermediate is inferred form the following experiments.

When the hydrolysis of the trifluorethyl ester of I is carried out in the presence of trifluoroethanol there is a marked decrease in k_{obs}, whereas, in the presence of equal molar amounts of methanol k_{obs} is little altered. This observation can be accounted for by Scheme V, which is based on a similar reaction observed with the half esters of maleate. This scheme predicts that a plot of $1/k_{obs}$ vs [ROH] should be linear

Scheme III

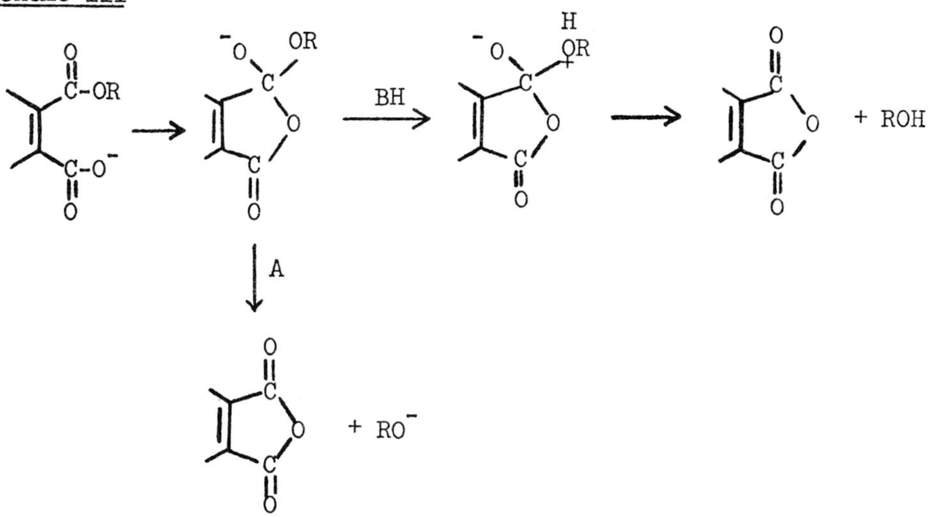

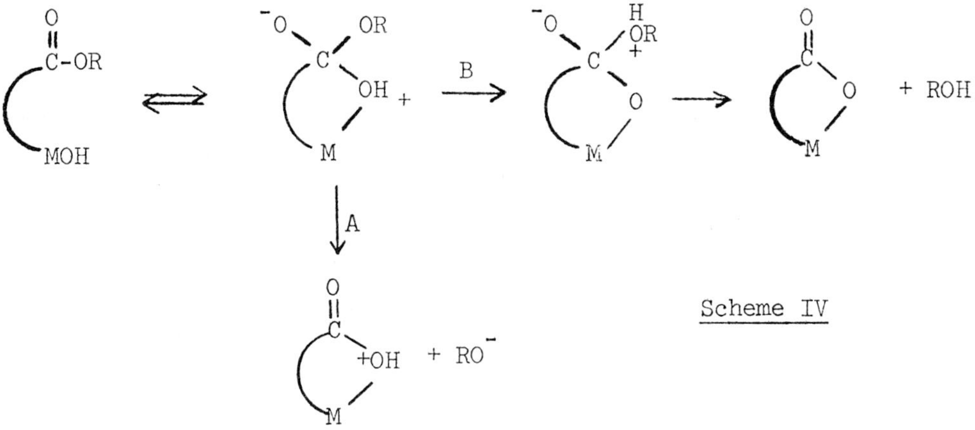

C-OR, C-O$^-$ → $^-$O, OR, C, O, C → (BH) $^-$O, H, OR$^+$, C, O, C → C, O, C + ROH

↓ A

C, O, C + RO^-

C-OR, MOH ⇄ $^-$O, OR, C, OH$^+$, M → (B) $^-$O, H, OR$^+$, C, O, M → C, O, M + ROH

↓ A

C, +OH, M + RO^-

Scheme IV

Scheme V

C-OR, MOH $\overset{K_{eq}}{\rightleftharpoons}$ $^-$O, OR, C, +OH, M $\underset{k_3[^-OR]}{\overset{k_1}{\rightleftharpoons}}$ C, +OH, M $\xrightarrow{k_2[OH^-]}$ C, O$^-$, MOH_2

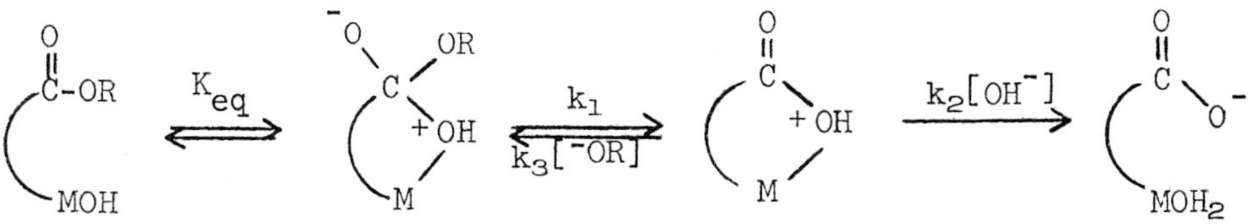

$$1/k_{obs} = 1/k_1 + \frac{k_3}{k_1 k_2}[OR^-]$$

with an intercept equal to the rate of hydrolysis of the trifluoroethyl ester in water. The results of such an experiment are shown in Figure 4

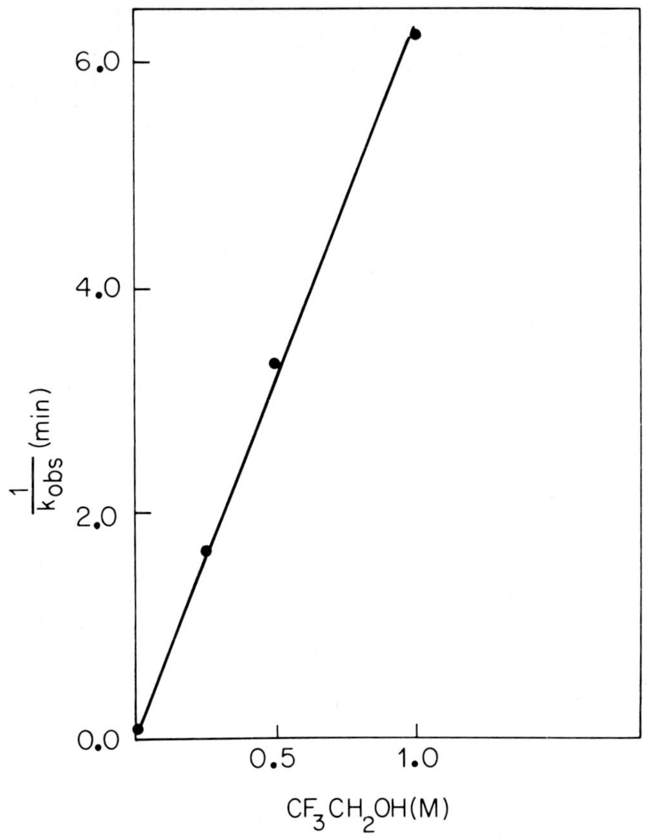

Fig. 4. Effect of trifluorothanol on the rate of hydrolysis of the Co(II) complex of the trifluoroethyl ester of I at pH 8.0.

and conform to the predictions of Scheme V. Such results could not arise from ROH attack on a carbonyl activated intermediate, since even if exchange occurred it would not alter the concentration of the ester and so the rate of hydrolysis would not be affected. Evidence that this

$$\text{Product} \xleftarrow{k[\text{OH}]} \text{C(OR)=O}\cdots\text{M} \overset{\text{ROH}}{\rightleftharpoons} \text{C(OR)=O}\cdots\text{M} \xrightarrow{k[\text{OH}]} \text{Product}$$

exchange process involves an intermediate before formation of the final product derives from the fact that the product-Co complex in the presence of 1 M trifluoroethanol gave no detectable trifluoroethyl ester in 8 h. This would give a rate constant much too small to account for the data of Figure 4.

The specific rate constants for the intramolecular catalysis by the metal bound hydroxide, $K_{eq}k_r$, for the various esters studied is given in Table I. The value for the trifluoroethyl ester is about 10% the value for k_{cat} of carboxypeptidase acting on its best ester substrate. If the enzyme were able to lower the apparent pK of the leaving group (ROH) and/or orient attack of the metal bound hydroxide on the carbonyl group more favourably than in the model system, the mechanism described above could account for the catalytic role of the metal in the enzyme catalyzed ester hydrolysis.

TABLE I
Specific rate constants for the Co-OH catalyzed hydrolysis of

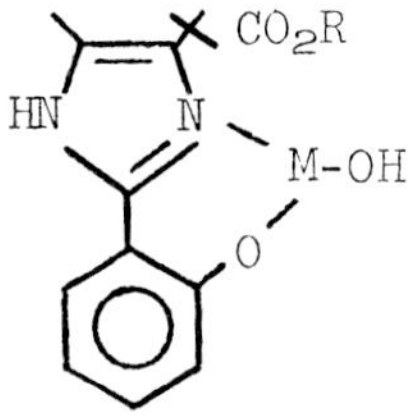

R	pK_a of ROH	k_rK_{eq} min^{-1}
CH_3O-	15.49	14.1
$ClCH_2CH_2O-$	14.31	21.7
$HC{\equiv}CCH_2O-$	13.55	81.0
F_3CCH_2O-	12.36	1500

REFERENCES

Aldersley, Michael F., Kirby, Anthony J., and Lancaster, Peter W.: 1974, 'Intramolecular Displacement of Alkoxide Ions by the Ionised Carboxy-group: Hydrolysis of Alkyl Hydrogen Dialkylmaleates', J. Chem. Soc., 1504-1510.

Auld, David S. and Holmquist, Barton: 1974, 'Carboxypeptidase A. Differences in the Mechanisms of Ester and Peptide Hydrolysis', J. Amer. Chem. Soc., 4355-4360.

Bruice, Thomas C. and Maley, J.R.: 1970, 'Equilibrium Titration and pH-Stat Cell for a Cary 15 Spectrophotometer', Analytical Biochemistry, 275-300.

Bruice, Thomas C. and Schmir, Gaston L.: 1959, 'The Influence of Mechanism on the Apparent pK_a' of Participating Groups in Enzymic Reactions', J. Amer. Chem. Soc. 81, 4552-4556.

Buckingham, D.A., Foster, D.M., and Sargeson, A.M.: 1969, 'Cobalt(III)--Promoted Hydrolysis of Glycine Esters. Kinetics, Product Analysis, and Oxygen-18 Exchange Studies of the Base Hydrolysis of $[Co(en)_2X(glyOR)]^{2+}$ Ions', J. Amer. Chem. Soc. 91, 4102-4112.

Buckingham, D.A., Foster, D.M., and Sargeson, A.M.: 1970, 'Cobalt(III)--Promoted Hydrolysis of Glycine Amides. Intramolecular and Intermolecular Hydrolysis Following the Base Hydrolysis of the cis-$[Co(en)_2Br(glyNR_1R_2]^{2+}$ Ions', J. Amer. Chem. Soc. 92, 6151-6158.

Buckingham, D.A. and Engelhardt, L.M.: 1975, 'Metal Hydroxide Promoted Hydrolysis of Carbonyl Substrates', J. Amer. Chem. Soc. Chem. Commun. 97, 5915-5917.

Rogers, Gary A. and Bruice, Thomas C.: 1973, 'Synthesis and Evaluation of a Model for the So-Called "Charge-Relay" System of the Serine Esterases', J. Amer. Chem. Soc. 96, 2473-2481.

Thanassi, John W. and Bruice, Thomas C.: 1966, 'Neighboring Carboxyl Group Participation in the Hydrolysis of Monoesters of Phthalic Acid. The dependence of Mechanisms on Leaving Group Tendencies', J. Amer. Chem. Soc. 88, 747-752.

COMPLEXES OF COBALT WITH POLYMERIC PHOSPHORAMIDES

J.JAGUR-GRODZINSKI AND Y.OZARI
Plastics Research Department, The Weizmann Institute of Science, Rehovot, Israel

ABSTRACT. The formation of Co(II)-thiocyanate-phosphoramide complexes in aqueous solutions was investigated spectroscopically. It was noted that in such systems the phosphoramides and thiocyanate are synergists. The complexing power of several monomeric phosphoramides and of their polymeric analogues was determined. It has been found that the polymers are much more effective than the monomers, and that the Co(II) affinity of the monomeric phosphoramides increases in the order of their increasing Donor Numbers. It has been suggested that the complexes formed have tetrahedral symmetry; Cobalt (II) ion is first surrounded by four thiocyanate ions, which form its first coordination shell, and then by four phosphoryl groups in the second coordination shell.

1. INTRODUCTION

Hexamethylphosphoramide (HMPA) is known to act as a powerful electron donor [1, 2]. Numerous complexes of HMPA with transition metal ions were desribed in the literature [1, 3]. Phosphoramides of donor strength even higher than that of HMPA were also recently discussed [4]. The complexing ability of a polydentate ligand is usually much greater than that of its monofunctional analogues [5]. It may be, therefore, expected that polydentate phosphoramides may be able to replace water molecules in the coordination sphere of certain cations.

Complexes of Cobalt (II) with HMPA in non aqueous media were described by several investigators [3, 6-8]. It has been pointed out that they tend to form tetrahedrally speced species [6,8]. A similar tendency was also observed in the case of isothiocyanates ions [9]. Cobalt (II) complex tetrahedrally surrounded by isothiocyanates groups may apparently be formed even in aqueous solutions [10]. However, measurable quantities of such complex have been observed only in the presence of an extremely large excess of thiocyanate ions. Apparently the formation constant of this complex in water is very low [10-12]. It has recently [13] been noted that the properties of the aqueous Cobalt thiocyanate solutions are profoundly affected by the presence of even small amounts of phosphoramides. A spectrophotometric investigation of such systems was therefore undertaken and its results are discussed in the present paper.

B. Pullman and N. Goldblum (eds.), Metal-Ligand Interactions in Organic Chemistry and Biochemistry, part 2, 285-298.

2.EXPERIMENTAL

NH_4SCN, NaSCN and Cobalt ammonium sulfate (Baker's, Analyzed) were used without further purification.

Hexamethyl phosphoramide (HMPA)(Fluka, Practical) was left overnight over Linde Molecular Sieves 4A and distilled in vacuum (76°C 10^{-3} Torr).

N-tetramethylene-N',N"-bis(dimethyl)phosphortriamide (MPPA),N,N' -bis(tetramethylene)-N"-dimethyl-phosphoramide (DPPA), N,N',N"-tris (tetramethylene)phosphortriamide (TPPA) and poly-TPPA, $\bar{M}n = 1.5x10^4$ were prepared and purified as described elsewhere (13).

Poly-hexamethylphosphoramide (poly-HMPA) was prepared as described in ref. [14]. It was fractionated by precipitating the higher molecular weight polymer from acetone solutions by adding trace amounts of water.

Cary 14 Spectrophotometer was used in the near IR and Cary 15 in all other spectral measurements. 0.2, 1.0 and 5.0 cm Quartz optical cells were used in absorbance measurements.

3.RESULTS AND DISCUSSION

The absorption band (λ_{max} = 616 nm) of the aqueous Co(II) solutions, containing a very large excess of thiocyanate, was attributed [10, 11] to the tetrahedral $[Co(NCS)_4]^=$ complex. However, it was reported [10], that the fraction of thus complexed Co(II) is low even in concentrated thiocyanate solutions, and it approaches 70% only in the extremely concentrated (20 M) solutions, which are obtained by refluxing Co(II) with an excess of solid NH_4SCN.

The addition of phosphoramide drastically changes this situation. It may be seen from Figure 1 that while addition of 0.5 mole of NaSCN to a diluted $Co(SO_4)$ solution does not induce any significant absorption in the 600-650 nm region, an intense blue absorption band (λ_{max} = 627 nm) does appear upon subsequent addition of 0.25 mole of TPPA. Absorbance measurements in the near IR region show that such solutions also have a broad absorption maximum at 1300 nm.

The electronic spectrum of the Co (II) complex formed in water in the presence of an excess of a phosphoramide and a thiocyanate salt is, therefore, essentially identical with that of the tetrahedral cobalt isothiocyanate complex (which was reproted [9, 10] to be formed in organic solvents in the absence of water molecules). Furthermore it must be pointed out that the observed spectrum is also quite similar (though more intense) to the spectra of Co(II) complexes which were isolated from organic solvents and which contained both NCS^- ions and HMPA molecules [3, 6, 8] or HMPA molecules only as ligand species [15, 16].

The observed absorption band is very intense and positions of its maxima correspond to the electronic transitions calculated on the basis of ligand field strength theory for the tetrahedral Co(II) complexes [15, 17]. The tretrahedral or pseudo-tetrahedral symmetry of the presently investigated complexes seems to be, therefore well

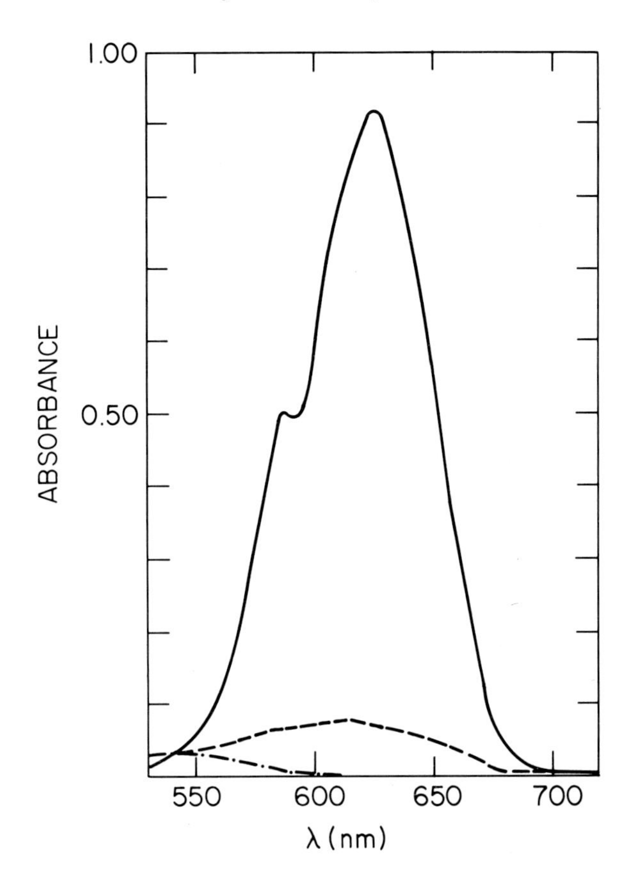

Fig. 1. Optical spectra, at the 500-700 nm range, of the aqueous 5.05×10^{-4} M Co(II) solutions, in the presence of a thiocyanate or a phosphoramide and a thiocyanate.
-·- 0.45M NaSCN; ---3.5M NH_4SCN; —— 0.45M NaSCN + 0.24M TPPA.

documented.To establish their exact composition and to determine their stability constants, spectroscopic measurements were taken at various concentrations of the reagents. Namely, the concentration of Co(II) and of the phosphoramides was kept constant, while the concentrations of NaSCN or NH_4SCN were changed or vice versa. It must be pointed out that addition of other anions than thiocyanates to the cobalt solutions had a negligible effect on their spectra. Obviously, the complex formed comprises both phosphoramides, P, and thiocyanate ions.

$$Co^{2+} + nP + m\,^{-}SCN \rightleftharpoons [CoP_n(NCS)_m]^{(m-2)-} \qquad (1)$$

Accordingly,

$$('C')/(Co^{2+})(P)^n(^{-}SCN)^m = K_{ov.} \qquad (2)$$

Where, ('C'), (P), (Co^{2+}) and (^{-}SCN) are activities of the complex of a phosphoramide, of free cobalt and of thiocyanate ions, respectively.

It follows from Equation (2) that at a constant $[P]_o > [Co^{2+}]_o$

$$('C')/(Co^{2+})(SCN)^m = \beta'_m \qquad (2a)$$

and for a constant (^{-}SCN) > $(Co^{2+})_o$

$$('C')/(Co^{2+})(P)^n = \beta''_n \qquad (2b)$$

Let us assume that: (1) the optical absorption, I, at 627 nm is due in such solutions to the complex, 'C' only and its maximum intensity I_{max}, is determined by the value of $[Co^{2+}]_o$; (2) two components only (e.g.'C' and Co^{2+}) must be considered in order to account for the total concentration of Co (II) present in the solution.

Accordingly, for a sereies of experiments in which only the concentrations of thiocyanate ions are changed, one could expect that a log-log plot of $I/(I_{max}-I)$ vs. (NaSCN) will yield a straight line with a slope = m and a intercept = log β'_m. Similarly, log-log plots of $I/(I_{max}-I)$ vs. P (at a constant $[NaSCN]_o > [Co^{2+}]_o$ should yield values of n and β''_n.

It may be seen from Figure 2 that straight and nearly parallel lines have indeed been obtained when values of $I/(I_{max}-I)$ were plotted against concentrations of various phosphoramides used in experiments in which a constant (3.5 M) concentration of NH_4SCN was maintained. For experiments in which the concentration of a phosphoramide was kept constant (e.g. 1.1 M HMPA or 0.5 M TPPA), the plot of $I/(I_{max}-I)$ vs. (^{-}SCN) has given an initial slope = 4. However, as may be seen from Figure 3, at concentrations of NaSCN higher than 0.1 M, the slope of the line changes and it seems to approach a value of 2.

Obviously, at least one of the previously outlined assumptions must be too simplistic. Let us reconsider them critically. The first assumption seems to be fully justified by our experimental findings and by their comparison with data taken from the literature; In the presence of phosphoramides the shape of the investigated blue absorption bond (λ_{max}=627 nm) has been found to be unaffected by changes in the concentration of the reagents or type of the phosphoramide used in an experiment. At higher (^{-}SCN) and phosphoramide concentrations, its intensity reached a plateau value from which ε_C^{627} = 1880 and ε_C^{1300} = 680 were calculated. The observed spectrum is therefore virtually identical with that of the $[Co(NCS)_4]^{2-}$ complex [9, 10]. Since the absorbances of other thiocyanate complexes of Co(II) at this wave length are at least three order of magnitude lower [10] than that of the tatrahedral complex, their contributions to the absorptions measured at 627 nm may be neglected even if their fractions are high.

Our second assumption seems to be, however, open to criticism. Cobalt thiocyanate complexes of various stoichiometry have been spectroscopically identified in aqueous solutions [10-12] and their presence was confrimed by nmr measurements [18]. The distributions of such

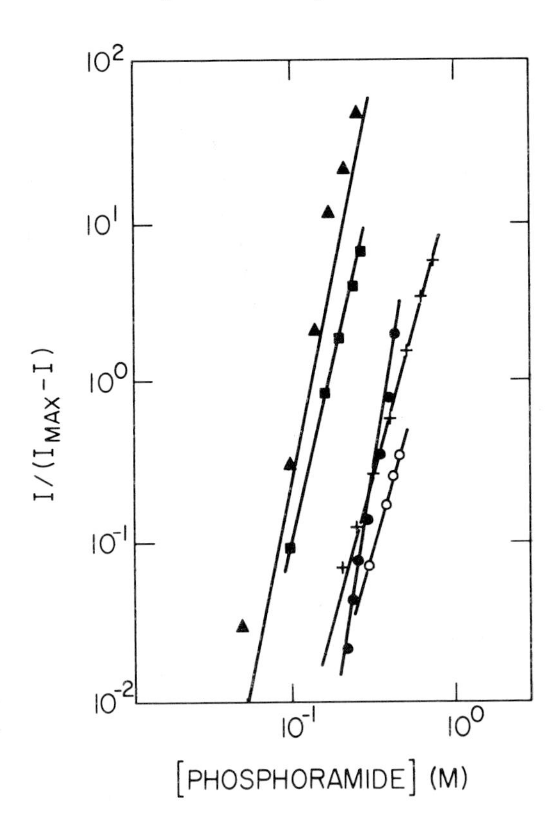

Fig. 2. Log-log plots of $I/(I_{max}-I)$ of the 10^{-3}M Co (II) solutions vs. phosphoramide concentration. I and I_{max} at 627 nm, 3.5M NH_4SCN.
▲ - TPPA; ■ - DPPA; ●,o - MPPA; + - HMPA.
Full and open signs represent experiments at 22 °C and 50 °C respectively.

complexes at various concentration of Na SCN were calculated [18]. It follows that though free Co^{2+} ions are predominant at low concentrations of NaSCN, their fraction decraeses rapidly as the concentration of NaSCN incraeses and comes down to 10% only at [NaSCN] ≃ 0.4 M. At such concentration of NaSCN, about 40% of Co(II) is converted into Co $(SCN)^+$, another 40% of Co(II) into $Co(SCN)_2$ and ~10% into $Co[SCN]_3^-$. The presence of the intermediate complexes must be, therefore, taken into account and it can explain the curvature noted in Figure 3. The overall stoichiometry of the complex is indicated by the initial slope of the curve, its value of 4 reveals that as much as four thiocyanate ions participate in the formation of the Cobalt-phosphoramide-thiocyanate complexes. The intercept of the initial line yields β'_4 = 4100 M^{-4}, as the apparent global formation constant of the complex.

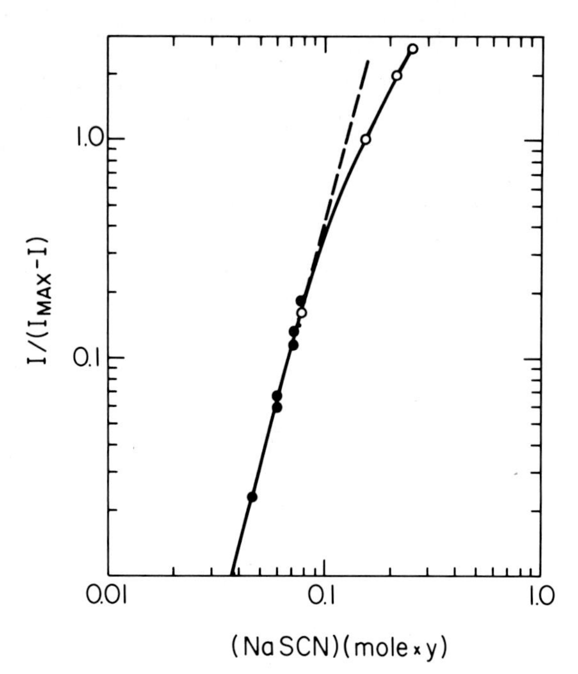

Fig.3. Log-log plot of $I/(I_{max}-I)$ of the cobalt solutions vs. NaSCN activity. I and I_{max} at 627 nm, HMPA = 1.1 M, temp.22 °C.

● - 5×10^{-3} M $[Co^{2+}]_o$; o - 2×10^{-3} M $[Co^{2+}]_o$.

A slope of about 2 is attained at higher concentrations of thiocyanate. This indicates that under such conditions the reaction Co $(SCN)_2$ + 2^-SCN + mP β_2* 'C' becomes the prevailing one. The extrapolated intercept of such line yields $\beta_2^* = 41.6\ M^{-2}$.

The ratio $\beta'_4/\beta_2^* \simeq 100$ which corresponds to the global formation constant, β_2, for the reaction $Co^{2+} + 2^-SCN \rightleftharpoons Co(SCN)_2$ is in excellent agreement with the value of $\beta_2 = 100\pm20$ reported in the literature [19].

It has been stated in the preceeding paragraph that plots of log $I/(I_{max}-I)$ vs. P do yield straight lines over a broad range of concentrations of the phosphoramides (cf. Figures 2 and 4). In these experiments, the ratio between various aqueous cobalt thiocyanate complexes has been fixed by a constant concentration of the thiocyanate salt. Co $(SCN)_2$ prevails at the 3.5 M NH_4CNS (before addition of a phosphoramide) and its conversion into the final tetrahedral complex represents, therefore, the main reaction reflected by the results of those experiments. With the exception of the MPPA at 22 °C (slope = 6) the slopes of the lines shown in Figure 2 are very close to a value of four. A slope = 4 is also obtained for MPPA in experiment conducted at 50 °C

At first glance, the overall stoichiometry revealed by the spectrometric measurements is somewhat puzzling. Namely, eight groups have

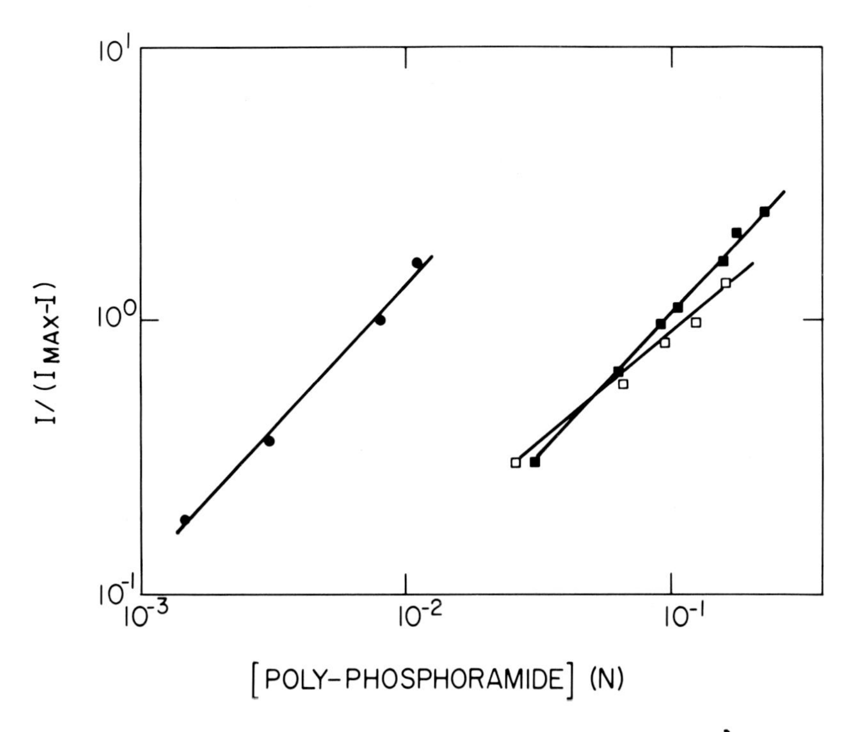

Fig. 4. Log-log plots of $I/(I_{max}-I)$ of the 10^{-3} M Co(II) solutions vs. normality (in respect to phosphoryl groups) of the poly-phosphoramide in the solution. 3.5M NH_4SCN.
● - poly-TPPA, $\bar{M}n = 1.5\times10^4$; ■,□ - poly-HMPA, $\bar{M}n = 1\times10^4$.
Full and open signs represent experiments at 22 °C and 50 °C respectively.

to be placed to form a tetrahedrally surrounded complex. It must be remembered that the mixed Co(II)-thiocyanate-HMPA complex which was isolated by several workers from organic solvents [3, 6-8] did contain two thiocyanate groups and two HMPA molecules only. On the other hand, tetrahedral Co(II) complexes containing $2\,^{-}NCS$ groups and 4 CH_3OH molecules were isolated from anhydrous methanol solutions [20]. Apparently in an aqueous medium the role of the phosphoramide molecules is restricted to the formation of the second coordination sphere arround the central Co(II) atom which stablizes the thiocyanate complex. Its stability may be due to a synergetic action of the two species. It has been noted [21] that the complexation of a transition metal ion results in the enhancement of the exchange rate of the remaining solvent molecules. Thus, partial complexation of Co(II) with ^{-}NCS may enhance the rate of exchange of water by a phosphoramide which in turn will enhance the exchange of remaining water molecules by ^{-}NCS groups.

The correctness of the mathematical approach used in the treatment of the experimental data should be re-examined at this point. This treatment was based on the assumption that the number of phosphoramide molecules participating in the complex is rigidly fixed. This may be an oversimplified picture. Nevertheless, it may be an admissible one if the contributions due to other forms may be neglected. An effort was, therefore, made to find out how extensive such contributions may actually be.

One can write down a series of equilibria:

$$A + P \overset{\beta_1}{\rightleftharpoons} AP_1$$
$$\vdots \qquad\qquad \vdots$$
$$A + nP \overset{\beta_n}{\rightleftharpoons} AP_n$$

and solve the resulting set of equations on the basis of the available experimental data.

A matrix composed of 12 equations, comprising terms corresponding to six elementary equilibria, was solved using a computer program based on a least square iteration procedure. Results of these computations summarized in Table I reveal that a complex comprising four HMPA molecules is the dominant one, indeed. Comparison of the K_i values corresponding to various steps, suggests that the introduction of the first phosphoramide entity into the coordination sphere of the central a atom brings about a powerful enhancement of the subsequent complexation step.

TABLE I
Equilibrium constants, K_i, global equilibrium constants, β_i, and molar extinction coefficients of Co(II)-thiocyanate-HMPA complexes formed in aqueous solutions in the presence of 3.5M NH_4SCN. Temp. 22 °C.

Number of attached HMPA molecules	1	2	3	4	5	6
ε_{627}	190	966	738	1872	2069	181
β_i	0.0009 M^{-1}	0.09 M^{-2}	0.3 M^{-3}	20.9 M^{-4}	0.027 M^{-5}	0.0106 M^{-6}
K_i (M^{-1})	9.0×10^{-4}	100	3.3	69.7	1.3×10^{-3}	0.4

In the investigated polydentate phosphoramide, the phosphoramide groups were built into the polymeric backbone and spaced at about 6 Å intervals. Because of a possible cooperative effect of the neighbouring phosphoryls, one could expect that the polymeric species might act as much more effective complexing agents than the monomeric ones. The log-log plots of $I/(I_{max}-I)$ vs. the normality of the polyphosphoramide solutions (in mole equivalents of phosphoryl groups per unit volume) are shown in Figure 4. Experiments conducted at 22 °C yield 45 °C plots for both poly-HMPA and poly-TPPA systems. This could be regarded as a sign of a 1:1 stoichiometry, in respect to the phosphoryl groups. However, the identity of their spectral characteristics with those of the monomeric phosphoramides seems to indicate that such interpretation of the experimental results is incorrect.

Let us consider a system in which the overall complexation equilibrium is determined by the rates of reactions involving the attachment of the first complexing entity and the removal of the last one from the cobalt center. Obviously, 1:1 apparent stoichiometry will be indicated by the concentration dependance of the equilibrium in such a system. It seems reasonable to postulate that exactly such behaviour characterizes the investigated polymeric systems. After the polydentate molecule attaches itself though one of its phosphoryl groups to the cobalt center, the probability of having additional phosphoryl groups in its vicinity is not determined by their overall concentration, but by their positions within the polymeric chain. Thus, the equilibria involving additional complexation steps become concentration-independent. Hence, an apparent 1:1 stoichiometry of the complex. Inspection of Figure 4 also reveals that the experiment conducted at 50 °C yields a slope = 0.8. The physical meaning of such a result is rather intriguing. A possible explanation may be based on a suggestion that a fraction of the cobalt-thiocyanate complexes, present at this temperature in the aqueous solution, prior to the addition of the phosphoramide, forms dimeric or oligomeric agglomerates such as for instance $[(SCN)_mCoSCNCo(SCN)_n]^{(3-m-n)+}$. Interaction of such a species with a phosphoramide group of a polydentate polymer molecule could trigger a 'Zipp' type mechanism leading to the formation of two or even more 'C' type complexes.

The reliability of the results obtained with poly-HMPA was further checked. A series of appropriately disigned experiments was conducted for poly-phosphoramide of $\bar{M}n = 2.5 \times 10^4$ M and their results were analyzed in the following way: for a 1:1 stoichiometry of complex formation and for a constant concentration of thiocyanate ions. Equation (2) can be written in the form:

$$\beta'' = \frac{I/\varepsilon\ell}{([Co^{2+}]_o - I/\varepsilon\ell)P} \tag{3}$$

where: ℓ is the optical path of the cell used in the experiments, ε the molar extinction coefficient of the complex and all other symbols retain their previous meaning.

Equation (3) can be conveniently rewritten as follows:

$$\frac{1}{P} = -\beta'' + (\beta''\varepsilon\ell[Co^{2+}]_o \frac{1}{I} \quad (4)$$

Thus, both β" and ε can be simultaneously derived from a series of experiments at a constant $[Co^{2+}]_o$.

A linear plot of 1/P vs. 1/I obtained for the poly-HMPA system at $[Co^{2+}]_o = 1\times10^{-3}$ M is shown in Figure 5. Its intercept yields β" = 80 M^{-1} and its slope, divided by $[Co^{2+}]_o \times \ell = 1\times10^{-3}$M x cm and by the intercept, yields ε = 1760 (in good agreement with the extinction coefficient of the 'C' complex derived from the plateau values of the experiments conducted with other phosphoramides).

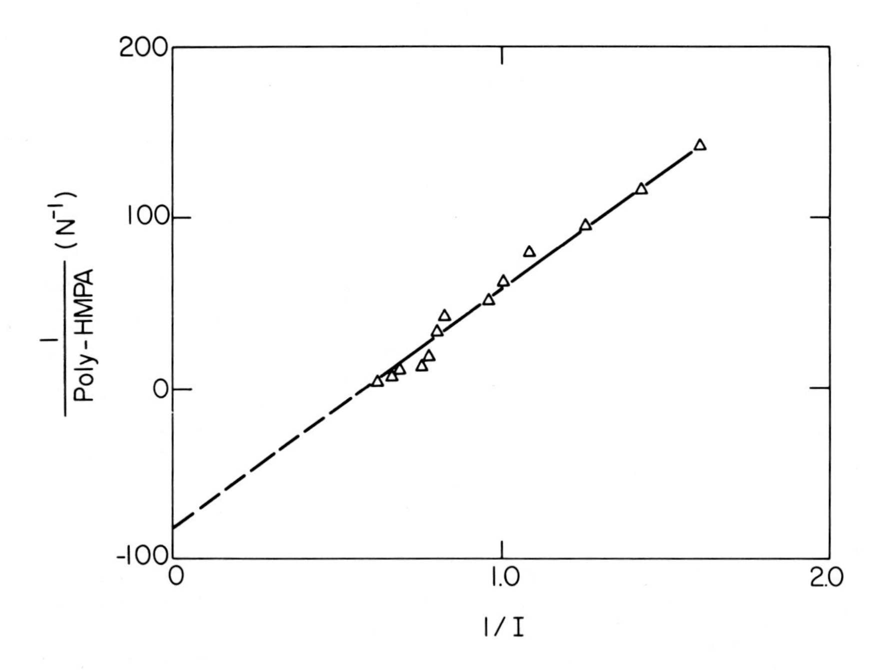

Fig. 5. Plot of the reciprocal concentration of poly-HMPA vs. reciprocal of the optical absorbtion at 627 nm of the 10^{-3} M Co (II) solutions. Temp. 22 °C; 3.5M NH_4SCN, ℓ = 1 cm; Δ - $\bar{M}n$ 2.5×10^4.

The equilibrium constants calculated for both monomeric and polymeric species are summarized in Table (II). The Donor Numbers of the investigated phosphoramides incraese in the order [4]: HMPA ≤ MPPA < DPPA < TPPA. As can be seen from the tabulated data, their respective abilities to complex Co(II) incraese in the same order.

A comparison between the complexing power of the monomeric and of the polymeric species, based on a direct comparison of the calculated global equilibrium constants would be greatly misleading, because they refer to different apparent stoichiometries. It seems,

TABLE II
Global equilibrium constants of complexation with Co (II) of various phosphoramides in aqueous solutions of 3.5M NH_4SCN.

Phosphoramide	Coordination number		β"	
	at 22 °C	at 50 °C	at 22 °C	at 50 °C
HMPA	4	-	21.4 M^{-4}	-
MPPA	6	4	1680.0 M^{-6}	166 M^{-4}
DPPA	4	3	1200 M^{-4}	204 M^{-3}
TPPA	4	-	4000 M^{-4}	-
Poly-HMPA $\bar{M}n=10^4$	1	0.8	10 M^{-1}	8.4 $M^{-0.8}$
$\bar{M}n=2.5x10^4$	1	-	80 M^{-1}	-
Poly-TPPA $\bar{M}n=1.5x10^4$	1	-	149 M^{-1}	-

however, that values of β" for the polymeric species, in conjunction with the value of K_1 for HMPA (see Table I) can provide the proper basis for camparison of the unidentate monomers with the polydentate polymers (cf. discussion in the preceeding paragraph). Such comparison reveals that the investigated systems are not different in this respect from other polydentate complexing agents, such as for instance polyethers [5] which have been shown to act as much more effective solvating agents than their monodentate analogues.

The effectiveness of the polymeric phosphoramides as complexing agents seems to be also dependent on their molecular weight. Thus, an eight-fold increase in the value of the formation constant was noted when the molecular weight of poly-HMPA was increased from 10 000 to 25 000. Similar to the monomer series, poly-TPPA is again a more powerful complexing agent than poly-HMPA of the same molecular weight.

It must be pointed out, however, that all three investigated polymers have been relatively low molecular weight. It seems reasonable to assume that with further increase in molecular weight of the polyphosphoramides, their complexing power would eventually reach a plateau value.

From the semilog plots of β" vs. reciprocals of absolute temperature shown in Figure 6 and 7 one can derive the apparent enthalpies and entropies of the complexation reaction. ΔH values were found to be -3.46 kcal/mole-1 for TPPA and -2.85 kcal/mole-1 for the poly-HMPA systems independently of their molecular weights.The larger value of ΔH for TPPA is in agreement with the high basicity of phosphoryl group of this compound. For TPPA the value of the entropy of complex formation is found to be positive (ΔS = 5.1 e.u.) Such value indicates that in this system the formation of the complex is accompanied by an increase of disorder in the bulk of the solution, probably disruption of TPPA clusters. On the other hand, for complexes involving poly-HMPA molecules the entropy values have been found to be negative (-1.55 e.u. and -4.44 e.u for $\bar{M}n = 2.5x10^4$ and for $\bar{M}n = 10^4$ respectively). The increase in complexing ability with the increase in the molecular weight

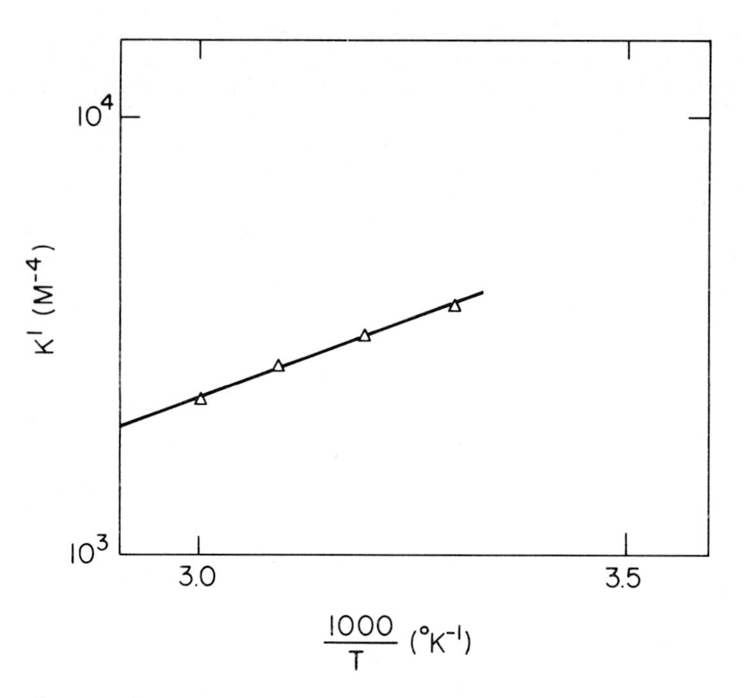

Fig. 6. Semi-log plot of the fourth order formation constant of the TPPA complex vs. reciprocal of the absolute temperature.

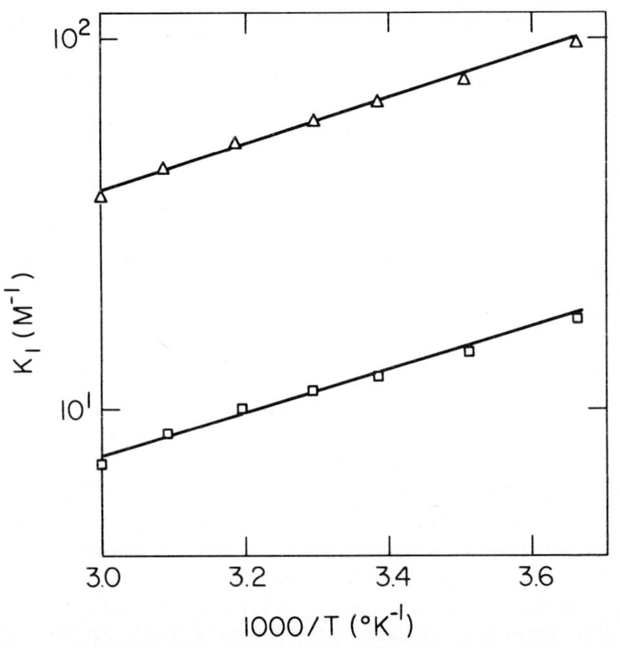

Fig. 7. Semi-log plots of the first order formation constants of the poly-HMPA complexes vs. reciprocal of the absolute temperature. □ - $\bar{M}n$ = $1x10^4$; Δ - $\bar{M}n$ = $2.5x10^4$.

of the polydentate polymer is therefore related to the decrease in the entropy losses due to complexation. Apparently, the phosphoryl groups in the longer polymer can organize themselves around the central atom in such a way that the resulting loss of segmental mobility is significantly reduced.

REFERENCES

1. (a) Normant, H.; Angew. Chem. Internat. Edit., 6, 1046, 1967; (b) Normant, H.: Russian Chem. Rev., 39, 457, 1970.
2. Gutmann, V. and Scherhaufer, A.: Monatsch., 99, 335, 1968.
3. (a) De Bolster, M.W.G. and Groeneveld, W.L.: Recueil Trav. Chim. Pays-Bas, 90, 477, 1971
   (b) idem, 91, 171, 1972; (c) idem, 91, 643, 1972.
4. Ozari, Y. and Jagur-Grodzinski, J.: J.C.S. Chem. Comm. 295, 1974.
5. Chan. L.L. and Smid, J.: J. Am. Chem. Soc., 89, 4547, 1967.
6. Donaghue J.T. and Drago, R.S.: Inorg. Chem., 2, 572, 1963.
7. Schafer, M. and Curran, C.: ibid., 4, 623, 1965.
8. Le Coz, E.,Guerchais, J.E. and Goodgame, D.M.L.: Bull. Soc. Chim. France, 3855, 1969.
9. Cotton, F.A., Goodgame, D.M.L., Goodgame, M. and Sacco, A.: J. Am. Chem. Soc., 83, 4157, 1961.
10. Tribalat, S.,and Zeller, C.: Bull. Soc. Chim. France 2041, 1962.
11. Lehné, M.: ibid., 76, 1951.
12. Babko, A.K. and Drako, O.F.: Zh. Obsh. Khim., 20, 228, 1950.
13. Ozari, Y.: Ph. D. Thesis, Weizmann Institute of Science, 1975.
14. (a) Bello, A., Bracke, W., Jagur-Grodzinski, J., Sckmann, G. and Szwarc, M.: Macromolecules 3, 98, 1970.
    (b) Ozari, Y. and Jagur-Grodzinski, J.: J.C.S. Dalton, 474, 1973.
15. Donaghue, J.T. and Drago, R.S.: Inorg. Chem., 1, 866, 1962.
16. Gutmann, V., Weisz, A. and Kerber, W.: Monatsch. 100, 2096, 1969.
17. Cotton, F.A. and Goodgame, M.: J. Am. Chem. Soc., 83, 1777, 1961.
18. Horrocks,Jr., W.D. and Hutchison, J.R.: J. Chem. Phys., 46 1703, 1967.
19. Tribilat, S. and Caldero, J.M.: J. Electroanal. Chem., 5, 176, 1963.
20. Pollak, P. and Cave, G.C.B.: Canadian J. Chem. 45, 1051, 1967.
21. Plotkin, K., Copes, J. and Vriesenga, J.R.: Inorg. Chem., 12, 1494, 1973.

DISCUSSION

Simon: In one of your systems of interaction of phosphoramides with ions you made a 12 parameter least square fit. Is there really significance attributable to the parameters obtained by such procedures?

The phosphoramides have very high dipole moments and should therefore interact strongly with group IA and especially 2A cations. Do you have information on the selectivity of such interactions?

Jagur-Grodzinski: (1) 12 parameters derived from a least square fit are of course quite sensitive to experimental errors. In our system the standard deviation from zero of the final F(i) values $\sigma_{F(i)/B}$ = 2%.

The extinction coefficient ε_4 obtained from the computation is also in excellent agreement with ε obtained from the plateau and cited in the literature for the tetrahedral complex. It seems, therefore, that our results are basically sound and possible errors are not significant.

(2) Indeed, phosphoramides interact strongly with group 1A and 2A cations. De Bolster et al. (Re Trav. Chim. Phys. 90, 1961) have published an extensive report on such systems. Radical-ions with Na, Li^+, Bu^{2+}, Cu^{2+} in HMPA were investigated by myself et al., by Johan Smid, Paul Sigvalt and several other investigators. However, detailed information on the respective selectivities of such interactions has not yet been reported (to the best of my knowledge).

INFRARED AND RAMAN STUDIES OF SOME ORGANOMAGNESIUM AND ALLYL COMPOUNDS

J. KRESS, C. SOURISSEAU and A. NOVAK
Laboratoire de Spectrochimie Infrarouge et Raman, C.N.R.S.
2, rue Henry Dunant, 94320 Thiais, France

1. INTRODUCTION

A study of organomagnesium and allylic compounds has been undertaken as a part of a program in investigation of the vibrational spectra of organometallic compounds. We were interested primarily in their structures, since there are few diffraction studies of these compounds in solid state and very little is known about liquids and solutions, and on metal-ligand interactions. The infrared and Raman spectra were examined generally in the 3600 to 50 cm^{-1} region. The high frequency spectrum was analysed in order to obtain data about the molecular geometry or conformation of polyatomic ligands and about the perturbation of intramolecular bonds by coordination. The low-frequency spectrum yielded information about the arrangement of ligands about metal atom and the metal-ligand force constants.

2. ORGANOMAGNESIUM COMPOUNDS

Three types of compounds were studied: (i) the symmetric type represented by $R_2Mg.n(C_2H_5)_2O$ (n = 0, 1, 2); (ii) the mixed type RMgX.n $(C_2H_5)_2O$ (n = 1, 2); (iii) the compounds of general formula $MgX_2.nR_2O$ (n =1, 2) (R =CH_3, C_2H_5; X = Cl, Br, I). Methods of preparation, details concerning the recording of infrared and Raman spectra and most assignments of the bands have been already published (Kress, 1974; Kress and Novak, 1974, 1975, pp. 23, 199, 281). We are reporting here the main spectroscopic results and the most important conclusions concerning these three types.

2.1. *Dimethylmagnesium $(CH_3)_2Mg$*

Dimethylmagnesium crystallises in the orthohombic system (Ibam) with four $(CH_3)_2Mg$ formula units per unit cell (Weiss, 1964). The crystal consists of (-Mg<CH_3, CH_3>Mg-) infinite chains the neighboring cycles being almost orthogonal.

B. Pullman and N. Goldblum (eds.), Metal-Ligand Interactions in Organic Chemistry and Biochemistry, part 2, 299-311.

The infrared and Raman spectra of $(CH_3)_2Mg$ and $(CD_3)_2Mg$ were obtained at ordinary and liquid nitrogen temperature and can be interpreted in terms of an isolated chain having C_{2h} factor group symmetry. The normal coordinate analysis supports this view rather than the D_{2h} symmetry of the chain suggested by Weiss (1964). The discrepancy between the X-ray and spectroscopic results is probably due to some disorder concerning the position of protons. The X-ray diffraction would give thus an average symmetry, higher than that seen by vibrational spectroscopy.

Most of the expected fundamentals have been observed and assigned to their symmetry species. The isotopic shifts and normal coordinate calculations show that the CH_3 vibrations can generally well be distinguished from skeletal vibrations except for a few cases in the 600-400 cm^{-1} region where some heavy mixing of Mg-C stretching and CH_3 rocking modes occurs.

Eighteen force constants have been determined and the most interesting one is related to the stretching of Mg-C bonds, K_{Mg-C} = 1.05 mdyne $Å^{-1}$. This is considerably less than the Mg-C stretching force constant of $C_2H_5MgBr.2(C_2H_5)_2O$ (Table I) of about 1.40 mdyne $Å^{-1}$ which

TABLE I
Force constants and interatomic distances of Mg-C, Mg-Br and Mg-O bonds

Compound	K_{Mg-C} (mdyne $Å^{-1}$)	r_{Mg-C} (Å)	K_{Mg-O} (mdyne $Å^{-1}$)	r_{Mg-O} (Å)	K_{Mg-Br} (mdyne $Å^{-1}$)	r_{Mg-Br} (Å)
$(CH_3)_2Mg$ Solid	1.05	2.24 [a]				
$C_2H_5MgBr.2(C_2H_5)_2O$ Solid	1.40	2.15 [b]	0.6	2.04 [b]	0.85	2.48 [b]
$MgBr_2.2(C_2H_5)_2O$ Solid			0.6	2.13 [c]	0.5	3.06 [c]
MgO gas			2.34 [d]	1.749 [d]		
$MgBr_2$ gas					1.48 [e]	2.34 [f]

[a] Weiss (1966); [b] Guggenberger and Rundle (1968); [c] Schibilla and Le Bihan (1967); [d] Herzberg (1950); [e] Randall et al. (1959); [f] Akisin and Spiridonov (1957).

can be explained by the electron-deficient character of $(CH_3)_2Mg$ bonds having 1/2 bond order. Nevertheless, the force constant of dimethylmagnesium is still larger than that of Li-C stretching of CH_3Li monomer which is 0.70 mdyne $Å^{-1}$ (Andrews, 1967). Another interesting fact is that the Mg-Mg force constant is different from zero and amounts to 0.07 mdyne $Å^{-1}$ which can be correlated with some direct bonding interaction between Mg atoms.

The methyl group frequencies are considerably different from those of a hydrocarbon like propane. The Table II shows that large modifications are observed for the C-H stretching, symmetrical CH_3 bending and CH_3 rocking vibrations. The δ_s CH_3 frequency appears to be useful for correlation purposes and its low value (1198 cm^{-1}) can be correlated with the electronegativity of the magnesium atom. The C-H stretching force constant (4.56 mdyne $Å^{-1}$) also reflects the electronegativity of Mg being intermediate between that of CH_3Li and CH_3Br, i.e. 4.32 (Andrews, 1967) and 4.95 (Linnett, 1940) mdyne $Å^{-1}$, respectively.

TABLE II
Some characteristic frequencies (cm^{-1}) of organomagnesium compounds

Compound	$\nu_a CH_3$	$\delta_s CH_3$	ρCH_3	ν_a C-O-C	δ C-O-C
$CH_3CH_2CH_3$ [a]	2968	1382	1134		
$(CH_3)_2Mg$	2897	1198	560		
$CH_3MgBr.2(C_2H_5)_2O$	2895	1108	517		
CH_3OCH_3	2922	1438	1185	1093	426
$MgBr_2.2(CH_3)_2O$	2961	1435	1178	1065	465

[a] Gayles and King, 1965.

2.2. *Mixed Organomagnesium Compounds* $C_2H_5MgBr.2(C_2H_5)_2O$ *and* $C_2H_5MgCl.(C_2H_5)_2O$

(a) Crystal

The structure of solid $C_2H_5MgBr.2(C_2H_5)_2O$ has been determined by Guggenberger and Rundle (1968) and can be described in terms of monomers having a quasitetrahedral arrangement of ligands (C_2H_5, Br, $C_2H_5OC_2H_5$) about the central magnesium atom. The conformation of ether molecules is gauche-gauche (Figure 1).

The infrared and Raman spectra of the crystalline solid (Figure 2) are very complex. More than sixty bands have been observed and their Raman and infrared frequencies are practically the same. It was necessary to study five isotopic species ($C_2H_5MgBr.2Et_2O$, $C_2H_5MgBr.2Et_2O$-d_{10}, $CH_3CD_2MgBr.2Et_2O$, $CD_3CH_2MgBr.2Et_2O$ and $C_2D_5MgBr.2Et_2O$) and ethylmagnesium iodide dietherate ($C_2H_5MgI.2Et_2O$) in order to give a suitable assignment (Kress and Novak, 1975).

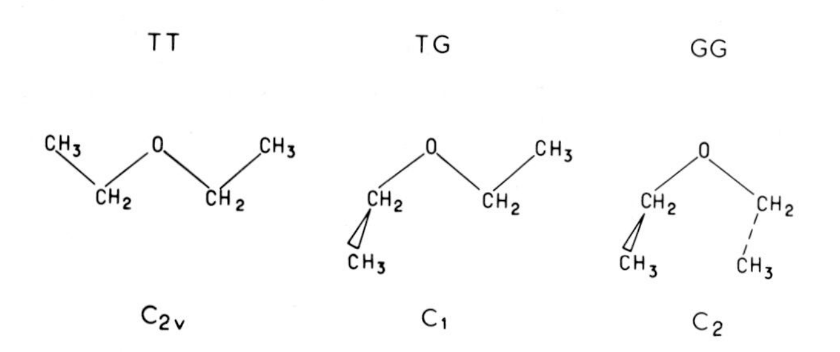

Fig. 1. Trans-trans (TT), trans-gauche (TG) and gauche-gauche (GG) conformations of diethylether molecule.

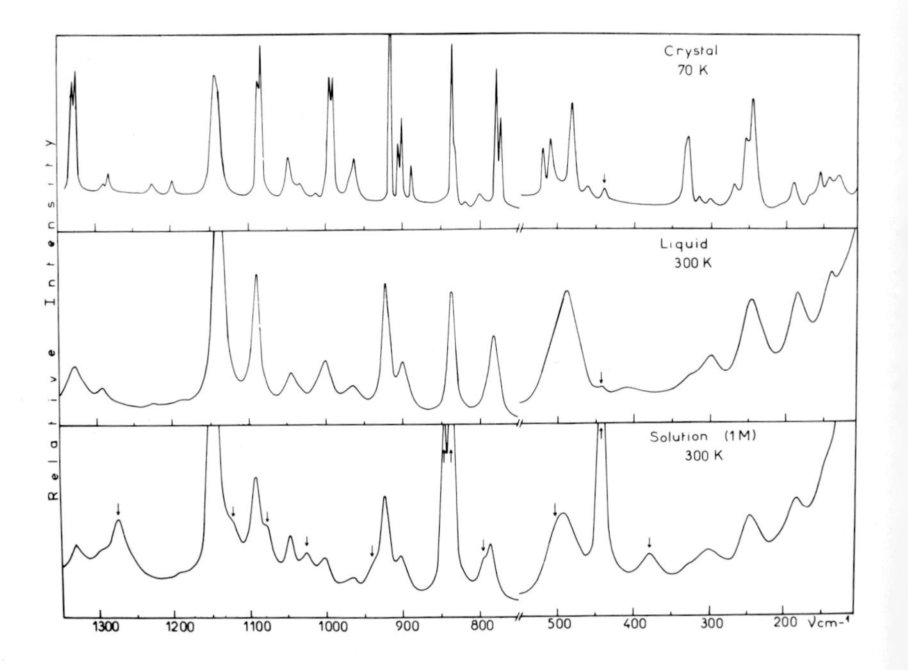

Fig. 2. Raman spectra of $C_2H_5MgBr.2(C_2H_5)_2O$ in solid and liquid phase and of a solution in diethylether.

Low frequency spectra: The region below 600 cm^{-1} contains both intramolecular and magnesium-ligand frequencies. The distinction between them is possible since the isotopic shifts for the former are much larger (1.14 to 1.28) than for the latter. Nine magnesium-ligand vibrations are expected and six of them have been observed and assigned using the shifts upon partial or total deuteration of ligands or chemical (Br → I) substitution. The four magnesium-ligand stretching frequencies are given in Table III and can be used to estimate the corresponding Mg-C, Mg-O and Mg-Br force constants (Table I).

TABLE III
Metal-ligand stretching frequencies (cm^{-1}) of some organomagnesium compounds

Compound [a]	νMg-C	$\nu_a MgO_2$	$\nu_s MgO_2$	νMg-X [b]
$EtMgBr.2Et_2O$	485	316	300	250
$EtMgI.2Et_2O$	481	313	294	228
$MeMgBr.2Et_2O$	508	304	289	247
$MgBr_2.2Et_2O$		320	286	208 161
$(EtMgCl.Et_2O)_2$	498 486	322 [c] 306 [c]		283 252
Et_2Mg	399			

[a] Et = C_2H_5, Me = CH_3

[b] X = Cl, Br, I; $\nu_a MgBr_2$ and $\nu_s MgBr_2$ for $MgBr_2.2Et_2O$

[c] νMgO

The Mg-C force constant is larger than that of symmetric compounds in agreement with shorter Mg-C distance. The Mg-O and Mg-Br force constants are comparable and much weaker than that of the Mg-C bond indicating a more ionic character of the former - at least in the investigated organomagnesium compounds. The Mg-O and Mg-Br force constants of gaseous MgO and $MgBr_2$ compounds on the other hand, are much higher, as expected on ground of shorter distances (Table I).

High frequency spectra: 30 fundamentals of coordinated $(C_2H_5)_2O$ molecules have been assigned and represent an excellent mean of identification of the gauche-gauche conformation of ether molecules unknown for pure ether.

All the fifteen fundamental vibrations of Mg bonded ethyl group have been assigned in terms of group frequencies. The methylene group frequencies decrease substantially with respect to hydrocarbon frequencies, especially the CH_2 stretching, scissoring, wagging, and rocking modes with the average frequencies of 2835, 1415, 920 and 520 cm^{-1}, respectively. The CH_3 stretching frequencies decrease too (2868 cm^{-1}) while the C-C stretching frequency increases (1146 cm^{-1}). These variations are ascribed to the electronegativity of magnesium in much the same way as those of methyl group of $(CH_3)_2Mg$.

(b) Liquid and ether solutions

A comparison of the Raman spectra of low temperature solid and room temperature liquid reveals a great similarity of the two phases (Figure 2). There are minor differences such as variation of relative intensities of several bands and some more band splitting in the crystal spectrum due doubtless to intermolecular coupling effects but practically no frequency shifts. It can thus be concluded that the structure of pure liquid, which can alternatively be considered as a 4 molar solution of 'C_2H_5MgBr' in ether is essentially the same as that of crystalline solid, i.e. a dietherate monomer with tetrahedral arrangement of ligands about Mg and gauche-gauche conformation of other molecules.

Solution spectra (1 to 4 moles l^{-1}) can be interpreted as a superposition of $C_2H_5MgBr.2Et_2O$ and of pure ether spectra and there is no dilution effect (Figure 2), which indicates that the same monomer compound predominates in diethylether solution and that all previous observations can be extended to the dissolved species. No bands attributable to Et_2Mg or $MgBr_2$ species could be detected which implies that they must be less than 3%. Schlenk's equilibrium is thus shifted strongly towards mixed species. It should also be pointed out that no evidence of C_2H_5MgBr polymers is found even at higher concentrations, contrary to the conclusions reached by Walker and Ashby (1969) on the ground of association measurements

$C_2H_5MgCl.(C_2H_5)_2O$

The infrared and Raman frequencies of liquid $C_2H_5MgCl.(C_2H_5)_2O$ (Kress and Novak, 1975) do not coincide unlike those of $C_2H_5MgBr.2(C_2H_5)_2O$ and the largest difference is observed for magnesium-chloride stretching vibrations: there is a 288-278 cm^{-1} doublet in infrared and a 263-241 cm^{-1} doublet in Raman.

Furthermore, it must be assumed that the Mg-X stretching force constant is lower for chloride than for bromide since the average νMg-Br frequency is but little lower (244 cm^{-1}) than the νMg-Cl one (265 cm^{-1}). These facts support a cyclic dimer model of C_{2h} symmetry for the monoetherate $C_2H_5MgCl.Et_2O$ (Figure 3). Moreover, the Mg-C and Mg-O stretch-

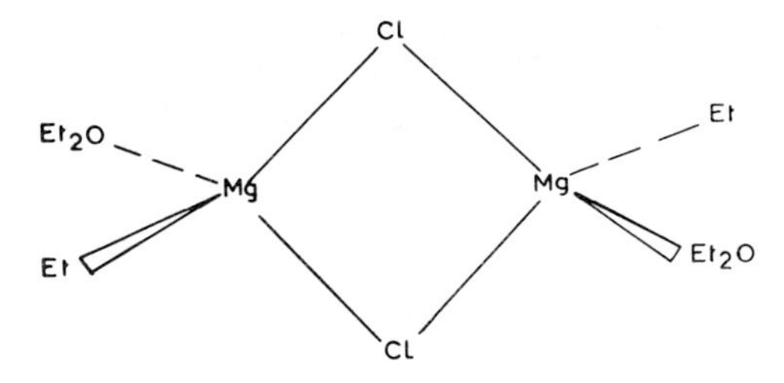

Fig. 3. Structural model proposed for ethylmagnesium chloride monoetherate, $C_2H_5MgCl.(C_2H_5)_2O$.

ing frequencies are very similar to those of $EtMgBr.2Et_2O$. As in the case of ethylmagnesium bromide, the same dimer species is shown to be predominant in diethylether solution.

3. $MgBr_2.2(C_2H_5)_2O$

The triclinic ($P\bar{1}$, Z=4) $MgBr_2.2Et_2O$ crystal has been studied by X-ray diffraction at -50° C (Schibilla and Le Bihan, 1967). The magnesium atom is tetracoordinated with quasitetrahedral arrangement of bromine and ether ligands. The conformation of $(C_2H_5)_2O$ molecules, however, could not be determined.

We investigated the infrared and Raman spectra of this compound at various temperatures (25° to -180° C) and various pressures up to 10 kbar. It has been shown that the $MgBr_2.2Et_2O$ crystal exists in two phases and that there is a reversible transition between them. This phase transition, which has also been determined calorimetrically, concerns primarily a conformational change of diethylether molecules and much less the arrangement of the ligands about the central magnesium atom.

There are two sets of ether bands in the infrared spectrum of $MgBr_2.2Et_2O$ at -40° C (Figure 4). One set disappears at - 180° C but is predominant at room temperature. The corresponding frequencies are practically identical with those of $C_2H_5MgBr.2(C_2H_5)_2O$ compound and are thus

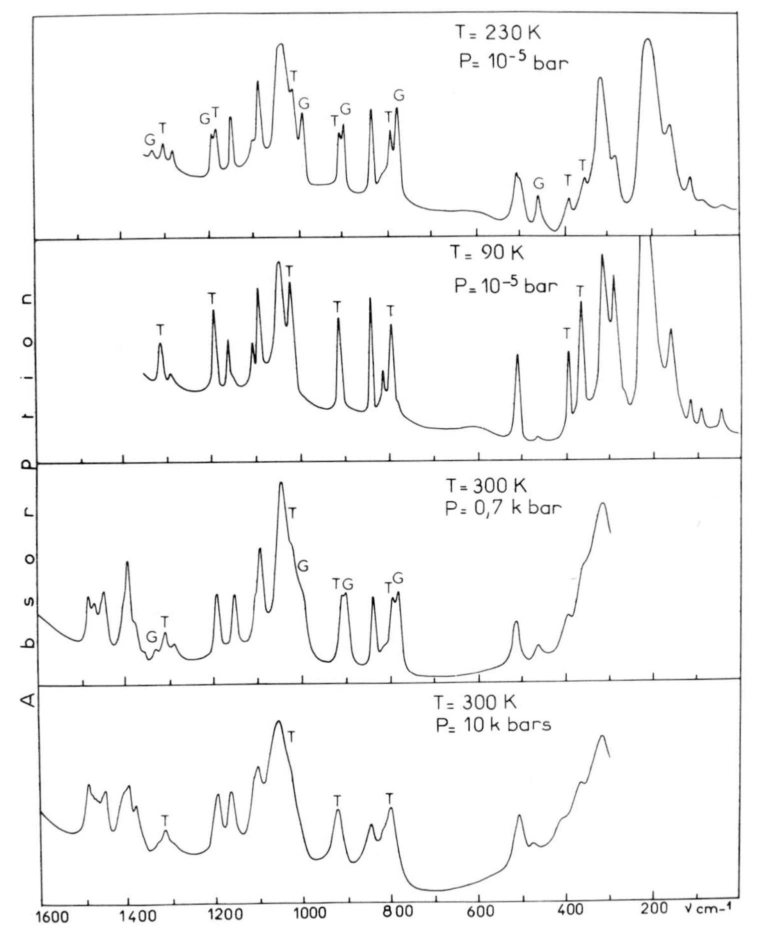

Fig. 4. Infrared spectra of crystalline $MgBr_2.2(C_2H_5)_2O$ at different temperatures and different pressures.

assigned to the gauche-gauche conformation of ether molecule. The other set alone persists at -180° C and is attributed to either trans-gauche or trans-trans conformation (Figure 1). The same behavior is observed in Raman spectra. The effect of the pressure is similar to the temperature lowering, i.e. when the pressure increases the contribution of the GG conformation diminished and at 10 kbars only one conformation (TG or TT) persists as shown by the corresponding infrared spectra (Figure 4). This can probably be interpreted in terms of intermolecular forces which become stronger at higher pressure and lower temperature. It is interesting to note that only one other compound, $CH_3MgBr.2Et_2O$, showed a similar behaviour which is likely to be due to similar volume of CH_3 and Br ligands.

The modification of the intramolecular frequencies of $(C_2H_5)_2O$ molecule upon coordination is due to two factors: conformational change and charge transfer from ether oxygen to magnesium. It can be shown qualitatively that the electronic effect decreases the ν_aC-O-C frequency and increases the skeletal bending and CH_2 stretching frequencies much in the same way as for the CH_3OCH_3 molecule for which the conformation remains inchanged in $MgBr.2(CH_3)_2O$ (Table II).

The metal-ligand frequencies (Table III) on the other hand, vary but little with temperature. The Mg-O stretching force constant estimated from the above frequencies is the same as that of $C_2H_5MgBr.2Et_2O$ while the Mg-Br force constant is significantly lower in agreement with longer Mg-Br distance of $MgBr_2.2Et_2O$ (Table I).

3. ALLYL COMPOUNDS

In organometallic compounds the allyl group can be bonded to metal atom in three different ways: σ, π or ionic. The corresponding C_3H_5 groups can thus be represented as shown in Figure 5. We have investigated some

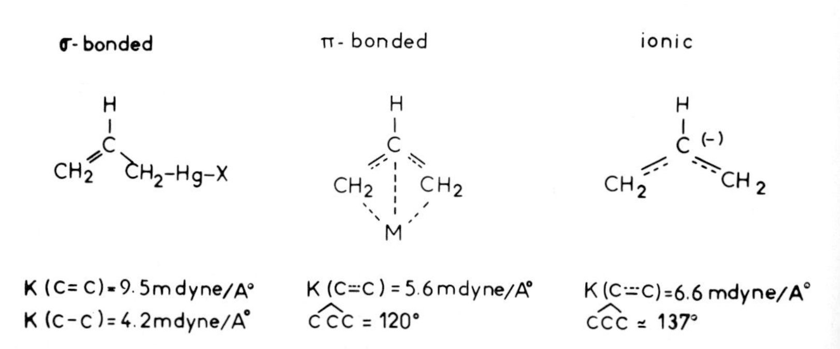

Fig. 5. Diagrammatical structures of σ-bonded, π-bonded and ionic allyl ligand.

σ bonded allylmercury compounds (Sourisseau and Pasquier, 1972, p. 51 and 65), π-bonded palladium and nickel complexes (Sourisseau and Pasquier, 1973 and 1974) and sodium, lithium and magnesium salts containing allyl anion (Sourisseau and Pasquier, 1975) in order to determine the main spectroscopic characteristics and to enquire about the geometry of C_3H_5 ligand as well as about intramolecular and metal-ligand force constants.

Infrared spectra of three different types of allyl compounds i.e. σ bonded C_3H_5HgCl, π bonded $(C_3H_5PdCl)_2$ and ionic NaC_3H_5 are illustrated in Figure 6. Their spectral features are very different and we shall discuss successively the skeletal, CH and metal-ligand vibrations.

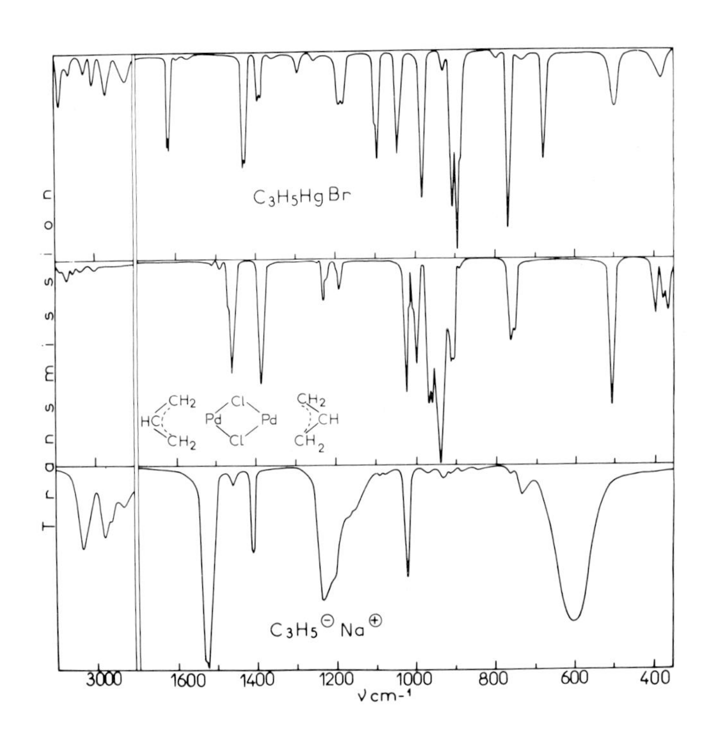

Fig. 6. Infrared spectra of (a) C_3H_5HgBr, (b) $(C_3H_5PdCl)_2$, (c) C_3H_5Na compounds in solid state.

3.1. *Intramolecular vibrations and structure of C_3H_5 ligand*

(a) Skeletal Vibrations

Two stretching and a bending skeletal vibrations are expected for a C_3H_5 entity. In the vibrational spectra of allylmercury chloride, they have been identified at 1627, 936 and 385 cm^{-1} (Table IV). The first two frequencies are but little different from those of allylhalides (Sourisseau and Pasquier, 1972, 1) and can be assigned to the C=C and C-C stretching vibrations. The above assignment implies a nearly localised double bond in these compounds which is corroborated by the calculated stretching force constants of 9.5 and 4.2. mdyne $Å^{-1}$ characteristic of double and single bonds, respectively (Figure 5).

TABLE IV
Some intramolecular frequencies of allylmetallic compounds

Compound	Skeletal stretch		Skeletal bend	$\overline{\text{CH}}$ [a] stretch	CH wagging	CH bend
σ C_3H_5HgCl	1627	936	385	3000	1125 [b]	1300
(π $C_3H_5PdCl)_2$	1383	1023	511	3049	963 945	1192
C_3H_5Na	1527	1022	608	2987	917 888	1090

[a] $\nu\overline{\text{CH}}$: average frequency of five CH stretching fundamentals.

[b] wCH_2 vibration of methylenic group.

The spectra of π complexes such as $(C_3H_5PdCl)_2$, $(C_3H_5PdBr)_2$, $(C_3H_5PdOCOCH_3)_2$ and $(C_3H_5NiOCOCH_3)_2$, on the other hand, show the skeletal stretching bands near 1383 and 1023 cm^{-1} and must be interpreted in terms of antisymmetrical (ν_aCCC) and symmetrical (ν_sCCC) stretching modes. This is consistent with the X-ray diffraction data showing, for $(C_3H_5PdCl)_2$ crystal for instance (Smith, 1965) that the C-C bonds are equivalent, their distance (1.377 Å) being intermediate between those of single and double bond.

Similar conclusions can be reached for ionic compounds exhibiting ν_aCCC and ν_sCCC modes near 1527 and 1022 cm^{-1}, respectively (Table IV). In order to explain the large differences between skeletal (ν_aCCC and δCCC) frequencies of palladium and sodium compounds, a valence force field calculation on C_3H_5 and C_3D_5 species using an XY_2 model was performed. The results indicate that the C-C stretching force constant increases from 5.6 to 6.6. mdyne $Å^{-1}$ and that the CCC angle opens from 120 to about 137° ∓ 2° when going from $C_3H_5^{\cdot}$ radical in a π complex to $C_3H_5^-$ anion in a salt. The 120° angle value obtained for Pd complex is in excellent agreement with the X-ray data (Smith, 1965) while the 137° angle calculated for the anion can be compared with the 133° angle estimated on the ground of π electron and all valence electron calculation methods (Meyer and Chrinovitzky, 1972).

(b) CH Vibrations

The CH stretching and bending vibrations corroborate the above results as far as the equivalency of CH_2 groups is concerned. In σ bonded allyli ligands methylenic and vinylic CH_2 groups can well be distinguished as shown by different frequencies of corresponding ν_aCH_2, ν_sCH_2, δCH_2 and ρCH_2 bands (Sourisseau and Pasquier, 1972, 51 and 65). The ν_a and ν_s CH_2 frequencies of vinylic group for instance, are much higher (3080 and 2985 cm^{-1}) than those of methylenic group (2963 and 2933 cm^{-1}) in

C_3H_5HgCl. In the vibrational spectra of Pd complexes, on the other hand, all the stretching fundamentals give rise to bands in the 3000 to 3090 cm^{-1} range (Sourisseau and Pasquier, 1973) according to the 'aromatic' character of these C-H bonds (Table IV). The CH stretching frequencies are quite sensitive to the distribution of electronic charges as shown by the average CH stretching frequency which decreases to 2987 cm^{-1} for the ionic compounds (Table IV).

3.2. *Metal-Ligand Vibrations and Force Constants*

The main results concerning the metal-ligand vibrations and the corresponding force constants for eight allylic compounds are in Table V.

TABLE V
Metal-ligand vibrations and force constants of some allylic compounds

Compound	Vibration	Frequency (cm^{-1})	Force constant (mdyne $Å^{-1}$)
C_3H_5HgCl	ν Hg-C	510	2.66
C_3H_5HgBr	ν Hg-C	502	2.62
$(C_3H_5)_2Hg$	ν_aC-Hg-C ν_sC-Hg-C	495 475	2.6
$(C_3H_5PdCl)_2$	ν Pd-C	407 401 380	2.9
$(C_3H_5NiCH_3COO)_2$	ν Ni-C	418 403 382	2.5
$(C_3H_5)_2Mg$	ν Mg-C	360	0.8
$(C_3H_5)Li$	ν Li-C	380	0.45
C_3H_5Na	ν Na-C	190	0.30

The following comments can be made:

(a) σ bonded compounds

The Hg-C stretching frequencies of allyl mercury compounds are observed near 490 cm^{-1} in agreement with previous works (Mink et al., 1970 and 1972). The mercury-carbon force constant, calculated using a XY_2 model, has been found to be close to 2.6 mdyne $Å^{-1}$ for both allyl mercury halides and diallyl mercury (Sourisseau and Pasquier, 1972, 61 and 65). This value is also similar to Hg-C force constants of aliphatic mercury derivatives (Mink et al., 1972; Meić and Randić, 1968).

(b) π Complexes

Comparison of the infrared and Raman spectra of $(C_3H_5PdCl)_2$, $(C_3H_5PdBr)_2$ and $(C_3D_5PdCl)_2$ helps to distinguish intramolecular C_3H_5, metal-halogen and metal-carbon vibrations. The intramolecular vibrations, in particular the low frequency modes, are more deuteration sensitive than the metal-carbon vibrations while the metal-halogen frequencies, unsensitive to deuteriation, shift upon halogen substitution. Six paladium--allyl frequencies have been identified (Sourisseau and Pasquier, 1973) and the most interesting appear to be three stretching modes near 400, 384 and 365 cm^{-1} yielding 1.08, 1.10 and 1.14 isotopic frequency ratios. The first one corresponds to a symmetric stretch derived from a translational motion of the Pd and allyl group while the other two are of librational origin. The average ν Pd-C frequency is but little sensitive towards anion substitution (Cl^-, Br^- or CF_3COO^-) and gives a force constant of 2.9 mdyne $Å^{-1}$ using a diatomic Pd-(C_3H_5) model approximation. A lower value (2.5 mdyne $Å^{-1}$) is obtained for nickel compounds. The metal-ligand forces appear thus to decrease in the series alkyl (Adams, 1967), allyl, ethylene (Grogan and Nakamoto, 1968, 918) for the same metal ion.

(c) Ionic compounds

The vibrational analysis of allylmagnesium compounds such as C_3H_5MgCl, C_3H_5MgBr and $(C_3H_5)_2Mg$ shows that the average Mg-C stretching frequency is at about 360 cm^{-1} (Sourisseau and Pasquier, 1975). The corresponding force constant amounts to 0.80 mdyne $Å^{-1}$ and appears thus to be significantly lower than those of allylmagnesium compounds of Table I.

The use of ^6Li and ^7Li isotopes in allyllithium derivatives makes the assignment of the Li-C stretching motion straightforward: the ν Li-C frequency is observed at 380 and 351 cm^{-1}, respectively. The resulting force constant of about 0.45 mdyne $Å^{-1}$ is much lower than that of CH_3Li monomer (0.78 mdyne $Å^{-1}$) (Andrews, 1967) and indicates a strongly ionic character of the former. Much the same conclusion is reached for sodium salt: a strong infrared absorption band at 190 cm^{-1} assigned to the ν Na-C vibration yields even a lower sodium-carbon force constant of 0.30 mdyne $Å^{-1}$ (Sourisseau and Pasquier, 1975).

The metal-carbon force constant of the investigated allylic compounds varies thus very strongly with the type of the metal-ligand interaction and decreases for an order of magnitude in going from palladium π complex to sodium derivative (Table V).

REFERENCES

Adams, D.M.: Metal-Ligand and Related Vibrations, Edward Arnold, London, 1967, p. 207.
Akisin, P.A. and Spiridonov, V.P.: 1957, Kristallografia 2, 475.
Andrews, L.: 1967, J. Chem. Phys. 47, 4834.
Gayles, J.N. and King, W.T.: 1965, Spectrochim. Acta 21, 543.
Grogan, M.J. and Nakamoto, K.: 1968, J. Amer. Chem. Soc. 90, 918.

Guggenberger, L.J. and Rundle, R.E.: 1968, J. Amer. Chem. Soc. 90, 5375.
Herzberg, G.: 1950, Spectra of Diatomic Molecules, Van Nostrand, New York.
Kress, J.: 1974, Thesis, Paris, C.N.R.S. AO 9349.
Kress, J. and Novak, A.: 1974, J. Mol. Struct. 23, 215.
Kress, J. and Novak, A.: 1975, J. Organometal. Chem. 86, 281.
Kress, J. and Novak, A.: 1975, J. Organometal. Chem. 99, 23.
Kress, J. and Novak, A.: 1975, J. Organometal. Chem. 99, 199.
Kress, J., Bougeard, D., and Novak, A.: Spectrochim. Acta, in press.
Linnett, J.W.: 1940, J. Chem. Phys. 8, 91.
Meić, Z. and Randić, M.: 1968, Trans. Faraday Soc. 64, 1438.
Meyer, A.Y. and Chrinovitzky, M.: 1972, J. Mol. Struct. 12, 157.
Mink, J., Bursics, L., and Vegh, G.: 1972, J. Organometal. Chem. 34, C4.
Mink, J. and Pentin, Y.A.: 1970, J. Organometal. Chem. 23, 293.
Randall, S.P., Greene, F.T., and Margrave, J.L.: 1959, J. Phys. Chem. 63, 758.
Schibilla, H. and Le Bihan, M.T.: 1967, Acta Cryst. 23, 332.
Smith, A.E.: 1965, Acta Cryst. 18, 331.
Sourisseau, C. and Pasquier, B.: 1972, J. Organometal. Chem. 39, 51.
Sourisseau, C. and Pasquier, B.: 1972, J. Organometal. Chem. 39, 65.
Sourisseau, C. and Pasquier, B.: 1973, Can. J. Spectry. 18, 91.
Sourisseau, C. and Pasquier, B.: 1974, Can. J. Spectry. 19, 11.
Sourisseau, C. and Pasquier, B.: 1975, Spectrochim. Acta 31A, 287.
Sourisseau, C. and Pasquier, B.: 1972, J. Mol. Struct. 12, 1.
Walker, F.W. and Ashby, E.C.: 1969, J. Amer. Chem. Soc. 91, 3845.
Weiss, E.: 1964, J. Organometal. Chem. 2, 314.

CONFORMATIONAL STUDIES OF FLEXIBLE MOLECULES USING LANTHANIDE IONS AS NMR PROBES

BERRY BIRDSALL, J. FEENEY AND A.M.GUILIANI
Molecular Pharmacology Division, National Institute for Medical Research, Mill Hill, London, NW7 1AA

Nuclear magnetic resonance spectroscopy has been used for several years to study the conformations of molecules in solution. Until relatively recently nearly all these studies relied on relating three-bond spin coupling constants to their appropriate dihedral angles in molecules. This method can only provide partial information about the overall molecular conformation since usually there are several torsional angles which are not characterised by suitable three-bond spin coupling constants. There is also the additional problem encountered when the molecules exist as a rapidly interconverting mixture of conformers and the observed coupling constants are averaged values. In such cases there is often insufficient coupling constant data to define the conformational state unequivocally. Thus, when the method of studying molecular conformations based on lanthanide induced shifts (LIS) and relaxation effects became available we were interested in assessing if additional information about the conformations of flexible molecules could be obtained by combining these measurements with the results from coupling constant data. Flexible molecules suitable for such an investigation must contain several torsional angles and be amenable to study by both methods. We chose to examine tryptophan (I) and nicotinamide mononucleotide (II), NMN (see Figure 1), because both molecules can be fairly well characterised from coupling constant measurements and both molecules have single cation-binding sites (carboxylate and phosphate groups respectively) for lanthanide ions in aqueous solutions. Furthermore, previous studies had indicated that the molecules exist as rapidly interconverting mixtures of conformers.

1. SHIFT AND RELAXATION THEORY

Pseudocontact shifts induced when a paramagnetic lanthanide binds to the molecule can be used to get conformational information because the shift arises from a through space dipolar interaction between the magnetic moments of unpaired electrons on the metal and the nucleus observed [1-4]. The magnitude of the pseudocontact shift is given by

B. Pullman and N. Goldblum (eds.), Metal-Ligand Interactions in Organic Chemistry and Biochemistry, part 2, 313-324.

$$\frac{\Delta\gamma}{\gamma_0} = D \frac{3\cos^2\theta_i - 1}{r_i^3} - D' \frac{\sin^2\theta_i \cos 2\phi_i}{r_i^3}$$

where D and D' are constants for each lanthanide complex at a given temperature, r_i is the distance vector between the metal ion and a specific nucleus, θ_i is the angle between the principal symmetry axis of the complexed ion and the distance vector r_i and ϕ_i defines the angular position in the plane around the symmetry axis. In an axially symmetric complex D' is zero and the ratios of the shifts for each nucleus in the molecule are the same for all lanthanide ions and depend solely on r_i and θ_i.

Gd^{3+} does not cause pseudocontact shifts since it has a symmetric distribution of electrons. However, it has a long electron relaxation time and thus causes relaxation effects which are observed in both the T_1 and T_2 relaxation times. The theory is reviewed elsewhere [5-7] but for this analysis it is only necessary to consider that the dipolar part of the paramagnetic contribution to the relaxation time is a function of the inverse of r_i^6 and the correlation time (τ_c) for the dipolar interaction

$$\frac{1}{T_1} \propto f(\tau_c) \frac{1}{r_i^6} .$$

Thus the ratios of $1/T_1$ effects for the different nuclei will give their relative distances from Gd^{3+} ion if the correlation times are the same. This is a reasonable assumption for the protons on a rigid ring system, for example the indole ring of tryptophan, but may not be valid for molecules with many flexible bonds.

2.CRITERIA FOR A GOOD SET OF LIS DATA

The LIS experiment is usually performed by titrating the ligand with increasing concentrations of many different lanthanide ions.[1]. The observed shifts are corrected for any diamagnetic effects of binding a metal ion by studying the binding of La^{3+} and Lu^{3+}. If the corrected shifts give constant shift ratios for the different lanthanide ions one can use the simplified pseudocontact shift equation. Large contact shifts are observed in ^{13}C experiments but for ^1H nmr contact effects are usually observed for protons only a few bonds away from the binding site. Corrections for contact shifts can sometimes be made [2] but, in general any protons suspected of having contact contributions are ignored in the conformational analysis. If the shift ratios, after correction, are not constant, the data is not amenable to analysis at the present time.

We have measured the lanthanide induced shifts for NMN at pH 4.4

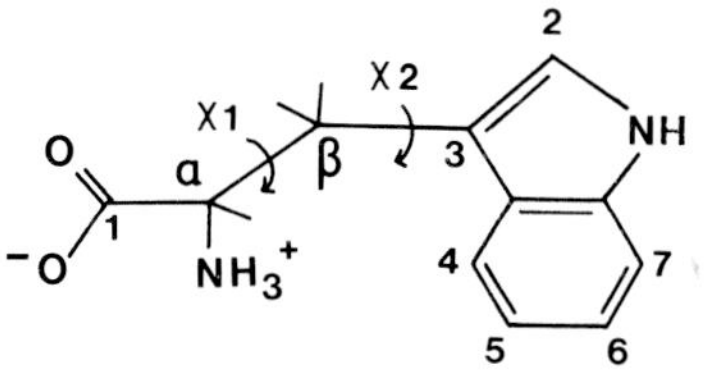

I Tryptophan

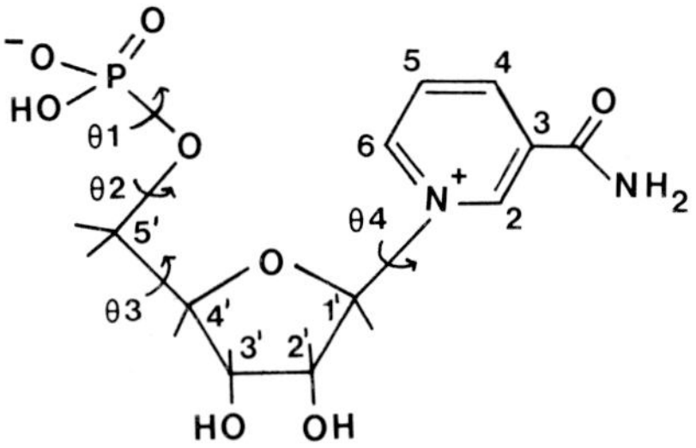

II NMN

Fig. 1. The structures of tryptophan and NMN. For torsion angles $\chi 1$ and $\chi 2$ in tryptophan (I) and $\theta 1$, $\theta 2$, $\theta 3$ and $\theta 4$ in NMN (II) positive angles (0-360°) are defined as right-handed rotations. When one looks along any bond the far atom rotates clockwise relative to the near atom. The zero angles for the flexible bonds in tryptophan are defined as follows: $\chi 1 = 0°$ when the $C_1 - C_\alpha$ bond is trans planar to the $C_\beta - C_3$ bond and $\chi 2 = 0°$ when the $C_\alpha - C_\beta$ bond is cis planar to the $C_3 - C_2$ bond. Thus for rotamer I, $\chi 1 = 240°$, rotamer II, $\chi 1 = 360°$ and rotamer III, $\chi 1 = 120°$. The zero angles for the flexible bonds in NMN are defined as follows: $\theta 1 = 0°$ when the $O_{5'}-C_{5'}$ bond is trans planar to the P-O (this O is that not bonded to Ln) bond, $\theta 2 = 0°$ when the $C_{5'}-C_{4'}$ bond is trans planar to the $P-O_{5'}$ bond, $\theta 3 = 0°$ when the $C_{4'}-C_{3'}$ bond is trans planar to the $O_{5'}-C_{5'}$ bond, and $\theta 4 = 0°$ when the $N-C_2$ bond is trans planar to the $C_{1'}$, ring O bond in the syn conformation of NMN, $\theta 4$ is in the range 320-40° and in the anti, depicted in II, $\theta 4$ is in the range 140- 250°.

[8] and the shift ratios are given in Table I. For tryptophan we used the ratios reported in the literature [2]](see Table II).

The relaxation data can be obtained either by measuring the changes in the line broadening or the T_1 relaxation time due to the bound Gd^{3+}. In the line broadening experiment the Gd^{3+} can cause chemical exchange spin-decoupling of proton multiplet signals resulting in errors in the apparent line-widths. However, by examining line broadening at high concentration of Gd^{3+} where the proton multiplets are completely decoupled these errors can be minimised. The relaxation data for NMN (Table I) was obtained by measurements of line broadening and this problem makes it difficult to get more than a qualitative order for the broadening.

By measuring T_1 relaxation times, the spin-decoupling problem is avoided. Spin-lattice relaxation time measurements for the tryptophan protons in the absence and in the presence of Gd^{3+} were made at 100 and 300 MHz at pH 4.5 using a conventional inversion recovery pulse sequence (180 - t - 90). Corrections for relaxation effects due to non-specific interactions with free Gd^{3+} were estimated from the T_1 relaxation time of dioxane measured under the same conditions. The T_1 relaxation data for tryptophan given in Table II does not agree with the data in the literature (2) which was obtained from broadening measurements.

3. NATURE OF LANTHANIDE ION COMPLEX

Before the shift and relaxation data can be analysed to give conformational information, the nature of the complex must be defined. One must consider the stoichiometry of the complex, the site or sites of binding, the ionisation states of the binding groups and whether the binding of lanthanide ion perturbs the conformation of the molecule in solution.

To determine the stoichiometry the ligand is titrated with varying concentrations of Ln^{3+}. For NMN one can fit the shift data to a 1:1 complex and obtain the same equilibrium constant for all protons [8]. At pH 4.4 lanthanide metals bind mainly to the dianionic form of the phophate group. Similar lanthanide shift ratios are observed for all protons at pH 1.7 where the lanthanide ion binds to the monoanionic form thus the metal is binding in a similar manner in both complexes and the conformations must be similar. For tryptophan the lanthanide ions have been shown to bind only to the carboxylate group [2].

The position of the metal and the direction of the principal symmetry axis for the metal complexes with phosphate and carboxylate groups have been previously determined by Williams and coworkers [1, 2] and we have used their values in this study.

4.EFFECTS OF LANTHANIDE BINDING ON THE CONFORMATION

If lanthanide ions are used as probes to determine the conformation of a molecule it is necessary to establish whether the binding of lanthanic ions perturbs the conformation in solution.

In some cases a consideration of the shift ratios at different lanthanide concentrations allows one to determine whether or not conformational changes accompany the binding of lanthanide ions. For exampl if lanthanide ions changed the composition of the equilibrium mixture of syn and anti conformers of the nicotinamide ring in NMN, this change would be reflected in the ratio of the shift observed for H_2 relative to that of H_6 at different lanthanide ion concentrations. This analysis assumes that the lanthanide ion binds to the phosphate group in both conformers under consideration. The ratio of lanthanide-induced shifts of H_2 and H_6 remains constant (within 5% over the total range of lanthanide ion concentration) thus showing that the binding of lanthanide ions has not changed the proportions of syn and anti conformers in solution.

A more general method of detecting conformational changes on metal binding is to monitor conformationally dependent vicinal coupling constants in the presence of lanthanide ions. For example, in NMN the proton coupling constants $J_{1'-2'}$ and $J_{4'-5'}$ do not vary (within 0.2 Hz) as the lanthanide concentration is changed over the range 0-10 mM (at 10mM concentration approximately 40% of the NMN molecules are complexed). This shows that the binding of lanthanide ions does not change the conformation of the ribose ring in NMN. In addition, the $J_{5'-P}$ coupling constants are unchanged in the presence of a 10 molar excess of La^{3+} showing that the conformation of the P-O-C-H backbone is also uneffected by the presence of lanthanide ions.

Similarly, tryptophan can be shown not to change conformation when a lanthanide ion binds to the carboxylate group; the $J_{\alpha C\underline{H}-\beta C\underline{H}}$ and $J_{\beta C\underline{H}-\delta C\underline{H}}$ spin coupling constants do not change in the presence of La^{3+} ions.

TABLE I
Observed and calculated shift and broadening ratios for NMN

Proton	Observed		Calculated			
			Solution I[a]		Solution II[b]	
	Shift	Broadening[c]	Shift	Broadening	Shift	Broadening
1'	11	6	15	4	12	5
2'	39		40		45	
3'	40		37		35	
4'	26		27		25	
5'	100	-[d]	100	182	100	100
2	19	30	23	28	23	62
4	-6	4	-4	5	-7	13
5	2	15[e]	-1	19	4	13
6	31	18	26	19	35	22

a. Solution I: θ1 rotated, θ2 = 360°, θ3 = 240°
67%(syn (θ4 = 10°),S), 33% (anti (θ4 = 190°), N)

b. Solution II: θ1 = 260°, θ2 = 350°, θ3 = 280
100% S, 50% syn (θ4 = 15°), 50% anti (θ4 = 245°).

c. The observed broadening data was normalized to the broadening of H_2 (taken as 30). This resonance was chosen for normalization because it exhibits only small proton spin couplings, and except for the 5' protons, has the largest observable broadening.

d. The broadening of the 5' protons is very difficult to measure because of the chemical nonequivalence of the 5' protons and the multiple spin-spin couplings on these resonances. The broadening ratio may be estimated to be in the range 50-200.

e. The H_5 proton has large spin-spin couplings to H_4 and H_6 and thus the broadening of this multiplet resonance is also difficult to measure.

TABLE II
Observed and calculated shift and relaxation ratios for tryptophan

Proton	Observed Shift[b]	Observed $1/T_1$	Calculated[a] Shift	Calculated[a] $1/T_1$
βCH_2	100	100	100	100
2	32	57	33	59
4	40	49	38	43
5	7	7	8	11
6	3	4	5	4
7	9	4	6	4

a. Calculated mixture
64% (rotamer II, 32% $\chi 2$ = 90, 32% $\chi 2$ = 270)
16% (rotamer I, 16% $\chi 2$ = 270)
20% (rotamer III, 10% $\chi 2$ = 280, 10% $\chi 2$ = 130).
b. Shift data taken from reference [2].

5. ANALYSIS OF DATA

Fitting of coupling constant, LIS and relaxation time data to a single fixed conformational structure is relatively straight-forward. However, when we consider a mixture of rapidly interconverting conformers the system is usually underdetermined and it is not possible to estimate directly the populations of the contributing conformations without making some assumptions about the potential energies of the various conformer For example, in the side-chain of an amino acid such as tryptophan it is usual to consider the three minimum energy staggered conformers for rotation about the C_α-C_β bond (rotamers I - III). Likewise, in NMN, a similar situation exists for rotations about the $C_{4'}$-$C_{5'}$ and the $C_{5'}$-$O_{5'}$ bonds. For other torsional angles such as the glycosidic bond in NMN and the C_β-C_γ bond in tryptophan the allowed angles also fall within well-defined regions. If we simplify the problem by considering only these likely minimum energy conformations then we can investigate the conformational equilibria as follows. The first step is to determine the coupling constants, pseudo-contact shifts and relaxation effects for the different nuclei in the minimum energy conformations. The coupling constants are obtained from model compound studies [9] and the lanthanide induced data are calculated for each minimum energy conformation using a modified Burlesk program [1]. The averaged values of the parameters are then calculated for various mistures of the minimum energy conformers and compared with the experimental data until a good fit is obtained.

When averaging conformations it is necessary to average values of $(3\cos^2\theta_i - 1)/r_i^{3}$ (for shifts) and r_i^{-6} (for relaxation) and not ratios of shifts and relaxation values. It should also be noted that this analysis averages point values representing low energy conformations rather than considering ensemble weighted averages within each conformation. These methods of analysis will now be applied to tryptophan and NMN.

6. TRYPTOPHAN - CONFORMATIONAL ANALYSIS

6.1. Conformations from Coupling Constants

The two three-bond coupling constants in the $\alpha CH\beta CH_2$ fragment of tryptophan can be calculated easily from its ABX type spectrum. The observed values (see Table III) can arise either (i) from a fixed rotamer with the appropriate dihedral angles for the observed coupling constants or (ii) from averaging of the component coupling constant values in a mixture of the three minimum energy staggered rotamers I to III. From a consideration of coupling constants in model cyclic compounds [9] we have estimated the component coupling constants in the rotamers to be as indicated on I to III.

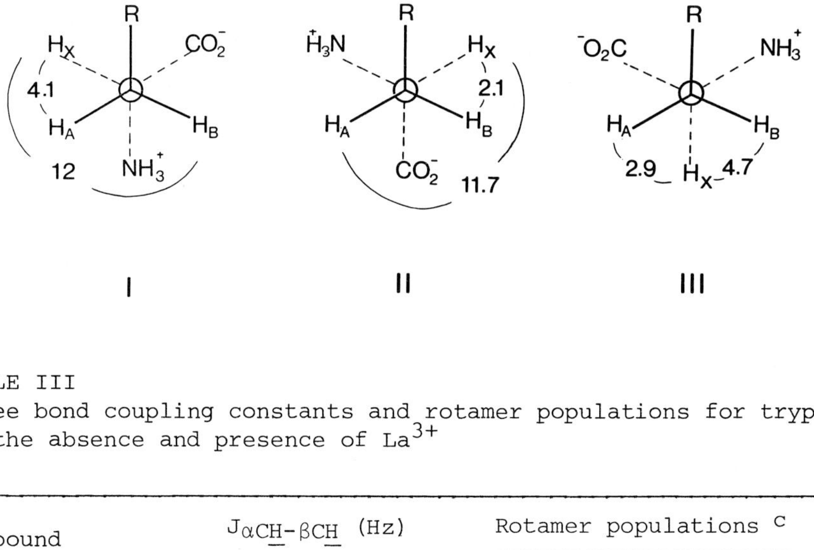

I II III

TABLE III
Three bond coupling constants and rotamer populations for tryptophan in the absence and presence of La^{3+}

Compound	$J_{\alpha C\underline{H}-\beta C\underline{H}}$ (Hz)		Rotamer populations [c]		
	J_{AX}	J_{BX}	P_I	P_{II}	P_{III}
Tryptophan [a]	8.5	4.4	0.18	0.61	0.21
Tryptophan - La^{3+} [b]	8.7	4.4	0.18	0.61	0.21

a. Tryptophan solution in D_2O, pH 4.5, 49 mM
b. 53 mM Tryptophan, 100 mM La^{3+}, pH 4.5
c. Errors $\pm$ 0.1

If the fractional populations are P_I, P_{II} and P_{III} then the averaged three-bond coupling constants are given by

$$J_{AX} = 4.1\ P_I + 11.7\ P_{II} + 2.9\ P_{III} \tag{1}$$

$$J_{BX} = 12.0\ P_I + 2.1\ P_{II} + 4.7\ P_{III}, \tag{2}$$

$$\text{where } P_I + P_{II} + P_{III} = 1. \tag{3}$$

From the measured values of J_{AX} and J_{BX} given in Table III it is possible to determine the fractional populations of the rotamers using Equations (1) to (3): these are given in Table III. The other possibility is that the side-chain has a single fixed non-staggered conformation with $\chi 1 \approx 20°$. This latter possibility seems unlikely in view of the results found for the side-chain conformations of other aromatic amino acids [10].

Unfortunately, there is no three-bond $^1H-^1H$ coupling constant to charaterise the $\chi 2$ torsion angle. However, by using van der Waal's criteria it is possible to place limits on the sterically allowed values of χ^2 for each of the rotamers I to III (Table IV). The ranges of χ^2 are different for each rotamer and for rotamer I one of these ranges is extremely restricted. Further restrictions of the possible conformations can be made from a qualitative analysis of the LIS and broadening data.

TABLE IV
Values of $\chi 2$ in Tryptophan allowed on Van der Waal's distance criteria

Rotamer	$\chi 2$	
I	0-60	260-280
II	0-90	240-360
III	90-130	280-360

6.2 Conformation from Lanthanide Data

In principle it is possible to determine the rotamer populations about $\chi 1$ from the shift and broadening ratios of the α and β protons. Unfortunately the possibility of contact shift contributions on the α proton complicates this form of analysis but the fact that the β protons both have the same LIS and relaxation effects implies that either $P_{II} = 1$ or that $P_I = P_{III}$.

We were able to fit the data satisfactorily with any of the conformational mixtures given in Table V and the best fit for the shift and relaxation data is given in Table II. The fractional population P_I, P_{II}

and P_{III} agree with the values obtained from the analysis of the three-bond coupling constants $J_{\alpha CH-\beta CH2}$. It should be noted that no solution could be obtained assuming a single conformation.

7. NMN - CONFORMATIONAL ANALYSIS

7.1. Conformations from Coupling Constants

The three-bond coupling constants $J_{5'a-p}$ and $J_{5'b-p}$ indicate that in NMN 90% of the molecules have the phosphorus atom gauche to both 5' protons [11] ($\theta 2 = 360°$). Similarly the coupling constants $J_{4'-5'a}$ and $J_{4'-5'b}$ show that more than 90% of the molecules have the 4' proton gauche to both 5' protons [11] ($\theta 3 = 240°$). Thus in NMN the phosphate backbone conformation is well defined although other nucleotides show more flexibility.

Again van der Waal's distance criteria are used to distinguish two distinct regions of allowed angles for the glycoside bond, syn $\theta 4 \sim$ 320-40 and anti $\theta 4 \sim$ 140-250). However, the range of sterically possible values of $\theta 4$ also depends on the puckering of the ribose ring. Semi-empirical energy calculations indicate that only small energy differences exist between syn and anti and one would expect a mixture of these conformers to exist [12]. There are essentially two classes of conformations of the ribose ring, S conformations (similar to $C_{2'}$-endo) and N conformations (similar to $C_{3'}$-endo) [13]. The observed ribose ring proton coupling constants of NMN indicate that there is an equilibrium mixture of N and S conformers. Using the table of ribose ring coupling constants, derived by Altona and Sundaralingam [13] for varying equilibrium mixtures of N and S conformations, we can estimate that for NMN in solution there is 70% S and 30% N conformers. The calculated coupling constants for such a mixture is $J_{2'-3'} = 5.2, J_{3'-4'} = 3.0$ Hz compared with the observed values of $J_{2'-3'} = 5.0$ Hz, $J_{3'-4'} = 2.6$ Hz. Energy calculations indicate the presence of 6 rotamers having approximately equal energy about $\theta 1$. These rotamers do not have well defined energy minima and would be able to interconvert rapidly [12]. Therefore to simplify the analysis we have assumed free rotation about $\theta 1$.

7.2. Conformation from Lanthanide Data

The problem of fitting the LIS and broadening data has now been simplified considerably. Free rotation is assumed about $\theta 1, \theta 2 = 360$, $\theta 3 = 240$ and the angles $\theta 4$ lie in the two regions, syn and anti. In addition, we consider a mixture of S and N conformations for the ribose ring. We have calculated the shifts and broadening for the S and N sugar conformations assuming free rotation about $\theta 1$. Excellent agreement between observed and predicted shifts for the ribose ring protons was obtained for a 2:1 mixture of S and N (Table I), which is very close to the proportion estimated for analysis of the ring proton coupling constants.

When the syn and anti ranges of both the S and N conformations were examined only one solution that fitted the observed and broadening data was found. This solution (Table I) was a 2:1 mixture of syn, S and anti,

N conformers. This particular solution is the simplest one that agrees with all the conformational information available. It should be emphasised that more complex averaged solutions which fit the data could possibly be found.

If no assumptions regarding the conformation of NMN are included in the analysis it is possible to fit the observed LIS and broadening data to different mixtures of conformers. One such solution is given in Table I. This consists of one conformation for the ribose ring and fixed values for $\theta 1$, $\theta 2$ and $\theta 3$. This solution is inconsistent with all the coupling constants data *and reveals that analysing the LIS and broadening data for a flexible molecule without considering other conformational information can lead to incorrect solutions.*

TABLE V
Mixtures of tryptophan conformers which fit the lanthanide shift and relaxation data

Fractional population	Rotamer	$\chi 2$
0.65 ± 0.10	II	240-270 or mixtures of 240-270 and 90
0.15 ± 0.10	I	270 or mixtures of 270 and 0-60
0.20 ± 0.10	III	260-280 or mixtures of 260-280 and 90-130

8.CONCLUSIONS

From the foregoing discussion it is seen that combined studies of lanthanide induced chemical shifts and relaxation effects and spin coupling constants can be used to provide information about the conformational states of flexible molecules. For tryptophan it has been shown that the LIS data cannot be explained in terms of a single conformation and can only be fitted to mixtures of rotamers about the C_α-C_β bond which were in good agreement with those found from spin-coupling constant studies. In principle, the LIS and relaxation method allows us to investigate parts of the molecule which are inadequately characterised by spin-coupling constants. However, in practice it is sometimes difficult to obtain unequivocal information because several different mixtures of conformers lead to equally acceptable fits of the observed data. Thus for the torsional angle about the C_β-C_γ bond in tryptophan we found several solutions corresponding to different mixtures of conformations (all of which were acceptable in terms of van der Waal's steric considerations).

For nicotinamide mononucleotide it proved to be much more difficult to find mixtures of conformers which fitted all the data and therefore many mixtures of conformations could be eliminated. From the LIS data it was possible to estimate the syn/anti ratio which is not accessible from spin-coupling constant data: our results are in agreement with previous studies where it was shown by Nuclear Overhauser experiments [14] that there are approximately equal amounts of the syn- and anti-forms in the equilibrium.

Spin coupling constants can only give localised conformational information concerning a particular fragment of the molecule and they cannot give any indication of the presence of two or more correlated conformational features in the same molecule. Such information can be obtained from the lanthanide studies and this is illustrated by the finding of the syn S and anti N conformations in NMN (see Table I).

The two methods of conformational analysis clearly complement each other. Spin coupling constant studies are very useful and simple to carry out for conformations about bonds where the minimum energy conformers can be well-characterised (such as a C-C bond between two sp^3 carbons) and where there are suitable nuclei to provide the necessary three-bond coupling constants. Lanthanide induced shifts and relaxation effects provide conformational information about parts of the molecule inaccessible to spin coupling constants. However, for flexible molecules where the contributing conformers have several different torsional angles it is much more difficult to analyse the data. Agreement between the observed and calculated LIS and relaxation data for all the nuclei is a necessary but not sufficient criterion for the correctness of the calculated conformational equilibrium.

REFERENCES

1. Barry, C.D., North, A.C.T., Glasel, J.A., Williams, R.J.P., and Xavier, A.V.: Nature (London) 232, 236, 1971.
2. Levine, B.A. and Williams, R.J.P.: Proc. Roy. Soc. London A. 345, 5, 1975.
3. Dobson, C.M. and Levine, B.A.: in Pain, R.H., Smith, B.J. (eds.), New Techniques in Biophysics and Cell Biology, John Wiley + Sons, London, 1976, vol. 3.
4. Reuben, J.: in Emsley, J.W., Feeney, J., Sutcliffe L.H. (eds.), Prog. Nucl. Magn. Reson. Spectroc., Pergamon Press, Oxford, 1973, p. 1.
5. Dwek, R.: NMR in Biochemistry, Clarendon Press, Oxford, 1973.
6. Solomon, I.: Phys. Rev. 99, 559, 1955.
7. Bloembergen, N.: J. Chem. Phys. 27, 595, 1957.
8. Birdsall, B., Birdsall, N.J.M., Feeney, J. and Thornton, J.: J. Am. Chem. Soc. 97, 2845, 1975.
9. Feeney, J.: J. Mag. Res., 21, 473, 1976.
10. Hansen, P., Feeney, J. and Roberts, G.C.K.: J. Mag. Res. 17, 249, 1975.

11. Sarma, R.H. and Mynott, R.J.: Chem. Comm. 975, 1972.
12. Thornton, J. and Bayley, P.M.: Biochem. J. 149, 585, 1975.

13. Altona, C. and Sundaralingam, M.: J. Am. Chem. Soc. 95, 2333, 1973.
14. Egan, W., Försen, S. and Jacobus, J.: Biochemistry 14, 735, 1975.

METAL IONS AS PROBES FOR NMR STUDIES OF STRUCTURE AND DYNAMICS IN SYSTEMS OF BIOLOGICAL INTEREST

JACQUES REUBEN
Isotope Department, The Weizmann Institute of Science, Rehovot, Israel

1. INTRODUCTION

The measurable quantities in nuclear magnetic resonance (NMR) spectra of molecules in solution are chemical shifts, spin-spin coupling constants, and nuclear relaxation times (or rates). All of them are related to molecular structure and the relaxation rates reflect in addition the molecular dynamics in solution. Thus, in principle, NMR studies provide direct information on molecular dynamics and structure. However, with macromolecular systems of biological interest this information is not always discernible, either because of extensive line overlap in the NMR spectra or due to the weakness of effects in the necessarily dilute solutions of the macromolecules. These difficulties can be circumvented in favourable cases by the use of metal ions as NMR probes. Presented in this paper is a discussion of some aspects of the application of metal ions in NMR studies of the dynamics and structure in systems of biological interest. It is based on recent results obtained in this laboratory and the original articles should be consulted for details and references not given here.

2. RELAXATION RATES OF QUADRUPOLAR NUCLEI AND MACROMOLECULAR DYNAMICS

The molecular dynamics in solution is reflected in principle in the nuclear relaxation rates. Nuclei with a spin quantum number $I > 1/2$ are particularly sensitive in this respect. Such nuclei posses an electric quadrupole moment which interacts with intramolecular electric field gradients resulting in enhanced relaxation rates. The molecular motion effectively modulates the quadrupole interaction and the nuclear relaxation rate is in fact related to the correlation time, τ_c, characteristic of the motion. It can be shown that to a good approximation the longitudinal relaxation rate, $1/T_1$, is given by

$$1/T_1 = C\tau_c/(1+2.67\,\omega^2\tau_c^2), \tag{1}$$

where ω is the nuclear resonance frequency and C is a constant for a given nucleus and chemical environment (Reuben and Luz, 1976). The constant C is larger for nuclei experiencing larger intramolecular electric field gradients. Equation (1) already suggests that measurements of relaxation times at several frequencies should permit an unequivocal determination of the correlation time.

In order to probe molecular dynamics by measuring the relaxation rates of metal ion nuclei, a sufficiently strong chemical interaction

B. Pullman and N. Goldblum (eds.), Metal-Ligand Interactions in Organic Chemistry and Biochemistry, part 2 , 325-330 .

is needed in order to ensure that the bound ions reside on the macromolecule for times longer than the tumbling time of the complex. Otherwise, if the residence time is shorter, it will constitute the effective correlation time. Experiments are normally carried out under conditions of a large excess of the metal ion over the concentration of macromolecular sites and the quantity of interest is the increment in the relaxation rate, $1/T_{1p}$, defined as

$$1/T_{1p} \equiv 1/T_1 - 1/T_1^{\circ} \qquad (2)$$

where $1/T_1^{\circ}$ is the relaxation rate in the absence of the macromolecule. This increment is related to the relaxation rate in the bound state, $1/T_{1b}$, by

$$1/T_{1p} = P_M/T_{1b}, \qquad (3)$$

where P_M is the fraction of bound ions and $1/T_{1b}$ obeys Equation (1). In writing Equation (3) it is assumed that chemical exchange is rapid, compared to the relaxation rate of the bound ions. By combining Equations (1) and (3) and rearranging, one obtains

$$T_{1p} = 1/(P_M C \tau_c) + 2.67\, \tau_c\, \omega^2/(P_M C) \qquad (4)$$

Thus, the correlation time is obtained from the square root of the ratio between the slope and the intercept on the ordinate of the plot of T_{1p} against ω^2.

2.1. Lanthanum - 139

A preliminary investigation of the effects of bovine serum albumin (BSA), a globular protein of molecular weight 69 000, on the relaxation rate of lanthanum - 139 in solutions of $LaCl_3$ has shown that the protein induced increment is frequency dependent and that ^{139}La could be a useful NMR probe for studying macromolecular dynamics (Reuben, 1975b). Some relevant properties of lanthanum - 139 are summarized in Table I.

TABLE I
Properties of lanthanum - 139

Natural abundance	99.91%
Nuclear spin, I	7/2
NMR freq. at 14 kG	8.43 MHz
NMR sensitivity (Rel. to ^1H)	5.92%

A large body of evidence from spectroscopic and magnetic resonance studies shows that the lanthanides form strong complexes with a number of proteins and nucleic acids (for recent reviews cf. Nieboer, 1975 and Reuben 1975a). In such studies the La^{3+} cation usually serves as a spectroscopically silent control. In the ^{139}La - NMR study of the interaction of La^{3+} with BSA it was found that the La^{3+} ions associate with virtually all of the free carboxylates of the protein. Very low protein concentrations (of the order of 1 mg ml^{-1}) produced easily measurable increments in the longitudinal relaxation rate of ^{139}La in a 0.4 M solution of $LaCl_3$. The increment was linear with the protein concentration and the reciprocals of the slopes of such curves obtained at different resonance frequencies were in turn linear with the squares of the resonance frequency in accord with Equation (4). The magnitude of this effect can be judged, e.g., from the ratio between the T_{1p} values at 8.4 and 4.2 MHz, which was found to be greater than 3. The analysis of the results yielded $\tau_c = 3.7 \times 10^{-8}$s (at 23±2°C), which is the correlation time charateristic of the isotropic rotational motion of the protein molecule (Reuben and Luz, 1976). This value is in excellent agreement with correlation times of BSA obtained by other methods.

2.2. *Sodium - 23*

Biological fluids are abundant in sodium ions and ^{23}Na, which is also a nucleus with a quadrupole moment, could serve as a natural probe in a manner similar to that described for ^{139}La. Some relevant properties of sodium - 23 are summarized in Table II.

TABLE II
Properties of sodium - 23

Natural abundance	100%
Nuclear spin, I	3/2
NMR freq. at 14 kG	15.8 MHz
NMR sensitivity (rel. to ^1H)	9.25%

The use of ^{23}Na as an NMR probe of macromolecular dynamics has been suggested by James and Noggle (1969) in their study of the interaction of sodium ions with soluble RNA (this is in fact tRNA with a molecular weight of ca. 27 000). However, in applying this approach to study the alkali ion- DNA interaction (using a sonicated DNA with an average molecular weight of 500 000) it was found that the DNA induced increments in the ^{23}Na relaxation rate showed a very small frequency dependence. The ratio between the increments at 4.3 and 15.7 MHz was only 1.25±0.2 from which an *upper limit* of 5.5×10^{-9} s was calculated for the correlation time (Reuben et al., 1975). A comparison of dissociation constants calculated per unit phosphate and relaxation rates, $1/T_{1b}$, obtained from ^{23}Na-NMR studies of tRNA and DNA is presented in Table III. This comparison shows that nucleic acids widely differing in molecular weight and size produce similar relaxation effects on the ^{23}Na nuclei.

TABLE III
Dissociation constants and ^{23}Na relaxation rates for sodium - Nucleic acid complexes

	tRNA[a]	DNA[b]
K_D, mM	3±1	11±1
$1/T_{1b}$, s^{-1}	222±19	170±14
Freq., MHz	15.0	15.7
Temp., °C	24.5±1	23±1

a. From James and Noggle (1969)

b. From Reuben et al. (1975)

Analogous similarities are observed with polyphosphates of varying chain length and with polyacrylic acid. Correlation times can be estimated from diffusion coefficients and the results are 2×10^{-8} and 10^{-4} s for tRNA and DNA, respectively, i.e. much longer than the above given upper limit obtained by NMR. Moreover, assuming that 'bound' sodium experiences the same electric field gradient as the aquo-ion and using $\tau_c = 5.5 \times 10^{-9}$ s to calculate a relaxation rate with Equation (1), the value $1/T_{1b} = 14\ 900\ s^{-1}$ is obtained. This should be regarded as a lower limit since any binding is expected to result in larger electric field gradients. It was therefore concluded, that only a small fraction, less than 1%, of the 'bound' sodium ions might come into a closer association with the DNA molecule. This interaction must be characteristic by a mean life time the upper limit of which is 5.5×10^{-9} s (Reuben et al., 1975). Thus, because of short life times of the macromolecular complexes of sodium, the relaxation rate of ^{23}Na cannot be used in studies of macromolecular dynamics in solution.

3. MANGANESE (II) AS A STRUCTURAL PROBE FOR A DNA INTERCALATION COMPLEX

The manganese aquo-ion has a characteristic and relatively sharp electron paramagnetic resonance spectrum, whereas the spectrum of bound manganese is usually much broader and of low intensity. Thus, the spectral intensity may serve to monitor the concentration of free Mn^{2+} in studies of the binding of Mn^{2+} to macromolecules. By competition the interaction of other cations can also be investigated. It may be of interest to point out that Na^+ competes, or interferes with the binding of Mn^{2+} to DNA. From the analysis of the competitive effects a dissociation constant of 10 mM has been obtained (Reuben and Gabbay, 1975) in good agreement with the one obtained by ^{23}Na-NMR (cf. Table III). In applying this approach to systems containing Mn^{2+}, DNA, and organic cations an important observation was made. The dissociation constants obtained for compounds that can intercalate between the base pairs of DNA were much higher than values obtained directly by equilibrium dialysis. On the other hand, for organic cations that cannot intercalate both methods yielded similar results (Reuben and Gabbay, 1975). Thus, with intercalating compounds conditions can be achieved such that Mn^{2+} and the organic cation are simultaneously bound to DNA in a way resembling that of metal ion and small ligand (substrate, inhibitor, etc.) bound to an enzyme. In the latter system, manganese (II) has successfully been used as a paramagnetic probe in NMR mapping studies (for reviews, cf. Cohn and Reuben, 1971; Dwek, 1973; Mildvan and Cohn, 1970). In such studies, the increments in the relaxation rates of the ligand nuclei are measured. The quantity P_M in Equation (3) is now the fraction of ligands bound in the vicinity of the Mn^{2+} ion in its macromolecular complex, effects in the absence of the macromolecule are appropriately accounted for, and $1/T_{1b}$ is the relaxation rate due to the electron-nuclear dipolar interaction given by

$$1/T_{1b} = Dr^{-6}f(\tau_c) \qquad (5)$$

where D is a known constant for given nucleus and electronic spin, r is the distance between the two, and $f(\tau_c)$ is a function of the correlation time.Thus, under favourable conditions the sixth root of the ratio between relaxation times gives the ratio between the distances.

This approach has been applied for the first time to study the structure of a DNA intercalation complex by Reuben, et al. (1976). The 3,8-dimethyl-N-methyl phenanthrolinium (DMP) cation was chosen as the intercalating compound on the basis of its easily interpretable proton NMR spectrum.

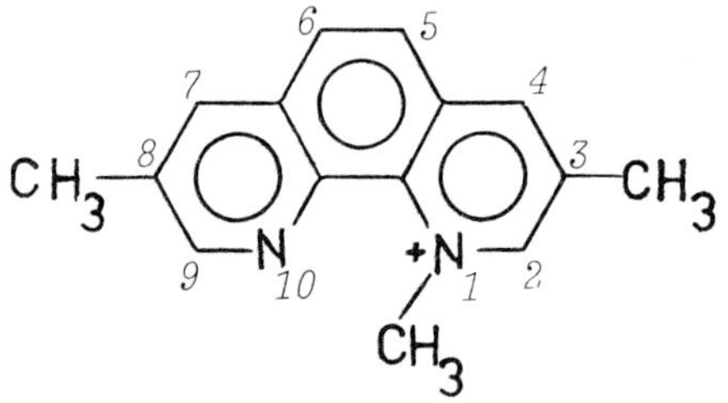

3,8-dimethyl-N-methyl phenanthrolinium (DMP)

The Mn^{2+} effected relaxation rates in the presence of DNA were markedly different for the different protons. Thus, the methyl group at position *1* was very strongly affected, whereas the relaxation rates of the methyls at positions *3* and *8* were unaffected at all. The effects on the ring protons were also different for different positions. Those at positions *5* and *6* were the most affected whereas the effect on that at position *2* was 7.6 times as small. Two structural models were examined in one of which the DMP molecule was placed with its long axis almost parallel to the hydrogen bonds. The base nitrogens (N-7) and the phosphate groups were taken as the likely manganese binding sites.On the basis of the relaxation data, one of the models was unequivocally discriminated. It was concluded that in the intercalation complex the long axis of the DMP molecule is almost perpendicular to the hydrogen bonds of the DNA base pairs (Reuben et al., 1976). The action of a number of drugs as well as of some mutagenic and carcinogenic agents is believed to result from the formation of DNA intercalation complexes and it seems that the NMR approach using manganese (II) as a paramagnetic probe should be fruitful in the study of such interactions.

REFERENCES

Cohn, M. and Reuben, J.: 1971, 'Paramagnetic Probes in Magnetic Resonance Studies of Phosphoryl Transfer Enzymes', Accounts Chem. Res. 4, 214-222.

Dwek, R.A.; 1973, 'Nuclear Magnetic Resonance in Biochemistry', Clarendon Press, Oxford.

James, T.L. and Noggle, J.H.; 1969, '^{23}Na Nuclear Magnetic Relaxation Studies of Sodium Ion Interaction with Soluble RNA', Proc. Natl. Acad. Sci. U.S. 62, 644-649.

Mildvan, A.S. and Cohn, M.: 1970, 'Aspects of Enzyme Mechanisms Studied by Nuclear Spin Relaxation Induced by Paramagnetic Probes', Adv. Enzymology 33, 1-70.

Nieboer, E.: 1975, 'The Lanthanide Ions as Structural Probes in Biological and Model Systems', Structure and Bonding 22, 1-47.

Reuben, J.: 1975a, 'The Lanthanides as Spectroscopic and Magnetic Resonanc Probes in Biological Systems', Naturwissenschaften 62, 172-178.

Reuben, J.: 1975b, 'Lanthanum-139 as a Nuclear Magnetic Resonance Probe of Macromolecular Dynamics', J. Am. Chem. Soc. 97, 3823-3824.

Reuben, J., Adawadkar, P. and Gabbay, E.J.: 1976, 'Structure of a DNA Intercalation Complex as Determined by NMR Using a Paramagnetic Probe', Biophys. Struct. Mechanism. 2, 13-19.

Reuben, J. and Gabbay, E.J.: 1975, 'Binding of Manganese (II) to DNA and the Competitive Effects of Metal Ions and Organic Cations. An Electron Paramagnetic Resonance Study', Biochemistry 14, 1230-1235.

Reuben, J. and Luz, Z.: 1976, 'Longitudinal Relaxation in Spin 7/2 Systems. Frequency Dependence of Lanthanum-139 Relaxation Times in Protein Solutions as a Method of Studying Macromolecular Dynamics', J.Phys. Chem. 80, 1357-1361.

Reuben, J., Shporer, M. and Gabbay, E.J.: 1975, ' The Alkali Ion - DNA Interaction as Reflected in the Nuclear Relaxation Rates of ^{23}Na and ^{87}Rb', Proc. Natl. Acad. Sci. U.S. 72, 245-247.

DISCUSSION

Haynes:

Does your conclusion that La^{3+} ions are rigidly attached to BSA molecules imply that the cation is liganded by two COO^- groups?

Reuben:

Since virtually all of the COO^- groups of BSA serve as binding sites for La^{3+} it is unlikely to have more than a few that could form such a cluster. It is very likely however that a COO^- group acts as a bidentate ligand in a fashion similar to that observed with small ligands like acetate.

SCF MO LCGO STUDIES ON ELECTROLYTIC DISSOCIATION: THE SYSTEM $LiF.nH_2O$ (n=1, 2)

GEERD H.F. DIERCKSEN* and WOLFGANG P. KRAEMER
Max-Planck-Institut für Physik und Astrophysik, 8 München 40, Föhringer Ring 6, Germany

ABSTRACT. Accurate single-determinant Hartree-Fock calculations have been performed, employing extended gaussian basis sets, in order to investigate the equilibrium geometrical structures and several binding potentials for a number of representative clusters of lithium fluoride with one and two water molecules. The SCF results are used to discuss qualitatively the energetics of the ionic LiF bond dissociation under the influence of the polar water molecules.

1. INTRODUCTION

Already in 1887 S. Arrhenius [2] stated that various experimental results could be explained from a common point of view, if it is assumed that molecules are able to dissociate in polar solvents into oppositely charged fragments, the so-called ions. The phenomenon of this dissociation of a large number of neutral and chemically stable molecules into solvated ionic fragments under the influence of electrostatic forces is of tremendous importance for chemistry, in particular for an understanding of chemical reactions in the gas phase and in solutions. Chemical reactions involving ion hydration are not restricted to terrestrial laboratory conditions but have also been found to be important in the chemistry of the Earth's atmosphere [3-8]. Though the interaction of neutral molecules and of ions with the surrounding solvent molecules plays a key rôle for the dissociation process, very little is known in detail about these interactions and the structure of ionic solutions.

Experiments on the solvation problem have accumulated a great amount of material both for gas phase and liquid phase processes. Among these data the solvation energies are of particular interest. They give some indication of the strength of the solvation shell formed and they can be most 'directly' compared to the stabilization energies calculated by quantum chemical methods.

Most experimental data are available for gas phase ion solvation.

---

* Part of this work has been submitted in partial fulfilment of the requirements of the Habilitation, Technische Universität München, December 1973 [1].

B. Pullman and N. Goldblum (eds.), Metal-Ligand Interactions in Organic Chemistry and Biochemistry, part 2 , 331-376 .

The pioneering and stimulating work has been done by P. Kebarle and coworkers [9-12], and by E.E. Ferguson and coworkers [5-7], mainly by high pressure mass spectroscopy. In the gas phase at low pressure the interaction of the complexes with the surrounding gas can be assumed to be negligible, and solvation energies can be determined as a function of solvation steps: $(Ion).Solv_{n-1} + Solv \rightarrow (Ion).Solv_n$. Such solvation energies have been determined for a large number of ions. The following conclusions hold in general: For equally charged ions the solvation energies decrease with increasing ion size. The solvation energies for a particular ion and solvent are found to decrease smoothly with increasing degree of solvation except for very few cases where a very sharp drop of solvation energy is observed at a given number of solvent molecules. Thus, in general, the change in the solvation energies cannot be used to determine the number of solvent molecules collected in the first or second hydration shell around the ion.

However, the results obtained for the gas phase solvation can only partially be transferred to the liquid phase, where the situation is obscured by influences arising from the interaction with the liquid, and where the dynamical nature of the solvation complexes becomes more important. Nevertheless it can be expected at least for strong ion solvent complexes, that the first hydration shells have a similar structure in both cases.

By combined X-ray and neutron diffraction techniques, Narten and coworkers [13] have investigated aqueous LiCl solutions of different concentrations, from nearly saturated solutions ($LiCl : H_2O = 1:3$) to very dilute solutions ($LiCl : H_2O = 1:166$). It has been found in this study that in sufficiently dilute solutions three different regions may be defined. In the first one predominantly tetrahedrally coordinated water clusters are present, similar to those clusters observed in the pure solvent. This region disappears when the molar ratio is 10 or less. The other two regions may be attributed to the hydration shells of the two ions. The Li cation appears in general to be tetrahedrally surrounded by 4 water molecules, the lone pair electrons of the water molecule pointing towards the cation. The Cl anion appears to be octahedrally hydrated by 6 water molecules on the average, the water molecules being linearly hydrogen bonded to the anion. These structures are also preferred in highly concentrated solutions. For all but the very concentrated solutions these coordination numbers are strongly influenced by dynamical effects and they can thus only be specified as 4 ± 1 and 6 ± 1, for the Li cation and the Cl anion, respectively. The coordination distances between the ions and water molecules vary with concentration. The fluctuations obtained are large and indicate that at any instant a considerable fraction of the molecules is present in largely distorted internal and intermolecular configurations.

From these results of Narten and coworkers a model *of ionic solutions* can be derived:

In the very concentrated solutions investigated ($LiCl.3H_2O$) the composition is similar to that of the crystalline dihydrate ($LiCl.2H_2O$). One therefore might expect to find some similarities between such solutions and the crystalline state. In the concentrated solution well--oriented hydration shells of 4 and 6 water molecules are observed around the cations and anions, respectively. The average structure of the hydrated complexes is already similar to that in the gas phase, but there are not enough water molecules to provide each ion with its own separate hydration shell and therefore many water molecules have to be shared between two hydration shells of the ions. This places a serious geometrical constraint on the structure, and a compromise between the competing interactions is necessary. This situation does not remind one so much of solvated ions, but is reminiscent of the crystal state. The OCl distance is just the value observed in crystals (R_{OCl} = 3.19 Å), and the OLi distance is particularly large (R_{OLi} = 2.25 Å). The liquid character of the solution is indicated by the high fluctuation of these values.

In more dilute solutions the strain on the complexes is lowered. An increasing number of water molecules becomes available and is used to build separate solvation shells around each ion. When the molar ratio has fallen to eight, the average distances have decreased to R_{OCl} = 3.1 Å and R_{OLi} = 1.94 Å. When the molar ratio has fallen to ten, enough water molecules are available in principle to guarantee each structural unit its individual existence: $Cl^{-}.(H_2O)_6$ and $Li^{+}.(H_2O)_4$. When the molar ratio falls even lower, experiment shows that most of the excess water is present in a state comparable to that of pure water. However, small amounts of water are found which are in random configurations and highly mobile.

It has thus been found from experimental measurements that for decreasing concentrations the exchange of water molecules by cooperative effects becomes significant. The fluctuations in the coordination numbers are estimated to be approximately ± 1. At very low concentrations, the ion-water correlation pattern becomes experimentally undetectable, indicating high fluctuations. But because of the strong interaction between the ions and water in the gas phase it is reasonable to assume that the ion-water complexes remain quite well defined in spite of the strong dynamical effects.

In the case of ion solvation, the strong electrostatic field of the ion leads to an interaction with polar solvent molecules, and hence the electrostatic energy is found to contribute most to the total interaction energy. A fairly simple model for ion solvation has been proposed by Frank and Wen [14]. Two different regions surrounding the solute are distinguished. In the inner sphere close to the ion the solvent molecules are more or less strongly bound and are therefore less mobile than the non-interacting free solvent molecules. In the outer region the solvent molecules are much less bound and there is a permanent exchange with free solvent molecules. Using this model, Frank and Wen were able to explain the dynamical behavior of ion-solvent complexes in aqueous solutions.

The structure of solvated ions has first been discussed by Buckingham [15] on theoretical grounds. According to these investigations, which are based on the classical electrostatic model, cations form solvation shells by building bonds directly to the atoms X of solute molecules such as HXR (where R stands for H or for an organic group), while anions form solvation shells by hydrogen bonds $A^{(-)}$---HXR to the solvent molecules. These structures derived by a purely classical model have in fact been proved to be the energetically most stable ones by a large number of ab initio calculations on various anions, cations and solvent molecules [16, 17, 18]. A water molecule is found to be bonded to the NO^+ molecular ion by an O-N bond, the N being the electropositive pole of the nitrosyl ion. Similarly, a water molecule is determined to be bonded to the cyanide anion CN^- by a hydrogen bond OH---N, the nitrogen being the electronegative pole of the cyanide ion. All other structures have been found to be energetically less stable, or even repulsive.

The solvation of polar neutral molecules, on the other hand, is more difficult to describe. The total interaction energy is much smaller and is mostly a sum of various contributions. Moreover the energy of interaction depends on the symmetry of the molecule, as in the case of molecular ions, and is different for different directions of interaction

During the last few years, particularly due to the progess in computational techniques, it has become possible to study the overall behavior of whole ensembles consisting of several hundred molecules, using dynamical [19-22] or statistical models [23-27]. These investigations have brought some new insight into the structures and properties of liquids and liquid mixtures, and some important characteristics could be explained.

However, any of these studies of the properties of molecular complexes needs *sufficiently* accurate interaction potentials between the molecular species involved. Such potentials have now become available for a number of ions and molecules, as for example for the hydrated lithium cation and fluorine anion [25, 28-40].

The structure of aqueous solutions of electrolytes of the type of LiCl can be reasonably well understood from the results of the experimental investigations of Narten [13]. But very little can be deduced from such experiments about the energetics of the actual electrolytic bond dissociation under the influence of polar solvent molecules. Thus the present study is aimed to study this question on a purely qualitative level utilizing quantum theoretical results for a number of representative clusters of the lithium fluoride-water system. The following properties have been investigated: (a) the equilibrium structures of the LiF-water complexes, (b) the hydration energies as a function of the lithium fluoride bond distance, and (c) the potential energy of the lithium fluoride bond dissociation for different hydration states of the system.

The clusters that have to be considered in order to take into account only the most important third body effects on the interaction potentials in the lithium fluoride-water system are *at the limits of feasibility of any reliable quantum-mechanical study*. The semiempirical methods developed so far and applied to ion solvent interactions do suffer from the unreliability of the calculated quantities, mostly geometries and binding energies. All ab initio calculations with small and inflexible basis sets are essentially semiepirical in character, and the results obtained depend strongly on the choice of the basis sets. The present investigations thus have been restricted to study the most important interaction potentials of a lithium fluoride system with one and two water molecules by ab initio calculations within the framework of the Roothaan closed shell self-consistent field method, using large and flexible basis sets and including polarization functions, which guarantee a near Hartree-Fock accuracy of the calculated properties. It is believed that for strong interactions the potentials calculated for these model systems do describe the actual interaction potentials in the electrolytic solution reasonably well, and that they can thus serve to build *model potentials* for dynamical and statistical calculations. The non-additivity of the different interaction potentials has been investigated previously [41-44], and may be accounted for qualitatively in modelling the surface.

2. DETAILS OF THE CALCULATIONS

The wave functions and the energy expectation values have been calculated using Roothaan's SCF-LCAO-MO expansion method [45]. All calculations have been carried out on an IBM 360/91 computer, using the program system MUNICH [46]. The molecular orbitals have been expanded in a set of cartesian gaussian functions of the form $\eta = x^l y^m z^n \exp(-\alpha r^2)$ (unnormalized). In particular, a (11s, 7p, 1d), (11s, 2p), and (6s, 1p) basis set has been employed for the fluorine and the oxygen, the lithium, and the hydrogen centers, respectively, contracted to [5s, 3p, 1d], [5s, 2p], and [3s, 1p] basis sets, respectively, to reduce the number of linear parameters in the SCF iteration procedure. The exponential parameters and contraction coefficients have been taken from the literature [47]. Only the polarization functions (d-type functions on the F- and O-centers, and p-type functions at the Li- and H-centers) have been optimized by SCF calculations on adequate systems (on the HF molecule for F [28], on the LiOH molecule for Li [31], and on the H_2O molecule for O and H, [48]).

The SCF energies calculated with these basis sets for the systems H_2O, Li^+, F^-, and LiF, as well as the best SCF energies reported in the literature for these systems, are compiled in Table I. For the H_2O molecule the Hartree-Fock limit has been estimated to be $E_{HF}(H_2O) = -76.068$ AU [35]. Thus for this representative system the SCF energy calculated with the basis sets described is about 0.014 AU (appr. 9 kal/ mol^{-1}) from the Hartree-Fock limit, and similar relative deviations from the Hartree-Fock limit should be expected for the other systems. This deviation from the Hartree-Fock limit is mainly due to an inadequate description of the wavefunction for inner shell electrons (of 1s-type) of the atoms O, and F. This energy contribution, however, is expected to change only insignificantly on bond formation. The basis sets used here can therefore be expected to yield reliable results for the geometrical structures and binding energies of the systems to be investigated within the accuracy of the Hartree-Fock approximation.

It is well established that the closed shell self-consistent field method leads to a qualitatively correct behaviour of the calculated interaction potential if the system separates into closed shell subsystems. In this case correlation effects on the potential energy curves are expected to be small, and the SCF interaction potentials are parallel to the non-relativistic potentials. Appropriate methods to study correlation effects of such systems are at hand now, and have recently been applied to ion-solvent interaction and hydrogen bond interaction [40, 51-54]. The correlation effects found for the interaction potentials are less than 10% of the hydration energy, and can thus be neglected in a first approximation.

TABLE I
SCF energies (E^{SCF}) for the systems H_2O, Li^+, F^-, and LiF.

System	E^{SCF} [AU]		ΔSCF [AU]
	This work	Best value	
H_2O	- 76.05118 d(OH) = 1.80887 [AU],	- 76.06587 [35] angle (HOH) = 104.53° (exp.)	0.01469
Li^+	- 7.23621	- 7.236413 [52]	0.00020
F^-	- 99.44068	- 99.45936 [52]	0.01868
LiF	-106.97485 d(LiF) = 2.95 [AU] (theor. equil. distance)	-106.99162 [53] d(LiF) = 2.9877 [AU]	0.02677

It is realized that since this project has been started, basis sets have become available which effectively yield the same energy with fewer basis functions [55, 56]. Energy optimized basis sets larger than those employed in the present study have been used in self-consistent field investigations on ion-solvation, as well. Larger basis sets necessarily do lead to lower total SCF energies than smaller ones. It is of importance to recall that for basis sets of comparable size the lower total energy computed is by no means a criterion for 'a more correct' SCF binding energy. It is obvious and well known that the energy optimisation of basis sets tends to a most accurate description of the inner shell electron distribution, connected with the largest gain of total energy. In contrast, the calculated binding energies depend critically on a correct description of the valence electron distribution mainly involved in the bond formation.

A serious problem in calculating binding energies from the SCF energies of the combined and the separated systems involves the so-called *superposition error*. The reason is that the SCF energies of the combined system and the separated subsystems are calculated in slightly different approximations. In the equilibrium geometry configuration of the total system, each subsystem can partly make use of the basis sets of the other subsystems involved in order to optimize the description of its 'own' electron density distribution. The energy effects connected with this phenomenon have been studied by B. Liu, and A.D. McLean [57] for the interaction of two helium atoms by stepwise increasing the basis set size until 'saturation' was achieved as far as the binding energy is concerned.

For practical reasons, the above procedure of studying the superposition error can only be applied to very small systems. In the case of the interaction of two hydrogen-bonded water molecules this phenomenon has been studied by supplementing the basis set of a separated water molecule by the basis set of a second water molecule, placed into the position which the second water molecule would have in the combined system. The energy decrease calculated in this way was found to be about 0.4 kcal $mole^{-1}$. For very obvious reasons, however, this value can not be considered to be a direct measure for the superposition error, because the flexibility of the total basis set is strongly reduced in the combined system compared to the supplemented separated subsystems.

The solvation energy can be calculated approximately as the difference between the equilibrium SCF energy of the total system and the SCF energies of the subsystems. This energy should in principle be corrected for correlation effects, as discussed above, and for the zero-point vibrational energy. Approximate values for these corrections have been computed for some ionhydrates in the gas phase, and the net effects have been found to be small [40]. There seems to be no chance of calculating the latter corrections in the liquid phase. In the following discussion the term solvation energy (hydration energy, binding energy) will therefore refer to the difference in the equilibrium SCF energies.

From previous theoretical studies it is known that water is bonded to the lithium cation by a direct Li-O bond and to the fluorine anion by a hydrogen bond F---HO. This can also be expected for the lithium fluoride molecule because LiF has a large permanent dipole moment with lithium as the electropositive pole and fluorine as the electronegative pole. Only the corresponding geometrical structures will be considered in the present invesitgation.

In all calculations reported here, the geometry of the water molecule has been kept fixed at the experimental values (d(OH) = 1.80889 AU, ∢ (HOH) = 104.52). This is justified in view of the results of a previous study on the influence of the Li^+ ion on the equilibrium geometry of the water in ($Li^+.H_2O$). It has been found there that the influence of the lithium cation on the equilibrium structure of the water molecule is negligibly small [31]. Because the lithium cation is the smallest singly charged ion known, except for the proton, and therefore should have the strongest polarisation effect on the ligands, it is safe to assume that singly charged ions have no significant effect on the equilibrium geometry of the solvent molecules.

The following clusters of 'lithium fluoride' with one and two water molecules have been investigated. The chemical formulas are written so as to indicate the position of the water molecule(s). The geometrical structure of the systems I-VII and the geometry parameters are defined in Figures 1 to 5.

(I) $H_2O.LiF$

(II) $LiF.H_2O$

(III) $H_2O.LiF.H_2O$

(IV) $Li.OH_2.F$

(V) $H_2O.Li.OH_2.F$

(VI) $Li.OH_2.F.H_2O$

(VII) $Li.OH_2.OH_2.F$

Potential energy curves have been calculated for the lithium-fluoride bond dissociations in the systems I-VII. Further, for the systems I and II a variation of the LiF-water bond angle has been performed. In addition, a very selected part of the energy hypersurface of the internal transition of cluster I onto cluster IV has been calculated. The total SCF energies are listed in Tables II-IX. Selected potential energy curves have been plotted in Figures 6-14. The numbers 1, 2, 3 in the structure formulas of these figures denote the attachment of one, two, or three water molecules, and the point in the structure formulas indicates the dissociated bond. From the potential energy surfaces the minimum geometry structures, the total SCF energies for all systems, and various hydration energies have been determined. The results are listed in Tables XI-XIII. For the most stable geometrical structures of the systems various electron density difference maps have been calculated and are displayed in Figures 16-26. The energy hypersurface for the interaction of lithium fluoride with one water molecule has recently been studied by Kress and coworkers [58].

In addition, the "non-bonding" interaction of two water molecules with their molecular planes perpendicular to each other and the two oxygen atoms pointing towards each other has been studied theoretically. The computed total SCF energies have been listed in Table X, and the potential energy curve has been displayed in Figure 15.

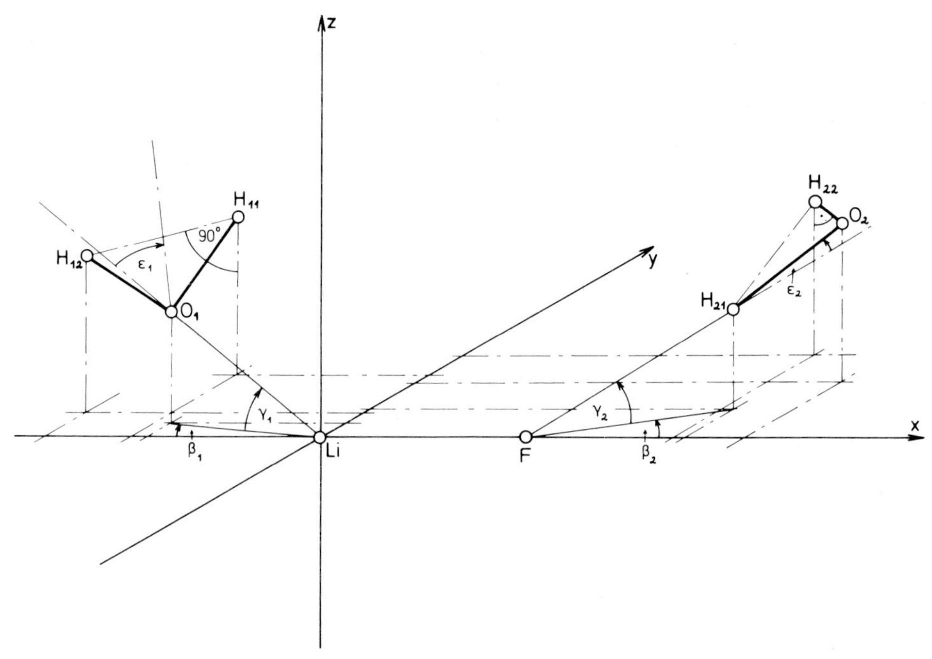

Fig. 1. Coordinate system for $H_2O.LiF$ and $LiF.H_2O$.

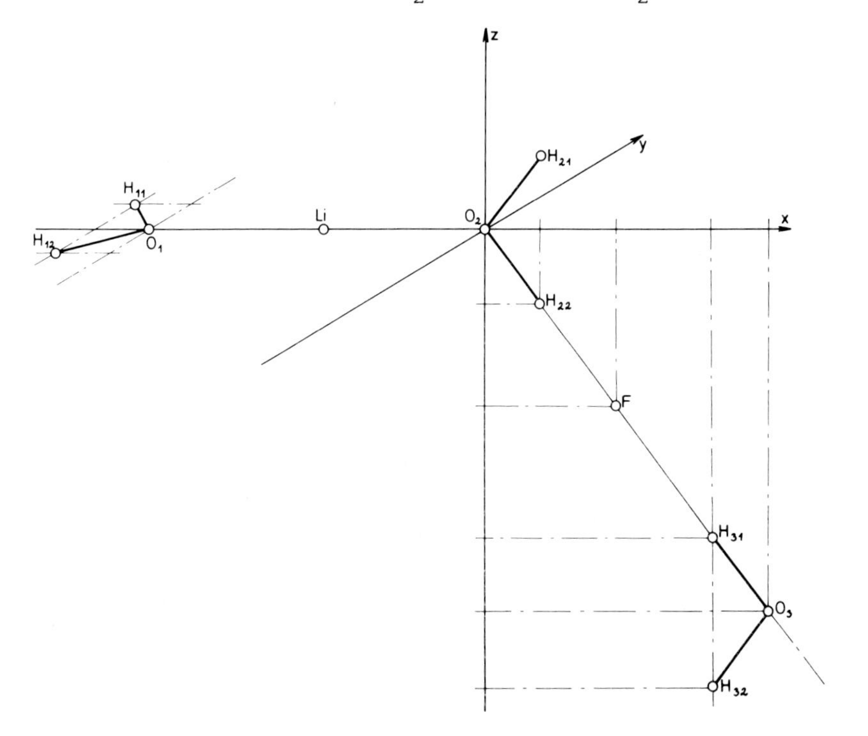

Fig. 2. Coordinate system for $H_2O.Li.OH_2.F$ and $Li.OH_2.F.H_2O$.

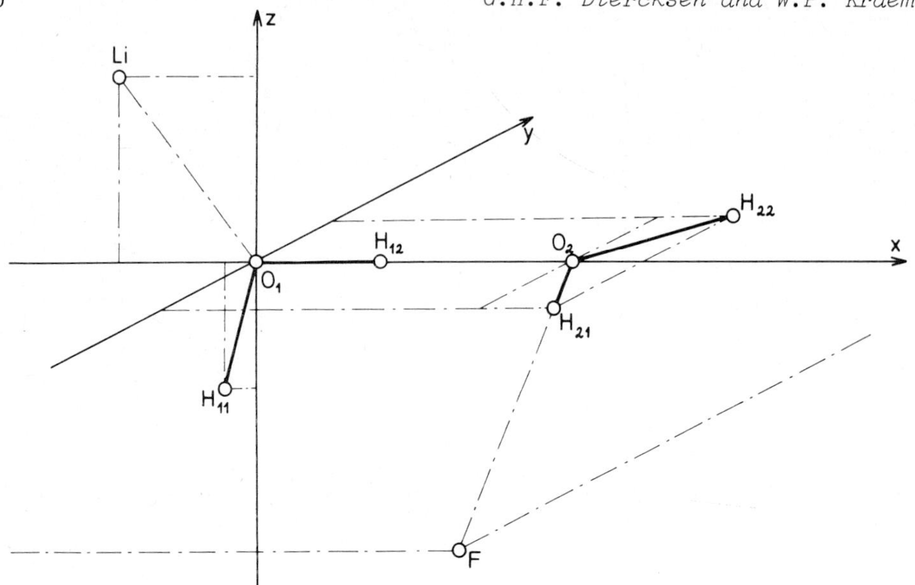

Fig. 3. Coordinate system for $Li.(OH_2)_2.F$.

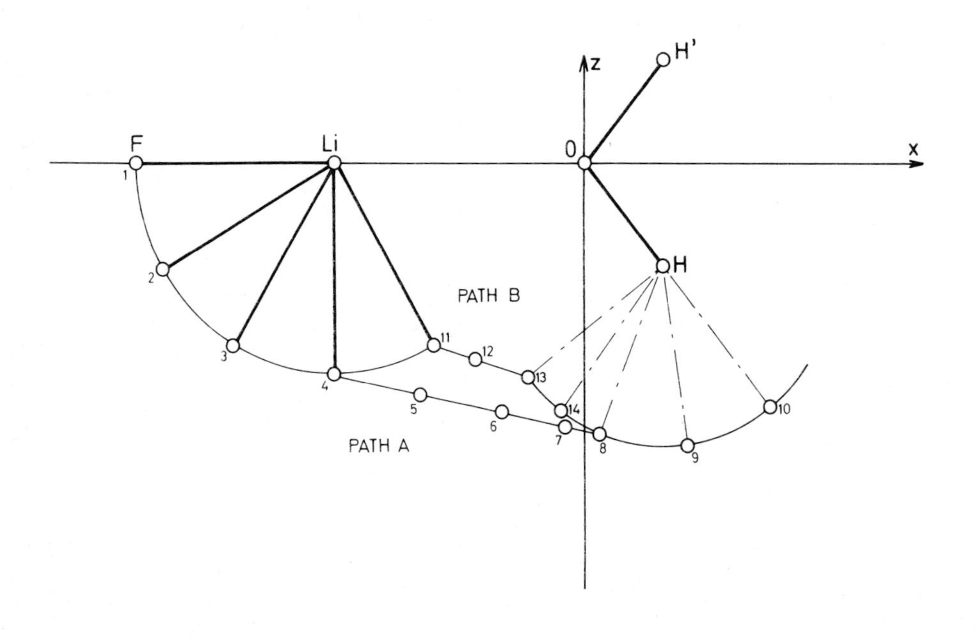

Fig. 4. Coordinate system for $LiF.H_2O$.

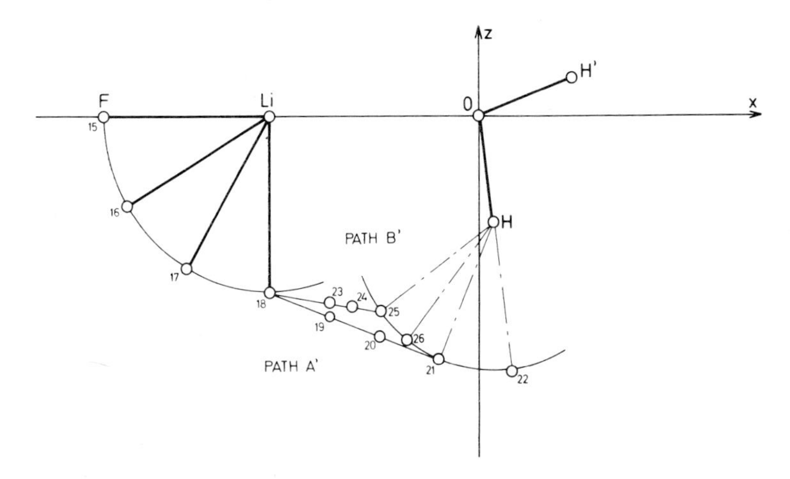

Fig. 5. Coordinate system for $LiF.H_2O$.

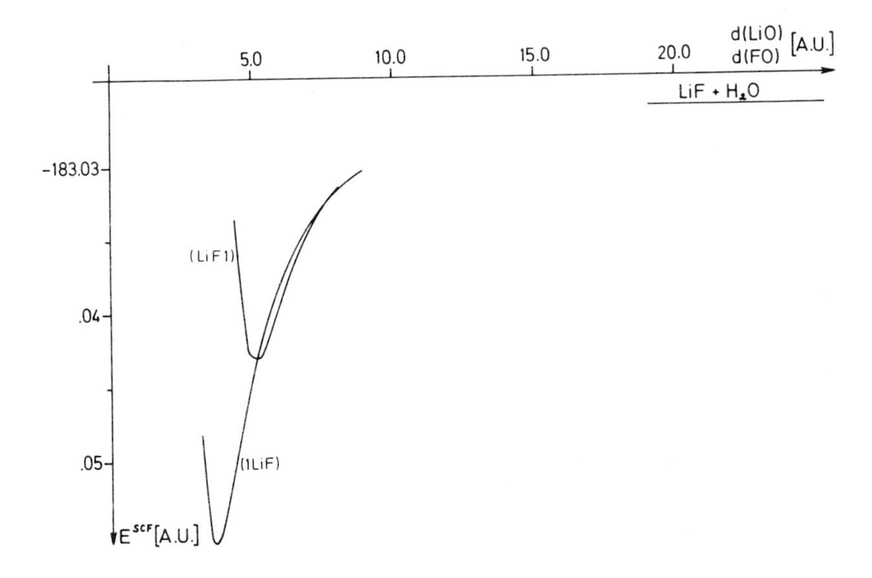

Fig. 6. Potential energy curves of $H_2O.LiF$ and $LiF.H_2O$ as function of the hydration bond distance. The number 1 in the chemical formula denotes the position of one water molecule.

TABLE II
Total SCF energies (E^{SCF}) of the lithium fluoride molecule (LiF) for different bond distances d(LiF).

	d(LiF) [AU]	D^{SCF} [AU]	E^{SCF} [a] [AU]
1	2.25	-106.89161	-106.89395
2	2.50	-106.94911	-106.95191
3	2.70	-106.96856	-106.97149
4	2.85	-106.97407	-106.97705
5	2.95	-106.97485	-106.97787
6	3.00	-106.97458	-106.97762
7	3.125	-106.97242	-106.97552
8	3.25	-106.96863	-106.97182
9	3.50	-106.95806	-106.96150
10	4.00	-106.93232	-106.93651
11	5.00	-106.88405	-106.89029
12	7.00	-106.82236	-106.83115
13	10.00	-106.77745	-106.78708
14	14.00	-106.74837	-106.75828

[a] Basis set specification: F,O (11.7.1), [5,4,1]
Li (11.2), [5,2]
H (6.1), [3.1]

TABLE III
Total SCF energies (E^{SCF}) of the system $H_2O.LiF$ for different geometrical configurations. (For a definition of the parameters, compare Figure 1.) All angles not explicitly listed are zero.

	d(LiF) [AU]	d(LiO) [AU]	β_1	γ_1	ε_1	E^{SCF} [AU]
1	2.95	3.25	0°	0°	0°	-183.05082
2	"	3.50	"	"	"	-183.05487
3(1)	"	3.68	"	"	"	-183.05541
4	"	3.75	"	"	"	-183.05528
5	"	4.00	"	"	"	-183.05390
6	"	4.50	"	"	"	-183.04934
7	"	5.00	"	"	"	-183.04474
8	"	6.00	"	"	"	-183.03800
9	"	8.00	"	"	"	-183.03168
10	"	11.00	"	"	"	
11	"	15.00	"	"	"	
12	2.50	3.68	"	"	"	-183.02567
13	2.75	"	"	"	"	-183.05000
14	3.25	"	"	"	"	-183.05141
15	3.50	"	"	"	"	-183.04253
16	5.00	"	"	"	"	-182.97625
17	7.00	"	"	"	"	-182.91986
18	10.00	"	"	"	"	-182.87891
19	15.00	"	"	"	"	-182.84794
20	2.95	3.68	10°	"	"	
21(2)	"	"	30°	"	"	-183.05494
22(3)	"	"	60°	"	"	-183.05303
23(4)	"	"	90°	"	"	-183.04384
24	"	3.88	"	"	"	-183.04398
25	"	3.48	"	"	"	-183.04223
26	"	3.28	"	"	"	-183.03808
27	2.95	3.68	0°	30°	"	-183.05442
28	"	"	"	60°	"	-183.05045
29	"	"	"	90°	"	-183.03576

TABLE III (Continued)

	d(LiF) [AU]	d(LiO) [AU]	β_1	γ_1	ε_1	E [AU]
(5)	3.435224	3.68	111.36	0°	0°	-183.01426
(6)	4.297474	"	126.41	"	"	-182.95803
(7)	4.890271	"	132.46	"	"	-182.94516
(8)	5.352571	"	136.01	"	"	-182.95189
(9)	6.423496	"	142.85	"	"	-182.97846
(10)	7.151367	"	151.75	"	"	-182.98292
(11)	2.95	"	120	"	"	-182.95721
(12)	3.492285	"	128.08	"	"	-182.89676
(13)	4.084979	"	133.91	"	"	-182.84507
(14)	4.729924	"	134.09	"	"	-182.91776
(15)	2.95	"	0	"	30°	-183.05051
(16)	"	"	30	"	"	-183.05303
(17)	"	"	60	"	"	-183.05567
(18)	"	"	90	"	"	-183.05460
(19)	3.475018	"	106.72	"	"	-183.03992
(20)	4.186656	"	118.06	"	"	-183.01235
(21)	5.056329	"	120.92	"	"	-182.99887
(22)	6.008598	"	135.10	"	"	-182.99049
(23)	3.286065	"	108.63	"	"	-183.03948
(24)	3.543604	"	114.77	"	"	-183.02057
(25)	3.847236	"	120.79	"	"	-182.99583
(26)	4.474823	"	122.85	"	"	-183.00031

TABLE IV
Total SCF energies E^{SCF} of the system LiF.HOH for different geometrical configurations. (For a definition of the parameters, compare Figure 1.) All angles not explicitely listed are zero.

	d(LiF) [AU]	d(FO) [AU]	β_2	γ_2	ε_2	E^{SCF} [AU]
1	2.95	4.00	0°	0°	0°	-183.00543
2	"	4.50	"	"	"	-183.03616
3	"	4.75	"	"	"	-183.04107
4	"	5.00	"	"	"	-183.04287
5	"	5.125	"	"	"	-183.04307
6	"	5.14	"	"	"	-183.04307
7	"	5.25	"	"	"	-183.04296
8	"	5.50	"	"	"	-183.04214
9	"	6.00	"	"	"	-183.03948
10	"	7.00	"	"	"	-183.03458
11	"	8.00	"	"	"	-183.03161
12	2.50	5.14	"	"	"	-183.01535
13	2.75	"	"	"	"	-183.03845
14	3.25	"	"	"	"	-183.03806
15	3.50	"	"	"	"	-183.02846
16	5.00	"	"	"	"	-182.95965
17	7.00	"	"	"	"	-182.90327
18	10.00	"	"	"	"	-182.86254
19	15.00	"	"	"	"	-182.83142
20	2.95	5.14	10°	"	"	
21	"	"	30°	"	"	-183.04168
22	"	"	60°	"	"	-183.03887
23	"	"	90°	"	"	-183.03266
24	"	"	- 30°	"	"	-183.04393
25	"	"	- 60°	"	"	-183.04421
26	"	"	- 90°	"	"	-183.04303
27	"	5.24	"	"	"	-183.04287
28	"	5.04	"	"	"	-183.04302
29	"	5.41	0°	30°	"	-183.04294
30	"	"	"	60°	"	-183.04220
31	"	"	"	90°	"	-183.03999

TABLE V
Total SCF energies of the system $H_2O.LiF.HOH$ for different geometrical configurations. (For a definition of the parameters, compare Figure 1.) All angles not explicitely listed are zero.

	d(LiF) [AU]	d(LiO) [AU]	d(OE) [AU]	E^{SCF} [AU]
1	2.75	3.50	4.75	-259.11683
2	2.95	"	"	-259.12334
3	3.05	"	"	-259.12373
4	3.15	"	"	-259.12280
5	3.05	"	4.55	-259.12049
6	"	"	4.85	-259.12458
7	"	"	5.05	-259.12525
8	"	"	5.25	-259.12499
9	2.95	"	5.05	-259.12493
10	3.15	"	"	-259.12425
11	3.05	3.60	"	-259.12563
12	"	3.70	"	-259.12557
13	"	3.90	"	-259.12459
14	2.50	3.64	5.09	-259.09378
15	2.75	"	"	-259.11918
16	3.02	"	"	-259.12574
17	3.25	"	"	-259.12255
18	3.50	"	"	-259.11458
19	5.00	"	"	-259.05334
20	7.00	"	"	-259.00220
21	10.00	"	"	-258.96519

TABLE VI
Total SCF energies (E^{SCF}) of the system LiO(H)HF in the geometrical configuration described in Figure 2 for different bond distances d(LiO_2) and d(FO_2).

	d(LiO_2) AU	d(FO_2) AU	E^{SCF} AU
1	2.55	4.35	-182.94452
2	3.00	"	-182.98859
3	3.10	"	-182.99090
4	3.20	"	-182.99166
5	3.30	"	-182.99122
6	3.50	"	-182.98777
7	3.60	"	-182.98516
8	4.00	"	-182.97176
9	4.50	"	-182.95334
10	6.70	"	-182.89489
11	10.90	"	-182.84843
12	3.30	3.46	-182.89706
13	"	3.75	-182.96137
14	"	4.00	-182.98229
15	"	4.25	-182.98794
16	"	4.30	-182.98799
17	"	4.50	-182.98589
18	"	4.75	-182.98009
19	"	5.10	-182.96938
20	"	5.40	-182.95972
21	"	7.40	-182.90831
22	"	11.60	-182.86405
23	3.10	4.28	-182.99133
24	3.20	"	-182.99204
25	3.30	"	-182.99155
26	7.044686	"	-182.88787
27	3.21	7.955603	-182.90085

TABLE VII
Total SCF energies (E^{SCF}) of the system H_2O LiO(H)HF in the geometrical configuration described in Figure 2 for different bond distances $d(LiO_2)$ and $d(O_1O_2)$.

	$d(LiO_2)$ [AU]	$d(FO_2)$ [AU]	$d(O_1O_2)$ [AU]	E^{SCF} [AU]
1	3.20	4.30	6.54	-259.07732
2	"	"	6.64	-259.07850
3	"	"	6.74	-259.07902
4	"	"	6.80	-259.07909
5	"	"	6.94	-259.07869
6	"	"	7.04	-259.07806
7	"	"	7.14	-259.07723
8	"	"	7.24	-259.07626
9	"	4.10	6.80	-259.07672
10	"	4.20	"	-259.07859
11	"	4.40	"	-259.07853
12	"	4.50	"	-259.07719
13	3.10	4.30	"	-259.07759
14	3.30	"	"	-259.07899
15	3.40	"	"	-259.07745
16	3.50	"	"	-259.07451
17	3.60	"	"	-259.07013
18	3.10	"	7.14	-259.07506
19	3.10	"	"	-259.07809
20	3.40	"	"	-259.07785
21	3.57	"	"	-259.07531
22	3.74	"	"	-259.07028
23	3.90	"	"	-259.06304

TABLE VIII

Total SCF energies (E^{SCF}) of the system LiO(H)HF HOH in the geometrical configuration described in Figure 2 for different bond distances $d(LiO_2)$, $d(FO_2)$, and $d(O_2O_3)$.

	$d(LiO_2)$ [AU]	$d(FO_2)$ [AU]	$d(O_2O_3)$ [AU]	E^{SCF} [AU]
1	3.21	4.28	8.8	-259.05936
2	"	"	9.0	-259.06268
3	"	"	9.2	-259.06401
4	"	"	9.35	-259.06417
5	"	"	9.6	-259.06347
6	3.11	"	9.35	-259.06344
7	3.31	"	"	-259.06372
8	3.21	4.13	"	-259.06179
9	"	4.43	"	-259.06367

TABLE IX
Total SCF energies of the system LiO(H)HO(H)HF in the geometrical configuration described in Figure 3 for different bond distances $d(LiO_1)$, $d(FO_2)$ and $d(O_1O_2)$.

	$d(LiO_1)$ [AU]	$d(FO_2)$ [AU]	$d(O_1O_2)$ [AU]	E^{SCF} [AU]
1	3.25	4.30	5.86925	-259.00667
2	"	"	5.66925	-259.01094
3	"	"	5.46925	-259.01524
4	"	"	5.26925	-259.01941
5	"	"	5.06925	-259.02322
6	"	"	4.86925	-259.02627
7	"	"	4.66925	-259.02796
8	"	"	4.60	-259.02808
9	"	"	4.46925	-259.02735
10	"	"	4.26925	-259.02293
11	"	4.20	4.60	-259.02681
12	"	4.35	"	-259.02829
13	"	4.40	"	-259.02826
14	"	4.50	"	-259.02763
15	3.15	4.35	"	-259.02759
16	3.35	"	"	-259.02792
17	3.27	4.37	4.055171	-259.01110
18	"	"	4.61	-259.02832
19	"	"	5.158263	-259.02212
20	"	"	6.777463	-258.99050
21	"	"	8.893842	-258.96466
22	"	"	12.014682	-258.94378

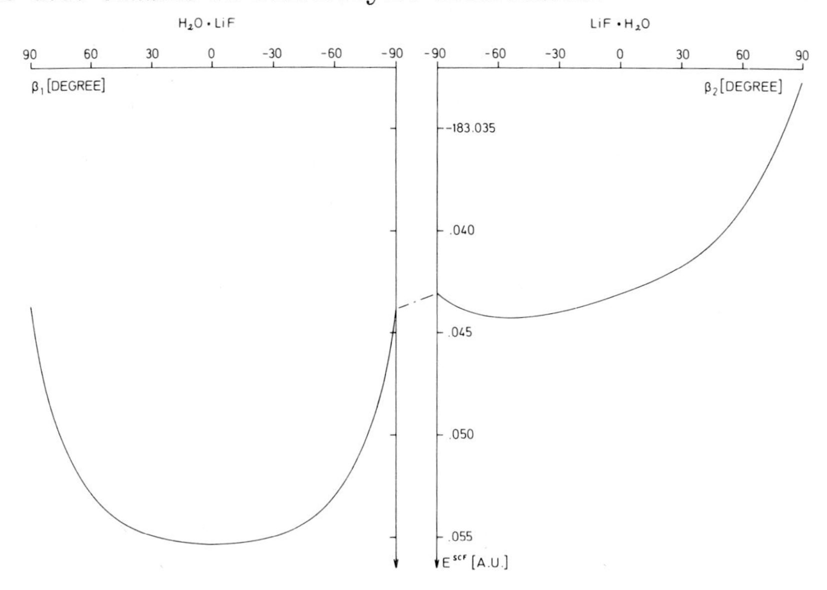

Fig. 7. Potential energy curves of $H_2O.LiF$ and $LiF.H_2O$ for a variation of the angle β (Figure 1).

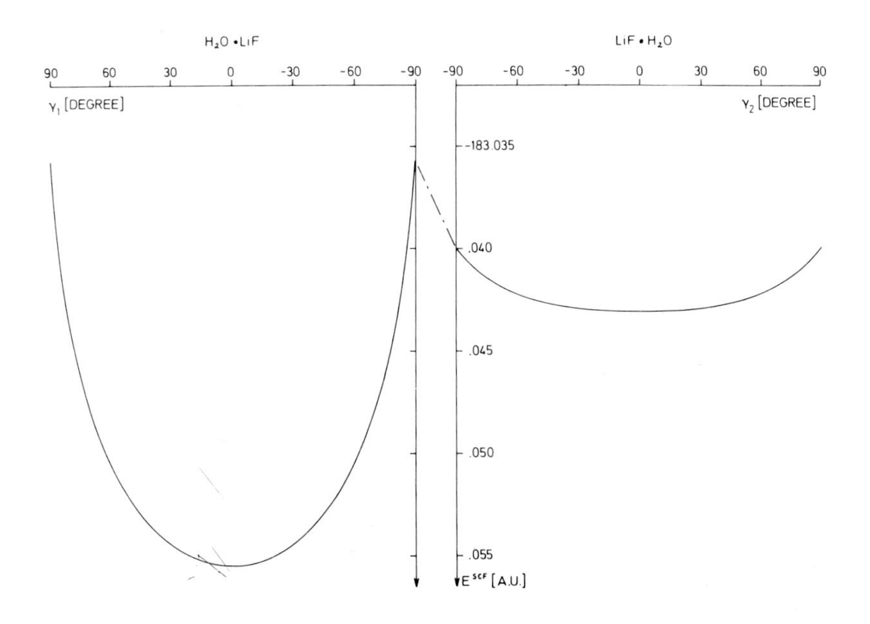

Fig. 8. Potential energy curves of $H_2O.LiF$ and $LiF.H_2O$ for a variation of the angle γ (Figure 1).

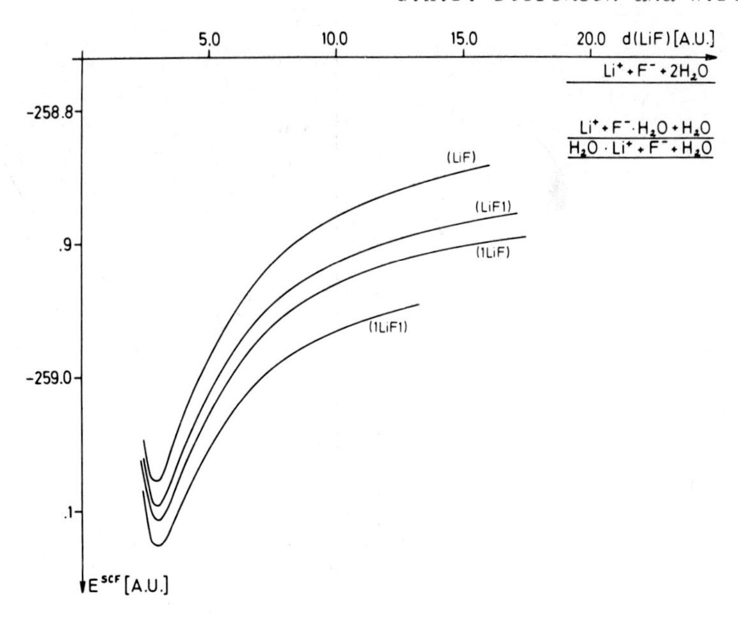

Fig. 9. Potential energy curves of LiF in different hydration states. The number 1 denotes the position of one water molecule. The point in the chemical formula denotes the bond dissociated.

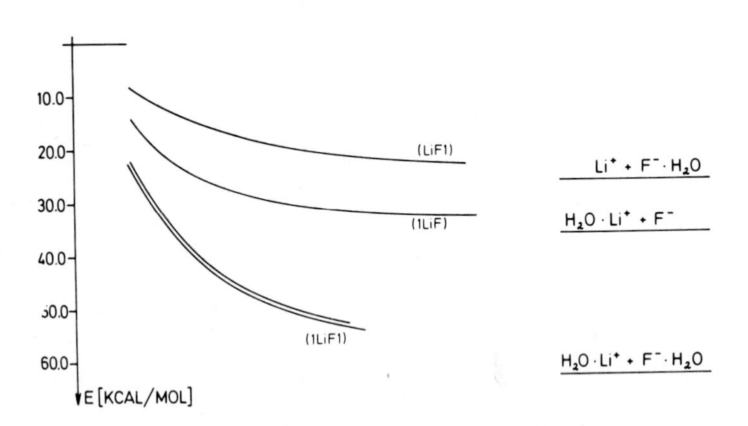

Fig. 10. Hydration energies as a function of the LiF distance. The number 1 denotes the position of one water molecule. The point in the chemical formula denotes the bond dissociated.

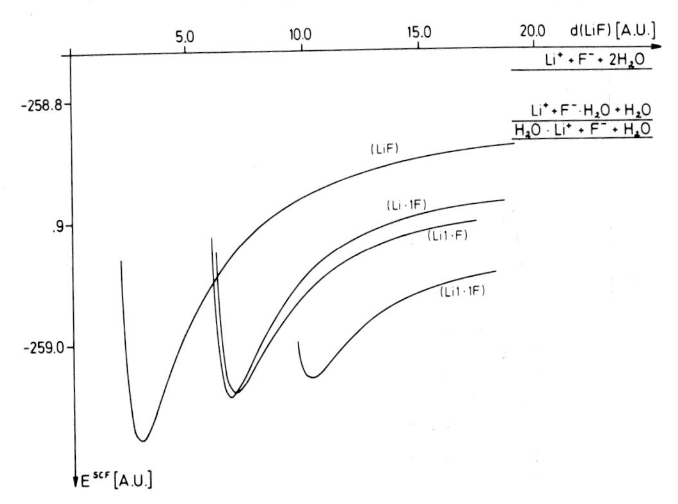

Fig. 11. Potential energy curves of LiF in different hydration states. The number 1 in the chemical formula denotes the position of one water molecule. The point in the chemical formula denotes the bond dissociated.

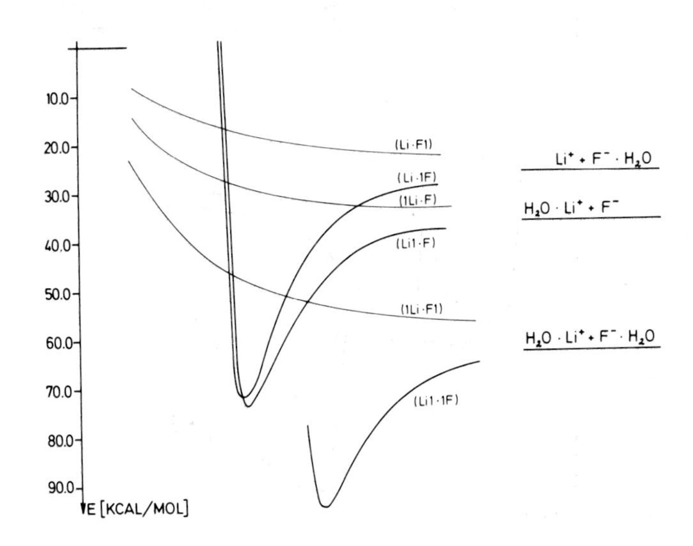

Fig. 12. Hydration energies as a function of the LiF distance in dif-fe ent hydration states. The number 1 denotes the position of one water molecule. The point in the chemical formula denotes the bond dissociated.

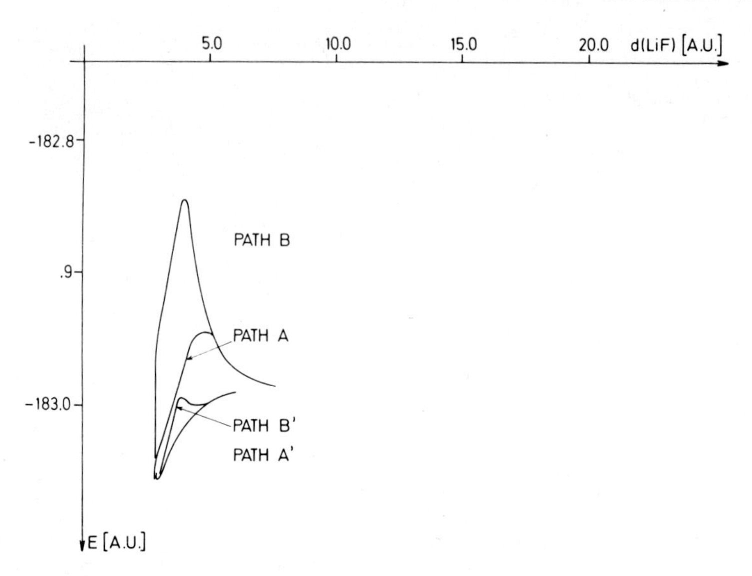

Fig. 13. Potential energy curves of $LiF.H_2O$ for different 'reaction paths' of the Lif bond dissociation (Figures 4 and 5)

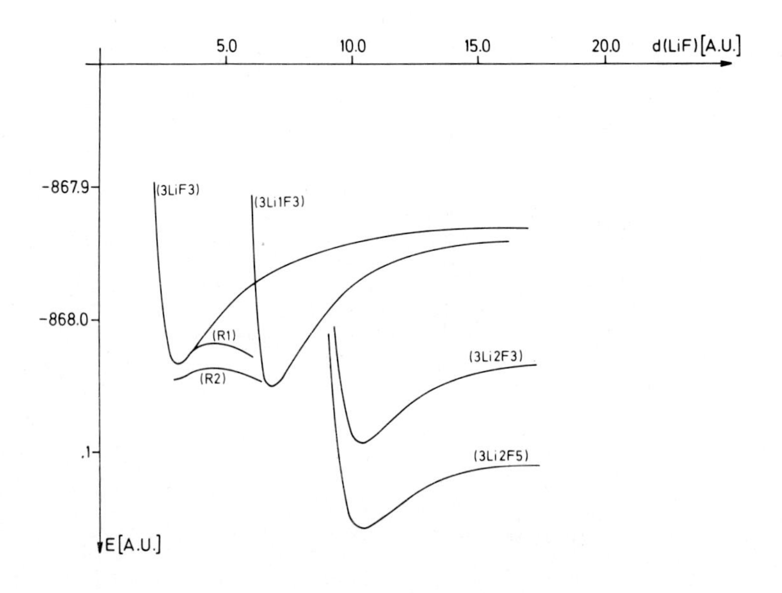

Fig. 14. 'Qualitative'potential energy curves for various LiF hydrates. The numbers 1, 2, 3,... in the chemical formula denotes the number and position of water molecules

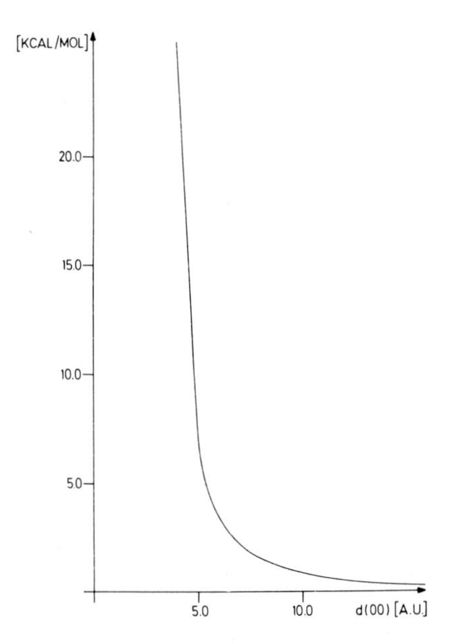

Fig. 15. Potential energy curve of $H_2O.OH$ (Table X).

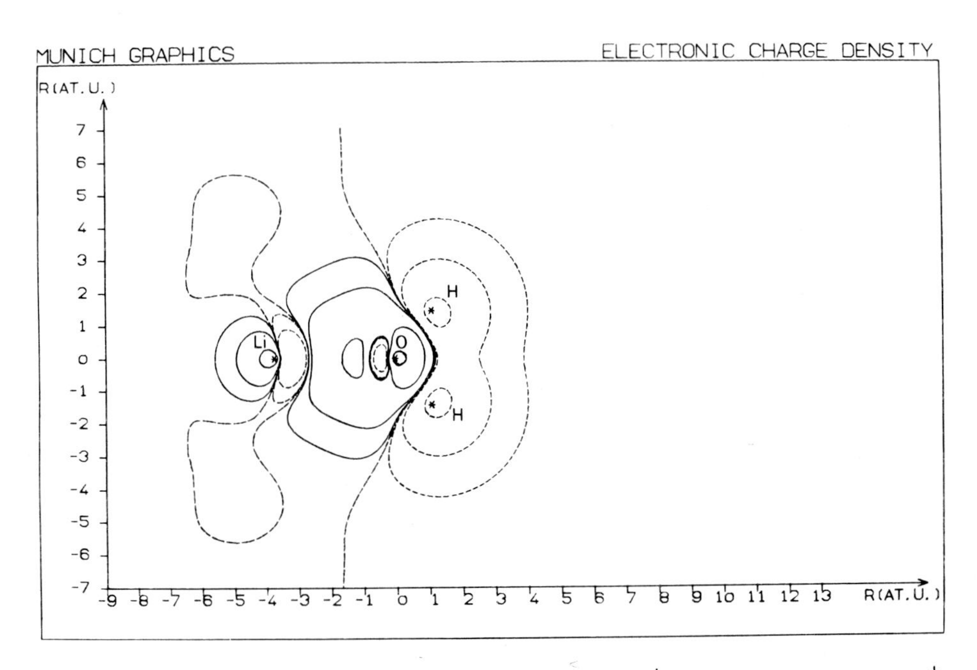

Fig. 16. Electron density difference map of $Li^+.H_2O$ relative to Li^+ and H_2O. (---- Positive-, -- -- Zero-, - - - Negative electron density differences. The contour lines correspond to density differences of 10^{-4}, 10^{-3}, 10^{-2} AU)

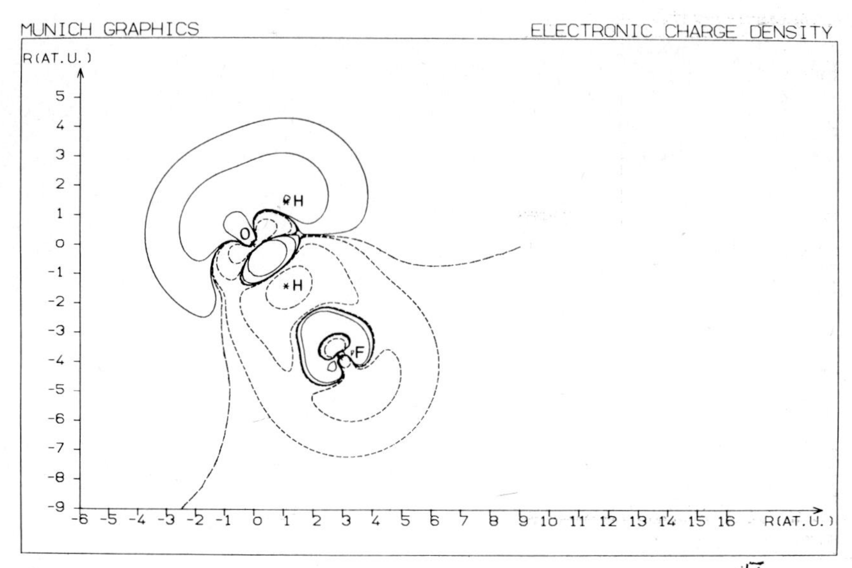

Fig. 17. Electron density difference map of $F^{-}.H_2O$ relative to F^{-} and H_2O. (---- Positive-, -- -- Zero-, - - - Negative electron density differences. The contour lines correspond to density differences of 10^{-4}, 10^{-3}, 10^{-2} AU)

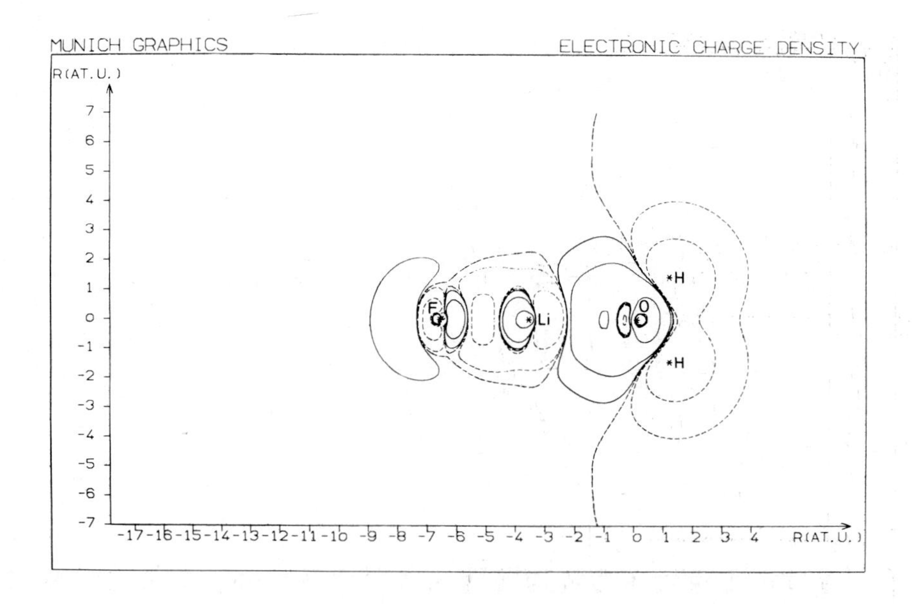

Fig. 18. Electron density difference map of $H_2O.LiF$ relative to LiF and H_2O. (---- Positive-, -- -- Zero-, - - - Negative electron density differences. The contour lines correspond to density differences of 10^{-4}, 10^{-3}, 10^{-2} AU)

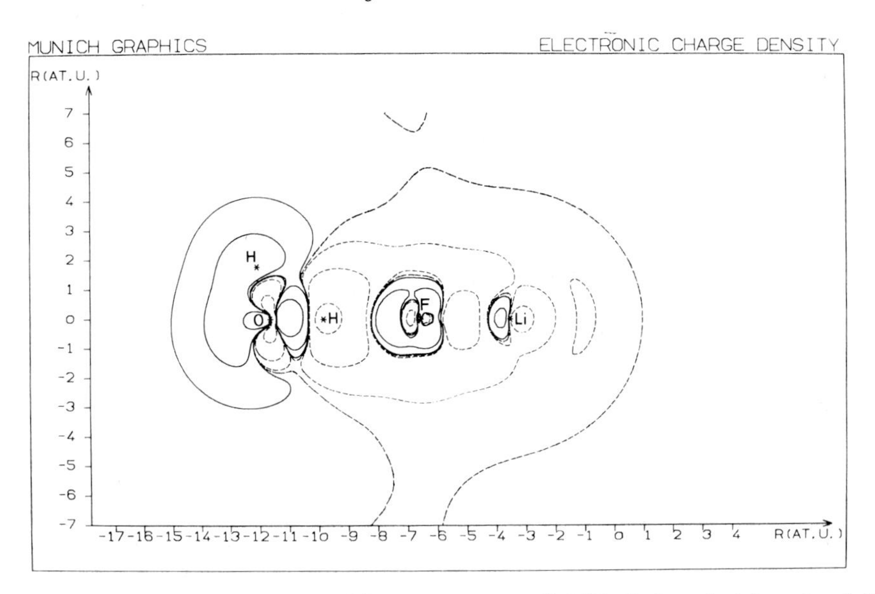

Fig. 19. Electron density difference map of $LiF.H_2O$ relative to LiF and H_2O. (---- Positive-, -- -- Zero-, - - - Negative electron density differences. The contour lines correspond to density differences of 10^{-4}, 10^{-3}, 10^{-2} AU)

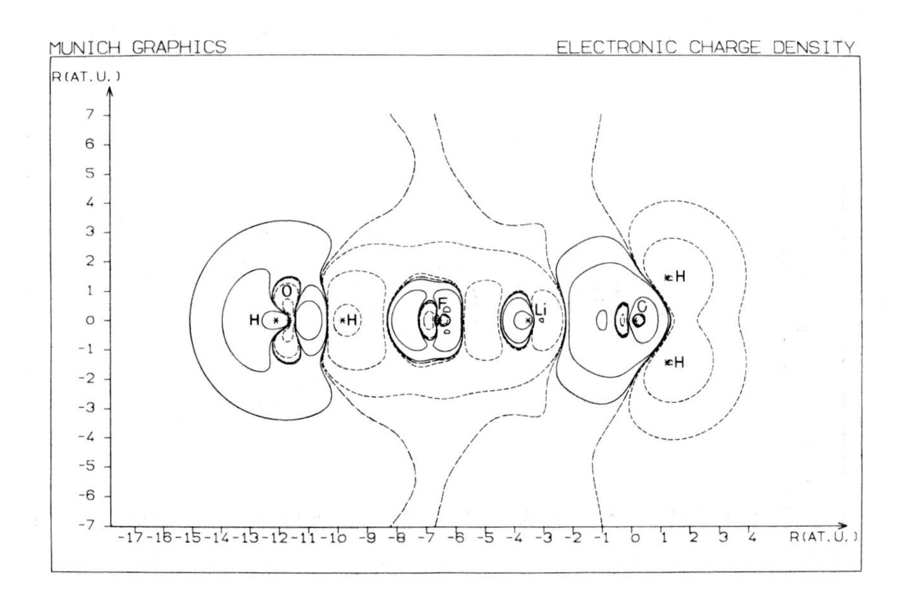

Fig. 20a. Electron density difference map of $H_2O.LiF.H_2O$ relative to LiF and $2H_2O$, in the xy plane (Figure 1). (---- Positive-, -- -- Zero-, - - - Negative electron density differences. The contour lines correspond to density differences of 10^{-4}, 10^{-3}, 10^{-2} AU)

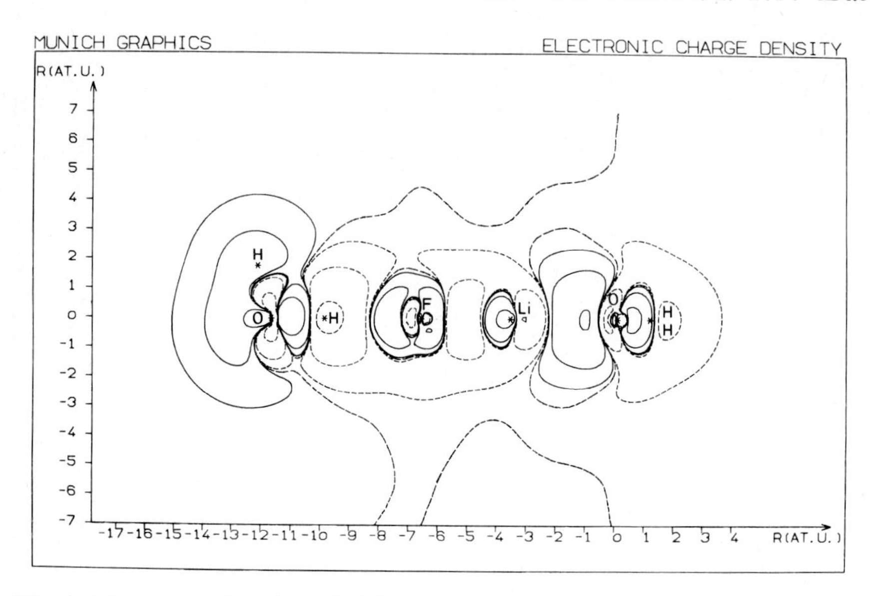

Fig. 20b. Electron density difference map of $H_2O.LiF.H_2O$ relative to LiF and $2H_2O$, in the xz plane (Figure 1). (---- Positive-, -- -- Zero-, - - - Negative electron density differences. The contour lines correspond to density differences of 10^{-4}, 10^{-3}, 10^{-2} AU)

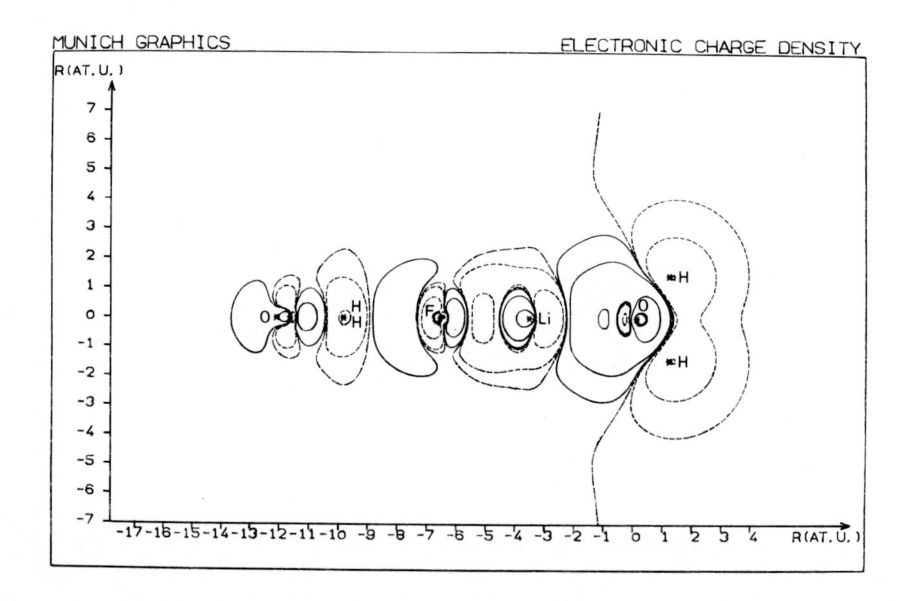

Fig. 21. Electron density difference map of $H_2O.LiF.H_2O$ relative to H_2O and $LiF.H_2O$. (---- Positive-, -- -- Zero-, - - - Negative electron density differences. The contour lines correspond to density differences of 10^{-4}, 10^{-3}, 10^{-2} AU)

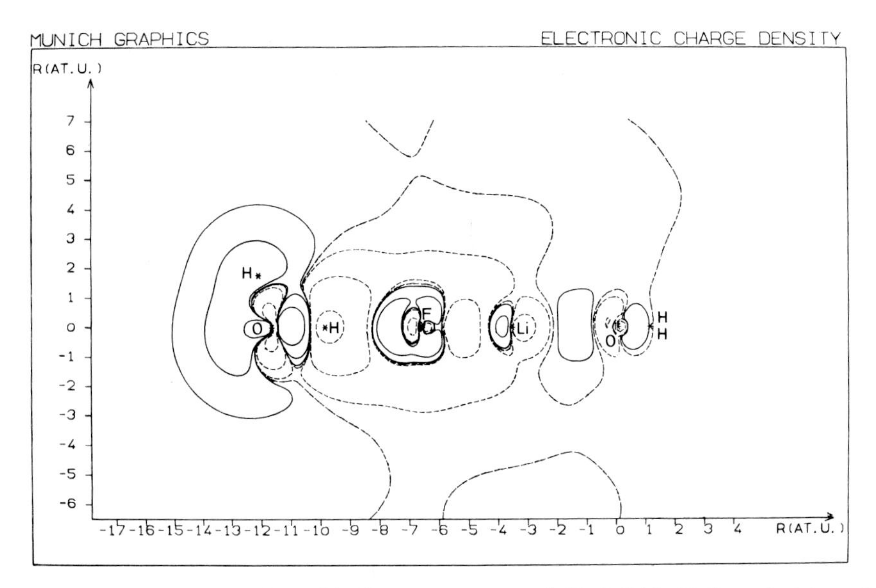

Fig. 22. Electron density difference map of $H_2O.LiF.H_2O$ relative to $H_2O.LiF$ and H_2O.)---- Positive-, -- -- Zero-, - - - Negative electron density differences. The contour lines correspond to density differences of 10^{-4}, 10^{-3}, 10^{-2} AU)

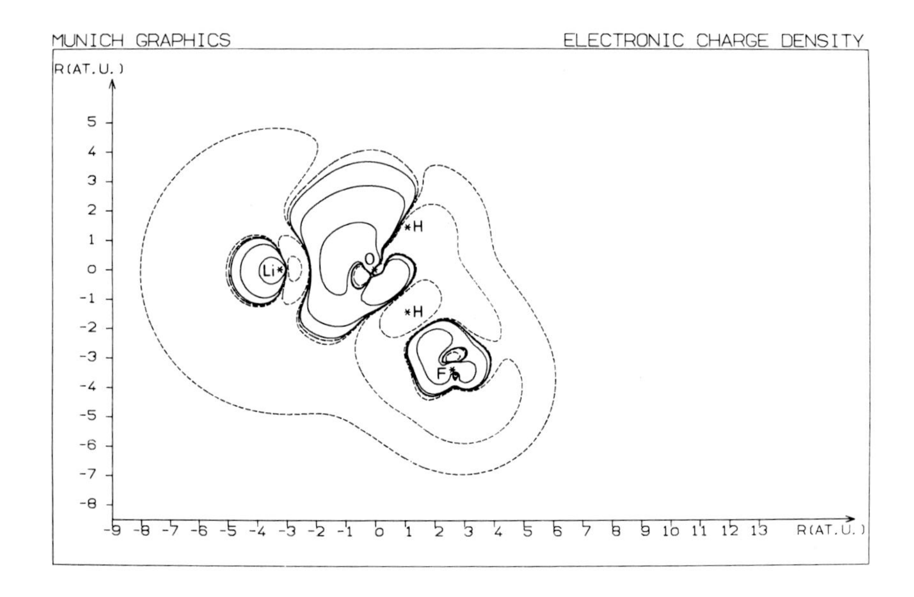

Fig. 23. Electron density difference map of $Li.OH_2.F$ relative to LiF and H_2O. (---- Positive-, -- -- Zero-, - - - Negative electron density differences. The contour lines correspond to density differences of 10^{-4}, 10^{-3}, 10^{-2} AU)

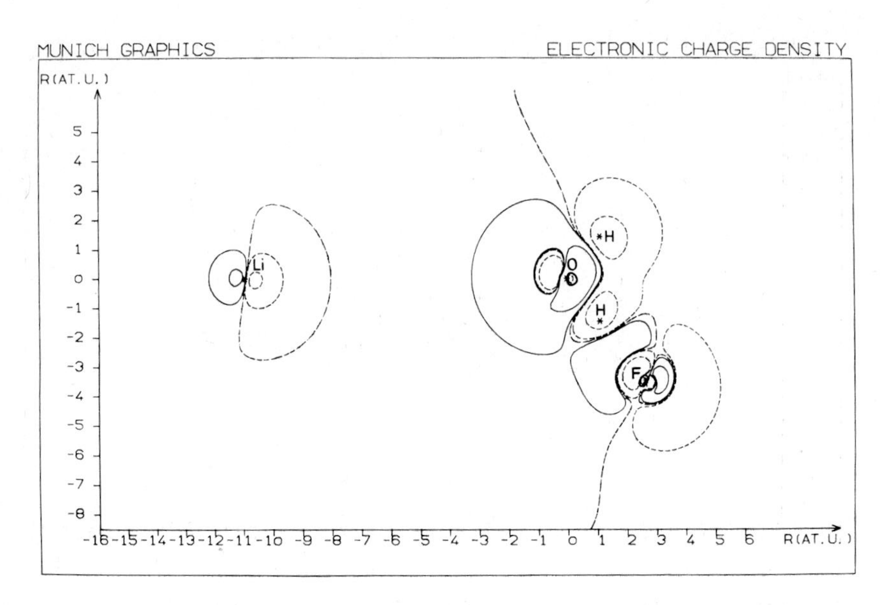

Fig. 24. Electron density difference map of Li.OH .F relative to Li^+ and $F.H_2O^-$. (---- Positive-, -- -- Zero-, - - - Negative electron density differences. The contour lines correspond to density differences of 10^{-4}, 10^{-3}, 10^{-2} AU)

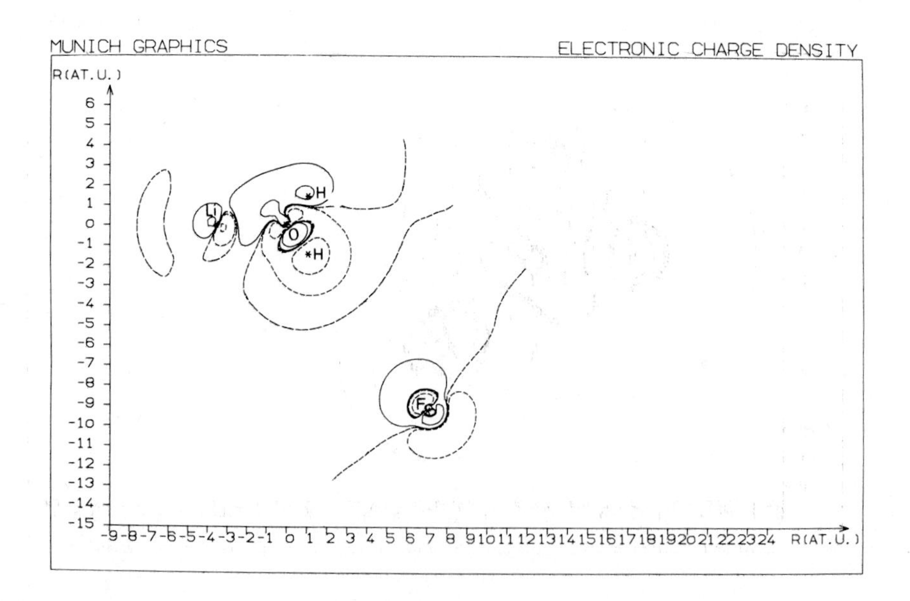

Fig. 25. Electron density difference map of $Li.OH_2.F$ relative to $Li.OH_2^+$ and F^-. (---- Positive-, -- -- Zero-, - - - Negative electron density differences. The contour lines correspond to density diffe-ences of 10^{-4}, 10^{-3}, 10^{-2} AU)

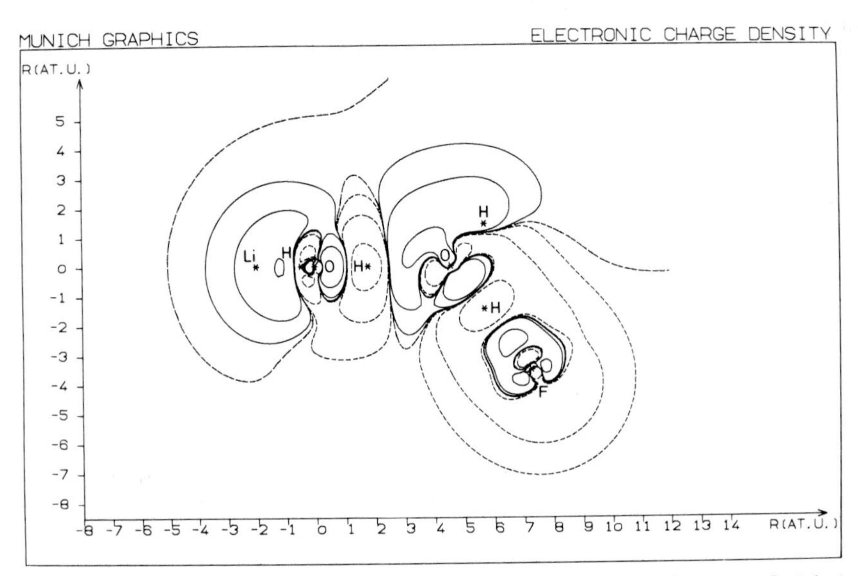

Fig. 26a. Electron density difference map of Li.$(H_2O)_2$.F relative to LiF and $(H_2O)_2$ in the xy-plane (Figure 3). (---- Positive, -- -- Zero-, - - - Negative electron density differences. The contour lines correspond to density differences of 10^{-4}, 10^{-3}, 10^{-2} AU)

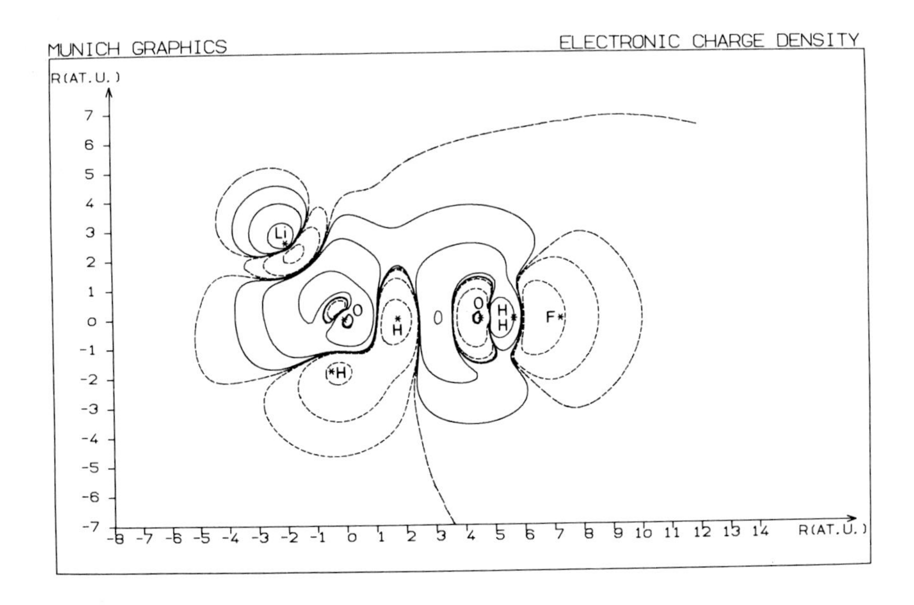

Fig. 26b. Electron density difference map of Li.$(H_2O)_2$.F relative to LiF and $(H_2O)_2$ in the xz-plane (Figure 3) (---- Positive-, -- -- Zero-, - - - Negative electron density differences. The contour lines correspond to density differences of 10^{-4}, 10^{-3}, 10^{-2} AU)

TABLE X

Total SCF energies of the system $H_2O.OH_2$ in C_{2v} symmetry, with the planes of the water molecules perpendicular to each other, and with the oxygen atoms pointing towards each other (d(OH) = 1.80887 AU ∡(HOH) = 104.52°)

	d(OO) [AU]	E^{SCF} [AU]
1	2.00	-151.338205
2	4.00	-152.063735
3	5.00	-152.092710
4	6.00	-152.098524
5	7.00	-152.100536
6	8.00	-152.101553
7	10.00	-152.102622
8	15.00	-152.103529
9	20.00	-152.103771

TABLE XI
Total SCF energies (E^{SCF}) for the hydrated systems Li^+, F^-, and LiF, and for various reference systems, in their equilibrium geometry (Compare Table XII).

	System (Structure formula)	E^{SCF}[541] [AU]	E^{SCF}[531] [AU]
1	H_2O	- 76.05199	- 76.05118
2	Li^+	- 7.23621	- 7.23621
3	F^-	- 99.45059	- 99.44068
4	LiF	-106.97787	-106.97485
5	H_2OHOH	-152.11167	-152.10989
6	$LiOH_2{}^+$	- 83.34573	- 83.34515
7	$H_2OLiOH_2{}^+$	-159.44869	
8	$LiO(H)HOH_2{}^+$	-159.42339	
9	$FHOH^-$	-175.54258	-175.53405
10	$HOHFHOH^-$		-251.62013
11	$FH(H)OHOH^-$		-251.60578
12	H_2OLiF		-183.05535
13	LiFHOH		-183.04307
12	LiO(H)HF		-182.99205
15	$H_2OLiFHOH$		-259.12578
16	$H_2OLiO(H)HF$		-259.07922
17	LiO(H)HFHOH		-259.06432
18	LiO(H)OH(H)HF		-259.02832

TABLE XII
Equilibrium geometry parameters for the hydrated systems Li^+, F^-, and LiF, and for various reference systems

	System (structure formula)	d(LiF) [AU]	d(LiO) [AU]	d(FO) [AU]	d(OO) [AU]
1	H_2O				
2	Li^+				
3	F^-				
4	LiF	2.95			
5	H_2OHOH				5.67
6	$LiOH_2^+$		3.57		
7	$H_2OLiOH_2^+$		3.62		
8	$LiO(H)HOH_2^+$				
9	$FHOH^-$			4.77	
10	$HOHFHOH^-$			4.83	9.66
11	$FH(H)OHOH^-$				
12	H_2OLiF	3.02	3.68		
13	LiFHOH	3.00		5.14	
14	LiO(H)HF	6.74	3.21	4.28	
15	$H_2OLiFHOH$	3.04	3.64	5.09	
16	$H_2OLiO(H)HF$	6.79	3.24	4.30	6.80
17	LiO(H)HFHOH	6.80	3.22	4.33	9.33
18	LiO(H)HO(H)HF	10.24	3.27	4.37	4.61

TABLE XIII
Hydration energies and equilibrium geometry parameters for the hydrated systems Li^+, F^-, and LiF. (For references compare Tables XI and XII)

	System (Structure formula)	d(LiO) AU	d(FO) AU	E^{Hydr} AU		E^{Hydr} AU	
				(LiO)	(F-H-O)	(Li-O)	(F-H-O)
1	$FLi.OH_2$	3.68		18.40			
2	$HOHFLi.OH_2$	3.64	5.09	19.26	11.20		
3	$H_2OLi^+.OH_2$	3.62		34.06			
4	$Li^+.OH_2$	3.57		36.11		34.03(36.08)	
5	$Li^+.OH_2$ (CI)					32.87(34.92)	
6	LiF.HOH		5.14		10.70		
7	$H_2OLiF.HOH$	3.64	5.09	19.26	11.20		
8	$HOHF^-.HOH$		4.83		24.20		
9	$F^-.HOH$		4.77		26.48		21.04(24.22)
10	$F^-.HOH$ (CI)						22.99(26.16)

Based on these results the following subjects are discussed: the equilibrium structure of the systems and the mutual interactions between the water ligands and the 'lithium fluoride' (section 3), the hydration energy as a function of the lithium fluoride bond distance, and the potential energy curve of the litium fluoride bond dissociation in different hydration states of the system (section 4).

3. EQUILIBRIUM STRUCTURES OF LITHIUM FLUORIDE-WATER COMPLEXES

The potential energy hypersurface for the interaction of a water molecule with a lithium fluoride molecule has two distinct minima. In the energetically lower one of the two geometrical configurations the water molecule is bonded to the lithium fluoride molecule by an OLi bond with the lone pair electrons of the oxygen atom pointing towards the lithium. This hydrate is symmetric with respect to the LiF axis. In the energetically less stable geometrical configuration, the water molecule is bonded to the fluorine atom of the lithium fluoride molecule by a linear hydrogen bond. This hydrate is planar with a lithium fluoride hydrate bond angle β_2 of 56°. The different bond angles of the water molecules at the lithium and at the fluorine centers are easily explained by the following considerations. In the strongly polar lithium fluoride molecule, the lithium has approximately a closed K shell of electrons, while the fluorine has a closed L shell. The K shell at the lithium is linearly polarized by the lithium fluoride bond, while the L shell at the fluorine is sp^3 hybridized, the maxima of the electron distributions pointing towards the edges of a tetrahedron. The potential energy curves for an approach of the water molecule along the LiF axis are attractive, and in particular there are no energy barriers against the formation of the hydrates from the reactants, LiF + H_2O. The potential energy curves for a variation of the angles β_1, β_2, γ_1, and γ_2 between 0° and 60° are rather flat, with ΔE^{SCF} = 0.5-3.0 kcal mole^{-1}, but they increase rather strongly for angles greater that 60°. Generally, these potential energy curves are more flat for the water bonded to the fluorine than to the lithium center. Thus the structure can be expected to be very sensitive to external perturbations through other ligands or molecules in a crystal or fluid state. For the two 'rectangular' geometrical configurations, with the hydrate bonds perpendicular to the LiF axis ($\beta_1 = 90^\circ$, and $\beta_2 = 90^\circ$, respectively) the potential energy hypersurface has been investigated more extensively. By a variation of the angles ε_1 and ε_2 (compare Figure 1) it is found that those configurations are energetically most stable in which the water molecule can interact with *both* atoms of the lithium fluoride molecule and forms appropriate bonds. The energy lowering in such structures, compared to those rectangular structures in which the water molecule preferably can interact with only *one* of the two centers of the lithium fluoride molecule, is considerable. From the present study of the energy hypersurface it can be concluded that the two minima on the surface are separated by energy barriers, and that therefore the associated geometrical structures are stable. They cannot be transformed into each other without activation energy.

A quite different result has been found for the hydrated molecular ions NO^+ and CN^- [18], and for the interaction of the lithium cation and a cyanide anion (LiCN) [59]. In all these cases, only one minimum has been found on the potential energy pypersurface.

The hydration energies of the monohydrated lithium fluoride molecule in the two stable geometrical configurations have been calculated to be $E^{Hydr.} = 18.40$ kcal mole^{-1} and $E^{Hydr.} = 10.70$ kcal mole^{-1}, respectively. The hydration energies computed for the strongly polar lithium fluoride molecule are larger than those of all other hydrated neutral molecules studied so far [16]. The energy of hydration at the lithium atom is larger than the energy of hydration at the fluorine atom of the lithium fluoride molecule. This is in agreement with the results for the Li^+ and F^- ions [16]. The ratio of the two hydration energy values is smaller for the lithium fluoride molecule than for the isolated ions.

In Table XIII are compiled the equilibrium bond distances and mono-hydration energies of those systems which include both Li and F. The hydration energies at the different centers of dihydrated systems have been calculated by subdividing the total di-hydration energy according to the ratio of the appropriate monohydrated systems.

The following general conclusions can be drawn from the results listed in Table XIII. The hydration energy increases and the bond length decreases with increasing (positive or negative) net charge on the atoms Li and F. In particular, the hydration energy decreases from $(Li.OH_2)^+$ to $FLi.OH_2$ by appr. 50%, while it decreases from $(F.H_2O)^-$ to $LiF.H_2O$ by more than 60%. Similar results hold for the equilibrium bond distances d(LiO) and d(FO). The bond distance d(LiO) increases for decreasing net charge on the Li atom by only 0.11 AU (appr. 3%), while the bond distance d(FO) increases with decreasing net charge on the F atom by appr. 0.37 AU (appr. 8%). This leads to the conclusion that the electronic charge of the center involved in the hydrogen bond greatly influences the hydrogen bond strength.

For the 'outer' lithium fluoride di-hydrate $H_2O.LiF.H_2O$ only the 'linear' structure has been studied, with the centers OLiFHO on the same axis. It is found that the LiF bond distance is slightly larger compared to the unhydrated lithium fluoride and its monohydrates, and that the hydration bond lengths are slightly smaller compared to those in the monohydrates. The total hydration energy of the lithium fluoride di-hydrate $H_2O.LiF.HOH$ is larger than the sum of the appropriate mono-hydration energies of the lithium fluoride molecule by about 1.0 kcal mole^{-1}. This shows that the hydration energies are not additive, but rather depend on the introduced charge shifts (polarization). The polarization introduced can either stabilize other bonds, as is found in the present example or in the chain and ring structures of water polymers and hydrogen fluoride polymers [60-62], or it can destabilize bonds or even lead to repulsive potential curves, as for example in the case of a fluorine ion interacting with a water dimer through the hydrogen atom of the proton donor system [33].

No calculations have been performed for the case in which two water molecules are both bonded either to the lithium atom or to the fluorine atom. Such dihydrates have been studied for the atomic ions Li^+ and F^-

[31, 32], as well as for the molecular ion NO^+ [17, 18]. In all cases it is found that the total hydration energy of such dihydrates is slightly smaller (by appr. 10%) than the sum of the monohydration energies of the corresponding systems. This is due to the fact that both bonds are polarizing the system in such a way that the formation of another bond of the same type is less favourable. The same behaviour can be expected to hold for the di-hydration at any center of the lithium fluoride molecule as well. No quantitative conclusion about the hydration energy and equilibrium bond length changes of the lithium fluoride molecule doubly hydrated at both centers *simultaneously* can be drawn from the present calculations. It can be expected that the two effects discussed, the polarization effects on double hydration at different centers and at the same center of the molecules, cancel to a large extent. In this case, the hydration energies and bond distances in higher hydrates, in general, should not be too different from those in the dihydrates of the type $H_2O.LiF.H_2O$, unless steric effects (hindrance) become important.

In the 'inner' hydrates of lithium fluoride, $Li.OH_2.F$, $Li.OH_2.OH_2.F$, and $H_2O.Li.OH_2.F$, $Li.OH_2.F.H_2O$, one or two water molecules are positioned between the lithium and fluorine centers and are bonded to both centers. This means the water molecule(s) are positioned in the field of the lithium fluoride dipole with the two poles near the lithium and fluorine centers. This can be expected to have a marked influence on the charge distribution, binding energies, and bond lengths in the system. There will be a considerable charge shift in the water molecule. Mutual polarization effects in the system will tend to increase the binding energy and to decrease the distance of the chemical bonds involved. In particular the energy of the chemical bonds between the water molecule and the lithium and fluorine centers, respectively, will be larger than those found for the systems $H_2O.LiF$ and $LiF.H_2O$. The total energy of such 'inner'lithium fluoride hydrates therefore is expected to be lower than the sum of the total energies of the subsystems water and lithium fluoride, with the same internuclear LiF separation. This result is of particular interest for the electrolytic dissociation process.

For the systems $Li.OH_2.F$ and $Li.OH_2.OH_2.F$, as well as $H_2O.Li.OH_2.F$ and $Li.OH_2.F.H_2O$, the energy hypersurface has been studied near the minimum energy configuration for a limited variation of a restricted number of parameters.

Two strong effects are observed for the equilibrium bond distances calculated for the lithium fluoride water complexes: First, the equilibrium bond distances d(LiO) and d(FO) for the 'inner' lithium fluoride hydrate are about 10% smaller than the corresponding distances in the pure ion hydrates $Li^+.OH_2$ and $F^-.H_2O$. This indicates that the water molecule(s) in the field of the lithium fluoride dipole are rather strongly polarized in such a way that the hydration bonds are strengthened. The 'inner' hydration energy therefore might well be expected to be larger than for the comparable atomic ions of lithium or fluorine. Further, for the same reason, a very strong decrease of the d(OO) distance of the water dimer placed in the field of lithium and fluorine atoms is observed, amounting to appr. 20% relative to the theoretically calculated distance in the isolated water dimer [44]. This also indicates a strong re-arrangement of the elctronic charge distribution of the water dimer when placed in the field of the lithium fluoride dipole. The charge distribution must be polarized in such a way as to increase the hydrogen bond energy in the water dimer.

The following characteristic changes in the equilibrium bond lengths are observed within the series of molecular complexes investigated: From $Li.OH_2.F$ to $Li.OH_2.OH_2.F$, the bond distances d(LiO) and d(FO) increase slightly by about 2-3%. In the series $Li.OH_2.F$, $H_2O.Li.OH_2.F$, and $Li.OH_2.F.H_2O$, the bond distances d(LiO) and d(FO) increase as well, in particular at those centers where double hydration takes place. This is well in agreement with the bond length increase observed for the mono- and for the dihydrates of the individual atomic ions, whereas the distances decrease for the double hydration of the lithium fluoride molecule at different centers.

The bond length effects found here for water molecules(s) in the field of two atomic ions of opposite charge can be expected to have a strong influence on the equilibrium bond length of the solvent clusters formed in ionic solution or in crystals, as compared with solvent clusters in the pure solvent.

The energy changes in the systems $Li.OH_2.F$ and $Li.OH_2.OH_2.F$, relative to the subsystems LiF and H_2O, in their minimum energy geometry, have been determined from the computed total energies of the systems in the appropriate configurations. It is found that the complexes are unstable with respect to dissociation into the subsystems LiF and H_2O, by 21 kcal $mole^{-1}$, and 31 kcal $mole^{-1}$, respectively. But the complexes are very stable, by 166 kcal $mole^{-1}$ and 156 kcal $mole^{-1}$, respectively, if the LiF distance in the LiF system is equal to that in the complexes. This shows that very stable bonds are formed in the 'inner' lithium fluoride-water complexes.

Electron density difference maps can be used to study polarization, or charge density changes, discussed above in connection with the hydration energies and bond length differences in the various states of hydration (degree of hydration). Such electron density difference maps are determined as the difference between the electron densities of the total system and of those subsystems whose mutual influence in forming the total system is to be studied. A number of electron density difference maps have been computed for the hydrates $H_2O.Li^+$, F^-H_2O, $H_2O.LiF$, $LiF.H_2O$, $H_2O.LiF.H_2O$, $Li.OH_2.F$, and $Li.OH_2.OH_2.F$ relative to the subsystems Li^+, F^-, LiF, $H_2O.H_2O$, $H_2O.LiF$ and $LiF.H_2O$ for bond distances near the theoretically determined equilibrium bond distances d(LiO) and d(FO). These maps are collected in Figures 16 to 26.

From these maps it is seen that a charge density redistribution generally takes place all over the system [16, 63-65], supporting the well-known fact that the hydration energies cannot be additive. A comparison of these maps shows that the charge shift is largest for the ion hydrates. For these systems the largest hydration energies and smallest equilibrium bond distances have been calculated. In particular it is found that the small lithium cation is very little polarized, while the large fluorine anion is very strongly polarized. From the plot of the electron density differences in the hydrated lithium cation it can be concluded that the charge distribution at the lithium cation has become more localized, and that the charge distribution of the water molecule has been polarized into the direction of the cation. This means that in both subsystems charge has been shifted towards the lithium center. From the plot for the hydrogen bonded fluorine anion hydrate it can be seen that the situation is more complicated. In particular, the charge density of the fluorine anion has become more localized than in the free ion, and the charge density in the water molecule has become polarized in such a way that the charge density at the hydrogen bridge atom has become more positive. At the water molecule, the charge has been shifted to the lone pair electron region and to the other hydrogen atom. Similar results have been found for the charge shifts in the water dimer [44]. In general, no quantitative results can be found from these plots concerning the charge shifts between the different subsystems involved, that is, between the ions and the water molecule.

In the monohydrates of the lithium fluoride molecule the charge shifts are smaller than in the hydrated ions. In particular, it is found that the lithium fluoride is polarized in such a way that the electron density at the lithium atom becomes more positive and at the fluorine atom more negative. The charge shifts induced in the water molecules are very similar, although smaller in magnitude, as those found in the hydrates of the atomic ions.

From the present results it can be concluded that the lithium fluoride molecule is polarized by the monohydration in such a way that the hydration energy at the other center of the molecule will be increased relative to the *monohydration* energy at the same center. This is in agreement with the theoretical results of the present investigation. Very similar effects have been found for the charge shifts and hydrogen bonding energies in the higher polymers of water, where the hydrogen bonding energy increases with the length of the chain or cycle until a certain limit is reached.

In this context it is of special interest to study the additional charge shifts introduced by binding a second water molecule at the other center of the molecule. The corresponding electron density difference maps have been computed and are displayed in Figures 21 and 22. The polarization in the *simultaneously* dihydrated lithium fluoride molecule is in agreement with the picture deduced from the results found for the monohydrated lithium fluoride molecules. In particular, the polarization of the lithium fluoride molecule is stronger than in the monohydrates, and the polarization effect of the monohydrates is amplified. The charge shifts induced in the monohydrates by binding a second water molecule are also in agreement with previous results. It is of particular interest to notice that these polarization effects are of very long range. They even influence the charge distribution at the other water molecule.

An analysis of the electron density difference maps of the 'inner' lithium fluoride hydrates displayed in Figures 23 to 26 and a comparison to the electron density difference maps of the lithium fluoride hydrates displayed in Figures 10 to 22 leads to the following results: In general agreement with the conclusions drawn in previous investigations it is found that there is a decrease of charge density at the hydrogen atoms acting as the hydrogen bridge in the various hydrogen bonds, while there is an increase of electronic charge in the region of the lone pair electrons of the hydrogen bonded water molecules. In general, the electron density difference maps of the system $Li.OH_2.F$, relative to the subsystems Li--F and H_2O, resemble very much a superposition of the electron density difference maps of the systems $Li^+.OH_2$ and $F^-.H_2O$ relative to the subsystem Li^+, F^-, and H_2O. In agreement with the conclusions drawn from an analysis of the interaction (hydration) energies, it is found that there are strong mutual polarization effects even for rather large internuclear distances d(LiO) and d(FO) of appr. 10 AU. The general pattern of the charge rearrangement for this long range interaction is very similar, although smaller in magnitude, to those found for the 'equilibrium' complex. A similar analysis holds for the electron density difference map of the system $Li.OH_2.OH_2.F$ relative to the subsystem LiF and $OH_2.OH_2$.

4. BOND DISSOCIATION IN LITHIUM FLUORIDE-WATER COMPLEXES

With regard to a purely qualitative discussion of the ionic dissociation of LiF under the influence of H_2O as polar solvent, the potentials of the lithium fluoride-water bonds and of the lithium fluoride bond itself in different hydration states will now be considered. The corresponding potential energy curves are displayed in Figures 9-14.

For the 'outer' hydrates $H_2O.LiF$, $LiF.H_2O$, and $H_2O.LiF.H_2O$, the hydration energy increases continuously with increasing internuclear separation between the lithium and fluorine centers, and it converges slowly to the hydration energy of the non-interacting ions. For $H_2O.LiF$, for example, at an internuclear distance of d(LiF) = 15 AU, the hydration energy differs from that of the lithium cation, $E^{Hydr}(Li^+.H_2O)$ = 38.08 kcal $mole^{-1}$, by approximately 5 kcal $mole^{-1}$ (15%). This means that the system becomes *completely* dissociated only for rather large distances. In the whole range of internuclear separations investigated, the hydration energy of the system $H_2O.LiF.H_2O$ is larger than the sum of the hydration energies of the systems $H_2O.LiF$ and $LiF.H_2O$ by about 5 kcal $mole^{-1}$. This indicates that there is a further stabilization due to an additional polarization of the system by the second water molecule.

The potential energy curves for a lithium fluoride bond dissociation in the 'outer' hydrates show a bond destabilisation with increasing number of water ligands. The increase in potential energy for the LiF bond dissociation energy decreases with increasing number of water ligands.

For the 'inner' hydrates, $Li.OH_2.F$ and $Li.OH_2.OH_2.F$, the hydration energy has a maximum for the equilibrium structure of the system and it converges for large LiF distances towards the value of the hydrated ions. This means that the hydration energy is larger than the sum of the hydration energies of the separated systems. The potential energy minima are shifted strongly upward with increasing number of internally bonded water molecules. The larger the number of internally bonded water molecules becomes, the smaller is the energy required to break the LiF bond. These results support the conclusion drawn earlier that there is a strong additional binding effect in the 'inner' hydration bonds.

The potential energy curves for the lithium fluoride bond dissociation in the 'inner' hydrates show even a stronger bond destabilisation with increasing number of water ligands than in the case of the 'outer' hydrates.

Summarizing, it can be said that a considerable amount of energy is gained if, in the course of the dissociation of a lithium fluoride bond, an internal hydrate with one or two water molecules is formed. This process may be considered as one of the first steps in the electrolytic dissociation process.

An analysis of the potential energy curves for the internal rearrangement $F.Li.H_2O \rightarrow Li.H_2O.F$ displayed in Figure 13 shows that there exists a reaction path (B') without energy barrier between the initial and final reactants. While in the present example the energy of the product is higher than the energy of the reactants, this situation is expected to change if completely hydrated reactants and products would be considered.

A discussion of the energetics of the LiF bond dissociation in solution has to take into account the influence of the surrounding medium. This is, however, not possible in any quantum-theoretical approach. Hence model treatments have to be used to simulate the effect of the surrounding medium. The change of total energy can be discussed qualitatively, based on the results of ab initio calculations of near Hartree--Fock accuracy. In the present investigation, the most important three body effects calculated for the representative model systems will be taken into account rigorously. Effects of higher order may be assumed to be reasonably small and not to change the results qualitatively.

Assuming that the electrolytic dissociation and hydration takes place stepwise, the following clusters may be expected as intermediate steps of the electrolytic dissociation process:

(VIII) $(H_2O)_3.LiF.(H_2O)_3$

(IX) $(H_2O)_3.Li.OH_2.F.(H_2O)_3$

(X) $(H_2O)_3.Li.OH_2.OH_2.F.(H_2O)_3$

(XI) $(H_2O)_3.Li.OH_2.OH_2.F.(H_2O)_5$

(XII) $(H_2O)_4.Li^+ + F^-.(H_2O)_6$

The total energy has been estimated as a function of the internuclear separation d(LiF) for the systems (VIII), (IX), and (X), assuming additivity of the energy components. Actually, the potential energy curves of these systems displayed in Figure 14 have been 'constructed' from the appropriate potential energy curves of the systems LiF, $Li.OH_2.F$, and $Li.OH_2.OH_2.F$ displayed in Figure 11 and from the hydration energy curves of the systems $H_2O.LiF$, $LiF.H_2O$, and $H_2O.Li^+$, $F^-.H_2O$ displayed in Figure 9. The potential energy curves for these systems include only the energy of the LiF system and the interaction energy with the given number of water molecules. The total energy of the water molecules is not taken into account because it is constant for the electrolytic dissociation process, under the assumption that no geometry changes in the water molecules take place.

From the potential energy curves of Figure 14 it can be seen that near an internuclear distance of d(LiF) = 3 AU the energetically most stable system is a lithium fluoride molecule surrounded by a hydration shell of six water molecules (curve 3LiF3 of Figure 14). At an internuclear distance of about d(LiF) = 6 AU there will be enough space for another water molecule to be placed between the two centers, without steric hindrance. As has been shown earlier, this water molecule will form strong hydration bonds with both nuclear centers in LiF. The LiF dissociation for this system (IX) is described by the potential energy curve (3Li1F3) of Figure 14. For internuclear separations larger than d(LiF) = 5 Au this curve is lower than that of the system (VIII). Similarly, at an internuclear distance of approximately d(LiF) = 10 AU, two water molecules can be positioned between the two centers Li and F, without steric hindrance. It may further be assumed that already at such an internuclear LiF distance, an octahedral hydration shell can be formed around the fluorine center, giving rise to the system (XI). The potential energy curve of this system (XI) is even lower than those of the systems (VIII) and (IX) for all LiF distances larger than about d(LiF) = 9 AU. This potential energy curve must finally converge to the energy of the totally hydrated lithium and fluorine ions. Within the present approximation, the potential energy curve of system (XI) approaches this limit from below (within the region investigated), indicating that the system (XI) is *energetically* stable against complete dissociation. This, however, does not contradict the experimental results showing complete dissociation of strong electrolytes at low concentrations, because the equilibrium state is not determined by the minimum enthalpies *but by the minimum free energy*. In addition only zero temperature is considered.

In the course of electrolytic dissociation transitions must take place between the different potential energy curves of Figure 14.

In the same way as for the potential energy curves of the lithium fluoride-water complexes (VIII), (IX), and (XI) displayed in Figure 14, potential energy curves have been constructed for the reaction B' of Figure 13. In aqueous solution this 'internal' reaction takes place in a complex of completely hydrated reactants (lithium fluoride) and reaction products. A complication in studying the potential energy surface is due to the fact that during the course of dissociation, while a water molecule of the hydration shell moves into the position between the two centers Li and F, another water molecule of the surrounding solvent moves into the (first) hydration shell. Therefore potential energy curves have been constructed for the following two systems

(XIII) $(H_2O)_2.H_2O.LiF.(H_2O)_3$ + H_2O (isolated)

(XIV) $(H_2O)_3.Li.OH_2.F.(H_2O)_3$

Both systems should be energetically stable in different regions of internuclear distance d(LiF) because of steric reasons, and there must be a transition between the two potential energy curves.

The potential energy curves of these systems, (XIII) and (XIV), displayed in Figure 14, have been 'constructed' as described previously from the potential energy curve of reaction path B', Figure 13. The main conclusion that can be drawn from these potential energy curves is that the potential energy barrier against dissociation of lithium fluoride is very small.

5. SUMMARY

Very little is known in detail about the structure of ionic solutions and the interactions in such solutions. On the other hand the solute-solvent clusters which have to be considered for this purpose are by far too large to be treated in a rigorous quantumchemical ab initio calculation. Due to this fact a number of relevant interaction potentials have been calculated in this study for the singly and doubly hydrated LiF molecule within the single determinant Hartree-Fock approximation employing rather extended gaussian basis sets to ensure a reasonable accuracy. Using these potentials a reaction pattern is discussed qualitatively along which the dissociation process could possibly proceed. It has to be stressed, however, that this is not necessarily the only possible scheme leading to complete dissociation of ionic solutions. More detailed informations should be obtained from further dynamical and statistical studies in which the calculated interaction potentials can be used.

REFERENCES

1. Diercksen, G.H.F.: Habilitationsschrift, Technische Universität München, 1973.
2. Arrhenius, S.: Z. phys. Chemie 1, 481 (1887).
3. Narcisi, R.S. and Bailey, A.D.: J. Geophys. Res. 70, 3687 (1965).
4. Narcisi, R.S. and Roth, W.: Advan. Electron. Electron Phys. 29, 79 (1970).
5. Fehsenfeld, F.C. and Ferguson, E.E.: J. Geophys. Res. 74, 2217 (1969).
6. Ferguson, E.E. and Fehsenfeld, F.C.: J. Geophys. Res. 74, 5743 (1969).
7. Fehsenfeld, F.C., Mosesman, M., and Ferguson, E.E.: J. Chem. Phys. 55, 2115 (1971); 55, 2120 (1971).
8. Howard, C.J., Rundle, H.W., and Kaufman, F.: J. Chem. Phys. 55, 4772 (1971).
9. Kebarle, P., Arshdi, M., and Scarborough, F.: J. Chem. Phys. 49, 817 (1968).
10. Dzidic, I. and Kebarle, P.: J. Phys. Chem. 74, 1466 (1970).
11. Arshdi, M., Yamdagni, R., and Kebarle, P.: J. Phys. Chem. 74, 1475 (1970).
12. Kebarle, P.: in Franklin, J.L. (ed.), Ion-Molecule Reactions, Butterworth, London, 1972.
13. Narten, A.H., Vaslow, F., and Levy, H.A.: J. Chem. Phys. 58, 5017 (1973).
14. Frank, H.S. and Wen, W-.Y.: Disc. Faraday Soc. 24, 133 (1957).
15. Buckingham, A.D.: Discuss. Faraday Soc. 24, 151 (1957).

16. Schuster, P., Jakubetz, W., and Marius, W.: 'Molecular Models for the Solvation of Small Ions and Polar Molecules', in Topics in Current Chemistry 60, Springer-Verlag, 1975.
17. Kraemer, W.P.: in 'Quantum Chemistry, The State of the Art', Atlas Computer Laboratory, Chilton (GB) May 1975, p.217.
18. Kraemer, W.P. and Diercksen, G.H.F.: Theoret. Chim. Acta (Berl.), to be submitted.
19. Rahman, A. and Stillinger, F.H.: J. Chem. Phys. 55, 3336 (1971); J. Am. Chem. Soc. 95, 7943 (1973).
20. Stillinger, F.H. and Rahman, A.: J. Chem. Phys. 57, 1281 (1972); 60, 1545 (1974); 61, 4973 (1975).
21. Heinzinger, K. and Vogel, P.C.: Z. Naturforsch. 29a, 1164 (1974).
22. Vogel, P.C. and Heinzinger, K.: Z. Naturforsch. 30a, 789 (1975).
23. Barker, J.A. and Watts, R.O.: Chem. Phys. Letters 3, 144 (1969).
24. Kistenmacher, H., Popkie, H., Clementi, E., and Watts, R.O.: J. Chem. Phys. 60, 4455 (1974).
25. Kistenmacher, H., Popkie, H., and Clementi, E.: J. Chem. Phys. 61, 799 (1974).
26. Watts, R.O., Clementi, E., and Fromm, J.: J. Chem. Phys. 61, 2550 (1974).
27. Watts, R.O.: Mol. Phys. 28, 1069 (1974).
28. Diercksen, G.H.F. and Kraemer, W.P.: Chem. Phys. Letters 5, 570 (1970).
29. Schuster, P. and Preuss, H.: Chem. Phys. Letters 11, 35 (1971).
30. Clementi, E.: in Physics of Electronic and Atomic Collisions VII ICPEAC, North-Holland, Amsterdam, 1971.
31. Diercksen, G.H.F. and Kraemer, W.P.: Theoret. Chim. Acta (Berl.) 23, 387 (1972).
32. Kraemer, W.P. and Diercksen, G.H.F.: Theoret. Chim. Acta (Berl.) 23, 393 (1972).
33. Kraemer, W.P. and Diercksen, G.H.F.: Theoret. Chim. Acta (Berl.) 27, 265 (1972).
34. Breitschwerdt, K.G. and Kistenmacher, H.: Chem. Phys. Letters 14, 288 (1972).
35. Clementi, E. and Popkie, H.: J. Chem. Phys. 57, 1077 (1972).
36. Kollmann, P.A. and Kuntz, I.D.: J. Am. Chem. Soc. 94, 9236 (1972).
37. Kistenmacher, H., Popkie, H. and Clementi, E.: J. Chem. Phys. 58, 5627 (1973).
38. Kistenmacher, H., Popkie, H., and Clementi, E.: J. Chem. Phys. 59, 5842 (1973).
39. Marius, W.: Thesis, Vienna, 1974.
40. Diercksen, G.H.F., Kraemer, W.P., and Roos, B.O.: Theoret. Chim. Acta (Berl.) 36, 249 (1975).
41. Morokuma, K. and Pedersen, L.: J. Chem. Physics 48, 3275 (1968).
42. Kollmann, P.A. and Allen, L.C.: J. Chem. Phys. 51, 3286 (1969).
43. Hankins, D., Moskowitz, J.W., and Stillinger, F.H.: Chem. Physics Letters 4, 527 (1970).
44. Diercksen, G.H.F.: Theoret. Chim. Acta (Berl.) 21, 335 (1971).
45. Roothaan, C.C.: Rev. Mod. Phys. 23, 69 (1951).
46. Diercksen, G.H.F. and Kraemer, W.P.: MUNICH Molecular Program System, Reference Manual, Special Technical Report, Max-Planck-Institut für Physik und Astrophysik.

47. Salez, C. and Veillard, A.: Theoret. Chim. Acta (Berl.) 11, 441 (1968).
48. Diercksen, G.H.F.: Chem. Physics Letters 4, 373 (1969).
49. Clementi, E.: Suppl. IBM Journal Res. and Dev., 1965
50. McLean, A.D. and Yoshimine, M.: J. Chem. Phys. 46, 3682 (1967); Suppl. IBM Journal Res. and Dev., 1967.
51. Meyer, W., Jakubetz, W. and Schuster, P.: Chem. Phys. Letters 21, 97 (1973).
52. Jeziorski, B. and Van Hemert, M.: preprint 1975.
53. Matsuoka, O., Clementi, E., and Yoshimine, M.: J. Chem. Phys. 64, 1351 (1976).
54. Roos, B.O., Kraemer, W.P., and Diercksen, G.H.F.: Theoret. Chim. Acta (Berl.), accepted.
55. Dunning, T.H.: J. Chem. Phys. 53, 2823 (1970); 55, 716 (1971; 55, 3958 (1971), Chem. Physics Letters 7, 423 (1970).
56. Huzinaga, S.: Approximate Atomic Functions, Vol. 1 and 2, Preprint, Division of Theoretical Chemistry, Department of Chemistry, The University of Alberta, 1971.
57. Liu, B. and McLean, A.D.: J. Chem. Phys. 59, 4537 (1973).
58. Kress, J.W., Clementi, E., Kozak, J.J., and Schwartz, M.E.: J. Chem. Phys. 63, 3907 (1975).
59. Clementi, E., Kistenmacher, H., and Popkie, H.: J. Chem. Phys. 58, 2460 (1973).
60. Hankins, D., Moskowitz, J.W., and Stillinger, F.H.: J. Chem. Phys. 53, 4544 (1970).
61. Lentz, B.R. and Scheraga, H.A.: J. Chem. Phys. 58, 5296 (1973).
62. Kraemer, W.P. and Diercksen, G.H.F.: Theoret. Chim. Acta (Berl.), to be submitted.
63. Kollman, P.A. and Allen, L.C.: J. Chem. Phys. 52, 5085 (1970).
64. Kollman, P.A., Liebman, J.F., and Allen, L.C.: J. Am. Chem. Soc. 92, 1142 (1970).
65. Kollman, P.A. and Allen, L.C.: J. Am. Chem. Soc. 92, 6101 (1970).

AB INITIO CALCULATIONS ON NICKEL-ETHYLENE COMPLEXES

BJØRN ÅKERMARK, MATS ALMEMARK, JAN-E. BÄCKVALL
Dept. of Organic Chemistry, Royal Institute of Technology, Stockholm, Sweden

JAN ALMLÖF, BJØRN ROOS
Institute of Theoretical Physics, University of Stockholm, Sweden

and

ÅSE STØGÅRD
Dept. of Chemistry, University of Bergen, Norway

1. INTRODUCTION

Semi-empirical MO calculations have been reported for a few complexes where the olefin is complexed to the nickel triad (Ni, Pd, Pt) (e.q., Elian et al., 1975). The calculations give a qualitative understanding of the interaction between the metal and the olefin. More accurate calculations have also been performed for π allylnickel (Veillard, 1969), bis π allylnickel (Rohmer et al., 1974), nickel carbonyl (Hillier et al., 1971), a nickel pentafluoride olefin complex (Baerends et al., 1972), Zeise's salt (Rösch et al., 1974), and a few other transition metal complexes.

Olefins complexed to transition metals are known to undergo a number of reactions, for example, forbidden thermal cycloaddition, polymerization reactions and reactions with weak nucleophiles like water, alcohols, and amines. Many empirical data are available, but a quantitative understanding of these reactions and of the interaction between the olefin and the metal is still lacking. This interaction can be described in a qualitative manner using the Dewar-Chatt-Duncanson model (Dewar, 1951 and Chatt et al., 1953). In this model the bond is described by: (i) charge transfer from the symmetric bonding π-orbital of the olefin to unoccupied symmetric metal orbitals and (ii) charge transfer from occupied antisymmetric metal orbitals into the antibonding π* orbital of the olefin ('back bonding'). The combined effect of these two interactions will determine the reactivity of a complexed olefin. If electron withdrawal from the olefin (i) predominates, the olefin becomes susceptible to nucleophilic attack. If both electron withdrawal (i) and back donation (ii) are significant, the olefin becomes relatively unreactive towards both nucleophiles and electrophiles.

B. Pullman and N. Goldblum (eds.), Metal-Ligand Interactions in Organic Chemistry and Biochemistry, part 2 , 377-387.

In cases where back donation dominates, the olefin is negatively charged and is susceptible to electrophilic attack and the olefin should undergo cycloaddition reactions.

In an attempt to gain further knowledge of transition metal interaction with olefins and the chemical consequences of these interactions, we have performed ab initio calculations on a series of nickel-ethylene complexes (Figure 1). Most of these complexes are not known as isolable compounds but may exist as intermediates in chemical reactions.

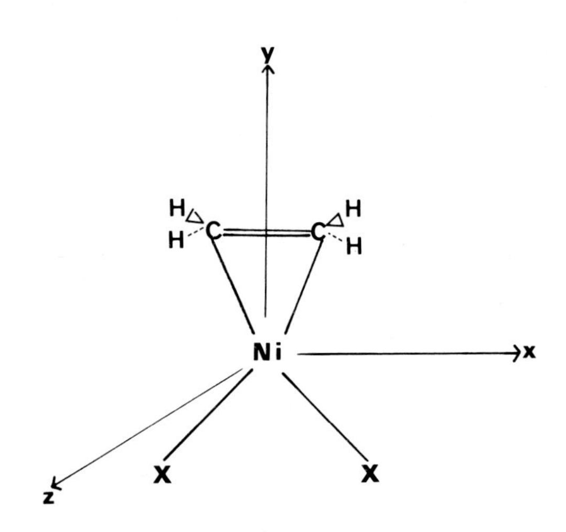

Fig. 1. $Ni(C_2H_4)X_2$
n = 0 X = F^-, Cl^-, NH_2^-, CN^-, CH_3^-, H_2O, NH_3, PH_3.
n = 2 X = H_2O, NH_3, PH_3.

2. DETAILS OF THE CALCULATION

The Ni atom is known to form square planar complexes. In the few $Ni(C_2H_4)X_2$ complexes for which the structure has been determined, the metal is approximately quadratically coordinated by the central atoms of the ligands and the carbon atoms of ethylene. The Ni-C distance and the /C-Ni-C were taken from X-ray determination on $(PPh_3)_2Ni(C_2H_4)$ (Cook et al., 1967). The metal-ligand distances and the coordination angles were taken from experimental data when available. The values used for these parameters are summarized in Table I. Standard values were used for the internal geometries of the various ligands.

For all systems the C-C double-bond distance was optimized, keeping all the other geometrical parameters constant. The calculations were performed in the MO-LCAO-SCF framework, using contracted gaussian functions for the expansion of the molecular orbitals. The choice of basis functions deserves some consideration. The basis set must be flexible enough to give a satisfactory description of the specific physical properties connected to the problem. At the same time the number of functions must be kept as low as possible to avoid excessive computation.

TABLE I
Geometrical parameter

X	R_{Ni-X} (Å)
F	1.986
Cl	2.38
C(cyanide)	1.86
C(methyl)	1.87
N	2.15
O	2.10
P	2.18
C(ethylene)	2.01

$\angle$ X-Ni-X = 112° in all cases

A(10(15), 7, 5) → [3(5), 4, 2] basis set was used for the Ni atom. The first-row atoms were described using a (7, 3) basis and the second-row atom by a (10, 6) basis set. These were contracted to minimal basis sets except for the carbon atoms of the ethylene molecules for which a double zeta type contraction was chosen in order to give some additional flexibility to the regions of particular interest. For the hydrogen atoms in ethylene a (4) → [2] basis was used, while a (3) → [1] basis was used for the hydrogens in the ligands.

Available experimental information indicated that a quadratically coordinated Ni complex has a low-spin ground state (Cotten et al., 1966). The calculations were therefore only performed on low-spin complexes. In addition to complexes of the type $Ni(C_2H_4)X_2$ the smaller systems $Ni(C_2H_4)$ and NiX_2 were also investigated.

All calculations were performed on the Univac 1110 Computer at the University of Bergen, using the program package MOLECULE (Almlöf, 1974).

3. RESULTS AND DISCUSSION

For the ligands F^-, Cl^-, CN^-, CH_3^-, and NH_2^- the charge distribution of the complexes are best described in terms of a divalent nickel ion and two negatively charged ligands. With the other ligands, H_2O, NH_3 and PH_3, the central atom is essentially nickel (O). Calculations on the complexes with these ligands were therefore repeated with a total charge of +2 for the systems, which would correspond to a Ni(II) complex.

The results of the calculated equilibrium C-C bond distances of ethylene are shown in Table II. In the case of Ni(II), the compounds have approximately the same C-C bond distance with the exception of the

TABLE II
Optimized C-C distances for $Ni(C_2H_4)X_2$.

X	Ni(II) R_{C-C}(Å)	Ni(O) R_{C-C}(Å)
-	1.360	1.390
F^-	1.352	
Cl^-	1.360	
NH_2^-	1.352	
CN^-	1.391	
CH_3^-	1.346	
H_2O	1.347	1.452
NH_3	1.349	1.455
PH_3	1.356	1.410

The C-C distance in C_2H_4: 1.320 Å (calculated)
1.338 Å (exp.)

CN^- complex. The Ni(O) complexes have longer C-C distances than the Ni(II) complexes. The difference is approximately 0.1 Å when the various ligands are added. The addition of neutral ligands causes further increase in the C-C distance for the Ni(O) complexes. The values for the H_2O and NH_3 complexes of 1.452 and 1.455 indicate that the olefin has lost about half of its double bond character.

The calculated C-C bond distances are in all cases longer than in the free ethylene molecule. This increase in bond length is explained by the model adopted for the nickel-ethylene interaction. Both the charge transfer from the bonding π system of ethylene and the back donation to the antibonding π^* orbital decrease the double bond character and consequently increase the length of the C-C bond.

A comparison between calculated C-C bond distances and experimental data is difficult since very few nickel (II) olefin complexes have been investigated by X-ray crystallography. However, much experimental evidence is available in the palladium and platinum series and although the bond strengths are probably somewhat greater than in the nickel complexes, the difference should be small enough to make a comparison relevant. A recent neutron diffraction study on Zeise's salt, $[Pt(C_2H_4)Cl_3]^-$ (Hamilton et al., 1969), gives a C-C distance equal to 1.354 Å which is in agreement with the value of 1.360 Å obtained from the calculation on $Ni(C_2H_4)Cl_2$. Other recent studies exist, which give C-C bond lengths for palladium and platinum compounds in the range of 1.30-1.40 Å, with a preference around 1.35 Å (see Åkermark et al., 1976 for further references). For the metal (O) complexes distinctly longer C-C bond lengths have been determined, ranging from 1.40 to 1.48 Å, with a preference around 1.43 Å. Most of these compounds contain phosphine as ligand X. The experimental results thus agree with the computed value of 1.41 Å for the $Ni(C_2H_4)(PH_3)_2$ complex.

The binding energies between the ethylene molecule and the fragments NiX_2 are given in Table III. Since the nickel-ethylene distance

TABLE III
Binding energies between the ethylene molecule and the fragments NiX_2.

X	Ni(II) Δ E(kcal mole^{-1})	Ni(O) Δ E(kcal mole^{-1})
-	80.53	27.18
F^-	- 5.84	
Cl^-	1.44	
NH_2^-	13.62	
CN^-	- 0.13	
CH_3^-	5.90	
H_2O	43.80	58.86
NH_3	34.39	53.97
PH_3	30.31	35.70

has not been optimized, the absolute values of the binding energies should not be taken too seriously. However, the trends shown by the relative values are probably relevant. The bond strength between nickel and ethylene is much greater in the ionic case. The difference must be interpreted as due to electrostatic interactions between the ionic charge and the polarizable π-electron system of the olefin. It is therefore not unexpected that the binding energies are very much lowered when ligands are added to the Ni(II) ethylene complex. This is especially true when X is a negatively charged ligand. The calculated low binding energies in the nickel(II) series derive indirect experimental support from the fact that no neutral nickel(II) complexes $Ni(C_2H_4)X_2$ have been isolated. In the neutral Ni(O) complexes the binding energies are increased on the addition of ligand. The higher binding energy in these cases should be viewed as a more pronounced charge transfer to the π^* orbitals of the olefin. Several complexes of zero-valent nickel have been isolated. The metal-ethylene binding energy for one such complex, $(R_3P)_2Ni(C_2H_4)$, R = O - o - tolyl, has been estimated at 33 kcal mole^{-1} (Tolman, 1974), which is in good agreement with the value of 36 kcal mole^{-1} calculated for $Ni(C_2H_4)(H_3P)_2$.

The validity of the concepts of σ symmetry donation to the metal and π back-bonding to the olefin can be checked by an inspection of the gross population of the basis functions of π and π* symmetry on ethylene. The present calculation indicates that for neutral nickel(II) complexes, donation to the metal and back-bonding to the olefin are of the same magnitude (Table IV). This means that the olefin is essentially neutral

TABLE IV
Gross charges on ethylene, Ni and the ligands X and total population of the π and π* orbitals in $Ni(C_2H_4)X_2$.

	X	$Q(C_2H_4)$	Q(Ni)	Q(X)	P(π)	P(π*)
	-	+0.61	+1.39	-	1.37	0.02
	F^-	+0.01	+0.89	-0.45	1.79	0.19
	Cl^-	+0.07	+0.93	-0.50	1.74	0.19
	NH_2^-	+0.02	+0.83	-0.43	1.76	0.21
Ni(II)	CN^-	-0.01	+1.20	-0.59	1.78	0.21
	CH_3^-	-0.02	+1.05	-0.51	1.77	0.26
	H_2O	+0.38	+1.15	+0.23	1.54	0.07
	NH_3	+0.38	+1.12	+0.25	1.53	0.09
	PH_3	+0.39	+0.93	+0.34	1.49	0.13
	-	-0.33	+0.33	-	1.81	0.52
	H_2O	-0.80	+0.56	+0.12	1.83	0.98
Ni(O)	NH_3	-0.78	+0.58	+0.11	1.85	0.93
	PH_3	-0.48	+0.35	+0.07	1.87	0.61

in these complexes and thus relatively unreactive towards both nucleophiles and electrophiles. Somewhat similar results have been obtained for ethylene complexes with TiF_5 and NiF_5 (Baerends et al., 1972). When the charged ligands are replaced by net neutral ligands like water, ammonia or phoshine, the donation from the olefin is approximately 0.5 electron while the back donation is about 0.1 electron. In these complexes, the olefin will carry a considerable positive charge, and is expected to be readily attacked by nucleophiles. The bonding situation is quite different in the nickel(O) complexes. Here back donation is a major factor in the metal-olefin bond. The configuration is close to $\pi^2\pi^{*1}$ and the complexed olefin may be expected to have properties similar to those of an excited olefin. Zero-valent metal should therefore catalyze the cycloaddition reactions characteristic of excited olefins, e.q. (2+2) additions.

The binding mechanism in three-membered heterocyclic ring systems of the type XC_2H_4, X = S, SO and SO_2, has been discussed (Rohmer et al., 1975). From these results is was possible to relate the C-C bond weakening to the donor-acceptor strength of the group X. In order to investigate whether such a quantitative correlation also exists for the nickel-olefin complexes, a number of calculations have been performed on the fragment NiX_2 with all the ligands. The results of these calculations are shown in Table V. The a-orbital is the lowest unoccupied

TABLE V
Energy of the highest occupied and the lowest unoccupied orbital (eV) of the fragment NiX_2.

X	Ni(II)		Ni(O)	
	a-orb.	b-orb.	a-orb.	b-orb.
-	16.35	-39.89	4.45	-6.44
F^-	1.25	-22.04		
Cl^-	- 0.20	-23.09		
NH_2^-	2.05	-20.65		
CN^-	- 0.23	-21.43		
CH_3^-	2.34	-20.50		
H_2O	-11.42	-33.90	8.72	-4.65
NH_3	-10.98	-33.36	9.10	-4.72
PH_3	-10.91	-30.53	7.17	-6.02

and is the acceptor orbital, while the b-orbital is the donor orbital. For the Ni(O) complexes the acceptor orbitals energies are very high and, therefore, the donation of electrons to Ni is unimportant in these complexes. Instead, the strength of the bond is almost entirely determined by the donor strength of the b-orbital which increases with increasing orbital energy. For the Ni(O) complexes there exists a close correlation between the orbital energy of the b-orbital in NiX_2 and the donated charge, the bond energy, and the length of the ethylene bond (Tables II-IV). In the case of the charged Ni(II) complexes, the donor orbital has too low an energy to be effective. The bond strength is therefore determined by the energy of the acceptor orbital. The situation in the neutral Ni(II) olefin complexes is more complicated. Here both the acceptor and the donor orbitals have unfavourable energies for strong interactions with the ethylene orbitals. Consequently, the nickel-olefin bond is weak in these complexes as is also indicated by experimental data.

Nucleophilic addition to olefins and acetylenes, which are the most extensively studied reactions, appear to be facilitated by electron withdrawal. The palladium promoted addition of water to ethylene has become an important industrial process for production of acetaldehyde. So far catalytic amination has only been achieved with ethylene and norbornadiene (see Åkermark et al., 1976 for further references). One of the goals of the calculations presented here was to attempt a theoretical description of the influence of ligands on the ability of the metals of the nickel triad to promote nucleophilic additions, particularly amination, and also to gain an insight into the mechanisms governing these reactions. Although the calculations have been carried out for nickel, the results can probably be extrapolated to Pd and Pt.

The calculations show that the olefin is almost neutral for the negatively charged ligands. The charge on the nickel atom varies from +0.83 to +1.20. Nucleophilic attack might therefore be expected to occur on the metal with displacement of a ligand rather than on the olefin. Nucleophilic attack on the olefin has not been observed with this type of nickel complex. For the neutral ligands the positive charge on the olefin is increased to +0.4 while the charge on the metal remains essentially as before. This will make nucleophilic attack on the olefin more probably. The charged Ni-complex(1) is known to undergo nucleophilic attack on the olefin with the formation of a σ complex (2)(Parker et al., 1974). In contrast to nickel(II), palladium(II) and platinum(II)

$$[\text{(norbornadiene)Ni Cp}]^+ \xrightarrow{CH_3O^-} \text{(2)}$$

(1) (2)

form stable olefin and diolefin complexes, which are readily attacked by nucleophiles. This is probably due to stronger electron withdrawal from the olefin by Pd(II) and Pt(II) as might be expected from an inspection of the ionization potentials: Ni(18.15 eV) $<$ Pt(18.56 eV) $<$ Pd(19.9 eV). The fact that rhodium(III) but not rhodium(I) catalyzes nucleophilic addition to olefins may also reflect the importance of the energy of the metal acceptor orbitals (rhodium first ionization potential 7.7 eV, third 31.0 eV).

Several questions concerning the mechanism of the amino-palladation reaction may perhaps be answered, using the results from the calculation for the nickel complexes. The addition of three equivalents of amine is necessary to convert the π complex (3) to an aminated σ complex, possibly (6) or (7). The first equivalent of amine serves to split the halogen bridge to yield a monomeric π complex (4), but the reason for the very limited amination (<5%) which takes place on addition of the second equivalent has remained obscure. A reasonable explanation now seems to be that the main intermediate in the aminations is the charged complex (5) rather than the neutral complex (4). If the displacement of chloride from the neutral complex (7) to yield (5) is rapid relative to amination, two equivalents of amine will be consumed before appreciable amination of the olefin can take place in accordance with experimental observations (Åkermark et al., 1974).

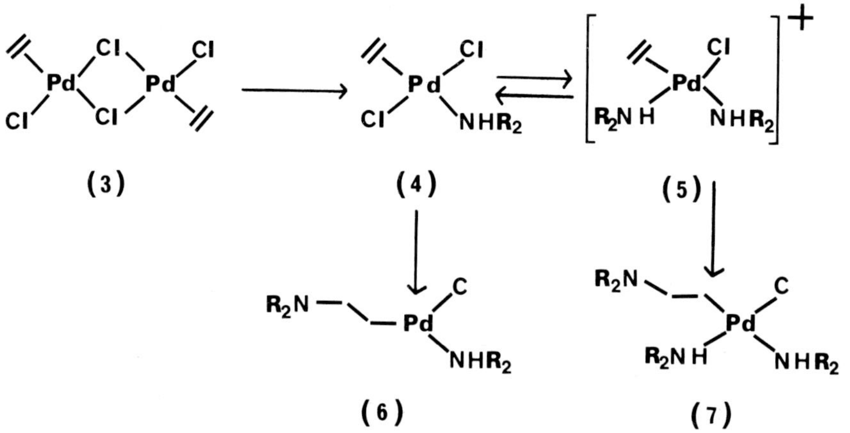

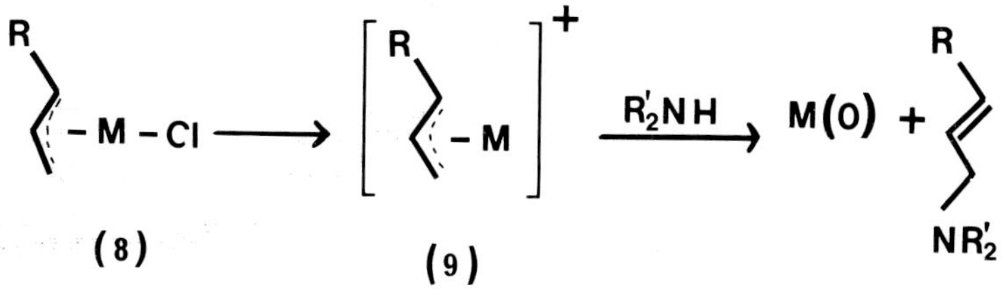

The relative reactivity of nickel and palladium complexes and the importance of the charge are further noticable in the aminations of π-allyl complexes. Palladium complexes of the general structure (8) and (9) are considerably more reactive than the corresponding nickel complexes. Furthermore, the ionized species (9) are much more reactive than the neutral ones (Akermark et al., 1975).

Concerning the cycloaddition reactions, excited olefins frequently participate in (2+2) cycloaddition reactions. Olefins complexed to neutral nickel(II) fragments have the approximate configuration $\pi^{1.75}$ $\pi^{*0.25}$ and thus have some excited state character. In general, the interaction between the metal and olefin might be expected to be too small to induce cycloaddition reactions. In the zero-valent metal complexes, a considerably stronger interaction may be anticipated and does indeed occur according to the present calculations and chemical evidence. The approximate configuration of the complexed olefin is $\pi^2\ \pi^{*1}$, indicating that close to one electron has been transferred from the metal to the olefin and it should act as donor in (2+2) cycloadditions. The Ni(O) catalyzed formation of cyclobutadienes from acetylenes and the addition of acrylonitrile to norbornadiene may well be examples of an essentially concerted (2+2) cycloaddition.

4. CONCLUSIONS

The σ-symmetry donation to the metal and π-symmetry back donation to the olefin (Dewar-Chatt-Duncanson model for the transition metal-olefin bond) have been quantitatively determined for a series of nickel complexes in order to determine the reactivity of a complexed olefin. In the neutral nickel(II) complexes, both effects are of the same magnitude resulting in an approximate olefin configuration of $\pi^{1.75}\ \pi^{*0.25}$. The complexed olefin is thus essentially neutral and unreactive towards nucleophiles. In related complexes of palladium and platinum, nucleophilic reactions should be more favourable due to the better acceptor properties of these elements in the two-valent state. This is confirmed by experiment. In the positively charged nickel(II) complexes, σ-symmetr donation dominates to give complexes where the olefin has an approximate configuration of $\pi^{1.5}\ \pi^{*0}$. This type of complexes should be readily attacted by nucleophiles as also shown by experiment. For the nickel(O) complexes, back donation dominates, the olefin configuration being close to $\pi^2\ \pi^{*1}$. The complexed olefin should therefore be able to participate in cycloaddition reactions but be unreactive towards nucleophiles, as also indicated by experimental results.

REFERENCES

Almlöf, J.: 1974, USIP Report 74-29, University of Stockholm.

Baerends, E.J., Ellis, D.E., and Ros, P.: 1972, Theoret. Chim. Acta 27, 339-354.

Chatt, J. and Duncansson, L.A.: 1953, J. Chem. Soc., 2939-2947.

Cook, C.D., Koo, C.H., Nyburg, S.C., and Shiomi, M.T.: 1967, Chem. Commun., 426-427.

Cotton, F.A. and Wilkinson, G.: 1966, Advanced Inorganic Chemistry, 2nd ed., Wiley-Interscience, New York, 1966, pp. 672, 884.

Dewar, M.J.S.: 1951, Bull. Soc. Chim. Fr. 18c, 79-85.

Elian, M. and Hoffmann, R.: 1975, Inorg. Chem. 14, 1058-1076.

Hamilton, W.C., Klander, K.A., and Spratley, R.: 1969, Acta Crystallogr. Sect. A 25, S172-S173.

Hillier, I. and Saunders, V.R.: 1971, Mol. Phys. 22, 1025-1034.

Parker, G., Salzer, A., and Werner, H.: 1974, J. Organometal. Chem. 67, 131-139.

Rohmer, M.-M., Demuynck, J., and Veillard, A.: 1974, Theoret. Chim. Acta 36, 93-102.
Rohmer, M.-M. and Roos, B.: 1975, J. Am. Chem. Soc. 97, 2025-2030.
Rösch, N., Messmer, R.P., and Johnson, K.H.: 1974, J. Am. Chem. Soc. 96, 3847-3854.
Tolman, C.A.: 1974, J. Am. Chem. Soc. 96, 2780-2789.
Veillard, A.: 1969, Chem. Commun., 1022-1023, 1427-1428.
Åkermark, B., Bäckvall, J.-E., Hegedus, L.S., Siirala-Hansén, K., Sjöberg, K., and Zetterberg, K.: 1974, J. Organometal. Chem. 72, 127-138.
Åkermark, B. and Zetterberg, K.: 1975, Tetrahedron Lett., 3733-3736.
Åkermark, B., Almemark, M., Almlöf, J., Bäckvall, J.E., Roos, B., and Støgård, Åse: 1976, J. Am. Chem. Soc., to be published.

DISCUSSION

A. Pullman: (1) In commenting the table where you gave the CC optimized distances you indicated that, in the $Ni^{II}X_2^-$ ethylene complexes this distance was about 1.35 Å. I noticed that when the X ion is CN^- the CC distance is 1.39 Å. Do you have any idea to the reason for this particular behavior?

(2) You mentioned that the distance from Nickel to ehtylene was not optimized. I would suspect that optimization of this distance would influence the results that you obtain.

(3) The last question concerns a possible biochemical application of one of your results. Your result shows that the $Ni^{O}X_2$ ethylene complex mimics an excited state of the olefin and thus favours reactions which normally occur in the excited states. Would it be possible to make such a $Ni^{O}X_2$ complex with the double bond of thymine and thereby favour the occurrence of thymine reactions which otherwise occur in the excited state of thymine that is under the effect of radiation?

Støgård: (1) We have not yet been able to explain the unexpected value of 1.39 Å for the C-C bond in the $Ni(C_2H_4)(CN)_2$ complex. Currently calculations on the same complexes with different metals are in progress. May be the results from these calculations can clarify this problem.

(2) The optimization of the distance between the Ni-atom and the ethylene will give other absolute values, but the relative trends should be retained.

(3) The dimerization of thymine brought about by UV radiation looks similar to the (2+2) cycloaddition reaction activated by Ni coordiation. Thus, in principle, metal binding to the double bond in thymine could stimulate the dimerization reaction of thymine.

METAL-LIGAND INTERACTIONS AND PHOTOELECTRON SPECTROSCOPY

HAROLD BASCH
Dept. of Chemistry, Bar-Ilan University, Ramat Gan, Israel

ABSTRACT. The importance of orbital relaxation in X-ray photoelectron spectroscopy is emphasized for the chemical shift effect in transition metal complexes. An assignment of metal core level ionization satellite bands in transition metal complex spectra to specific shake-up processes is detailed and related to the appearance/non-appearance of the corresponding bands in the ligand core level ionization spectra. Valence shell relaxation effects are included in a model that relates multiplet splitting to the unpaired orbital spin density distribution giving direct information on the orbital distribution of the radical electron(s).

1. INTRODUCTION

Photoelectron Spectroscopy offers great promise as an analytical and research tool in elucidating the electronic structural properties of matter. Of particular interest are the results of looking at core electron binding energy spectra using an X-ray photon source; called X-ray photoelectron spectroscopy (XPS) [1]. The very appearance of a primary photoelectron line in a spectrum can be the source of qualitative, quantitative, and chemical bonding type information. In addition, observed multiplet structure and satellite bands can potentially add to the chemical information content of an XPS experiment. The primary purpose of this paper is to present models, including ion state relaxation effects, capable of giving insight into the orbital nature of the metal-ligand bond from these multiplet structure and satellite bands.

2. CHEMICAL SHIFT

The most direct source of chemical bonding type information in an XPS experiment has commonly been considered to come from the chemical shift effect [1]. Thus core electron binding energies exhibit apparently systematic changes with molecular environment on a scale that is small compared with the electron binding energies themselves, but large (full scale ∿12 eV in the carbon atom, for example [2, 3]) compared to the resolving power capabilities of today's instrumentation. For sulfur, whose compounds have been extensively investigated, a binding energy spread of ∿15 eV has been observed spanning a formal oxidation state

B. Pullman and N. Goldblum (eds.), Metal-Ligand Interactions in Organic Chemistry and Biochemistry, part 2 *,* 389-400.

range of -2 to +6. Generally it is found that there is a consistent correlation relating electron binding energy with formal oxidation state, or, more accurately, with charge on the sulfur atom [4, 5]. A more refined picture requires inclusion of the ground Madelung potential due to all the other atoms in the system [1].

However, there are specific cases which do not seem to fit neatly into this general picture. In an interesting study on a series of square planar nickel-dithiolate complexes, $\{Ni[S_2C_2R_2]_2\}^{n-}$ with n = 0, 1, and 2 and R = C_6H_5 or CN, the values of the sulfur 2p binding energies were found to be very close to that of S^{2-} (in Na_2S, for example) while the nickel $2p_{3/2}$ binding energies were essentially independent of n or R in the ranges studied, and very near the values found for known zero-valent nickel complexes [6, 7]. This latter observation was interpreted as evidence for a formal zero nickel oxidation state in all these compounds, independent of charge, n. The sulfur binding energies seem to tell a different tale, however. They, of course, are indicative of large electron charge densities on sulfur and therefore favor a higher nickel oxidation state in these compounds.

Analogously, it was concluded [8] that olefin complexes of nickel can be properly regarded as Ni(O) complexes on the basis of the relatively low $2p_{3/2}$ electron binding energies found for these compounds. Roughly speaking there seems to be a generally inverse correlation between Ni metal core electron binding energies and size of the ligand system [7, 8]. Analogous trends are also to be found in the corresponding classes of Cu complexes [9-11].

Table I and II give metal $2p_{3/2}$ binding energies for some selected nickel and copper complexes, respectively. Although it is possible to

TABLE I
Nickel $2p_{3/2}$ binding energies[c]

Compound	Ni $2p_{3/2}$ binding energy[a]
Ni[b]	852.8
$Ni[S_2C_2(C_6H_5)_2]_2$	852.9
$Ni[S_2C_2(C_6H_5)_2]_2^-$	852.5
$Ni[S_2C_2(C_6H_5)_2]^{2-}$	852.8
$C_2H_4Ni(P\phi_3)_2$	855.6
$(\pi-C_5H_5)_2Ni$	856.8
NiF_6^{4-}	856.7
NiF_6^{2-}	861.2

[a] in eV

[b] nickel metal

[c] from references [6-8]

TABLE II
Copper $2p_{3/2}$ binding energies

Compound	Cu $2p_{3/2}$ binding energy[a]
$Cu[S_2C_2(CN)_2]_2^-$	933.7
$Cu(DEDC)_2$[c]	934.0
$CuBr_2$	935.8
$CuCl_2$	936.4
CuF_2	939.7

[a] in eV

[b] from references [9-11]

[c] DEDC = Diethyldithiocarbamate

present cogent arguments in favor of the concept that the extent of metal-ligand covalency determines the metal core level binding energy by determining the charge division among metal and ligand systems it should be realized that this approach takes into account only the initial (unionized) state properties of the complex and ignores the corresponding situation in the final (ionized) state. This emphasis on ground state properties, although well based theoretically [2], can be an incomplete basis for interpreting core electron binding energy chemical shifts.

As has been repeatedly emphasized [12-17] in the electronic relaxation accompanying photoionization of a core electron there is a compensatory net transfer of charge density on to the ionizing atom. The extent of such a charge transfer will clearly depend on the relative abilities of the surrounding atoms to release electron density to the ionized atom and the stability of the resultant ion state will, in turn, depend on the amount of charge transferred to the ionized atom. Clearly a more extended ligand system, especially one with a highly mobile network such as is found in the (planar) dithiolate ligand, will both more readily release charge density to an ionized metal and be able to disperse the partial positive charge more effectively than small ligand systems.

Thus in correlating the trends in metal $2p_{3/2}$ electron binding energies for nickel and copper complexes it should be realized that the interpretation of these trends in terms of metal-ligand interactions by interpolating known data for simple ligand systems (such as halides, for example) to obtain information about extended ligand systems is not unique. A very different picture can be obtained by considering relaxation effects in the positive ions which may very well be the dominating effect in determining the metal core level binding energy chemical shift. Clearly in interpretive work a more quantitative model is called for which takes into account final state properties. Such models have recently been discussed [12-17] and represent a quantitative basis for a more accurate description of the chemical shift effect.

3. SATELLITE BANDS

Satellite bands have been observed in the core electron spectra of a large number of substances. These bands are generally attributed to the excitation of a valence electron to a previously unoccupied molecular orbital simultaneous with the primary core electron photoejection process. These 'shake-up' bands are typically of low intensity ($\sim$1/20) relative to the primary photoline except in certain metal core level spectra of transition metal complexes where the relative intensity can reach as high as 50% of the primary photoline [18].

This last observation is both interesting and significant. Within the framework of the sudden approximation [19-21] the relative intensity of a photoelectron peak is proportional to the (square of the) overlap of the N-1 electron part of the initial state wave function with the N-1 electron part of the final (ion) state wave function. This overlap will clearly be maximal for a final state wave function whose principle electronic configuration component is the same as in the initial state, and whose molecular orbitals overlap maximally with the initial state corresponding orbitals in that principle electronic configuration. All other possible final (ion) state wave functions, where the principle electronic configurations are necessarily different from that of the N-1 electron part of the initial state, will have low intensity and constitute a set of shake-up states giving rise to satellite bands.

Since satellite band intensity can only come at the expense of the photoionization intensity to form the primary ion state, if satellites appear prominently in the XPS spectrum then there must be a corresponding diminution of intensity that would otherwise signal the presence of the primary ion. This diminution comes as a result either of (1) a significant change in orbital character of the corresponding occupied molecular orbitals relative to the initial state, or (2) the inadequacy of the essentially single electronic configuration description of the initial and/or primary ion state wave function. Both causes indicate a drastic electronic rearrangement or relaxation effect in the positive ion relative to the neutral (unionized) state and probably occur simultaneously. Thus in light of the high intensities observed for satellite bands it can safely be concluded that relaxation effects play an important role in the XPS spectra of transition metal complexes.

A good deal of insight can be brought to bear on the nature of satellite bands from the simple consideration that an electron hole created in a core level of a given atom has as its major effect a stabilization of all the electrons on that atom external to the ionized core level [22]. In a molecular system the energy of a given molecular orbital should be lowered proportionately to the extent of ionizing atom character in that molecular orbital. This simple model has been used successfully to interpret the appearance of satellite bands in small organic molecules [21-24].

Transition metal complexes can be conveniently divided into two types; those with ligands containing empty, low-lying antibonding molecular orbitals (typically of the π^* variety), such as CO, CN^- and maleonitrile^{2-}, and those ligands with no available unoccupied orbitals such as the halides and chalcogenides. In the latter case two types of satellite bands have been observed in the 5-12 eV range to the higher binding energy side of the metal 2p core level ionization energy [7, 9, 10, 18, 25-34]. A great deal of confusion has arisen in the published literature as a result of the failure to recognize and distinguish between these two satellites with their different characteristics.

The satellite line closest to the primary photoline decreases in energy and increases in intensity with increasing covalency of the ligand. Its intensity also increases in going from manganese to copper complexes [7, 25-27]. The origin of this satellite line is best explained as a simultaneous ligand to metal (L → M), $np_L \rightarrow 3d^*_M$ charge transfer excitation accompanying the metal 2p level photoionization. This assignment explains all the known characteristics and trends of this satellite line within monopole selection rules as follows:

(1) Satellites are not observed to appear in the pure metals or in $3d^{10}$ complexes [such as those of Cu(I) and Zn(II)] because in the latter case there are no available terminating $3d^*_M$ orbitals, and in the former case due to the absence of ligand levels.

(2) Satellite bands decrease in energy relative to the primary photoline with increasing covalency in accord with the known ligand dependence of $np_L \rightarrow 3d^*_M$ charge transfer transitions in ordinary optical absorption spectroscopy [35]. The analogous decrease in excitation energy is also observed for these type transitions in going across the transition metals, Ti to Cu [35].

(3) Satellite line intensity increases with the covalency of the ligand-metal bond apparently because the greater the covalency of the complex the larger the degree of valence orbital relaxation upon hole state formation. Covalent bonding allows greater valence electron relaxation because of the increased polarizability of the metal-ligand bond as compared with more ionic bonding situations. This point has been stressed by Robert [28].

(4) No low energy (5-10 eV) satellites are observed accompanying ligand core electron ionization simply because a core electron hole in a ligand atom will preferentially stabilize the ligand level and shift $np_L \rightarrow 3d^*_M$ transitions to higher energy. In contrast, a core electron hole in the metal atom will preferentially stabilize the primarily metal 3d level leading to low energy $np_L \rightarrow 3d^*_M$ type shake-up states, as observed. This explanation differs substantially from that proposed by Ikemoto et al. [29] based on intensity considerations but is in accord with the experimental observations of Kim and Winograd [34] on metal d^0 oxide complexes.

(5) The apparent absence of low energy satellites in diamagnetic Ni(II) complexes can be attributed to a large ligand field affect pushing the unoccupied metal 3d level to a higher energy than if it were partially occupied. This also explains the apparently discontinuous decrease in satellite energy for octahedral halide complexes [27] in going from Cr^{3+} to Fe^{3+}; in the latter complex the $3e^*_g(3d^*_M)$ level begins to be

partially occupied. Thus in simple octahedral complexes this satellite line is clearly $2e_g(np_L) \rightarrow 3e_g^*(3d_M^*)$.

The second (higher energy displacement from the primary photoline) satellite band in the halide and chalcogenide metal complexes has been somewhat of a mystery and variously assigned [9, 10, 28]. In Cu(II) complexes, for example, the $np_L \rightarrow 3d_M^*$ shake-up transition fromally gives rise to both a singlet and triplet state (a sort of shake-up multiplet splitting) whose energy separation and relative intensities, unlike the situation in the pure hole state, do not necessarily depend on simple spin density and degeneracy considerations. Thus the second satellite band in halide and chalcogenide metal complexes could correspond to a final state electronic configuration where the metal core hole and ligand np_L level are spin coupled to give a singlet state, distinct and displaced from the corresponding lower energy triplet shake-up state.

In the absence of detailed configuration interaction calculations we can only speculate in order to draw further conclusions here. Nontheless, in contrast to the lower energy satellite, the second satellite bands are observed to increase in energy (relative to the primary photoline) with increasing covalency of the ligand and therefore the energy separation between the two types of satellites increases with increasing covalency. This trend accords with increasing exchange interaction between the metal core and np_L orbitals. As the latter level increases its metal valence orbital character, both because of increasing covalency and because of the metal core hole, the singlet-triplet splitting energy, which in the case of Cu(II) is approximately equal to twice the exchange interaction integral between metal core and np_L levels, will increase. Thus both satellite bands in halide and chalcogenide metal complexes can be assigned to the same $np_L \rightarrow 3d_M^*$ excitation process and their observed behavior rationalized on this basis.

Another type of satellite line is observed in metal carbonyl complexes, for example, where the ligand has an available π^* molecular orbital. Here, low energy satellites are observed both in the metal and ligand (carbon and oxygen) core state spectra [36, 37] but are not found in the uncomplexed ligand XPS spectra. For example, in $Cr(CO)_6$ satellite lines are observed at 6.5 eV, 5.4 eV, and 5.4 eV from the $2p_{3/2}$, 1s, and 1s hole states of the Cr, C, and O atoms respectively. These satellite bands could also be assigned as σ_L or $\pi_L \rightarrow 3d_M^*$ except for their simultaneous appearance in the ligand core hole spectra and the higher relative energy of the chromium satellite line compared to carbon and oxygen. Both these characteristics suggest a $3d_M \rightarrow \pi_L^*$ assignment since photoionizing a ligand core level stabilizes the terminating orbital in the shake-up process (π_L^*) while ionization of a metal core electron lowers the initial orbital ($3d_M$). This $M \rightarrow L^*$ assignment is, in fact, the calculated result [38]. In all fairness it is probably true that the observed satellite line also has a significant $L \rightarrow M^*$ component. However, this analysis suggests that the appearance of low energy satellite lines in the complexed ligand core electron photoelectron spectra can be used as a diagnostic indication of a $3d_M \rightarrow \pi_L^*$ shake-up process.

4. MULTIPLET SPLITTING

Satellite structure found in the XPS spectra of paramagnetic transition metal complexes accompanying the $2p_{3/2}$ core level ionization can actually also arise from multiplet splitting in the primary core hole state. Since the extent of such multiplet splitting is related to the unpaired orbital spin density on the ionizing atom [15, 39-41] it is clear from the energies of the satellite line relative to the primary photoline that the satellite bands found in nickel and copper complexes, where the number of unpaired spins is small, could only be due primarily to shake-up processes. For the earlier high spin transition metal complexes this question of multiplet splitting versus shake-up has been taken up by Carlson, et al. [42] who demonstrate convincingly in general favor of the latter mechanism. That multiplet splitting is not observed for the primary metal $2p_{3/2}$ core ion state is apparently due to the lower spin multiplet component having its intensity distributed over a number of other electronic states through configuration interaction [43]. The net result is to considerably broaden the primary photoline rather than leading to distinctive multiplet peaks. A similar effect is expected for the metal 3s core level [44, 45]. Here, however, the predicted shake-up intensity is much less [42] so that multiplet splitting is clearly observed [41] but with much reduced splitting compared to free ion values for the same number of unpaired spins.

Although it is clear that 3s core level multiplet splitting in transition metals cannot be interpreted in simple terms the same is not expected for the metal 2s level where the special quasi-degeneracy effects of the n=3 level are not present and thus correlation effects are not so important. [44] the simplest model for interpreting XPS spectra involves ground state properties only but it would be naive to expect that any model valid on a semi-quantitative level could be proposed that does not account also for relaxation effects. We therefore here apply such a model, derived previously [15] to some first row atom containing radicals as test cases. The results are expected to be equally applicable to metal 2s core level multiplet splitting which, unfortunately, to date has not been extensively studied.

It has previously been shown [15, 39] that the multiplet splitting energy, δ_i, for ionization of core level electron i in a radical of initial space and spin symmetry $^{2S+1}\Gamma$ to give two multiplet states $^{2S+2}\Gamma$ and $^{2S}\Gamma$ can be written as,

$$\delta_i = \frac{2S+1}{2S} \sum_\mu (N_\mu^\alpha - N_\mu^\beta + \tfrac{1}{2}\bar{N}_\mu^\alpha) K_{\mu i} ,$$

or as,

$$\delta_i = \frac{2S+1}{2S} \sum_\mu (n_\mu + \tfrac{1}{2}\Delta_\mu) K_{\mu i} , \qquad (1)$$

where N_μ^α and N_μ^β are the ground state gross α and β spin orbital populations of basis orbital μ, respectively, and the bar represents the corresponding quantities in the ion state. n_μ is just the unpaired orbital spin density (UOSD) in orbital μ and Δ_μ is the α spin density migration into or out of basis orbital μ consequent upon photoionization of a core electron. $K_{\mu i}$ is the exchange interaction (integral) between core level i and basis orbital μ.

For first row atoms Equation (1) can be simplified by restricting the sum over μ to same-atom orbitals as core level i [39] and grouping n_μ and Δ_μ values for 2s and 2p electrons together. The result for first row atoms is,

$$\delta_X = \frac{2S+1}{2S} \; (n_X + \tfrac{1}{2}\Delta_X) K^X \qquad (2)$$

where X represents the atom, n_X is the total UOSD on atom X, and Δ_X is the total shift of α spin density from or to atom X as a result of core electron ionization.

Equation (2) was tested on the series of radicals NF_2, NO_2, NO, O_2, HCO and CN. Since experimental values for the core level multiplet splitting are not available for the last 2 species the multiplet splitting was calculated as the difference in Hartree-Fock ion multiplet energies of proper space and spin symmetry. The expansion basis used was of the (gaussian) triple-zeta plus polarization type where the atom s,p bases have been described previously for carbon [46] and oxygen [21]. Both the polarization p functions on hydrogen and d functions on the first row atoms (carbon to fluorine) were taken as two term gaussian fits to an exponential function [47]. The total ground state energy result for NO_2, for example, is -204.08648 AU compared to the previous double-zeta result of -203.94461 AU [15]. The calculated and observed multiplet splittings are displayed in Table III.

TABLE III
Values of multiplet splitting for some small radicals

Molecule	Atom	δ_{1s} [a]		
		DZ [b]	TZ+P [c]	Experiment [d]
HCO	C	0.87	0.90	-
	O	0.38	0.36	-
NF_2	N	1.42	1.55	1.93
	F	0.08	0.09	0.72
NO_2	N	-	0.71	0.70
	O	0.33	0.33	0.67
NO	N	-	1.26	1.42
	O	-	0.51	0.55
O_2	O	0.96	1.02	1.12
CN	C	-	0.74	-
	N	-	1.60	-

[a] Multiplet splitting in eV.

[b] Double-zeta gaussian basis from ref. [15]. Note that DZ values for NO reported in ref. [15] are in error.

[c] Triple-zeta plus polarization gaussian basis as described in text.

[d] Experimental value from ref. [40].

The experimental values [40] and in their absence the best theoretical results from Table III were combined with Equation (2) to obtain values of n_X, the UOSD on each atom X. Firstly, it is necessary to decide on values for K^X, the valence-core exchange interaction (integral). This was done by ignoring the Δ_K term in Equation (2) and fitting the resultant simple expression, $[(2S+1)/2S]n_X K^X$, to calculated multiplet splittings for the ground ^3P states of carbon and oxygen with extrapolation and interpolation for nitrogen and fluorine. The results (in eV) are $K^C = 1.37$, $K^N = 1.82$, $K^O = 2.34$, and $K^F = 2.93$.

The next problem is how to treat the relaxation term Δ_X. Actually, for a diatomic system (X-Y) with no hydrogen atoms and where the multiplet splittings for each atom are available there 4 unknowns (n_X, n_Y, Δ_X, and Δ_Y) and 3 equations (Equation (2) twice and $n_X + n_Y = 1$ for a doublet state). A simple assumption would be to set $\Delta_X = k\Delta_Y$ and look at the results for various values of k. This, in fact is what has been done here where possible. However, in addition, it would be advantageous to be able to predict, in advance, the direction of the majority spin density flow (i.e. the signs of Δ_X and Δ_Y) in order to compare with the calculated result and thus increase confidence in the values of n_X produced by the model.

The direction of majority spin density flow upon K-shell photoionization can be anticipated on the following basis. Consider the NO molecule in which the first π level (π_1) is fully occupied and the second π level (π_2) contains a single electron. In ordinary NO the π_1 orbital has predominant oxygen character and, accordingly, π_2 is mainly nitrogen. Upon core electron ionization, whether from nitrogen or oxygen, the π_1 level will increase its proportion of the ionizing atom valence orbital character simply as a result of that atom's stabilization. This is simply an expression of charge density flow on to an ionizing atom. Thus the antibonding π_2 orbital containing the unpaired electron will automatically decrease its orbital component on the ionizing atom. These qualitative arguments applied to the NO molecule are borne out by the numerical application of Equation (2) as described above, shown in Table IV for k=1.

Analogously for O_2, another two π level system, this time with 2 unpaired spins in the antibonding π_2 level, a large migration a majority spin density away from the ionizing atom is predicted. The direction and relative magnitudes of spin density flow in both NO and O_2 agree with the results of Davis, et al. [38]. For the CN radical the unpaired electron is in the π_1 bonding orbital and the expectation therefore is of a migration of spin density on to the ionizing atom; precisely as calculated and shown in Table IV.

In NO_2 the odd electron is in a σ orbital which is mainly nitrogen in character and antibonding (carbon dioxide has all its bonding σ orbitals fully occupied). Thus a net decrease in orbital spin density is predicted for both nitrogen and oxygen. In NF_2 the radical electron is in the third highest (π_3) of a set of three π-type molecular orbitals. Here, although the electron is in an antibonding orbital there are two lower occupied molecular orbitals within which the electron density rearrangement upon core level ionization can take place, making an unambiguous prediction of the resultant orbital characteristics of π_3 very

TABLE IV
Calculated unpaired orbital spin densities and relaxation spin density flow

Molecule	Atom	δ_X [a]	n_X	Δ_X
HCO	C	0.90	0.587	-0.183
	O	0.38	0.213	-0.183
	H	-	0.200[b]	-
NF_2	N	1.93	0.964	-0.334
	F	0.72	0.018	+0.334
NO_2	N	0.67	0.383	-0.188
	O	0.70	0.309	-0.188
NO	N	1.42	0.705	-0.237
	O	0.55	0.295	-0.237
O_2	O	1.12	1.000	-1.042
CN	C	0.74	0.373	+0.068
	N	1.60	0.627	+0.068

[a] Multiplet splitting in eV; see Equation (2).

[b] Assumed value.

difficult. It is also, therefore, not automatically true that the spin density migration will have the same sign for both atoms. Assuming, for example, that $\Delta_N = -\Delta_F$ gives the results in Table IV. The alternate assumption of $\Delta_N = \Delta_F$(k=1 always) gives $n_N = 0.741$, $n_F = 0.129$, and $\Delta_N = \Delta_F = +0.112$. The former numbers seem to be closer to the corresponding values derives from electron spin resonance (ESR) experiments [48-50].

The HCO molecule presents the new problem in how to treat an atom whose multiplet splitting is unavailable; here the hydrogen atom on which XPS measurements cannot be made. The simultaneous equations were solved for various assumed values of n_H and the resultant n_C and n_O values closest to Mulliken [51] and Lowdin [52] population analyses [53] of the ab initio wave function for ground state HCO are shown in Table IV. As expected from the radical electron being in an essentially antibonding σ molecular orbital, both Δ_C and Δ_O are negative. In fact, the simultaneous equations turn out to be incompatible with $\Delta_C = -\Delta_O$ for n_H less than 0.30. The ESR results on HCO [54] favor somewhat larger values of the UOSD on oxygen and hydrogen than those shown in Table IV.

The relaxation model for multiplet splitting detailed above can, of course, readily be applied to transition metal complexes where the unpaired spins are typically located in an antibonding metal-ligand molecular orbital. Thus majority spin density should shift away from each atom upon core level ionization. A more detailed application requires additional experimental and theoretical work which hopefully will be available in the near future.

ACKNOWLEDGEMENT

This work was supported by a grant from the Israel Commission for Basic Research.

REFERENCES

1. Siegbahn, K. et al.: 'ESCA - Atomic, Molecular and Solid State Structure studied By Means of Electron Spectroscopy', Almqvist and Wiksells, Uppsala, 1967.
2. Basch, H.: Chem. Phys. Letts. 5, 337 (1970).
3. Davis, D.W., Banna, M.S., and Shirley, D.A.: J. Chem. Phys. 60, 237 (1974).
4. Gianturco, F.A., and Coulson, C.A.: Mol. Phys. 14, 223 (1968).
5. Kramer, L.N. and Klein, M.P.: J. Chem. Phys. 51, 3618 (1969).
6. Grim, S.O., Matienzo, L.J., and Swartz, W.E.: J. Am. Chem. Soc. 94, 5116 (1972).
7. Matienzo, L.J., Yin, L.I., Grim, S.D., and Swartz, W.E.: Inorg. Chem. 2, 2762 (1973).
8. Tolman, C.A., Riggs, W.M., Linn, W.J., King, C.M., and Wendt, R.C.: Inorg. Chem. 12, 2770 (1973).
9. Frost, D.C., Ishitani, A., and McDowell, C.A.: Mol. Phys. 24, 861 (1972).
10. Frost, D.C., McDowell, C.A., and Tapping, R.L.: J. Electron Spectrosc. 6, 347 (1975).
11. Frost, D.C., McDowell, C.A., and Tapping, R.L.: J. Electron Spectrosc. 7, 297 (1975).
12. Shirley, D.A.: Chem. Phys. Letts. 16, 220 (1972).
13. Davis, D.W. and Shirley, D.A.: Chem. Phys. Letts. 15, 185 (1972).
14. Davis, D.W., Banna, M.S., and Shirley, D.A.: J. Chem. Phys. 60, 237 (1974).
15. Basch, H.: J. Electron Spectrosc. 5, 463 (1974).
16. Citrin, P.H. and Hamann, D.R.: Phys. Rev. B10, 4948 (1974).
17. Bossa, M., Ramunni, G., and Gianturco, F.A.: Chem. Phys. Letts. 30, 235, (1975).
18. Brisk, M.A. and Baker, A.D.: J. Electron Spectrosc. 7, 197 (1975).
19. Aberg, T.: Phys. Rev. 156, 35 (1967).
20. Aberg, T.: Ann. Acad. Sci. Fenn. A6, 308 (1969).
21. Basch, H.: Chem. Phys. 10, 157 (1975).
22. Snyder, L.C.: J. Chem. Phys. 55, 95 (1971).
23. Basch, H.: J. Am. Chem. Soc. 97, 6047 (1975).
24. Basch, H.: Chem. Phys. Letts. 37, 447 (1976).
25. Rosencwaig, A., Wertheim, G.K., and Guggenheim, H.J.: Phys. Rev. Letts. 27, 479 (1971).
26. Yin, L., Adler, I., Tsang, T., Matienzo, L.J., and Grimm, S.O.: Chem. Phys. Letts. 24, 81 (1974).
27. Carlson, T.A., Carver, J.C., Saethre, L.J., Santibanez, F.G. and Vernon, G.A.: J. Electron Spectrosc. 5, 247 (1974).
28. Robert, T.: Chem. Phys. 8, 123 (1975).
29. Ikemoto, I., Ishii, K., Kuroda, H., and Thomas, J.M.: Chem. Phys. Letts. 28, 55 (1974).

30. Robert, T. and Offergeld, G.: Chem. Phys. Letts. 29, 606 (1974).
31. Sen, K.: 'Shake-up Satellites in the 2p X-Ray Photoelectron Spectra of the Transition Metal Ion Complexes', preprint.
32. Larsson, S.: Chem. Phys. Letts. 32, 401 (1975).
33. Kim, K.S.: Phys. Rev. B11, 2177 (1975).
34. Kim, K.S. and Winograd, N.: Chem. Phys. Letts. 31, 312 (1975).
35. Basch, H., Viste, A., and Gray, H.B.: J. Chem. Phys. 44, 10 (1966).
36. Barber, M., Conner, J.A., and Hillier, I.H.: Chem. Phys. Letts. 9, 570 (1971).
37. Pignitaro, S., Fofani, A., and Distefano, G.: Chem. Phys. Letts. 20, 350 (1973).
38. Hillier, I.H. and Kendrick, J.: Farad. Trans. II, 71, 1369 (1975).
39. Basch, H.: Chem. Phys. Letts. 20, 233 (1973).
40. Davis, D.W., Martin, R.L., Banna, M.S., and Shirley, D.A.: J. Chem. Phys. 59, 4235 (1973).
41. Carver, J.C., Schweitzer, G.K., and Carlson, T.A.: J. Chem. Phys. 57, 973 (1972).
42. Carlson, T.A., Carver, J.C., and Vernon, G.A.: J. Chem. Phys. 62, 932 (1975).
43. Gupta, R.P. and Sen, S.K.: Phys. Rev. B10, 71 (1974).
44. Bagus, P.S., Freeman, A.J., and Sasaki, F.: Phys. Rev. Lett. 30, 850 (1973).
45. Viinikka, E.K. and Ohrn, Y.: Phys. Rev. B11, 4168 (1975).
46. Basch, H.: J. Chem. Phys. 55, 1700 (1971).
47. Stewart, R.F.: J. Chem. Phys. 52, 431 (1970).
48. McDowell, C.A., Nakajima, H., and Raghunathan, P.: Canad. J. Chem. 48, 805 (1970).
49. Kasai, P.H. and Wipple, E.B.: Mol. Phys. 9, 497 (1965).
50. Brown, R.D., Burden, F.R., Godfrey, P.D., and Gillard, I.R.: J. Mol. Spectry. 25, 301 (1974).
51. Mulliken, R.S.: J. Chem. Phys. 46, 497 (1949).
52. Lowdin, P.O.: J. Chem. Phys. 18, 365 (1950).
53. Basch, H.: Chem. Phys. Letts. 12, 110 (1971).
54. Pederson, L.: J. Molec. Structure 5, 21 (1970).